Mastering
Technical
Mathematics

ABOUT THE AUTHORS

Stan Gibilisco is one of McGraw-Hill's most prolific and popular authors. His clear, reader-friendly writing style makes his books accessible to a wide audience, and his experience as an electronics engineer, researcher, and mathematician makes him an ideal editor for reference books and tutorials. Stan has authored several titles for the McGraw-Hill *Demystified* library of home-schooling and self-teaching volumes, along with more than 30 other books and dozens of magazine articles. His work has been published in several languages. *Booklist* named his *McGraw-Hill Encyclopedia of Personal Computing* one of the "Best References of 1996," and named his *Encyclopedia of Electronics* one of the "Best References of the 1980s."

Norman Crowhurst (deceased) originated the concept of *Mastering Technical Mathematics*. He authored the first edition.

Mastering
Technical
Mathematics

Third Edition

Stan Gibilisco

Norman Crowhurst

New York Chicago San Francisco Lisbon London Madrid Mexico City
Milan New Delhi San Juan Seoul Singapore Sydney Toronto

The McGraw·Hill Companies

CIP Data is on file with the Library of Congress

Copyright © 2008, 1999 by The McGraw-Hill Companies, Inc. All rights reserved. Printed in the United States of America. Except as permitted under the United States Copyright Act of 1976, no part of this publication may be reproduced or distributed in any form or by any means, or stored in a data base or retrieval system, without the prior written permission of the publisher.

The first edition of this book, with the title *Basic Mathematics*, © 1961 by Norman H. Crowhurst, was published by John F. Rider Publisher Inc., a division of Hayden Publishing Company, Inc., New York, New York.

McGraw-Hill books are available at special quantity discounts to use as premiums and sales promotions, or for use in corporate training programs. For more information, please write to the Director of Special Sales, McGraw-Hill Professional, Two Penn Plaza, New York, NY 10121-2298. Or contact your local bookstore.

1 2 3 4 5 6 7 8 9 0   DOC/DOC   0 1 3 2 1 0 9 8 7

ISBN 978-0-07-149448-9
MHID 0-07-149448-0

Sponsoring Editor
Judy Bass

Editing Supervisor
David E. Fogarty

Project Manager
Vastavikta Sharma, International
Typesetting and Composition

Copy Editor
Patti Scott

Proofreader
Shruti Pandey

Indexer
Stan Gibilisco

Production Supervisor
Pamela A. Pelton

Composition
International Typesetting and Composition

Art Director, Cover
Jeff Weeks

This book was printed on acid-free paper.

To Tim, Samuel, and Tony

Contents

22 Complex Numbers 323

Part 4 Tools of Applied Mathematics

23 Trigonometry in the Real World 339

24 Logarithms and Exponentials 359

25 Scientific Notation Tutorial 381

26 Vectors 395

27 Logic and Truth Tables   411

28 Beyond Three Dimensions   433

29 A Statistics Primer   455

30 Probability Basics   473

Final Exam   495

Appendices

Introduction

This book is intended as a refresher course for scientists, engineers, and technicians. It begins with a review of basic calculation techniques and progresses through intermediate topics in applied mathematics. This edition contains new or revised material on scientific notation, geometry, trigonometry, vectors, coordinate systems, logarithms, exponential functions, propositional logic, truth tables, statistics, and probability.

You'll notice that some of the discussions, especially in the early chapters, deal with archaic units such as feet and pounds. This isn't sheer madness or nostalgia for the olden days; there is a reason for it! The unfamiliarity (and in some cases strangeness) of these "English" units can help you more fully grasp the principles relating the phenomena the units describe.

Ideally, you'll have worked with all the material you'll see in this book, or at least seen it in some form. If something here is alien to you—for example, propositional logic—consider taking a formal course on that subject, using this book as a supplement. What if you took logic from your Alma Mater 20 years ago? Are the concepts still in your mind but no longer fresh? In that sort of situation, this book can bring things back into focus, so you can again work easily with concepts you learned long ago.

Each chapter ends with a Questions and Problems section. Refer to the text when solving these problems. Answers are in an appendix at the back of the book. In some cases, descriptions of the problem-solving processes are given in the answer key. Of course, many problems can be solved in more than one way. If you get the right answer by a method that differs from the scheme in the answer key, you might have found a better way!

In recent years, electronic calculators have rendered much of the material in this book academic. To find the sine of an angle, for example, you can punch it up on a calculator you bought at a grocery store and get an answer accurate to 10 significant figures. Personal computers have "calculator" programs that can go to many more digits than that. Nevertheless, it's helpful to understand the theory involved. You should at least read (if not painstakingly study) every chapter in this book.

Most people are strong in certain areas of mathematics and weak in others. If your job involves the use of math, you'll need proficiency in some fields more than in others. When you use this book as a refresher course, keep in mind that you might need intensive work on subjects that you don't like or that you have trouble grasping.

The material here is presented in a condensed format. You'll sometimes think that your progress must be measured in hours per page. If you get stuck someplace, skip ahead, work on something else for awhile, and then come back to the "trouble spot." Of course, you can always refer to more basic or subject-intensive texts to reinforce your knowledge in those areas where you are not confident.

Suggestions for future editions are welcome.

Stan Gibilisco

Part 1

WORKING WITH NUMBERS

1 From Counting to Addition

We've all seen or heard people count. You put a number of things in a group, move them to another group one at a time, and count as you go. "One, two, three ..." We learn to save time when counting by making up groups. Figure 1-1 shows four different ways in which seven items, in this case coins, can be grouped.

NUMBERS AND NUMERALS

You'll often hear the terms "number" and "numeral" used interchangeably, as if they mean the same thing. But there's a big difference (although subtle to the nonmathematician). A number is a quantity or a concept; it's abstract and intangible. You can think about a number, but you can't hold one in your hand or even see it directly. A numeral is a symbol or set of symbols that represents a number; it is concrete. In this book, numerals are arrangements of ink on the paper that represent numbers.

COUNTING IN TENS AND DOZENS

When you have a large number of things to count, putting them into separate groups of convenient size makes the job easier. People in most of the world use the quantity we call *ten* as the *base* for such counting. How can we show the idea of ten in an unambiguous way? Here's one method:

• • • • •

• • • • •

There are ten dots in the above pattern. The numbering scheme based on multiples of this number is called the *decimal system*. The number ten is represented by the numeral

Figure 1-1

These are some of the ways seven things can be arranged.

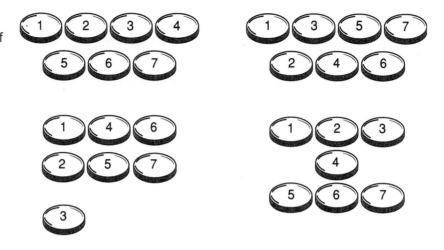

10 in that scheme. Two groups of ten make *twenty* (20); three groups of ten make *thirty* (30); four groups of ten make *forty* (40); and so on, as shown in Fig. 1-2.

Tens aren't the only size of base (also called *radix*) that people have used to create numbering systems. At one time, many things were counted in *dozens*, which are groups of *twelve*. This method is called the *duodecimal system* (Fig. 1-3). We can show the concept of twelve unambiguously, as a set of dots, like this:

· · · · · ·
· · · · · ·

WRITING NUMERALS GREATER THAN 10

When we have more than ten items and we want to represent the quantity in the decimal system, we state the number of complete tens with the extras left over. For example, 35 means three tens and five ones left over. The individual numerals or

Figure 1-2

It's convenient to count big numbers in tens.

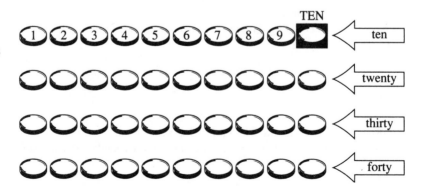

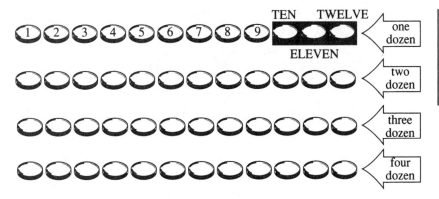

Figure 1-3
Items can be counted by twelves, but this is not done much nowadays.

digits are written side by side. The left-hand digit represents the number of tens, and the right-hand digit represents the number of ones. Figure 1-4 shows how this works for 35.

WHY ZERO IS USED IN COUNTING

If we have an exact count of tens and no ones are left over, we need to show that the number is in "round tens" without any ones. To do this, we write a digit zero (0) in the ones place, farthest to the right. This tells us that we have an exact number of tens

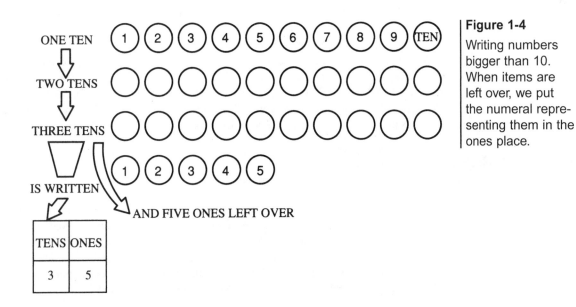

Figure 1-4
Writing numbers bigger than 10. When items are left over, we put the numeral representing them in the ones place.

Figure 1-5

When there aren't any ones left over, we write zero (0) in the ones place.

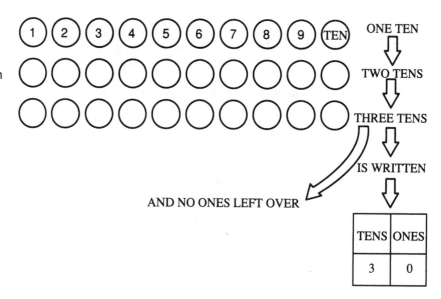

because there are no ones left over. "Zero" means "no" or "none." Figure 1-5 shows how this works for 30.

BY TENS AND HUNDREDS TO THOUSANDS

One indirect way to visualize numbers is to think of packing things into boxes. A box might hold 10 rows of 10 apples, for example. The total would be 10×10, or 100 apples. Then you might stack 10 layers of 100, one on top of the other, getting a cube-shaped box holding $10 \times 10 \times 10$, or a *thousand* (1000) apples.

If you imagine packing things this way, you'll find it easy to comprehend large numbers. You might have two full boxes holding 1000 apples each, and then a third box holding five full layers, six full rows on the next layer, and three ones in an incomplete row. This would add up to two thousand (2×1000, or 2000) plus five hundred (5×100, or 500) plus sixty (6×10, or 60) plus three (3), written as 2563. Figure 1-6 shows how this works. You can use the same sort of "boxing-up" concept to envision huge numbers of items.

DON'T FORGET THE ZEROS

When a count has leftover layers, rows, and parts of rows with this systematic arrangement idea, you will have numbers in each column. But if you have no complete hundreds layers (Fig. 1-7A), there will be a zero in the hundreds place. You might have no ones left over (as at B), no tens (as at C), or no tens and no hundreds (as at D). In each case, it's important to write a zero to keep the other numerals in their proper places. For this reason, zero is called a *placeholder*. Never forget to use zeros!

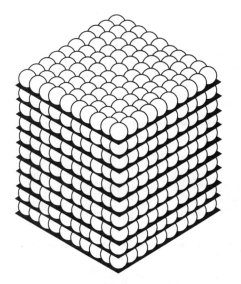

TEN ROWS OF TEN
IN EACH LAYER IS
1 HUNDRED

TEN LAYERS OF
ONE HUNDRED IS
TEN HUNDRED OR

1 THOUSAND

Figure 1-6
Ten rows of 10
items in each layer
is a hundred (100).
Ten layers of 100
is a thousand
(1000). We can
build up large
numbers such as
2563 in this way.

thousands	hundreds	tens	ones
2	**5**	**6**	**3**

Two thousand five hundred sixty three

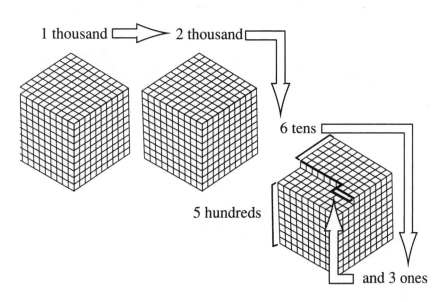

1 thousand ➡ 2 thousand

6 tens

5 hundreds

and 3 ones

Figure 1-7

Using rows, columns and layers to represent numbers that contain zeros. At A, there are 3065 items; at B, there are 4370 items; at C, there are 2504 items; at D, there are 3008 items.

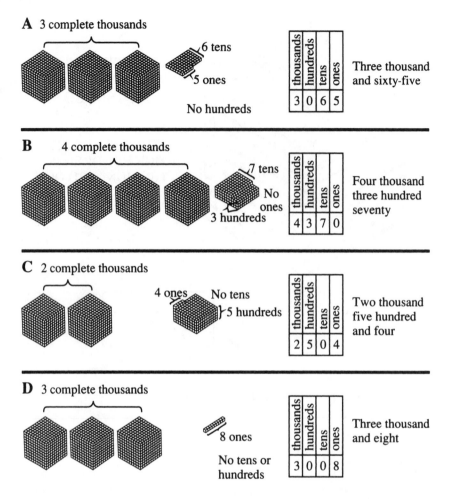

A 3 complete thousands

6 tens

5 ones

No hundreds

thousands	hundreds	tens	ones
3	0	6	5

Three thousand and sixty-five

B 4 complete thousands

7 tens

No ones

3 hundreds

thousands	hundreds	tens	ones
4	3	7	0

Four thousand three hundred seventy

C 2 complete thousands

4 ones   No tens

5 hundreds

thousands	hundreds	tens	ones
2	5	0	4

Two thousand five hundred and four

D 3 complete thousands

8 ones

No tens or hundreds

thousands	hundreds	tens	ones
3	0	0	8

Three thousand and eight

MILLIONS AND MORE

Imagine stacking thousands of "boxes" to get a powerful way of counting. In Fig. 1-8, one complete box that contains 1000 apples is magnified in a "stack" that has 999 other identical boxes. Think about it: 1000 boxes, each with 1000 apples in it! Each layer of 10 by 10 boxes contains a hundred thousand (100,000) apples. Each row or column of ten boxes in a layer contains ten thousand (10,000) apples. The whole stack contains a thousand thousand or a *million* (1,000,000) apples.

You can go on with this. Suppose each of the tiniest boxes in the magnifying glass is really a stack containing 1,000,000? Then the 10-by-10-by-10 "superbox" in the glass has a thousand million (1,000,000,000) apples. In the United States this is called a *billion*. The entire stack in this case has a million million (1,000,000,000,000) apples. In the United States it is called a *trillion*, but some people in England call it a billion. If you go to another multiple of a thousand, you get a *quadrillion* (written as a one with fifteen

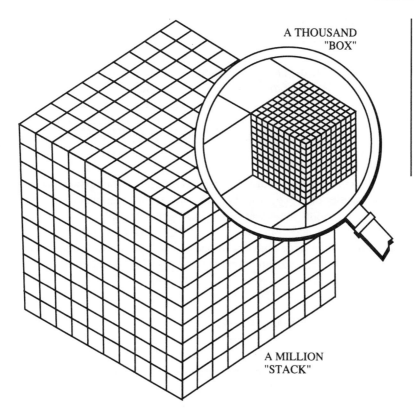

A THOUSAND "BOX"

A MILLION "STACK"

Figure 1-8
Rows, columns, and layers of 10 items can make a thousand "box." Rows, columns, and layers of 10 such "boxes" can make a million "stack."

zeros after it). Going onward by multiples of a thousand, you get a *quintillion* (a one with eighteen zeros after it), a *sextillion* (a one and then twenty-one zeros), a *septillion* (a one and then twenty-four zeros), an *octillion* (a one and then twenty-seven zeros), a *nonillion* (a one and then thirty zeros), and a *decillion* (a one and then thirty-three zeros).

WHAT ABOUT INFINITY?

A decillion decillion would be written as a one followed by 66 zeros. A decillion decillion decillion would go down as a one and then 99 zeros. Multiply that huge number by 10, and you get a *googol*, which is written as a one and then 100 zeros. Of course, we can go on for years with this game, but no matter how long we keep it up, the number we get will be *finite*. That means that you could count up to it if you had enough time. We can never get an *infinite* number this way.

No matter how big a number is, you can always get a bigger number by adding one. In that sense, infinity isn't a number at all. Some people say there isn't even such a thing as infinity. But enough about that! Now let's get back to the practical stuff and see how *addition* relates to counting.

ADDITION IS COUNTING ON

Now that we've developed a method of counting, we can start to work on a scheme for calculating. The first step is addition. Suppose you've already counted five items in one group and three items in another group. What do you have when you put them together? The easiest way to picture this addition is to count on. Have you seen children doing this using their fingers? They haven't memorized their *addition facts* yet!

If you memorize your addition facts, it's convenient. But nothing is theoretically wrong with counting on. It's just cumbersome and tedious. You can make an addition table like the multiplication tables you first saw in elementary school. Or you can use a calculator! Calculators add by counting on, but they're a lot faster than people. Calculators are great for adding large numbers to one another. But for the single-digit numbers, it's a good idea to memorize all your addition facts. You should know right away, for example, that seven and nine make 16 ($7 + 9 = 16$).

ADDING THREE OR MORE NUMBERS

Here's a principle that the people who invented the "new math" gave a fancy name: the *commutative law for addition*. Put simply, it says that you can add two or more numbers in any order you want. Suppose you want to add three, five, and seven. No matter how you do it, you always get 15 as the answer. Figure 1-9 shows two examples with coins. The commutative law applies to as few as two *addends* (numbers to be added), up to as many as you want.

ADDING LARGER NUMBERS

So far, we have added numbers with only a single figure in the ones place. Bigger numbers can be added in the same way, but be careful to add only ones to ones; tens to tens, hundreds to hundreds, and so on. Just as $1 + 1 = 2$, so $10 + 10 = 20$, $100 + 100 = 200$, and so on. We can use the counting-on method or the addition table for any group of numbers, as long as all the digits in the group belong. That is, they all have to be the same place: one, tens, hundreds, or whatever.

So, let's add 125 and 324. Take the ones first: $5 + 4 = 9$. Next the tens: $2 + 2 = 4$. Last the hundreds: $1 + 3 = 4$. Our result is four hundreds, four tens, and nine ones, which total 449. This process is shown in Fig. 1-10. Notice that we are taking shortcuts. We no longer count tens and hundreds one at a time, but in their own group, tens or hundreds. If you added all those as ones, you'd get sick and tired of it a long time before you were done. You might get careless, and you would have 449 chances of skipping one, or of counting one twice. The shortcuts not only make the process go by quicker, but they reduce the risk of making a mistake.

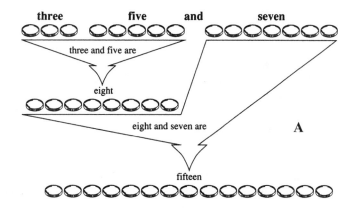

Figure 1-9

When adding three numbers, it doesn't matter which two are added first. At A, we add three and five, and then add seven more. At B, we add seven and five, and then add three more. Either way we get fifteen.

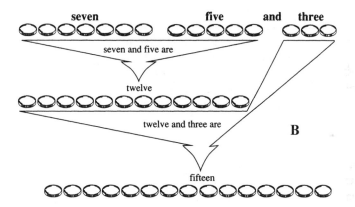

Figure 1-10

Big numbers are added in the same way as single-digit numbers. Here, we see that 125 added to 324 equals 449.

CARRYING

In the preceding example, we deliberately chose numbers in each place that did not add up to over 10, to make it easy. If any number group or place adds to over 10, we must "carry" it to the next higher group or place. Three examples follow; it will help if you write the digits down in the columns for *powers of 10* (that is, the ones place, the tens place, the hundreds place, and so on) as you read these descriptions.

Suppose we want to add 27 + 35. We take the ones first: 7 + 5 = 12. The numeral 1 belongs in the tens place. Now, instead of just 2 and 3 to add in the tens place, you have an extra 1 that appears in the tens place from adding 7 and 5. This extra 1 is said to be *carried* from the ones place. This *carrying* process goes on any time the total at a certain place goes over 10. The final result of this addition process, called the *sum*, is therefore 27 + 35 = 62, because we have 1 + 2 + 3 = 6 in the tens place.

Now what if we want to add 7,358 and 2,763? Start with the ones: 8 + 3 = 11, so we write 1 in the ones place and carry 1 to the tens place. Now the tens: 5 + 6 = 11, and the 1 carried from the ones makes 12. We write 2 in the tens place and carry 1 to the hundreds place. Now the hundreds: 7 + 3 = 10 and 1 carried from the tens makes 11 hundreds. We write 1 in the hundreds place and carry 1 to the thousands place. Now the thousands: 7 + 2 = 9 and 1 carried from the hundreds make 10 thousands. The final sum is therefore 10,121.

Another example: suppose we have to add 7,196 and 15,273. We start with the ones: 6 + 3 = 9. We write 9 in the ones place and nothing is left to carry to the tens. Next, 9 + 7 = 16. We write the 6 and carry the 1 to the hundreds. Now the hundreds: 1 + 2 = 3, and the 1 carried makes 4. We have nothing to carry to the thousands. So in the thousands: 7 + 5 = 12. Now we carry 1 to the ten-thousands place, where only one number already has 1. We add 1 + 1 to get 2 in the ten-thousands place. The final sum in this case is 22,469.

CHECKING ANSWERS

Before calculators made things easy, bookkeepers would use two methods to add long columns of numbers. First, they would add starting at the top and working down. Then they would add the same numbers starting at the bottom and working up. They kept right on doing this once they got calculators because it was a good way for them to check their work! Adding long lists of numbers can get tedious, and it's easy to make a mistake. You don't want to do that when you're dealing with important things such as tax returns.

You see the advantage of using more than one method. The *partial sums* that you move through on the way are different when you go upward, as compared to when you work down. But you should always reach the same answer at the end. It's not likely that you would enter the same wrong number twice under these conditions. If you get different answers, work each one again until you find where you made your mistake. (And be happy you found it before the tax man did!)

WEIGHTS

Now let's get away from pure numbers look at some real-world examples of addition and measurement. How about weight?

Modern scales read weight in digital format. They "spit out the numbers" at you. But scales weren't always so simple. You might have seen another kind of scale that uses a sliding weight. You have to balance it and read numbers from a calibrated scale. The old-fashioned *grocer's balance* worked in an even more primitive way. This device is now considered an antique, but knowing how it worked can help you understand addition in a practical sense.

The simplest grocer's balance, like the thing you've seen the blindfolded "Lady of Justice" holding, had two *pans* supported from a beam pivoted across a point or *fulcrum*. The pans were equally far away from the fulcrum on opposite sides. When the weights in both pans were equal, the scales balanced and the pans were level with each other. When the weights were unequal, the pan with the heavier weight dropped and the other one rose. To use such scales, the grocer needed a set of standard weights, such as those shown in Fig. 1-11.

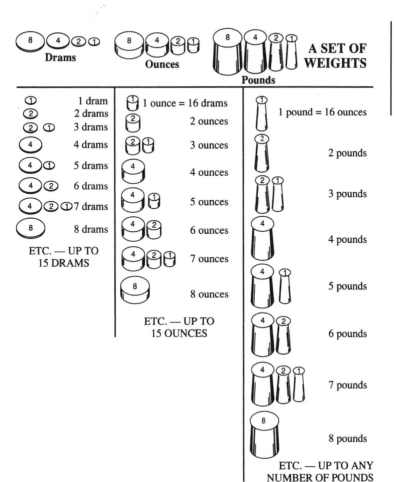

Figure 1-11

Standard weights were combined for use on an old-fashioned grocer's balance.

Standard or *avoirdupois* weight, still used in some English-speaking countries, does not follow the more sensible power-of-10 *metric system*. Instead, it defines 16 *drams* to an *ounce*, 16 ounces to a *pound*, 28 pounds to a *quarter*, four quarters (112 pounds) to a *hundredweight*, and 20 hundredweights (or 2240 pounds) to the *long ton*. (A *short ton* of 2000 pounds is used by most laypeople in "nonmetric" countries).

USING THE GROCER'S BALANCE

A set of weights for use with a grocer's balance consisted of those shown at the top of Fig. 1-11. It was only necessary to have 12 of them, unless the grocer wanted to measure more than 15 pounds. With these weights, if the scale was sensitive enough, the grocer could weigh anything to the nearest dram.

Suppose we want to weigh a parcel using a grocer's balance. Figure 1-12 shows how this process might go. First, we put the parcel in the pan on the left. Then we put standard weights on the other pan until the scale tips the other way. If a 1-pound weight doesn't tip it, we try a 2-pound weight. Suppose it still doesn't tip! But then the 2-pound and 1-pound weights together, making 3 pounds, do tip it. So we know that the parcel weighs more than 2 pounds, but less than 3 pounds.

Now we leave the 2-pound weight in the pan and start using the ounce weights. Suppose 8 ounces don't tip the scale. If 4 ounces are added to make 12 ounces, it still doesn't tip. But when we add a 2-ounce weight, which brings the weight up to 2 pounds 14 ounces, it tips. If the 1-ounce weight is used instead of the 2-ounce weight, the scale doesn't tip. Now we know that the parcel weighs more than 2 pounds 13 ounces, and less than 2 pounds 14 ounces. If we want to be more accurate, we can follow this method until it balances with 2 pounds, 13 ounces, and 3 drams.

QUESTIONS AND PROBLEMS

This is an open-book quiz. You may refer to the text in this chapter when figuring out the answers. Take your time! The correct answers are in the back of the book.

1. Does it make any difference in the final answer whether you count objects (a) one by one, (b) in groups of ten, or (c) in groups of twelve?

2. Why do we count larger numbers in hundreds, tens, and ones, instead of one at a time?

3. Why should we bother to write down zeros in numerals?

4. What are (a) 10 tens and (b) 12 twelves?

5. What are (a) 10 hundreds, (b) 10 thousands, and (c) 1000 thousands?

PARCEL WEIGHS...

Figure 1-12
An example of how we can weigh a parcel using a grocer's balance.

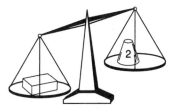

more than 2 pounds

less than 3 pounds

more than 2 pounds
12 ounces

less than 2 pounds
14 ounces

more than 2 pounds
13 ounces

less than 2 pounds
13 ounces and 4 drams

more than 2 pounds
13 ounces and 2 drams

**BALANCES AT 2 POUNDS
13 OUNCES AND 3 DRAMS**

6. By counting on, add the following groups of numbers. Then check your results by adding the same numbers in reverse order. Finally, use your calculator.

    (*a*) 3 + 6 + 9   (*b*) 4 + 5 + 7   (*c*) 2 + 7 + 3

    (*d*) 6 + 4 + 8   (*e*) 1 + 3 + 2   (*f*) 4 + 2 + 2

    (*g*) 5 + 8 + 8   (h) 9 + 8 + 7   (*i*) 7 + 1 + 8

7. Add together the following groups of numbers. In each case, use a manual method (without using a calculator) first, and then verify your answer with a calculator.

    (*a*) 35,759 + 23,574 + 29,123 + 14,285 + 28,171

    (*b*) 235 + 5,742 + 4 + 85,714 + 71,428

    (*c*) 10,950 + 423 + 6,129 + 1 + 2

    (*d*) 12,567 + 35,742 + 150 + 90,909 + 18,181

    (*e*) 1,000 + 74 + 359 + 9,091 + 81,818

8. How does adding money differ from adding pure numbers?

9. Add together the following weights: 1 pound, 6 ounces, and 14 drams; 2 pounds, 13 ounces, and 11 drams; 5 pounds, 11 ounces, and 7 drams. Check your result by adding them in at least three ways.

10. What weights would you use to weigh out each of the quantities in question 9, using the system of weights for a grocer's balance? Check your answers by adding up the weights you name for each object weighed.

11. In weighing a parcel, suppose the 4-pound weight tips the pan down, but the 2- and 1-pound weights do not. What would you do next to find the weight of the parcel (a) if you wanted it to the nearest dram; (b) if you had to pay postage on the number of ounces or fractions of an ounce?

12. The yard is a unit of length commonly used in the United States. It has 3 feet, and each foot has 12 inches. How many inches are in 2 yards?

Figure 1-13

Non-metric standard units of liquid measure commonly used in the United States. Illustration for problems 13 and 14.

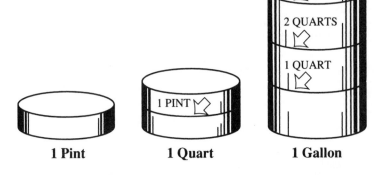

3 QUARTS

2 QUARTS

1 QUART

1 PINT

1 Pint   **1 Quart**   **1 Gallon**

13. In the United States, common liquid measures are the *pint*, the *quart*, and the *gallon*. There are 2 pints to a quart and 4 quarts to a gallon (Fig. 1-13). Suppose that a fleet of cars need oil changes. Three cars require 5 quarts each, two cars require 6 quarts each, and four cars require 1 gallon each. How many gallons of oil does the owner need?

14. If the owner of the cars in the previous problem can buy quarts of oil at 90 cents and gallons at $3.50, how should he buy the oil to be most economical?

15. Suppose a woman buys three dresses at $12.98 each, spends $3.57 on train fare to get to town and back, and spends $5.00 on a meal while she is there. How much did she spend altogether?

2 Subtraction

Just as addition is counting on, subtraction is counting away. We start with the total number, count away the number to be subtracted, and then see how many of the original items remain.

TAKING AWAY

Figure 2-1 shows an example of how subtraction works by counting away. When you were in grade school you might have said, "Eight take away three is five." You would write this as $8 - 3 = 5$. The teacher would say, "Three from eight is five."

Many people, when first learning arithmetic, have more trouble with subtraction than with addition. Some people keep having this problem for years! If you are one of these folks, you can make a subtraction table, just as you can make an addition table. But it really is a good idea to memorize your *subtraction facts* for all the single-digit numbers. Then you'll know right off, for example, that $7 - 4 = 3$.

Note that when we subtract in practical situations involving objects such as the coins in Fig. 2-1, the number "taken away" is never bigger than the number "taken away from." So we don't get expressions such as $8 - 9$. We can't have eight pennies and then take away nine! But in more advanced mathematics, we can have a *difference* such as $8 - 9$. When the number "taken away" is larger than the number "taken away from," we get a *negative number*. You will learn about negative numbers, and even stranger ones, later in this book.

CHECKING SUBTRACTION BY ADDITION

It is most important, all through mathematics, to be sure that we arrive at the right answers when we're done calculating. No matter how sophisticated the method might be, if the result is wrong, the method isn't worth a thing! That is why we use at least

Figure 2-1

An example of subtraction as counting away with coins. Three from eight equals five. We write this as $8 - 3 = 5$.

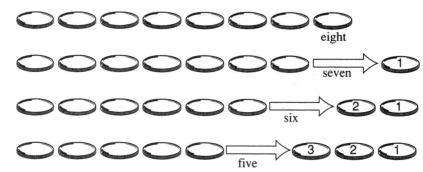

two ways of adding, one to check the other. In subtraction, an easy way to check the answer is to reverse the process by addition, to see if we get the number we began with. An example is shown in Fig. 2-2.

BORROWING

When we work out addition problems, if the digits in the ones column add up to 10 or more, the tens digit in the answer carries over into the tens column. If the tens figures in the whole sum add up to 10 or more, the tens digit in the answer carries over into

Figure 2-2

Subtraction can be checked by adding back. To be sure that $8 - 3 = 5$, we take five coins and add back the three we took away, getting the original eight coins.

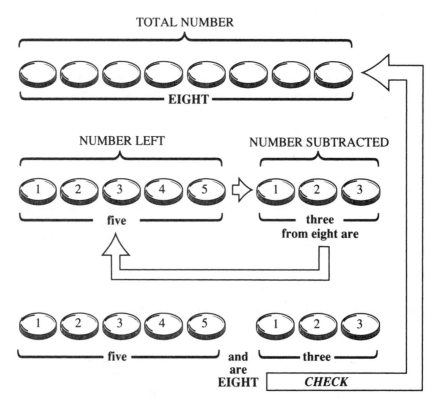

the hundreds column, and so on. In subtraction, this process is reversed. Instead of *carrying* digits from one column to another, we can *borrow* them from one column and give them to another if we get what seems to be a negative number in any particular column.

Suppose we must subtract 17 from 43. That means we want to find the difference between the two numbers, written as $43 - 17$. First, we subtract the numbers in the ones column. But then we get $3 - 7$. How do we deal with this? We can't attack this problem head on, but we can go at it indirectly. We can subtract 7 from 13. To get 13 instead of 3, we "borrow" a 1 from the tens column. That's like adding 10 to the 3. Now we have $13 - 7 = 6$. Taking away the 1 that was borrowed from the tens column leaves only 3 in that column from which we take away 1, getting 2. So the final answer is $43 - 17 = 26$. This process is shown in Fig. 2-3.

Now let's check the result by adding 26 to 17. In the ones column, $6 + 7 = 13$. We write down the 3 and carry the 1 to the tens column. In the tens column, 1 carried plus 2 equals 3, plus 1 more equals 4. So $26 + 17 = 43$, which checks back with the number we began with.

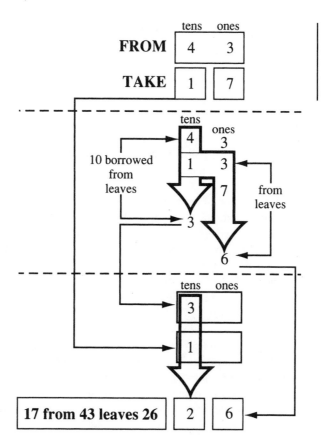

Figure 2-3

An example of borrowing in order to subtract 17 from 43.

When you check subtraction by adding back in this way, you can be confident that your answer is correct. You would have to make two "mirror-image" mistakes for things to check out as if they were right when they really weren't. Nearly always, if you happen to make one mistake when subtracting and another mistake when adding back, you don't get the original number at the end. That's a "red flag" that says you had better work the problem out all over again!

SUBTRACTING WITH LARGER NUMBERS

Now that you have the idea, try some subtraction problems involving big numbers. How about 17,583 from 29,427? Work through it yourself. You will have to borrow from the hundreds for the tens, and again from the thousands for the hundreds.

When borrowing, some people like to cross out the original figure and reduce it by 1. In this example, at the hundreds figure you are subtracting 5 from 3, which, with 1 borrowed from the thousands, is 13. Then, in the thousands, subtract 7 from 8, not 9. You should get 11,844 for the difference.

Now turn things around and find the sum 17,583 + 11,844 to check your work. Note that you'll have to carry in the same places that you borrowed when subtracting. You should return to 29,427, which was the number in the top line of the original subtraction.

WORKING WITH MONEY

Numbers that express money are no more difficult to add and subtract than are other numbers. The only difference is the presence of a decimal point and a change in the positions of the commas.

In an expression of cash, the decimal point, which separates dollars from cents, always stays immediately to the left of the hundreds digit. That is, the decimal point should always have two digits after it. If you're working with large amounts of money, you will have to put commas at multiples of three digits to the left of the decimal point. For example, a new bridge across the river in a town might cost $34,456,220.05 to design and build. That's thirty-four million, four hundred fifty-six thousand, two hundred twenty dollars and five cents. Look carefully at the positions of the decimal point and the commas here.

When you add and subtract cash amounts, just remember that you're always adding or subtracting numbers of cents, no matter how large the dollar figures happen to be. And again, that old reminder: Never forget to include all the zeros, whether they occur before the decimal point or after it. In the subtraction example of the previous section, if you were to put in decimal points to make the numbers into cash amounts, you would have had $294.27 − $175.83 = $118.44.

MAKING CHANGE

This idea of counting on, or using addition to check subtraction, is often used by sales-people when making change. Suppose you buy something for $3.27 and use a $5.00 bill for payment. Subtraction will show that you should get $1.73 in change. The salesper-son figures the bill (or maybe a computer does it!) and then "proves" that the bill is correct by giving you change and counting it back, as shown in Fig. 2-4.

Putting three pennies in your hand, the salesperson says, "$3.27, 28, 29, 30." Then he puts two dimes in your hand, saying "$3.40, 50." Next two quarters, saying

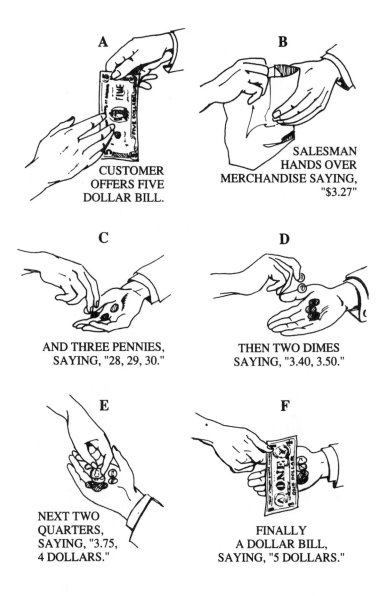

A

CUSTOMER
OFFERS FIVE
DOLLAR BILL.

B

SALESMAN
HANDS OVER
MERCHANDISE SAYING,
"$3.27"

C

AND THREE PENNIES,
SAYING, "28, 29, 30."

D

THEN TWO DIMES
SAYING, "3.40, 3.50."

E

NEXT TWO
QUARTERS,
SAYING, "3.75,
4 DOLLARS."

F

FINALLY
A DOLLAR BILL,
SAYING, "5 DOLLARS."

Figure 2-4

Making change for a $3.27 purchase done with a $5.00 bill. The customer gives the salesper-son the $5.00 bill (A) and receives the goods (B). Then the salesper-son hands over and "counts back" pennies (C), dimes (D), quarters (E) and a dollar (F).

"$3.75, $4.00." Finally he gives you a dollar bill, saying "$5.00," which was the amount you tended. During this process, you and the salesperson were both checking the change by adding it to the cost of what you bought, in order to get back to the $5.00.

SUBTRACTING WEIGHTS

Suppose a mother wants to weigh her baby, who is too big for baby scales and too "wriggly" for ordinary scales. The mother weighs herself holding the baby, and then weighs herself without the baby. The difference is the baby's weight. For example, if the mother weighs 156 pounds holding the baby (Fig. 2-5A) and 121 pounds without the baby (Fig. 2-5B), then the baby weighs 156 − 121, or 35 pounds.

Now let's look at an example of how weights can be subtracted with an old-fashioned grocer's balance. Suppose you are weighing something that you eventually learn weighs 3 pounds 14 ounces. In the traditional method, you put the parcel in one pan and a selection of weights in the other pan. You have 1-pound and 2-pound standard weights, along with 8-ounce, 4-ounce, and 2-ounce weights. After some trial and error, you can determine the weight of the parcel down to the nearest ounce (Fig. 2-6A).

The other method, shown in Fig. 2-6B, uses subtraction (or backward addition, if you'd rather think of it that way). You find that the parcel weighs just under 4 pounds. So you put small weights in the pan with the parcel, and you find that it balances with the 2-ounce weight in the parcel pan and the 4-pound weight in the weight pan. So you know that the parcel weighs 2 ounces less than 4 pounds. That, of course, is 3 pounds 14 ounces. Alternatively you can say that the weight of the parcel, plus 2 ounces, equals 4 pounds.

Figure 2-5

Weighing by subtraction. A mother weighs 156 pounds holding her baby (A) and 121 pounds without the baby (B). The baby weighs 156 − 121, or 35 pounds.

156 pounds   121 pounds

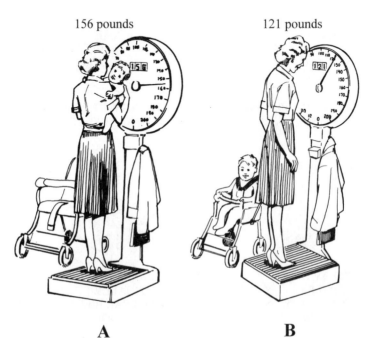

A   B

A

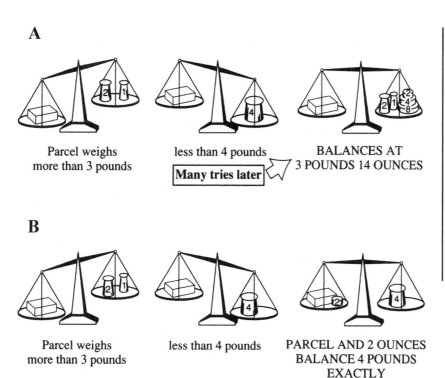

Parcel weighs
more than 3 pounds

less than 4 pounds

Many tries later

BALANCES AT
3 POUNDS 14 OUNCES

B

Parcel weighs
more than 3 pounds

less than 4 pounds

PARCEL AND 2 OUNCES
BALANCE 4 POUNDS
EXACTLY

Figure 2-6
Two methods of weighing a parcel. At A, add weights to the right-hand (non-parcel) side until the pans balance. At B, note that the parcel weighs less than a certain amount, and then add weights to the left-hand (parcel) side until the pans balance.

SUBTRACTING LIQUID MEASURES

Another application of subtraction is figuring out how much gas you use on a car trip, or figuring out how much gas you have left in the tank after a trip.

Suppose you know that your car's gas tank can hold 20 gallons when full. At the beginning of a journey, you fill the tank until the gas pump stops automatically, indicating a full tank. At the end of the trip, you fill the tank again in the same way. The pump delivers 9 gallons and then stops automatically. That means you used 9 gallons during the trip, assuming both pumps detect "fullness" in the same way. You also know that there were $20 - 9 = 11$ gallons left in the tank at the end of the trip.

QUESTIONS AND PROBLEMS

This is an open-book quiz. You may refer to the text in this chapter (and Chap.1, too, if you want) when figuring out the answers. Take your time! The correct answers are in the back of the book.

1. Make the following subtractions by hand, and check your results by addition.
(a) $69 - 46$ (b) $123 - 81$ (c) $543 - 37$
(d) $762 - 371$ (e) $509 - 410$ (f) $263 - 74$
(g) $4,321 - 1,234$ (h) $6,532 - 2,356$ (i) $11,507 - 8,618$

2. A refrigerator has a list price of $659.95. A local discount store offers a $160 discount on this item. How much would you pay at this store? Check your result by addition.

3. A lady purchases items priced at $2.95, $4.95, $3.98, $10.98, and $12.98. After adding up the bill, the store clerk offers to knock off the extra cents, so she would pay only the dollar amount. The lady has a better idea. Why not knock the extra cents off each item? How much more will she save if the store clerk accepts this suggestion?

4. A child wants to weigh her pet cat. The cat won't stay on the scales long enough for her to get a reading. So she weighs herself while holding the cat, and then without it. With the cat, she weighs 93 pounds. Without it, she weighs 85 pounds. How much does the cat weigh?

5. A recipe calls for 1 pound 12 ounces of rice. You have a grocer's balance to measure it with. All the pound weights are there, but the only ounce weights that have not been lost are the 1-ounce and 4-ounce ones. How can you use this balance to be sure you have 1 pound 12 ounces of rice? Prove it by showing that the scale balances.

6. You are studying a map on which town B is between towns A and C on highway X. The map shows only selected distances. Between A and B, the map shows 147 miles along highway X. Between A and C, it shows 293 miles along highway X. If you want to go from B to C along highway X, how far will you have to drive?

7. A freight company charges partly on weight and partly on mileage. The distance charge is based on direct distance, even if the company's handling necessitates taking it farther. Suppose that a package addressed to town B travels 1,200 miles from A to C and then 250 miles back to B on the same route. On what distance is the charge based?

8. A man has a parcel of land along a 1-mile frontage of highway. He has sold pieces with frontages of 300 yards, 450 yards, 210 yards, and 500 yards. A mile is 1,760 yards. How much frontage does he have left to sell?

3 Multiplication

Suppose you go into a store and buy seven articles for $1.00 each. The total cost is $7.00; you count $1.00 seven times. What if the articles cost $3.00 apiece? In that case, to find the total, you must count $3.00 seven times. This type of problem leads to the next step in calculating: *multiplication*.

IT'S A SHORTCUT

Multiplication is a shortcut for repeated addition. At one time, children learning arithmetic memorized huge *multiplication tables* without knowing why they'd ever need such facts. If you memorized, say, 7 times 3 is 21 (written $7 \times 3 = 21$), you could do calculations more quickly than if you had to go to a table and look up every multiplication fact. But not many people could tell you the reason why $7 \times 3 = 21$! Figure 3-1 shows why this is true: you must add up seven "threes" to get 21.

You won't regret it if you memorize multiplication facts up through $9 \times 9 = 81$, as shown in Table 3-1. These facts will be a great convenience if you know them "by heart." Then, along with the material you will learn in the rest of this chapter, when you need to multiply two numbers together and you don't happen to have a calculator handy, you will be able to do it without any trouble.

MAKING A TABLE

The multiplication table was one of humankind's earliest computers: a ready way of getting answers without having to add things over and over. To understand it, you can set up a multiplication table for yourself, as shown in Table 3-1. Start with the numbers along the top and down the left side. Now, count in twos, and write the results in the next column, under "2." Each next figure down the column is 2 more than the one above it. Now, do the same thing with the "3" column, adding 3 for each next figure down the column. Continue until you have done the "9" column.

Figure 3-1

Multiplication is a form of repeated addition. In this case, seven sets of three items, all added up, gives us 21 items. We write $7 \times 3 = 21$.

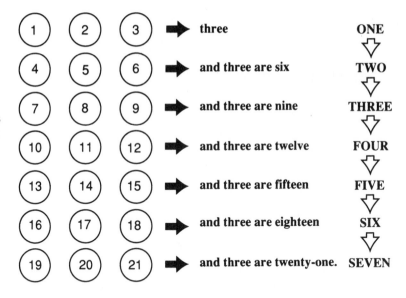

PATTERNS IN MULTIPLICATION TABLES

If you count carefully enough, your table should turn out right. The idea behind making your own table is to convince yourself that the table is true. That way, using it will not be "cheating," which is what some teachers used to say about students who used printed tables to solve their multiplication problems. You will be using what you have already done and verified. So, how do you verify it?

The simplest check involves looking at the odds and evens. An even number is one whose ones figure is 2, 4, 6, 8, or 0. An odd number has a ones figure of 1, 3, 5, 7, or 9. Notice that the only places in the table (called *cells*) where you have odd numbers are where both the number at the top of the column and the number at the left-hand end of the row are odd. If either of those numbers is even, the number in the cell is even.

Table 3-1

Multiplication table from 2×2 up through 9×9. This is the sort of table you used in grade school. It's not a bad idea to memorize it!

	2	3	4	5	6	7	8	9
× 2 =	4	6	8	10	12	14	16	18
× 3 =	6	9	12	15	18	21	24	27
× 4 =	8	12	16	20	24	28	32	36
× 5 =	10	15	20	25	30	35	40	45
× 6 =	12	18	24	30	36	42	48	54
× 7 =	14	21	28	35	42	49	56	63
× 8 =	16	24	32	40	48	56	64	72
× 9 =	18	27	36	45	54	63	72	81

In Fig. 3-2A, examine the numbers in the cells whose column or row header is equal to 5. All these numbers have a ones figure of either 0 or 5. Another thing you should notice is that any multiplication answer, called a *product*, can be found in two places, except where a number is multiplied by itself. For example, 3 × 7 is the same as 7 × 3, as shown at B. You see 21 in two places. This rule is true for every combination of two different numbers. It is called the *commutative law for multiplication*.

Another interesting thing happens in the column or row against 9, as shown at C. Notice that each successive number has one more in the tens and one less in the ones, and that adding the two digits in the product together always makes 9.

Now take the diagonal where numbers are multiplied by themselves. These numbers are called *squares*, and they are shown at D. You can complete the table by adding the digit 1 at the extreme upper left. Put a circle around it to indicate that 1 × 1 = 1, so you remember that the number 1 is its own square. Notice the difference between successive places down the *squares diagonal*:

$$4 - 1 = 3$$
$$9 - 4 = 5$$
$$16 - 9 = 7$$

A

	2	3	4	5	6	7	8	9
				times				
2				10				
3				15				
4				20				
5	10	15	20	25	30	35	40	45
6				30				
7				35				
8				40				
9				45				

B

	2	3	4	5	6	7	8	9
				times				
2								
3						21		
4								
5								
6								
7		21						
8								
9								

C

	2	3	4	5	6	7	8	9
				times				
2								18
3								27
4								36
5								45
6								54
7								63
8								72
9	18	27	36	45	54	63	72	81

D

	①	2	3	4	5	6	7	8	9
					times				
2		4		8					
3			9		15				
4		8		16		24			
5			15		25		35		
6				24		36		48	
7					35		49		63
8						48		64	
9							63		81

Figure 3-2

Some patterns in numbers make useful checks. At A, checking the fives; at B, checking by symmetry; at C, checking the nines; at D, checking by diagonals.

$$25 - 16 = 9$$
$$36 - 25 = 11$$
$$49 - 36 = 13$$
$$64 - 49 = 15$$
$$81 - 64 = 17$$

The differences start at 3 and then keep on increasing by 2 each time: 3, 5, 7, 9, 11, 13, 15, and 17. If you want to expand this table, you can verify that the progression continues on like this for all squares, that is, for all numbers that are equal to some smaller number multiplied by itself.

Here's another interesting quirk in the multiplication table. Notice what happens when you move upward and to the right ("northeast") or downward and to the left ("southwest") by one cell from a square. Going either way from 9 gives you 8. Going either way from 16 gives you 15. Going either way from 25 gives you 24, and so on. Moving one cell "northeast" or "southwest" away from the square diagonal always causes a decrease of 1.

HOW CALCULATORS MULTIPLY

The tables we have just examined only go up to 9 × 9. Years ago, children had to learn the tables up to 12 × 12, or even up to 20 × 20. What a chore! In the decimal number system, all we really need to memorize is the multiplication facts up to 9 × 9. Multiplying by 10 merely adds a zero to the end. We can expand on other facts, too. For example, 5 × 6 = 30, so it follows that 5 × 60 must be 300 and then 50 × 60 must be 3000. Whichever number you multiply by 10, the product also gets multiplied by 10 (using an extra 0).

Calculators use this fact. Actually a calculator does not multiply. It keeps on adding. But because it goes so much faster than any human can, it seems as if it arrives at the product instantly. If you enter 293 × 135 in a calculator, for example, it follows the process shown in Fig. 3-3 to arrive at the product 39,555. The calculator performs repeated addition operations to do what we humans call multiplication.

Sometimes, people have to multiply big numbers by hand. To do this, you must multiply every part of one number, the *multiplicand* (as it used to be called), by every part of the other number, the *multiplier*. You can multiply in either order as long as you do it systematically. Figure 3-4 shows how the product 293 × 135 can be determined manually, going by hundreds, tens, and ones. We must take things in order, from left to right in each number, working with the multiplicand (293) first in each case. Most people did not learn to multiply large numbers by using this method. Maybe they would understand multiplication better if they had. At least, they'd have a better idea of how calculators work!

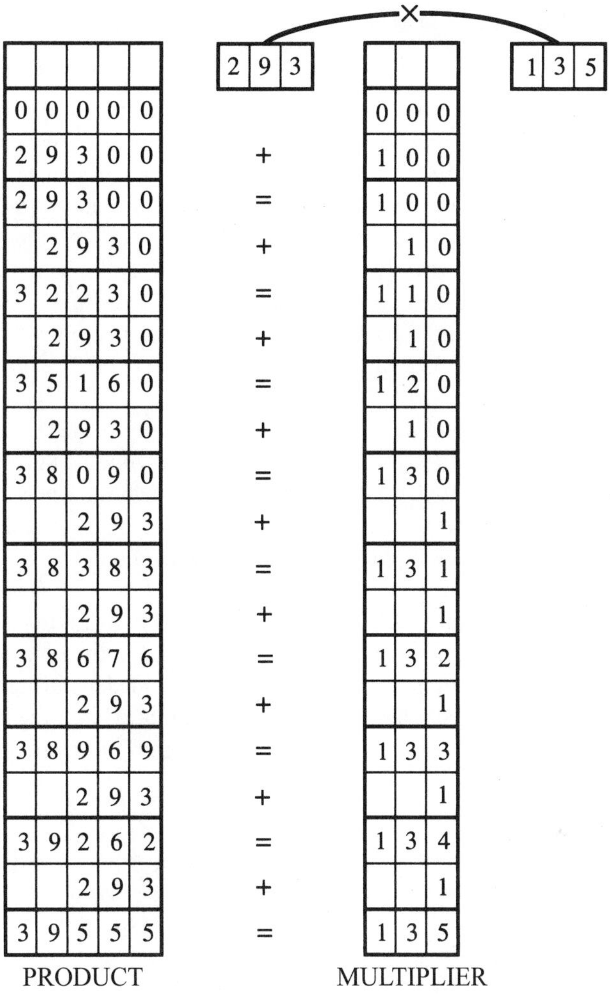

Figure 3-3
An example of the process by which a calculator multiplies two numbers.

0	0	0	0	0
2	9	3	0	0
2	9	3	0	0
	2	9	3	0
3	2	2	3	0
	2	9	3	0
3	5	1	6	0
	2	9	3	0
3	8	0	9	0
		2	9	3
3	8	3	8	3
		2	9	3
3	8	6	7	6
		2	9	3
3	8	9	6	9
		2	9	3
3	9	2	6	2
		2	9	3
3	9	5	5	5

PRODUCT   MULTIPLIER

Figure 3-4

Here is a way to find the product 293 × 135 manually by hundreds, tens, and ones.

Multiplicand		Multiplier		Subproduct	Cumulative Total
200	×	100	=	20 000	20 000
90	×	100	=	9 000	29 000
3	×	100	=	300	29 300
200	×	30	=	6 000	35 300
90	×	30	=	2 700	38 000
3	×	30	=	90	38 090
200	×	5	=	1 000	39 090
90	×	5	=	450	39 540
3	×	5	=	15	39 555

CARRYING IN MULTIPLICATION

When we are performing addition, carrying is a way of leaving out an extra digit, saving it for awhile, and then adding it back in later. We can do the same thing in multiplication. Figure 3-5 shows how carrying is done when we want to multiply 3,542 × 27. This can be broken down into a sum of two products: 3,542 × 7 and 3,542 × 20.

To get the first *partial product*, namely, 3,542 × 7, we first do the ones. That is 7 × 2 = 14; write 4 and carry 1. The tens: 7 × 4 = 28 with 1 carried to make 29; write 9 and carry 2. The hundreds: 7 × 5 = 35 with 2 carried to make 37; write 7 and carry 3. The thousands: 7 × 3 = 21 with 3 carried to make 24. This gives us 24,794.

To get the second partial product, we multiply 3,542 × 20 in the same fashion to get 70,840. Sometimes you don't write the zero; you just move the last digit (in this case 4) over one place to the left, so it's in the tens column. The whole thing is written down in one "piece" or *algorithm*, as the professional mathematicians call it. This is how people performed multiplication before they had computers.

At one time, the "new math" consisted in multiplying from the other direction. In Fig. 3-6, the same multiplication is performed in the reverse order: first the 20 and then the 7. The answer is the same either way, provided no mistakes are made. If the multiplier (that is, the second number or the lower number) has three or more digits, we must work consistently, either from left to right or from right to left.

Figure 3-5

A multiplication problem that involves finding the product in two parts.

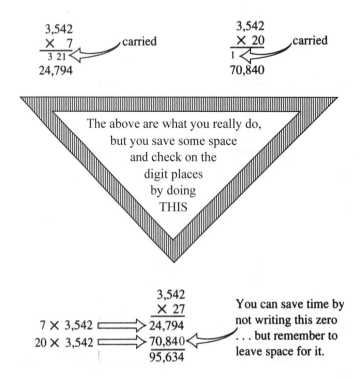

```
   3,542                        3,542
 ×    7      carried          ×  20      carried
 3 21                         1
 24,794                         70,840
```

The above are what you really do, but you save some space and check on the digit places by doing THIS

```
                3,542
              ×  27
7 × 3,542  ⟹  24,794
20 × 3,542 ⟹  70,840
              95,634
```

You can save time by not writing this zero . . . but remember to leave space for it.

Figure 3-6

It doesn't matter which part of the problem we do first. We get the same answer either way.

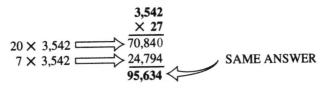

$$
\begin{array}{r}
3{,}542 \\
\times\ 27 \\
\hline
70{,}840 \\
24{,}794 \\
\hline
95{,}634 \\
\end{array}
$$

20 × 3,542 ⟹ 70,840
7 × 3,542 ⟹ 24,794   SAME ANSWER
95,634 ⟸

USING YOUR CALCULATOR TO VERIFY THIS PROCESS

When you have a digital calculator, it is easy to punch in one number, then the "times sign" (×), then the other number, and finally the equals sign (=). Bingo, you have the answer, all complete! But this doesn't help you see how the calculator actually does the work. Figure 3-7 can help you see that for the product 3,542 × 27.

Suppose we have a calculator with a single memory, which is the simplest type. Multiplying 7 by 2 gives us 14, which we enter in memory with a button labeled something like "MS" for "memory save." Then we multiply 7 by 40, which gives us 280. Next, we add this number to the 14 with the "M+" or "memory add" button. We can read what we already have by pressing the "MR" or "memory recall" button. Finish multiplying by 7 and then press the MR button to display the first partial product of 24,794.

After that we can go on and multiply by 20. With a single memory, we don't see the "times 20" part separately, as we do in longhand. But the final answer is the same. If you have a calculator with more than one memory, you can store each partial product in a separate memory and then add the contents of the two memories.

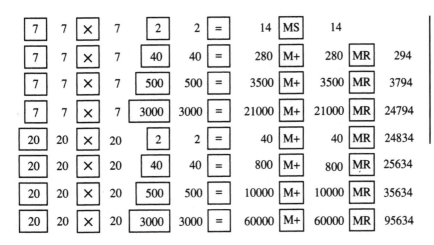

Figure 3-7

Here is how you can use a calculator to find the product 3,542 × 27 in two parts, as shown previously by the longhand method.

Figure 3-8

In multiplication, zeros can be used to keep figures in their proper places.

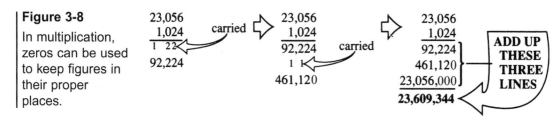

PUT THOSE ZEROS TO WORK!

When you multiply by longhand, don't forget the zeros, if your multiplier has a zero in it. When you pass zero, there is no point in writing a line of zeros, because zero times anything is zero. But don't forget that, in this case, after multiplying by 20 (the tens figure), the next one is the thousands figure, which moves to the left two places instead of one, because the multiplier has no hundreds figure. It's a good idea to write down all the zeros in a partial product that ends with them. This can help you keep the digits in their proper places, as shown in Fig. 3-8.

When we multiply two numbers by the longhand method, we tend to use the shorter number as the multiplier because it looks like less work. Of course it really isn't. We still have to multiply every digit in one number by every digit in the other. If you use a calculator to solve a multiplication problem straightaway by entering one number in full, then the times button, then the other number in full, and finally the equals button, you can multiply a short number by a long one or vice versa. The answer, as well as the amount of work you do, will be the same either way.

USING SUBTRACTION TO MULTIPLY

Sometimes it's convenient to make complicated calculations in your head, instead of relying on a calculator. Although calculators don't need shortcuts or tricks, they can be a big help when you do things in your head. Figure 3-9 shows two examples of using subtraction to help do multiplication problems. People are almost always amazed when I show them this trick, which works best when the multiplier ends in 8 or 9.

In Fig. 3-9A, we multiply 47,392 by 29. At left, the multiplier 29 is just 1 less than 30. So we multiply by 30, which is much easier than multiplying by 29. Then we subtract 1 times the original number, which is itself. Check it in the usual way as shown at right, and you will find the same answer. In Fig. 3-9B, we multiply 63,257 by 98. At left, 98 is 2 less than 100. Multiplying by 100 puts two zeros to the right of the original number.

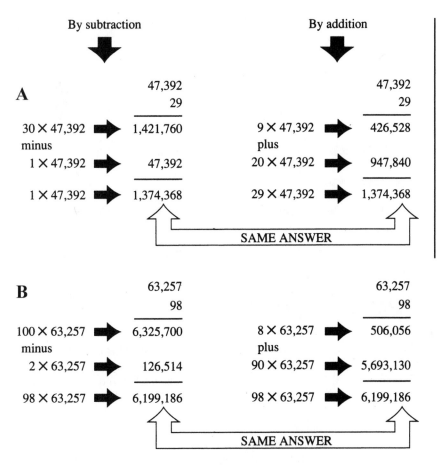

By subtraction

By addition

A
47,392
29

30 × 47,392 ➡ 1,421,760
minus
1 × 47,392 ➡ 47,392

1 × 47,392 ➡ 1,374,368

47,392
29

9 × 47,392 ➡ 426,528
plus
20 × 47,392 ➡ 947,840

29 × 47,392 ➡ 1,374,368

SAME ANSWER

B
63,257
98

100 × 63,257 ➡ 6,325,700
minus
2 × 63,257 ➡ 126,514

98 × 63,257 ➡ 6,199,186

63,257
98

8 × 63,257 ➡ 506,056
plus
90 × 63,257 ➡ 5,693,130

98 × 63,257 ➡ 6,199,186

SAME ANSWER

Figure 3-9

Subtraction can help us do certain multiplication problems. At A, we multiply 47,392 by 29. At B, we multiply 63,257 by 98. The examples at left show subtraction. The examples at right show the traditional addition method for comparison.

Subtract 2 times the original number in the multiplier, and you have the answer, which is verified at right.

MULTIPLYING BY FACTORS

Figure 3-10 illustrates a trick that works well with numbers that can be broken into single-digit *factors*. What is a factor? When two numbers are multiplied by each other to get a third number, then the first two numbers are factors of the third number. Some numbers can be split into whole-number factors, and others cannot.

Suppose we have a multiplication problem in which the multiplier is 35. That happens to be 5 × 7. The factors of 35 are therefore 5 and 7. Instead of multiplying the whole original number by 5 and by 30 and adding the results, you can multiply first by 5 and then multiply that result by 7. As is shown here, both ways give the same answer. You can verify this on your calculator, too.

Figure 3-10

An example of multiplication by factors. Here, we multiply 23,657 by 35. At left, the addition method is shown. At right, we multiply 23,657 by 5 and then multiply that product by 7, because 5 and 7 are factors of 35.

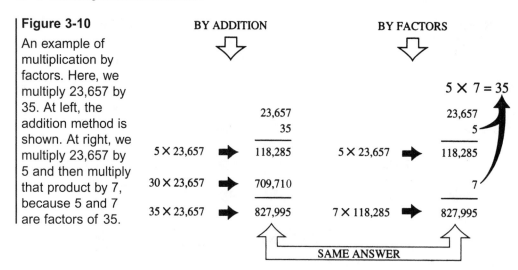

BY ADDITION

BY FACTORS

$5 \times 7 = 35$

	23,657			23,657
	35			5
$5 \times 23,657$ →	118,285	$5 \times 23,657$ →	118,285
$30 \times 23,657$ →	709,710			7
$35 \times 23,657$ →	827,995	$7 \times 118,285$ →	827,995

SAME ANSWER

MULTIPLYING IN NONDECIMAL SYSTEMS

When you use systems that are *nondecimal* (not based on the number we call ten), multiplication is a little bit complicated. Some calculators are equipped to make conversions to nondecimal systems, but you need to know what to do if your calculator can't do that and you encounter such a problem.

The "pounds and ounces" or "English" weight measurement system is nondecimal, with 16 ounces to the pound. Suppose you want to find 25 times 1 pound 3 ounces. You can multiply the pounds by 25, then multiply the ounces by 25, and then just stick them together and say that the product is 25 pounds 75 ounces. But that isn't the way you'd normally express weight! The ounces part should always be less than 16. If it goes to 16 or more, you should add 1 to the pounds part of the figure and subtract 16 from the ounces part. You might have to do this more than once to get an ounces figure of less than 16. As you find the product 25 times 1 pound 3 ounces, doing this procedure four times converts the 75 ounces to 4 pounds and 11 ounces (Fig. 3-11). Then you add the 4 pounds to the 25 pounds, so the total is 29 pounds 11 ounces.

Figure 3-11

A multiplication problem involving weight in the "pounds-and-ounces" system. Here, we multiply 25 times 1 pound 3 ounces. Note that a pound contains 16 ounces.

25 times   3 ounces is

$$\begin{array}{r} 75 \text{ ounces} \\ -16 \\ \hline \end{array}$$

or 1 pound 59 ounces

$$\begin{array}{r} -16 \\ \hline \end{array}$$

or 2 pounds 43 ounces

$$\begin{array}{r} -16 \\ \hline \end{array}$$

or 3 pounds 27 ounces

$$\begin{array}{r} -16 \\ \hline \end{array}$$

or 4 pounds 11 ounces

25 times 1 pound is   25 pounds

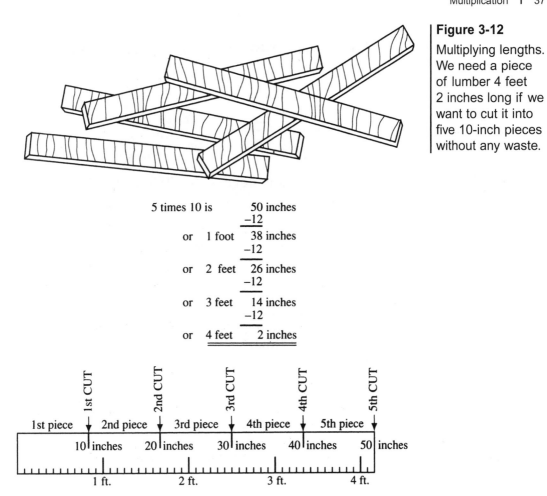

Figure 3-12
Multiplying lengths.
We need a piece
of lumber 4 feet
2 inches long if we
want to cut it into
five 10-inch pieces
without any waste.

5 times 10 is		50 inches
		−12
or	1 foot	38 inches
		−12
or	2 feet	26 inches
		−12
or	3 feet	14 inches
		−12
or	4 feet	2 inches

Figure 3-12 shows another nondecimal multiplication problem. The "yards-feet-inches" or "English" length measurement system defines 12 inches to the foot and 3 feet to the yard. Where necessary, we have to make conversions between inches, feet, and yards. How long must a piece of lumber be in order to get five 10-inch pieces without any waste when we cut it into equal pieces? The answer is 50 inches, but we would usually say it is 4 feet 2 inches. We could also say it is 1 yard 1 foot 2 inches.

In the same way, if you multiply a measure by a large number, it is usually convenient to change the unit of measure in which we express it, as shown by the example in Fig. 3-13. Suppose a motor crankcase takes 3 pints of oil. How much oil will you need for 250 motors of this same kind? You can multiply 3 pints by 250 and get an answer of 750 pints. But quantities this large are usually given in gallons, not in pints. Remembering that 8 pints make a gallon, you can proceed to count backward in eights. From 750, 8 can be subtracted 93 times, and when you are all done with this process, you'll have 6 pints (or 3/4 of a gallon) left over. But that is a tedious way to do it! You must be thinking there's an easier way, and you are right. We can use *division*. We'll learn about that in Chap. 4.

Figure 3-13

Multiplying liquid measures. This requires that we subtract 8 pints over and over and over, dozens of times! In the next chapter we'll learn how division can make it easier to do problems like this.

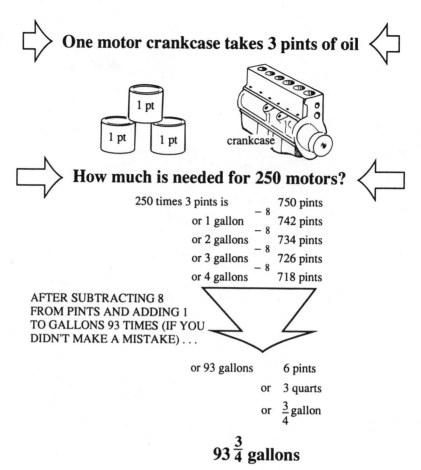

One motor crankcase takes 3 pints of oil

1 pt
1 pt   1 pt
crankcase

How much is needed for 250 motors?

250 times 3 pints is		750 pints
or 1 gallon	− 8	742 pints
or 2 gallons	− 8	734 pints
or 3 gallons	− 8	726 pints
or 4 gallons	− 8	718 pints

AFTER SUBTRACTING 8
FROM PINTS AND ADDING 1
TO GALLONS 93 TIMES (IF YOU
DIDN'T MAKE A MISTAKE) . . .

or 93 gallons   6 pints

or   3 quarts

or $\frac{3}{4}$ gallon

$93\frac{3}{4}$ **gallons**

QUESTIONS AND PROBLEMS

This is an open-book quiz. You may refer to the text in this chapter (and earlier ones, too, if you want) when figuring out the answers. Take your time! The correct answers are in the back of the book.

1. Multiply the following pairs of numbers as shown. Check your results by using the left-hand number as the multiplier in each case.

    (*a*) 357 × 246   (*b*) 243 × 891   (*c*) 24 × 36

    (*d*) 37 × 74   (*e*) 193 × 764   (*f*) 187 × 263

2. Multiply the following pairs by using subtraction to make the work simpler. Verify your results by the more usual method.

    (*a*) 2,573 × 19   (*b*) 7,693 × 28   (*c*) 4,497 × 18

    (*d*) 5,396 × 59   (*e*) 7,109 × 89

3. Multiply the following pairs by using factors of the second number. Check your results by the more usual method.

    (a) 1,762 × 45   (b) 7,456 × 32   (c) 8,384 × 21

    (d) 9,123 × 63   (e) 1,024 × 28

4. An airline runs four flights per day between two cities, except on Sundays when it runs only two flights. How many flights per week is this?

5. The flight described above is also made only twice on 12 public holidays during the course of a year. How many flights are made in a period of a year based on exactly 52 weeks?

6. A mass-produced item costs 25 cents each to make and $2.00 to package. The packet cost is the same, no matter how many items are in it. Determine the cost for packets of

    (a) 1   (b) 10   (c) 25   (d) 100

    (e) 250   (f) 1,000   (g) 5,000   (h) 10,000

7. If you ignore the manufacturing cost in problem 6, how much is saved by packing 500 items in packets of 250 as opposed to packets of 100?

8. On a commuter train, single tickets sell for $1.75. You can buy a 10-trip ticket for $15.75. How much does this save over the single-ticket rate if you take 10 trips?

9. On the same train, a monthly ticket costs $55. If a commuter makes an average of 22 round trips a month, how much will he save by buying a monthly ticket as compared with paying for every trip individually?

10. Suppose that a corporation offered a 10-hour-a-week part-time employment contract beginning at $500 a month, with a raise of $50 a month every year for five years. The contract expired after the sixth year. Employees bargained for a starting figure of $550 a month, with a raise of $20 a month every six months. Which rate of pay was higher at the beginning of the sixth year? Which rate resulted in the greatest total earnings per employee for six years? By how much?

11. A manufacturer prices parts according to how many are bought at once. The price is quoted per 100 pieces in each case, but the customer must take the quantity stated to get a particular rate. The rates are $7.50 apiece for 100, $6.75 apiece for 500, $6.25 apiece for 1,000, $5.75 apiece for 5,000, and $5.50 apiece for quantities over 10,000. The rate for any in-between quantity is based on the next-lower number. What is the difference in total cost for quantities of 4,500 and 5,000 parts?

12. Small parts are counted by weighing. Suppose 100 of a particular part weigh 2-1/2 ounces and you need 3,000 of the parts. What is the total weight?

13. A bucket that is used to fill a tank with water holds 4 gallons. To fill the tank, 350 bucketfuls are required. What is the tank's capacity?

14. A freight train has 182 cars each loaded to the maximum which, including the weight of the car, is 38 tons. What is the total weight that the locomotive has to haul?

15. A car runs 260 miles on one tank of gas. It has an alarm that lets the driver know when it needs refilling. If a particular journey required 27 fillings and at the end, the tank was ready for another, how long was the journey?

16. Two railroads connect the same two cities. The first railroad company charges 10 cents per mile, and the second company charges 15 cents per mile. The distance between the cities is 450 miles by the first railroad, but only 320 miles by the second. Which company offers the cheaper fare? By how much?

17. One airline offers rates based on 14 cents per mile for first-class passengers. Its distance between two cities is 2,400 miles. Another airline offers 10 cents a mile for coach, but uses a different route, marking the distance at 3,200 miles. Which fare is cheaper? By how much?

18. The first airline in the scenario of problem 17 offers a family plan. Each member of a family, after the first member, pays a rate that is based on 9 cents per mile. Which way will be cheaper for a family (a) of 2 and (b) of 3? By how much?

19. A health specialist recommends chewing every mouthful of food 50 times. One ounce of one food can be eaten in 7 mouthfuls. A helping of this food consists of 3 ounces. How many times will a person have to chew this helping to fulfill the recommendation?

20. An intricate pattern on an earthenware plate repeats nine times around the edge of the plate. Each pattern has seven flowers in the repetition. How many flowers are around the edge of the plate?

4

Division

Before the days of electronic calculators and computers, division was a shortcut for counting out. To divide 28 items among 4 people, you would keep giving one to each of the four until the items were all gone, and then you would see how many each person had. Figure 4-1 shows two ways of how this could be done. Division is symbolized by using the division sign ($\div$), and the equation in this case is written $28 \div 4 = 7$. In a division problem written in this way, the first number (in this case 28) is called the *dividend*, the second number (in this case 4) is called the *divisor*, and the answer (in this case 7) is called the *quotient*.

LET'S THINK LIKE A CALCULATOR

A calculator uses a repeated process of subtraction to perform division. Let's think like a calculator for a moment and see how this works when we want to divide 45,355 by 193. You can follow along here by reading down the columns in Fig. 4-2.

First, note that $100 \times 193 = 19,300$. Subtracting 193 from 45,355 over and over a total of 100 times leaves us with 26,055. Now 26,055 is greater than 19,300, so we can subtract 193 over and over, 100 more times, leaving us with 6,755. So far, we have subtracted 193 from 45,355 a total of 200 times. Now, let's start taking away 193×10, which is 1,930, over and over. When we begin with 6,755 and do this, we get, in succession, 4,825, 2,895, and 965, while adding 10 to the quotient column (called a *register* in calculator parlance) each time. So far, we have taken away 193 from 45,355 a total of 230 times. To finish, we subtract 193 five more times, leaving us with 0. Now we are all done with the process. We have subtracted 193 from 45,355 a total of 235 times, and there's nothing left over. That means $45,355 \div 193 = 235$.

Some students say that 193 "goes into" 45,355 exactly 235 times. When I did that in elementary school, my teachers scolded me, saying that "goes into" was a bad expression. But it helped me to understand division. You might prefer to say that 193 "comes out of" 45,355 exactly 235 times.

41

Figure 4-1

Two ways in which 28 items can be divided among four people so each person gets seven items.

Divide 28 items among 4 people

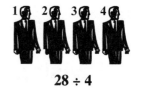

$28 \div 4$

First Method

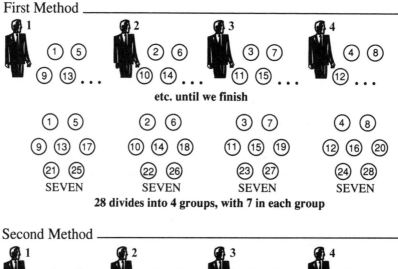

etc. until we finish

28 divides into 4 groups, with 7 in each group

Second Method

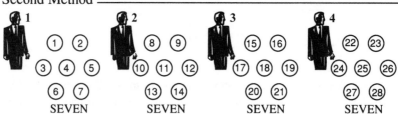

DIVISION "UNDOES" MULTIPLICATION

Just as a calculator performs multiplication by repeated addition, it also performs division by repeated subtraction. In fact, this pattern in mathematics is useful to follow through. Each process that we learn has a reverse, and each reverse process provides a way to check the one it reverses:

$$- \text{ can "undo" } +$$

$$+ \text{ can "undo" } -$$

$$\div \text{ can "undo" } \times$$

$$\times \text{ can "undo" } \div$$

We can say that subtraction is the opposite of addition in some ways, and division is the opposite of multiplication in some ways. But that doesn't mean that these pairs of

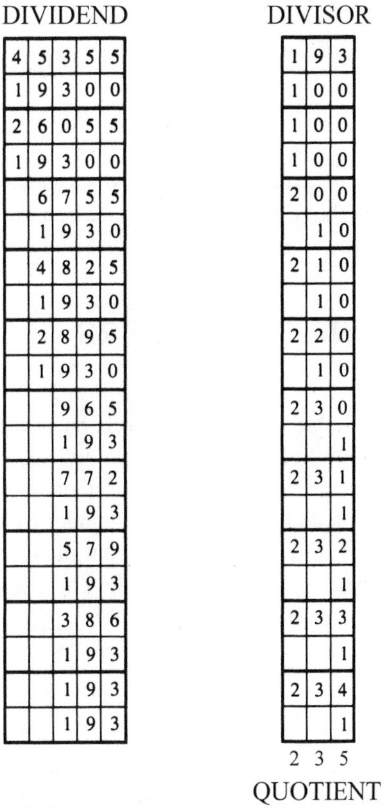

DIVIDEND

4	5	3	5	5
1	9	3	0	0
2	6	0	5	5
1	9	3	0	0
	6	7	5	5
	1	9	3	0
	4	8	2	5
	1	9	3	0
	2	8	9	5
	1	9	3	0
		9	6	5
		1	9	3
		7	7	2
		1	9	3
		5	7	9
		1	9	3
		3	8	6
		1	9	3
		1	9	3
		1	9	3

DIVISOR

1	9	3
1	0	0
1	0	0
1	0	0
2	0	0
	1	0
2	1	0
	1	0
2	2	0
	1	0
2	3	0
		1
2	3	1
		1
2	3	2
		1
2	3	3
		1
2	3	4
		1

2 3 5

QUOTIENT

Figure 4-2

A pictorial representation of the way a calculator divides 45,355 by 193 to get a quotient of 235.

operations have identical properties. They are not mirror-image operations; some big differences exist! You can add or multiply two numbers in either order, and it doesn't matter. But the order is critical in subtraction and division. For example, the following statements are *true*:

$$4 + 11 = 11 + 4$$
$$4 \times 11 = 11 \times 4$$

But these statements are *false*:

$$4 - 11 = 11 - 4$$
$$4 \div 11 = 11 \div 4$$

ALTERNATIVE NOTATION

Mathematicians, scientists, and engineers often use a forward slash to represent division, and you should get used to this notation. Here are a few examples:

$$10 \div 5 = 10/5$$
$$25 \div 5 = 25/5$$

$$333 \div 11 = 33/11$$

$$5,280 \div 176 = 5,280/176$$

$$49,000 \div 7,000 = 49,000/7,000$$

If an expression looks too "scrunched up" without spaces between the slash and the numbers, you can add spaces on either side of the slash. This is not a bad idea, especially if either or both of the numbers are long.

DIVIDING INTO LONGER NUMBERS

Now let's try another division problem: 1,743/7. You might find it helpful to have a multiplication table for the divisor, in this case 7, at the side. This table enables you to subtract the number in the quotient all in one "bite," rather than one piece at a time as a calculator does. The remainder each time is less than the divisor, so "bring down" the next digit or figure and continue for the next place in the quotient. Figure 4-3 shows what you really do, and then how it is usually written.

Figure 4-3

In division, we start with the larger number and work toward smaller ones. In this example, we divide 1,743 by 7 to get a quotient of 249.

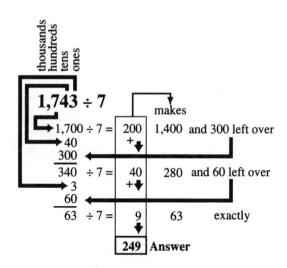

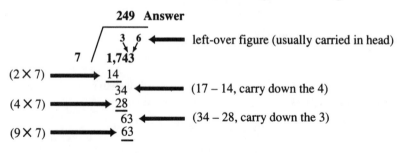

MULTIPLICATION CHECKS DIVISION

You can always check division by multiplication (Fig. 4-4), whether you do it the old-fashioned way or on a calculator. Using a calculator, you punch in the number 1,743, then the "divided by" button (marked "÷" or "/"), then the number 7, and finally the equals button (marked "="). The calculator reads 249. Now, with 249 still reading, punch the "times" button (marked "×"), then the number 7, and finally the = button. You should get back the original number: 1,743.

This procedure confirms that you hit the correct keys. If the last figure isn't the one you began with, you probably hit a wrong button somewhere. Try again!

MORE ABOUT HOW A CALCULATOR DOES IT

Look back to the section "Let's think like a calculator." How does the calculator "know" that its first subtraction must be 100 times the divisor? To figure this out, it begins with 1 times the divisor (or the divisor itself), then tries 10 times the divisor (1,930), and then 100 times the divisor (19,300). If it finds that the dividend is larger, it will increase to 1000 times the divisor (193,000). If it finds that the divisor is larger, it will drop back to 100 times. After subtracting 100 times the divisor twice, it tries the third time, finds that the division is bigger, so it drops back to 10 times the divisor. The same thing happens when it has subtracted 10 times the divisor a total of three times. The fourth time

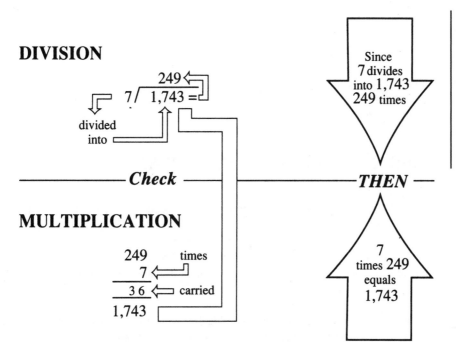

Figure 4-4
You can check your division by multiplying. When you multiply the quotient by the divisor, you should get the dividend (the original number) back.

Figure 4-5

Here is how a calculator figures out which numbers to take away as it divides 45,355 by 193.

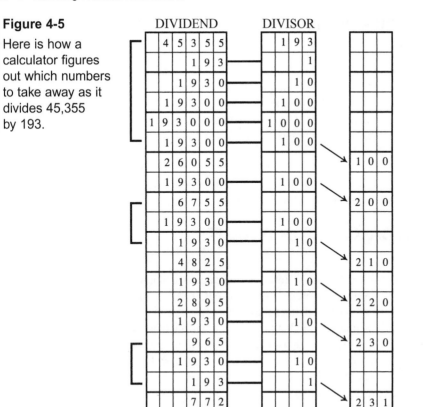

it finds that the divisor is bigger, so it drops back to the divisor itself, as shown in Fig. 4-5. It "grinds away" at the problem, putting the quotient together in a register called the *quotient accumulator*.

DIVIDING BY LARGER NUMBERS (THE PEOPLE WAY)

Figure 4-6A shows an example of why we suggested having a multiplication table handy when doing division "by hand," a process that has traditionally been called *long division*. In this example, we divide 14,996 by 23. The divisor, 23, is not in your regular multiplication table. But you can make one up, getting something like the table at left. Then you can see at a glance which digits you should use to do the long division.

In this example, we might first try to divide 149 by 7. But when we multiply 23 by 7, we get 161. That number is bigger than 149, so we try multiplying 23 by 6, which is 138.

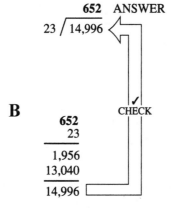

A

23 × 2	46
23 × 3	69
23 × 4	92
23 × 5	115
23 × 6	138
23 × 7	161
23 × 8	184
23 × 9	207

Figure 4-6
At A, we can make a table of multiples of the divisor to help us do long division when the divisor has more than one digit. At B, we use long multiplication to be sure we did the division problem correctly.

Then 138 is larger than 119, so we try 5 times. Finally, the last digit (conveniently) is exactly twice.

With multiple-digit numbers such as 23 as divisors, it's more important to check our techniques because we have more chances of making a mistake. These mistakes occur both when we do it "by hand" and when we use a calculator. Figure 4-6B illustrates how multiplication "by hand" (or *long multiplication*) can be used to check the long division problem we just finished.

DIVISION BY FACTORS

You can sometimes solve multiplication problems by using factors. It also works in division, if the divisor can be factored into a product of two whole numbers. To see an example of how this can be done, examine Fig. 4-7, where we find the quotient 37,996/28. In this case, the divisor, 28, factors into 4 × 7. So you can divide the dividend by 4, and then divide the result by 7 to get the final quotient. (You could divide by 7 first and then by 4, and things would end up the same, although the intermediate quotient would be 5,428 instead of 9,499.) You can use long division, and also multiplication by factors, to check the answer.

Figure 4-7

An example of division by factors, and two methods of checking the answer. Here, we divide 37,996 by 4 to get 9,499, and then divide that result by 7 to get 1,357.

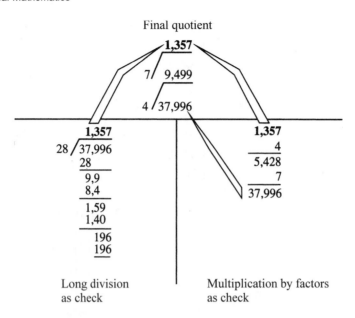

Final quotient

1,357

7 / 9,499

4 / 37,996

1,357
28 / 37,996
28
9,9
8,4
1,59
1,40
196
196

Long division as check

1,357
  4
5,428
  7
37,996

Multiplication by factors as check

WHEN A REMAINDER IS LEFT

What happens when division doesn't produce a whole number? Try it on a calculator and you'll see. Instead of a whole number, you get some digits, then a decimal point, and then some more digits. If you do long division, you get a *remainder*.

Figure 4-8 shows how we find the quotient 10,050/37 by using long division. We get a whole number, 271, and a remainder of 23. Now you ask, "What does that mean? My calculator doesn't show things that way!" Well, the remainder indicates how many parts of the divisor are left over in addition to the whole-number part of the quotient.

Figure 4-8

Division doesn't always produce a whole-number quotient. In this example, dividing 10,050 by 37 gives us 271 with a remainder of 23.

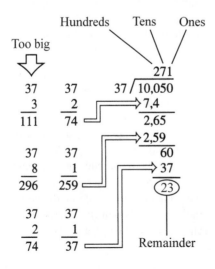

Hundreds   Tens   Ones

Too big

271
37 / 10,050

37   37
 3   2
111   74

37   37
 8   1
296   259

37   37
 2   1
 74   37

7,4
2,65
2,59
60
37
(23)

Remainder

In the example of Fig. 4-8, we have a remainder of 23 and a divisor of 37. That translates to a fraction: 23/37. Our quotient is therefore 271 and 23/37, written like this:

271-23/37

The dash here is not a minus sign! It just serves to separate the whole-number part from the fractional part. Sometimes the dash is left out and a space is put there instead. Fractions can also be written with the numerator on top, then a horizontal line under the numerator instead of a slash after it, and then the denominator underneath the line. The first or top number in the fraction (in this case 23) is called the *numerator*. The second or bottom number (in this case 37) is called the *denominator*.

If you use your calculator, you will see that dividing 23 by 37 gives you a bunch of numbers after a decimal point, like this:

23/37 = .621621621621 . . .

On some calculators you will see a 0 in front, like this:

23/37 = 0.621621621621 . . .

Now try dividing the original dividend, 10,050, by 37 on a calculator. You will get

10,050/37 = 271.621621621 . . .

That means 271 plus 0.621621621 . . . . The remainder of 23 means 23 parts out of 37, which is equivalent to the decimal number 0.621621621 . . . . The three dots mean that the numbers "621" keep repeating in that order, over and over, forever! That's called a *recurring decimal*. Some texts call it a *repeating decimal*.

HOW A CALCULATOR HANDLES FRACTIONS

Now, let's look more closely at what a calculator does with fractions. If you divide 25 by 6, it will read out 4.16666 . . . . Figure 4-9 shows what goes on inside the little thing's "brain." As you move down until the quotient accumulator reads 4, it follows what you already know. However, it doesn't stop. Had you punched in 2500 divided by 6, it would read 416.6666 . . . . Expressed as a whole number and a fraction, that's 416-2/3.

A division problem such as 25/6, when done in decimal form, never ends. Each recurring 4 in the quotient accumulator can be divided by another 6. A real calculator stops when it runs out of digits to display. Depending on the type of calculator you have, it will either leave the last digit as a 6 or round it up to a 7. If the calculator didn't know better, it would get "hung up" in the calculation process and spit out 6s until you switched it off or its battery died!

Figure 4-9

Here is the process a calculator goes through when you tell it to divide 25 by 6.

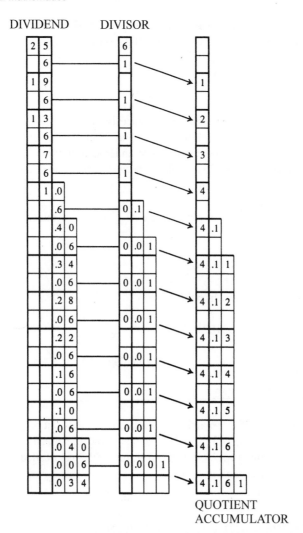

DIVIDEND   DIVISOR

QUOTIENT
ACCUMULATOR

DECIMAL EQUIVALENTS OF FRACTIONS

We have seen how a calculator turns a fraction into a decimal. That's what it does naturally, if you want to believe that there is anything "natural" about a calculator! Now, let's look at what a calculator does with some simple fractions. Here are a few that convert to relatively simple decimals. You can use your own calculator to verify these:

$$1/2 = 0.5$$

$$1/4 = 0.25$$

$$3/5 = 0.6$$

$$5/8 = 0.625$$

$$7/10 = 0.7$$
$$17/20 = 0.85$$
$$39/40 = 0.975$$
$$49/50 = 0.98$$

MORE DIFFICULT FRACTIONS

The above fractions are easy. The harder ones are those that always run off the end of the calculator's space. The whole number and fraction 416-2/3 is a good example. Here's another one that looks simple: 1/3.

When you divide 3 into 1, after bringing down zero after zero, you get an unending string of 3s. Earlier generations had a useful trick to avoid having to keep writing 3s; they put a dot over the 3 to indicate that it keeps repeating. Sometimes they put a line over it instead of a dot. Here are some more fractions that end up with one number that keeps repeating. Again, you can use your calculator to check them, as follows:

$$1/9 = 0.11111\ldots$$
$$1/6 = 0.16666\ldots$$
$$4/9 = 0.44444\ldots$$
$$7/12 = 0.583333\ldots$$
$$5/6 = 0.83333\ldots$$

WHERE GROUPS OF DIGITS REPEAT

In all the fractions in the last section, the decimal equivalent, which is what your calculator always gives you, ends up with one figure that kept repeating. Another kind of fraction first shows up in the decimal equivalent for 1/7. Here, a sequence of six digits repeats. Try dividing it out with a calculator that displays plenty of digits, such as the one on your computer! You should get this:

$$1/7 = 0.142857142857142857\ldots$$

The sequence that repeats here is 142857, always in the same order. Try some others now. See what happens when you divide out the following quotients. How many digits are in the sequence that repeats? You might need a calculator that can display a lot of digits to see the pattern, but there will always be one!

$$3/11 = ?$$
$$5/13 = ?$$

$$7/17 = ?$$

$$20/21 = ?$$

$$21/22 = ?$$

$$26/27 = ?$$

$$97/99 = ?$$

$$851/999 = ?$$

$$6{,}845/9{,}999 = ?$$

$$25{,}611/99{,}999 = ?$$

Do you notice something in common about the last four examples above?

CONVERTING RECURRING DECIMALS TO FRACTIONS

Although recurring decimals are easier to handle when you know what they mean, using old-fashioned fractions is often easier. How can we convert a recurring decimal to a fraction? There's a neat little trick that works for decimal numbers less than 1— that is, numbers that you write down as a zero followed by a decimal point and then a group of digits that keeps recurring. We can call this general rule the "law of the nines."

First, find the pattern that repeats, and write it down. For example, you might see the following decimal:

$$0.673867386738 \ldots$$

Here, the sequence of digits that repeats is 6738. Now put this in the numerator of a fraction, and then put an equal number of 9s in the denominator. In this case, you get 6,738/9,999. This is the fractional equivalent of the above decimal number. Check it out with your calculator. Punch in the digits 6, 7, 3, and 8; then press the "divide by" key, then the digits 9, 9, 9, and 9, and finally the "equals" key. Here are some more examples:

$$0.979797 \ldots = 97/99$$

$$0.851851851 \ldots = 851/999$$

$$0.684568456845 \ldots = 6{,}845/9{,}999$$

$$0.256112561125611 \ldots = 25{,}611/99{,}999$$

You will recognize these as the last four numbers from the previous section.

QUESTIONS AND PROBLEMS

This is an open-book quiz. You may refer to the text in this chapter (and earlier ones, too, if you want) when figuring out the answers. Take your time! The correct answers are in the back of the book.

1. Perform the following divisions and check your answers by multiplication.
(*a*) 343 ÷ 7	(*b*) 729 ÷ 9	(*c*) 4,928 ÷ 8
(*d*) 3,265 ÷ 5	(*e*) 6,243 ÷ 3	(*f*) 7,862 ÷ 2
(*g*) 3,936 ÷ 4	(*h*) 3,924 ÷ 6	(*i*) 3,081 ÷ 13
(*j*) 16,324 ÷ 11	(*k*) 6,443 ÷ 17	(*l*) 8,341 ÷ 19

2. Perform the following divisions by successive (division by factors) and long division. If your answers do not agree, check them with long multiplication.
(*a*) 3,600 ÷ 15	(*b*) 15,813 ÷ 21
(*c*) 73,625 ÷ 25	(*d*) 10,136 ÷ 28

3. Perform the following divisions and write the remainder as a fraction, using whichever method of conversion you like best.
(*a*) 3,459 ÷ 7	(*b*) 23,431 ÷ 8	(*c*) 13,263 ÷ 9
(*d*) 14,373 ÷ 3	(*e*) 29,336 ÷ 6	(*f*) 8,239 ÷ 17
(*g*) 34,343 ÷ 28	(*h*) 92,929 ÷ 29	

4. A profit of $14,000,000 has to be shared among holders of 2,800,000 shares of stock. What is the profit per share?

5. The total operating cost for a commuter airline flight between two cities is estimated as $8,415. What fare should be charged so that a flight with 55 passengers just meets the operating cost?

6. A machine needs a special tool that costs $5,000 to make a certain part. When equipped with this tool, the machine makes parts for 25 cents each. But the price must also pay for the tool. If the tool cost is to be paid for out of the first 10,000 parts made, what will be the cost of each part?

7. A freight car carries 58 tons (a ton here is defined as 2000 pounds), including its own weight, and runs on 8 wheels. Its suspension distributes the weight equally among the wheels. What is the weight on each wheel?

8. A man makes 1,200 production units of a certain part in 8 hours. How much time is spent making each part?

9. A package of 10,000 small parts weighs 1,565 pounds. The empty package weighs 2.5 pounds. How much does each part weigh? (*Hint:* convert pounds to ounces.)

10. Another package weighs 2,960 pounds full and 5 pounds empty. One part weighs 3 ounces. How many parts are in the full package?

11. A narrow strip of land exactly 1 mile long is to be divided into 33 lots of equal width. How wide is each lot?

12. On a test run, a car travels 462 miles on 22 gallons of gas. Assuming performance is uniform, how far does it go on each gallon of gas?

13. A particular mixture is usually made up 160 gallons at a time. It uses 75 gallons of ingredient 1; 50 gallons of ingredient 2; 25 gallons of ingredient 3; and 10 gallons of ingredient 4. If only 1 gallon is required, what amounts of each ingredient should be used?

14. Find the simplest fractional equivalents of the following decimals. By "simplest," we mean that the numerator and denominator should both be as small as possible so the fraction "divides out" into the decimal expression shown.
    (*a*) 0.125   (*b*) 0.7
    (*c*) 0.375   (*d*) 0.95

15. Find fractional equivalents of the following decimals. Check each by dividing back to decimal form again. You don't have to find the simplest expressions here.
    (*a*) 0.416416416 . . .   (*b*) 0.212121 . . .
    (*c*) 0.189189189 . . .   (*d*) 0.489248924892 . . .

5
Fractions

Chapter 4 presented some concepts involving fractions that you might not yet fully understand. In this chapter, you'll "catch up" by learning some important techniques for dealing with fractions.

SLICING UP A PIE

If you imagine fractions as slices of a pie, it can help you see how fractions work. Notice that the fraction 1/4 can be cut into smaller pieces without changing its value as part of the whole, as shown in Fig. 5-1. You can multiply 1/4 by various fractions that are always equal to 1, such as 2/2, 3/3, or 4/4, like this:

$$1/4 \times 2/2 = 2/8$$

$$1/4 \times 3/3 = 3/12$$

$$1/4 \times 4/4 = 4/16$$

When the dividend and the divisor (or numerator and denominator) are identical, the value of the quotient or fraction is equal to 1.

Whenever we multiply or divide a fraction by some other fraction that's really equal to 1, we have the same number in a different form. Imagine, for example, that we see the fraction 20/80. We can divide this by 20/20 and get 1/4, like this:

$$(20/80) / (20/20) = (20/20) / (80/20) = 1/4$$

We divide the numerator by 20 to get 1, and the denominator by 20 to get 4. Presto! We have just done a trick known as *reducing* or *simplifying* a fraction. Here, 1/4 is the *lowest form* or *simplest form* of this particular number, because we can't find smaller whole-number numerators and denominators for it.

Figure 5-1

The fraction 1/4 can also be represented as 2/8, 3/12, 4/16, or any other expression of whole numbers where the denominator is exactly 4 times as big as the numerator.

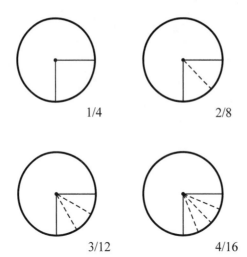

1/4   2/8

3/12   4/16

FACTORS HELP SIMPLIFY FRACTIONS

A fraction can sometimes contain huge numbers for both the numerator and denominator. When you see a fraction like that, you should wonder whether the fraction can be reduced to a simpler form so that the numerator and denominator are smaller but the quotient is still the same. A calculator that handles fractions will automatically find and present them in their simplest forms. If your calculator cannot handle fractions and you want to reduce a fraction to its simplest form, you can do it "by hand."

If the numerator and denominator both end in zeros, you can strike off the same number of zeros from each. If both are even, you can divide each number by 2. (We did both of those things when we reduced 20/80 to 1/4 a moment ago.) These rules are obvious, and it's easy to spot situations that call for them. But often the factors are harder to find. For example, in the fraction 455/462, both the numerator and the denominator can be divided by 7 and the results are whole numbers again. We can reduce 455/462 down to 65/66 that way. It isn't immediately apparent to most people that both 455 and 462 are divisible by 7. How can we find factors in situations like this?

SPOTTING THE FACTORS

Here are a few tricks that can help you spot factors. If the last digit of a number divides by 2 without a remainder, then the whole number does. If the number consisting of the last 2 digits divides by 4 without a remainder, then the big number does. If the number consisting of the last 3 digits divides by 8 without a remainder, then the big number does.

This line of reasoning leads to a set of checks for powers of 2. A *power* is defined as a certain number of times that a number is multiplied by itself. For example, 2 to the second power is 4; this is written $2^2 = 4$. The number 2 to the third power is 8; this is written $2^3 = 8$. Continuing, we have $2^4 = 16$, $2^5 = 32$, $2^6 = 64$, $2^7 = 128$, $2^8 = 256$, and so on.

A similar set of facts works for powers of 5. Starting with the first power and working up, we have $5^1 = 5$, $5^2 = 25$, $5^3 = 125$, $5^4 = 625$, $5^5 = 3,125$, and so on.

Rules also exist for 3s and 9s. Add all the digits of a number together. If this sum divides by 3 without a remainder, then the whole number does. If the sum of digits divides by 9 without a remainder, then the whole number does.

The check for dividing by 11 is more complicated. Add alternate digits in two sets. If the sums are identical, differ by 11, or differ by any whole-number multiple of 11, then the original number divides by 11 without a remainder.

There is no easy rule that works if you want to test a number to see if it will divide by 7 without a remainder. But there's another way to find factors when none of the above rules work and you seem to be "stuck."

PRIME NUMBERS

A *prime number* is a whole number larger than 1 that can only be factored into a product of 1 and itself. (For some reason, most mathematicians do not consider 1 a prime number, even though it can be "factored" into a product of 1 and itself.) The first few prime numbers, often called simply *primes*, are 2, 3, 5, 7, and 11. Table 5-1 lists the first 24 prime numbers.

Prime numbers have a property that makes them ideal for finding factors. This property is so important in mathematics that it deserves a "bullet"!

- **Any nonprime whole number can be factored into a product of primes. The numbers in this product are called the *prime factors*.**

Order	Prime	Order	Prime	Order	Prime
1st	2	9th	23	17th	59
2nd	3	10th	29	18th	61
3rd	5	11th	31	19th	67
4th	7	12th	37	20th	71
5th	11	13th	41	21st	73
6th	13	14th	43	22nd	79
7th	17	15th	47	23rd	83
8th	19	16th	53	24th	89

Table 5-1

The first 24 prime numbers. By convention, the number 1 is not considered prime.

FACTORING INTO PRIMES

When you want to find the prime factors of a large number, first use a calculator to find the *square root* of the number, and then divide the original number by all the primes less than or equal to that square root. If you ever get a whole-number quotient as you grind your way through this process, then you know that the divisor and the quotient are both factors of the original number. Sometimes the quotient is prime, and sometimes it isn't. If it isn't, then it can be factored down further into primes. Don't stop dividing the original by the primes until you get all the way down to 2.

Usually, the square root of a number is not a whole number. Don't worry about that when you're trying to find its prime factors. Just round the square root up to the next whole number, and then look up all the primes less than or equal to that. If Table 5-1 doesn't go far enough, you can find plenty of lists of primes on the Internet, some of which go to extremes far beyond anything you will ever need!

PRIME FACTORS OF 139

Suppose you want to find the prime factors of 139. When you find the square root of 139 using a calculator, you get a number between 11 and 12. Round it up to 12. Now find all the primes less than or equal to 12. From Table 5-1, you can see that these are 2, 3, 5, 7, and 11. Divide 139 by each of these and look for whole-number quotients:

$$139/11 = 12.\text{xxx} \ldots \text{(Not whole)}$$

$$139/7 = 19.\text{xxx} \ldots \text{(Not whole)}$$

$$139/5 = 27.8 \text{ (Not whole)}$$

$$139/3 = 46.333 \ldots \text{(Not whole)}$$

$$139/2 = 69.5 \text{ (Not whole)}$$

Here, "xxx" means a string of numbers that has a recurring pattern too long to write down easily. (It doesn't matter what the pattern is anyway, if the number is not whole.) You know from this that the only factors of 139 are 1 and itself. By definition, that means 139 is a prime number.

PRIME FACTORS OF 493

Now imagine that you are told to find the prime factors of 493. Using a calculator to find the square root and then rounding up, you get 23. From Table 5-1, you can see that the primes less than or equal to 23 are 2, 3, 5, 7, 11, 13, 17, 19, and 23. Now divide 493 by each of these and look for whole-number quotients:

$$493/23 = 21.\text{xxx} \ldots \text{(Not whole)}$$

$$493/19 = 25.\text{xxx} \ldots \text{(Not whole)}$$

$$493/17 = 29 \text{ (Whole)}$$

$$493/13 = 37.\text{xxx} \ldots \text{(Not whole)}$$

$$493/11 = 44.818181 \ldots \text{(Not whole)}$$

$$493/7 = 70.\text{xxx} \ldots \text{(Not whole)}$$

$$493/5 = 98.6 \text{ (Not whole)}$$

$$493/3 = 164.333 \ldots \text{(Not whole)}$$

$$493/2 = 246.5 \text{ (Not whole)}$$

You know from this that the prime factors of 493 are 17 and 29, because 29 appears in Table 5-1. You can't factor 493 down any further than that.

PRIME FACTORS OF 546

Now suppose you want to factor 546 into primes. Using a calculator to find the square root and then rounding up, you get 24. Again, from Table 5-1, the primes less than or equal to 24 are 2, 3, 5, 7, 11, 13, 17, 19, and 23. So you divide by them all:

$$546/23 = 23.\text{xxx} \ldots \text{(Not whole)}$$

$$546/19 = 28.\text{xxx} \ldots \text{(Not whole)}$$

$$546/17 = 32.\text{xxx} \ldots \text{(Not whole)}$$

$$546/13 = 42 \text{ (Whole)}$$

$$546/11 = 49.636363 \ldots \text{(Not whole)}$$

$$546/7 = 78 \text{ (Whole)}$$

$$546/5 = 109.2 \text{ (Not whole)}$$

$$546/3 = 182 \text{ (Whole)}$$

$$546/2 = 273 \text{ (Whole)}$$

Now you know that 546 has prime factors of 2, 3, 7, and 13. But is that all? There's a way to test and see. Using your calculator, you hit 2, then $\times$ (the "times" key), then 3, then $\times$, then 7, then $\times$, then 13, and finally $=$ (the "equals" key). The result, if you hit all the right buttons and none of the wrong ones, is 546. So you know that the prime factors of 546 are 2, 3, 7, and 13.

The business of prime numbers and prime factorization can get incredibly complicated, and this chapter could go on about it for a long time. But that would be a diversion. You should now have the general idea of how prime factorization works, and that's all you need for most practical problems.

ADDING AND SUBTRACTING FRACTIONS

When you want to add or subtract fractions "by hand," they must all have the same denominator. For instance, to add 1/2, 2/3, and 5/12, both 1/2 and 1/3 can be changed to 12ths. You can figure out that 1/2 = 6/12 and 2/3 = 8/12. Once you have found a *common denominator* in this way, you can add the numerators straightaway because the denominators are all 12ths: 6 + 8 + 5 = 19. Then you get 19/12, which is more than 1. Subtract 12/12 (which is equal to 1) from this and you get 1-7/12 as the final answer in its proper form.

Now suppose you want to subtract 3-3/5 from 7-5/12. You can always find a common denominator in problems like this if you multiply the two denominators together, although the result might not be the smallest common denominator. In this case, multiplying the denominators gives you 5 × 12 = 60. Now proceed like this:

$$3/5 = (3/5) \times (12/12)$$
$$= (3 \times 12) / (5 \times 12)$$
$$= 36/60$$

and

$$5/12 = (5/12) \times (5/5)$$
$$= (5 \times 5) / (12 \times 5)$$
$$= 25/60$$

Note that when you multiply two fractions, the product of the whole fraction is equal to the product of the numerators divided by the product of the denominators. The subtraction problem now looks like this:

$$7\text{-}25/60 - 3\text{-}36/60 = ?$$

Change the whole-number parts of these numbers into fractions to get

$$7 = (7 \times 60) / 60 = 420/60$$

and

$$3 = (3 \times 60) / 60 = 180/60$$

The numbers in the original problem can be found by addition:

$$7\text{-}25/60 = 420/60 + 25/60$$
$$= (420 + 25) / 60$$
$$= 445/60$$

and

$$3\text{-}36/60 = 180/60 + 36/60$$
$$= (180 + 36) / 60$$
$$= 216/60$$

Now subtract the larger of these two numbers from the smaller:

$$(445/60) - (216/60) = (445 - 216) / 60$$
$$= 229/60$$

In the "olden days" this sort of expression was called an *improper fraction* because the numerator is larger than the denominator, and it "doesn't look right." There's nothing technically wrong with it. But to keep to "proper" form, you can divide out 60 from 229 three times and have a remainder of 49. That means the above improper fraction is equal to 3-49/60. You can't reduce 49/60 to a lower form (try it and see!), so this is the final answer in its simplest form. Stated from the first to the last, the "subtraction fact" looks like this:

$$7\text{-}5/12 - 3\text{-}3/5 = 3\text{-}49/60$$

Again, don't be confused by the little dashes that separate the whole-number parts of each value from the fractional parts. They aren't minus signs! The long dash, with a space on either side of it, is the minus sign here.

FINDING THE COMMON DENOMINATOR

How do you find a reasonable common denominator when you have a long sum of fractions? Sometimes it's easy, but sometimes it's quite a hard trick. You can multiply all the denominators together and get a number that will work, but it will probably be a lot larger than what you need.

Older textbooks had a routine for the job, but it was difficult to follow and it confused a lot of folks. Here's a way you can understand. Figure 5-2 illustrates an example. Suppose you have this addition problem involving fractions:

$$1/4 + 1/3 + 2/5 + 1/6 + 5/12 + 3/10 + 7/30 + 4/15 = ?$$

Figure 5-2

Here's a way to find a reasonable common denominator in a long sum of fractions.

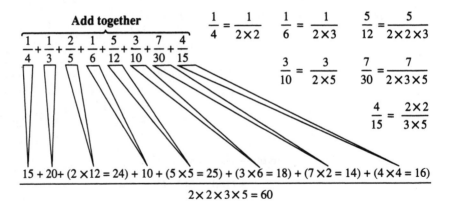

$$\frac{142}{60} = \frac{71}{30} = 2\frac{11}{30}$$

same answer

$$\text{or } \frac{142}{60} = 2\frac{22}{60} = 2\frac{11}{30}$$

To begin, find the prime factors of each denominator in order:

$$4 = 2 \times 2$$

3 is already prime

5 is already prime

$$6 = 2 \times 3$$

$$12 = 2 \times 2 \times 3$$

$$10 = 2 \times 5$$

$$30 = 2 \times 3 \times 5$$

$$15 = 3 \times 5$$

Our "ideal common denominator" must contain every prime factor that is in any denominator. These are 2, 3, and 5. The common denominator is then found by multiplying all these primes together: $2 \times 2 \times 3 \times 5 = 60$. Notice that the prime number 2 is used twice here. That sort of thing is not unusual when you are finding the prime factorizations of numbers.

Now convert all the fractions to 60ths:

$$1/4 = 15/60$$

$$1/3 = 20/60$$

$$2/5 = 24/60$$

$$1/6 = 10/60$$

$$5/12 = 25/60$$

$$3/10 = 18/60$$

$$7/30 = 14/60$$

$$4/15 = 16/60$$

These fractions add up to 142/60. You can reduce this to 71/30. That's an improper fraction, so you divide out 30 twice and get a remainder of 11. Therefore, the final answer is 2-11/30. Stated fully, here's the fact:

$$1/4 + 1/3 + 2/5 + 1/6 + 5/12 + 3/10 + 7/30 + 4/15 = 2\text{-}11/30$$

SIGNIFICANT FIGURES

Devices that give you numerical information can be either *analog* or *digital*. An example of an analog device is an old-fashioned spring-loaded scale. A grandfather clock, with its hour and minute hands, is another example of an analog display. Calculators have digital displays, as do numeric-readout clocks, numeric-readout thermometers, and automotive odometers. Some people think that digital displays are more accurate than analog ones. But that isn't always true.

Suppose a 150-pound man steps on a scale to weigh himself (Fig. 5-3A). Does he weigh exactly 150 pounds, and not an ounce more or less? Is he between 149.5 and 150.5 pounds, or is he between 145 and 155 pounds? The answer depends on which

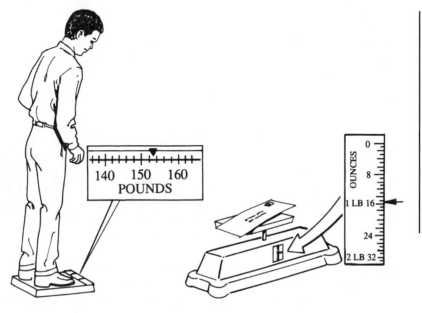

A   **B**

Figure 5-3
At A, a man weighs himself and sees that he weighs 150 pounds, more or less. How much more or less? At B, an envelope weighs 16 ounces, more or less. How much more or less? Accuracy is relative!

figures in that number 150 are "significant." It also depends on what the man is wearing and how well-made the scale happens to be.

If the 0 were just a placeholder, then if he weighed less than 145 or more than 155, the number would be stated as 140 or 160. If the 0 were "significant," then 150 would mean that he weighed from 149.5 to 150.5 pounds. If he weighed less than 149.5 or more than 150.5, the number would be 149 or 151. For it to mean not an ounce more or less than 150 pounds, the weight should be written as 150 pounds, 0 ounces. You might even specify the weight to greater accuracy, such as 150.00 pounds or even 150.000 pounds.

The same sort of situation occurs when you weigh an envelope (Fig. 5-3B) on an analog postal scale. It looks like this envelope weighs just about 16 ounces, or maybe a little less. But what if the scale is not a good one, and the envelope actually weighs a little more than 16 ounces? How much "leeway" should you allow? (And, you might wonder, what does this envelope contain that makes it so heavy for its size?)

APPROXIMATE DIVISION AND MULTIPLICATION

Suppose you divide an "approximate" 150 by an "exact" 7. If your calculator can display 10 digits, it gives you 21.42857142 (if it *truncates*, or simply cuts off, the digits after it runs out of space) or 21.42857143 (if it *rounds off* to the nearest significant digit). You are tempted to believe all those figures. But can you?

Apply what was just said about significant figures. If 150 means more than 145 and less than 155, that means only two figures are significant (the 1 and the 5). In that case, dividing by exactly 7 could yield anything between 145/7 and 155/7. Check these out on your calculator. Now, if the 0 of 150 is significant, the result can be anything between 149.5 / 7 and 150.5 / 7. Check out this range of values on your calculator. It's a lot narrower. Now what if you had 150.000 as the specified number? That means a range of 149.995 to 150.005. When you divide these numbers by exactly 7, you get a range of 149.995 / 7 to 150.005 / 7. Check this out on your 10-digit calculator. Or better yet, use a computer calculator that can show a lot more digits.

In these division problems, an "exact 7" means a theoretically perfect number 7, which you can call 7.0, or 7.00, or 7.000, or 7.0000—with as many zeros after the decimal point as you want. In this context, 7 is a *mathematical constant*. But the quantity 150 (more or less), as used here, refers to an observed or measured thing. It can never be exact, because no machine or device is perfect, and if the device is analog, there can be human error in the reading, too. But if you want to find exactly 1/7 of some figure, you can divide by 7 and consider that value to have no margin of error at all.

Suppose that you want to divide an "approximate" 23,500 by an "exact" constant of 291, and you believe that the two end zeros in the dividend aren't significant. You could assume it is between 23,450 and 23,550, perform both divisions, and then decide what is significant. But longhand, that's a lot of work! In the olden days, the practice was to draw a vertical line where figures began to be doubtful. You could divide between the "limiting values" as the possible errors, because only so many figures are significant, and then you could guess at the most reasonable value. An example of this procedure is illustrated in Fig. 5-4.

If you are willing to play around with a calculator, it can show you how to do things that aren't practical by using longhand methods. Take values that represent the greatest possible variation on either side of the stated value, and then deduce how accurate the

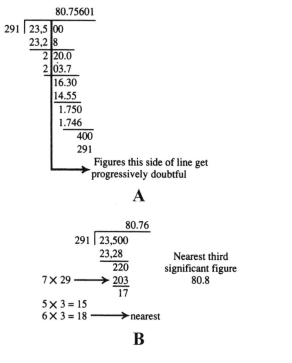

Figure 5-4

Division of 23,500 by 291. At A, conventional long division; the digits get more and more doubtful as we churn them out. At B, an approximate method of long division. At C, a way to find the limits of accuracy.

Figure 5-5

An example of approximation in long multiplication. As in division, the figures become more and more doubtful as we churn them out.

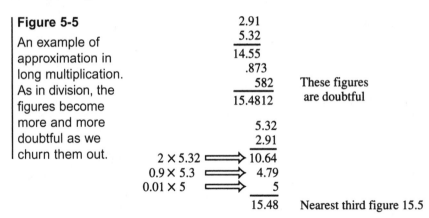

$$
\begin{array}{r}
2.91 \\
\underline{5.32} \\
14.55 \\
.873 \\
\underline{582} \\
15.4812
\end{array}
$$

These figures are doubtful

$$
\begin{array}{r}
5.32 \\
2.91 \\
\end{array}
$$

$2 \times 5.32 \implies 10.64$
$0.9 \times 5.3 \implies 4.79$
$0.01 \times 5 \implies \underline{5}$

15.48 Nearest third figure 15.5

answers will be by dividing them out with the calculator. Both the dividend and the divisor can have a certain number of significant figures unless one of them is known to be mathematically exact, as 7 and 291 are in the examples above.

The approximation methods you use for long division will also work for long multiplication. Figure 5-5 shows this process for the product of 2.91 and 5.32. Here, both values are expressed to three significant figures. So that's as far as we can reasonably go with the product.

APPROXIMATE ADDITION AND SUBTRACTION

Suppose you add "exactly 55" to "roughly a million." What does "roughly a million" mean here? It could be greater or less than an exact million (1,000,000) by 500,000, or 50,000, or 5,000, or 500. If you say "pretty close to a million," it could be greater or less than 1,000,000 by 50, or by 5, or maybe even by 0.5. It depends on how many of the zeros after the 1 you take seriously! Mathematicians and engineers call such give-or-take by a technical name, *plus or minus*, symbolized as a plus sign with a minus sign underneath ($\pm$). Adding 55 doesn't make a practical difference when we think of an approximate million that could vary by 50,000 either way, that is, 1,000,000 $\pm$ 50,000. But it makes a big difference if we take all six of the zeros seriously, in which case we mean a rather precise million—a number between 999,999.5 and 1,000,000.5!

In this chapter, you've had a taste of how significant figures work in practical mathematics. We'll look more closely at this subject, and how significant figures are dealt with by scientists and engineers, in Chap. 25. For now, here is the fundamental rule for determining significant figures in any practical problem that involves quantities subject to error. It deserves another "bullet"!

- **An answer cannot have more significant figures than the number with the least number of significant figures used as "input" for the problem.**

QUESTIONS AND PROBLEMS

This is an open-book quiz. You may refer to the text in this chapter (and earlier ones, too, if you want) when figuring out the answers. Take your time! The correct answers are in the back of the book.

1. Arrange the following fractions into groups that have the same value.

    1/2   1/3   2/5   2/3   3/4   3/6   4/6

    4/8   3/9   4/10   4/12   8/12   9/12   5/15

    6/15   6/18   9/18   8/20   10/20   15/20   7/21

2. Reduce each of the following fractions to its simplest form.

    7/14   26/91   21/91   52/78   39/65   22/30

    39/51   52/64   34/51   27/81   18/45   57/69

3. Without actually performing the divisions, indicate which of the following numbers divide exactly by 3, 4, 8, 9, or 11.

    (*a*) 10,452   (*b*) 2,088   (*c*) 5,841

    (*d*) 41,613   (*e*) 64,572   (*f*) 37,848

4. Find the prime factors of the following numbers.

    (*a*) 1,829   (*b*) 1,517   (*c*) 7,387

    (*d*) 7,031   (*e*) 2,059   (*f*) 2,491

5. Add the following groups of fractions and reduce each answer to its simplest form.

    (*a*) 1/5 + 1/6 + 4/15 + 3/10 + 2/3

    (*b*) 1/8 + 1/3 + 5/18 + 7/12 + 4/9

    (*c*) 1/4 + 1/5 + 1/6 + 1/10 + 1/12

    (*d*) 4/7 + 3/4 + 7/12 + 8/21 + 5/6

6. Find the simplest fractional equivalents of the following decimals.

    (*a*) 0.875   (*b*) 0.6   (*c*) 0.5625

    (*d*) 0.741   (*e*) 0.128

7. Find the decimal equivalents of the following fractions.

    (*a*) 2/3   (*b*) 3/4   (*c*) 4/5   (*d*) 5/6

    (*e*) 6/7   (*f*) 7/8   (*g*) 8/9

8. Find the decimal equivalents of the following fractions.

    (*a*) 1/3   (*b*) 1/4   (*c*) 1/5   (*d*) 1/6

    (*e*) 1/7   (*f*) 1/8   (*g*) 1/9

9. Find the fractional equivalents of the following recurring decimals.

    (*a*) 0.416416416 . . .

    (*b*) 0.212121 . . .

    (*c*) 0.189189189 . . .

(*d*) 0.571428571428571428 . . .
(*e*) 0.909909909 . . .
(*f*) 0.090090090 . . .

10. To define "significant figures," show the limits of possible meaning for measurements given as 158 feet and 857 feet.

11. Using the approximate method, divide 932 by 173. Then by dividing 932.5 by 172.5 and 931.5 by 173.5, show how many of your figures are justified. Noting that 932 and 173 have three significant figures, what conclusion can you draw?

12. Divide 93,700 by 857, using an approximate method. Then by dividing 93,750 by 856.5 and 93,650 by 857.5, show how many of your figures are justified. Can you shorten your method still further to avoid writing down meaningless figures?

13. (a) List all the prime numbers less than 60. (b) If you use this list, how can you test a given number to determine whether or not it is prime? (c) What is the largest number you can test in this way, using this list?

14. Find the differences of the following pairs of fractions. Reduce the results to standard form with the smallest possible denominators.
    (*a*) 3/4 − 1/16   (*b*) 11/13 − 1/7   (*c*) 16/20 − 3/8
    (*d*) 255/100 − 1/10   (*e*) 23/17 − 1/34

6
Plane Polygons

So far, this book has dealt with counting and calculation in one dimension. Now you will learn how math can help to relate different measures and dimensions by working with *plane polygons*: flat objects with straight sides.

LENGTH TIMES LENGTH IS AREA

Suppose someone asked you to multiply 15 oranges by 23 pears (Fig. 6-1). What would you do? You can multiply 15 by 23 and get 345—but of what? Not oranges! Not pears! You can do it if you call them all "fruit," but that changes the context of the problem. (In science and engineering, units are formed by multiplying different quantities together, such as feet and pounds to get *foot-pounds*. But let's not worry about that yet.)

Multiplying oranges by oranges, or pears by pears, makes sense in any context. In the same way, you can multiply a length by a length. When you do that, you get *area*.

If a piece of wallpaper is 27 inches wide and 108 inches long, its area is 27 inches times 108 inches. The answer is in *square inches*. (Scientists sometimes call them *inches squared* instead.) The answer is deduced from the way you began counting: laying objects out in rows and columns.

If you lay 27 little pieces of paper, each measuring 1 inch by 1 inch and having an area of 1 square inch, in a row and then line up 108 rows just like it, the total area is $108 \times 27 = 2,916$ square inches (Fig. 6-2). The same thing would happen if you laid 108 of the squares in a row, and then lined up 27 rows just like it. Can you think of any other ways you can arrange 2,916 little 1-inch squares into a whole number of complete rows and a whole number of complete columns?

Figure 6-1

Suppose we multiply 15 oranges by 23 pears. What do we get?

Figure 6-2

When we multiply inches by inches, we get square inches. That makes sense.

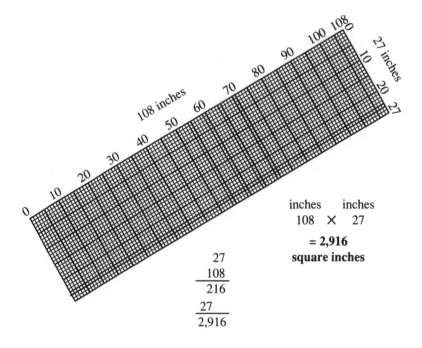

inches   inches
108 × 27

= 2,916
square inches

$$\begin{array}{r} 27 \\ 108 \\ \hline 216 \\ 27 \\ \hline 2,916 \end{array}$$

WHAT IS A SQUARE?

You are so accustomed to the shape called a *square* that you've probably never bothered to precisely define it. Counting those square inches, you most likely think of them as measuring 1 inch "each way." But if you construct a four-sided polygon that happens to measure 1 inch along each of its four sides, you will not necessarily end up with a square. What makes a polygon a true square? For one thing, the fourth side must end where the first side began. All the sides must join at their ends. No two sides may cross each other. Opposite pairs of sides must be parallel. Even then, the figure still might not be square. In addition to all the above requirements, the four *interior angles* must all be *right angles*, as shown at the top left in Fig. 6-3. None of the other objects in that drawing are squares.

Suppose a carpenter makes a table. She must attach its four legs to the tabletop. All four legs should be attached at one particular angle so that the table stands securely (unless you use some fancy reinforcements to hold the table). In the olden days, this angle was called the "right" angle, and that is how the term originated. Any other angle was a "wrong" angle because it did not provide the best possible support when the table was loaded down with oranges, pears, or whatever else people might put on it.

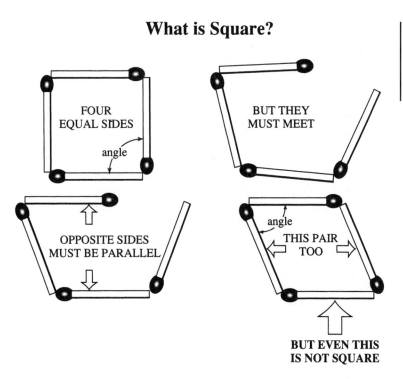

What is Square?

FOUR EQUAL SIDES

angle

BUT THEY MUST MEET

OPPOSITE SIDES MUST BE PARALLEL

angle

THIS PAIR TOO

BUT EVEN THIS IS NOT SQUARE

Figure 6-3

A square always has four equal sides and four right angles.

Figure 6-4

All of these shapes
have 660 square
inches.

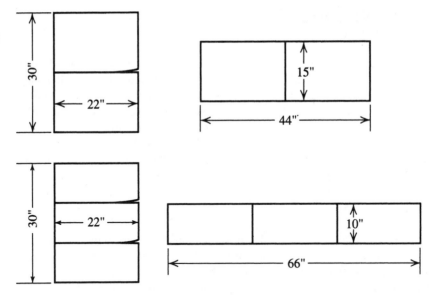

DIFFERENT SHAPES WITH THE SAME AREA

Take a piece of drawing paper measuring 22 inches by 30 inches, cut it across, and rejoin it to be 44 inches by 15 inches. The area is the same both times because it's the same paper, just rearranged. The product 22 inches times 30 inches is 660 square inches. So is the product 44 inches times 15 inches. The same original piece of drawing paper can be cut up and arranged in other ways, too, as you can see by looking at Fig. 6-4. The area is always 660 square inches, as long as you don't throw any paper away or let it overlap.

SQUARE MEASURE VS. LINEAR MEASURE

In *linear measure*, or determining length in one dimension, each foot has 12 inches and each yard has 3 feet. But in *square measure*, or determining area in two dimensions, how many square inches are in a square foot? How many square feet are in a square yard?

When you count, 12 rows of 12 give you 144, which is 12 × 12. This means there are 12 × 12 = 144 square inches in a square foot. Similarly, 3 rows of 3 give you 9, which is 3 × 3. This means there are 3 × 3 = 9 square feet in a square yard. If you like to use the *metric system*, where there are 100 centimeters in a meter by linear measure, you can count out 100 rows of 100 little 1-centimeter squares to get 10,000, which is 100 × 100. That means there are 100 × 100 = 10,000 square centimeters in a square meter.

The important thing to remember is this: When you are making conversions with areas, things work differently than they do with linear measures. Note that an area of 12 square inches is not the same thing as the area of a polygon that measures 12 inches square! Each side of a 12-inch by 12-inch square is 12 inches long, so it has $12 \times 12 = 144$ square inches. But an area of 12 square inches could be inside an oblong that measures 12 inches long by 1 inch wide, or 6 inches long by 2 inches wide, or 4 inches long by 3 inches wide. Check these products out; they all equal 12 square inches.

Here is another way to think of this. If you have a square and you double the length of each side, you increase the area by $2 \times 2 = 4$ times. If you triple the length of each side, you increase the area by $3 \times 3 = 9$ times. If you quadruple the length of each side, you increase the area by $4 \times 4 = 16$ times, and so on.

RIGHT TRIANGLES

All the polygons that have been considered so far have had four sides with right angles between the sides. Mathematicians call such shapes *rectangles*, a word of Latin origin, which means "having right angles."

An easy way to understand areas of a *triangle*, or three-sided polygon, is to think of it as a rectangle cut in half diagonally. When you cut a rectangle in this way, you get two triangles, each of which has one corner with a right angle. This type of triangle is called, appropriately enough, a *right triangle*. The two sides that come together at a right angle are called the *adjacent sides*. The longest side, opposite the right angle, is technically called the *hypotenuse*.

Figure 6-5 shows an example of a rectangle cut in half diagonally to get two right triangles. Notice that these triangles are identical; one of them is just turned by one-half a revolution with respect to the other. When you have two triangles that have the same size and shape so that one of them can be "pasted over" the other, the two triangles are said to be *congruent*.

Figure 6-5

The area of this rectangle is $6 \times 8 = 48$ square inches. The area of the triangle we get when we cut the rectangle in half diagonally is half the area of the rectangle, or 24 square inches.

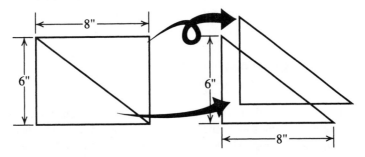

Figure 6-6

At A, examples of a square, a rectangle, a rhombus, and a parallelogram. At B, as a rectangle is flattened into a parallelogram while leaving all four of its sides the same length, its area decreases.

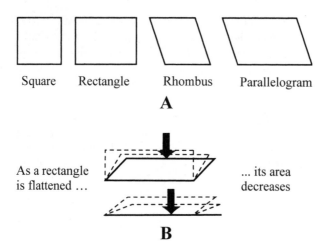

Square   Rectangle   Rhombus   Parallelogram

A

As a rectangle is flattened ...   ... its area decreases

B

The rectangle in Fig. 6-5 measures 8 inches wide by 6 inches high. Its area is therefore $6 \times 8 = 48$ square inches. The area of either one of the right triangles is exactly one-half that, or 24 square inches. From this, you should easily be able to see that the area of a right triangle must be equal to one-half its width (or *base length*) times its vertical height. As things turn out, this is true of all triangles, not only right triangles. We'll look at other types of triangles later in this chapter.

PARALLELOGRAMS

So far, squares, rectangles, and right triangles have been covered. What about four-sided shapes that have parallel sides, but not right angles? Geometry distinguishes two kinds, just as squares and rectangles have right-angle corners. If the four sides are equal, it's called a *rhombus*, or in common terms, a *diamond*. If the sides are unequal, it can be called a *rhomboid*, or more commonly, a *parallelogram* (Fig. 6-6A).

When you look at Fig. 6-6B, you will see that if a polygon starts out as a rectangle and then you change it into a parallelogram by flattening it out while keeping all four sides the same length, its area decreases. If you keep "squashing" it this way, it eventually becomes a straight line with an area of zero!

AREA OF A PARALLELOGRAM

One way to find the area of a parallelogram is to change it into a rectangle that has one pair of sides the same as the parallelogram, but the other two sides shorter. The *perpendicular* (straight-across) distance between the "top" and "bottom" of the rectangle

A

Sides 10" and 15"
Distance squarely
between 15" sides is 8"

B

Same parallelogram
Sides 10" and 15" Distance
between 10" sides is 12"

Figure 6-7

Two ways of deter-
mining the area of
a parallelogram
with pairs of
sides measuring
10 inches and
15 inches. At A,
the perpendicular
distance between
the 15-inch sides is
8 inches. At B, the
perpendicular dis-
tance between the
10-inch sides is
12 inches. Either
way we look at it,
the area turns out
the same.

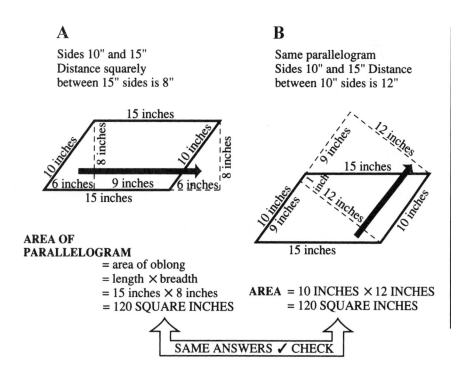

**AREA OF
PARALLELOGRAM**
= area of oblong
= length × breadth
= 15 inches × 8 inches
= 120 SQUARE INCHES

AREA = 10 INCHES × 12 INCHES
= 120 SQUARE INCHES

SAME ANSWERS ✓ CHECK

is the same as the distance between those two sides in the parallelogram, as shown in Fig. 6-7A. This rectangle has the same area as the parallelogram because you add in a triangle on the right-hand side and then take out a congruent triangle from the left-hand side. From this, you should be able to see that the area of a parallelogram is always equal to the product of its base length and its height.

Figure 6-7B illustrates another way of figuring out the area of the same parallelogram by this method. The process is just like that used in Fig. 6-7A, except that the slanting pair of parallel sides, rather than the horizontal pair, is used to make the "top" and "bottom" of the rectangle. (If you have trouble envisioning this, rotate the page clockwise until the slanting sides appear horizontal.)

AREA OF AN ACUTE TRIANGLE

Any angle smaller than a right angle is called an *acute angle*. A triangle in which all three angles are acute is called an *acute triangle*. Any acute triangle can be split into two right triangles, each of which is one-half of a corresponding rectangle. These two rectangles, put together, form a larger rectangle (Fig. 6-8). The line where the two smaller rectangles join becomes the vertical height of the acute triangle. The "bottom" sides of the two smaller right triangles, joined together, form the base of the acute triangle.

Figure 6-8

An acute triangle can be divided into two right triangles, and its area determined by adding the areas of the right triangles together.

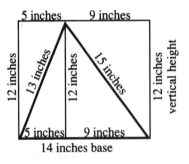

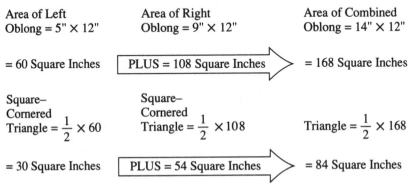

Area of Left
Oblong = 5" × 12"

= 60 Square Inches

Area of Right
Oblong = 9" × 12"

PLUS = 108 Square Inches

Area of Combined
Oblong = 14" × 12"

= 168 Square Inches

Square–Cornered Triangle = $\frac{1}{2}$ × 60

= 30 Square Inches

Square–Cornered Triangle = $\frac{1}{2}$ × 108

PLUS = 54 Square Inches

Triangle = $\frac{1}{2}$ × 168

= 84 Square Inches

In Fig. 6-8, the large acute triangle breaks into two right triangles, each with a vertical height of 12 inches. The right triangle on the left-hand side has a base length of 5 inches, so its area (as we learned earlier) is $1/2 \times 5 \times 12 = 1/2 \times 60 = 30$ square inches. The right triangle on the right-hand side has a base length of 9 inches, so its area is $1/2 \times 9 \times 12 = 1/2 \times 108 = 54$ square inches. The total area, which is the area of the large acute triangle, is the sum of the areas of the two smaller right triangles, or $30 + 54 = 84$ square inches.

Here's another way to look at it. Note that the large rectangle breaks into two smaller ones, and each of these smaller rectangles is split in half diagonally by the slant side of a right triangle. One-half of the left-hand rectangle is inside the left-hand right triangle. One-half of the right-hand rectangle is inside the right-hand right triangle. It follows that one-half of the large rectangle must lie within the large acute triangle. The area of the large rectangle is $14 \times 12 = 168$ square inches. One-half of this, which must be the area of the large acute triangle, is 84 square inches. That's the same answer as we got using the other method.

Now the general rule emerges: The area of any acute triangle is equal to one-half the base length times the vertical height. Figure 6-8 does not give us a *rigorous proof* of this fact, because it is only one example and an airtight proof must be done for the general case. But you should be able to see from this example how the general principle works.

AREA OF AN OBTUSE TRIANGLE

Any angle larger than a right angle, but smaller than two right angles combined, is called an *obtuse angle*. A triangle with an obtuse angle inside it is, as you should be able to guess, called an *obtuse triangle*. An example is shown in Fig. 6-9. Here, the vertical height "overhangs" the base. You can envision the obtuse triangle as part of a larger right triangle from which a smaller right triangle has been taken away. Each triangle is one-half the corresponding rectangle and the formula holds good, even though the vertical height is measured outside the base of the main triangle.

You can turn that same triangle around, using its longest side, which is the one opposite the obtuse angle, as the base. Then you can find the area in the same way as you would for an acute triangle. As long as you take the height from the base to the corner (or *vertex*) where the obtuse angle is, the formula is true and gives the same answer for the same triangle (Fig. 6-10).

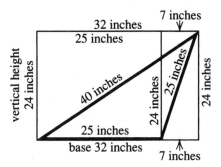

Figure 6-9

Determining the area of an obtuse triangle. The area is equal to half the base length times the vertical height, just as with an acute triangle.

Area of Large Rectangle	= 24" × 32"	Area of Small Right Rectangle	= 24" × 7"	Area of Left Rectangle	= 24" × 25"

= 768 Square Inches MINUS = 168 Square Inches → = 600 Square Inches

Area of Large Square-Cornered Triangle	$= \frac{1}{2} \times 768$	Area of Small Square-Cornered Triangle	$= \frac{1}{2} \times 168$	Area of Left Obtuse Triangle	$= \frac{1}{2} \times 600$

or .... *or*

= 384 Square Inches MINUS = 84 Square Inches → = 300 Square Inches

Figure 6-10

Another method of finding the area of an obtuse triangle. Take the side opposite the obtuse angle as the base.

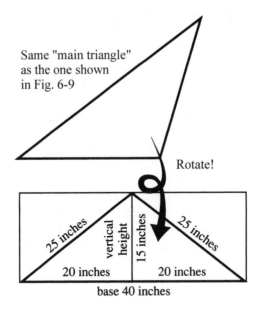

Same "main triangle" as the one shown in Fig. 6-9

Rotate!

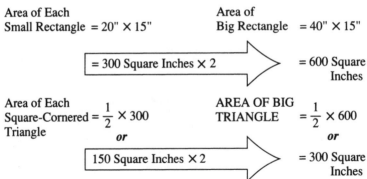

25 inches
vertical height
15 inches
25 inches
20 inches
20 inches
base 40 inches

Area of Each
Small Rectangle $= 20'' \times 15''$

Area of
Big Rectangle $= 40'' \times 15''$

$= 300$ Square Inches $\times 2$

$= 600$ Square Inches

Area of Each
Square-Cornered $= \frac{1}{2} \times 300$
Triangle

or

AREA OF BIG
TRIANGLE $= \frac{1}{2} \times 600$

or

150 Square Inches $\times 2$

$= 300$ Square Inches

METRIC MEASURE

For many years, inconsistency in systems of measurement made learning arithmetic difficult and frustrating. There are 12 inches to the foot, 3 feet to the yard, 220 yards to the *furlong*, and 8 furlongs to the mile. In liquid measure there are 4 ounces to the *gill*, 2 gills to the *cup*, 2 cups to the pint, 2 pints to the quart, and 4 quarts to the gallon. The common measure of weight is known as *avoirdupois*, in which there are 16 *drams* to the ounce, 16 ounces to the pound, 14 pounds to the *stone*, 2 stones to the *quarter*, 4 quarters to the *hundredweight*, and 20 hundredweights to the ton. *Troy weight*, used for jewelry, has 24 *grains* to the *pennyweight*, 20 pennyweights to the ounce, and 12 ounces to the pound! It's almost as if the people who made up these systems centuries ago wanted to make them as confusing as possible (Fig. 6-11).

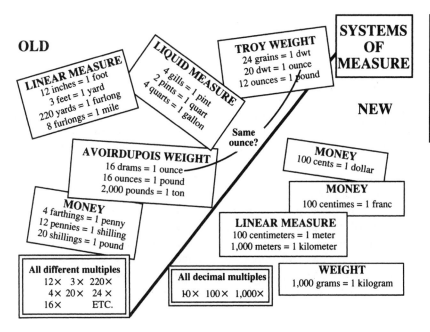

Figure 6-11

Some of the old units of measurement are confusing and inconsistent. The simplest new systems are based on powers of 10.

Some countries have adopted the metric system, in which most measures are based on powers of 10. This system is certainly easier to learn than other systems. But so many people have learned to use the old systems that the new one, though simpler to learn and work with, seems strange. Nevertheless, it's not too hard to find a *meter stick* that shows centimeters as well as inches (Fig. 6-12) and is a little longer than the conventional *yardstick*. Perhaps as the 21st century unfolds, the metric system will be adopted throughout the world.

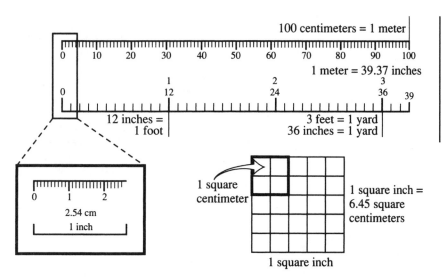

Figure 6-12

Conversion between inches and centimeters, and between square inches and square centimeters.

AREA PROBLEMS

In construction and engineering, *area problems* involve figuring out the area of a physical surface using whatever square units are most appropriate. Often these problems are a matter of "fitting" something into a space of stated dimensions. The regions to be fitted do not always have a simple shape. Then you must figure out the total area by adding and/or subtracting the areas of various other regions. Two such schemes, for calculating the total floor space in a room of irregular shape, are shown in Fig. 6-13.

Problems such as papering a wall can be complicated by a pattern, which must be matched on adjoining strips. Tiles or sections of material that come in standard square or rectangular pieces give rise to another kind of "matching" problem. How do you cut pieces to fill the space so as to minimize the amount of material that goes to waste? These problems must be worked out by making detailed trial calculations. An old carpenter's saying goes, "Measure twice, cut once." When covering a complicated surface with standard pieces of material, a mathematician might say, "Calculate three times, measure three times, and glue the stuff down once!"

Figure 6-13

Two ways to find the surface area of the floor in an irregular room.

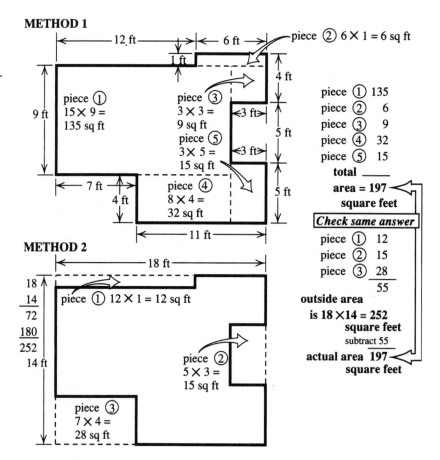

QUESTIONS AND PROBLEMS

This is an open-book quiz. You may refer to the text in this chapter (and earlier ones, too, if you want) when figuring out the answers. Take your time! Consider all values given as exact, so you don't have to worry about significant figures. The correct answers are in the back of the book.

1. Find the areas of the following rectangles.
   (a) 54 inches by 78 inches
   (b) 13 feet by 17 feet
   (c) 250 yards by 350 yards
   (d) 3 miles by 7 miles
   (e) 17 inches by 5 feet
   (f) 340 yards by 1 mile

2. What is a right angle? Why is it so named?

3. How many square feet are in (a) 5 square feet and (b) a 5-foot square?

4. Find the areas of square-cornered triangles, where the sides by the square corner have the following dimensions.
   (a) 5 inches by 6 inches
   (b) 12 feet by 13 feet
   (c) 20 yards by 30 yards
   (d) 3 miles by 4 miles
   (e) 20 inches by 2 feet
   (f) 750 yards by 1 mile

5. A field has four straight sides. Opposite pairs of sides measure 220 yards and 150 yards. But the field does not have square corners. A measurement between the opposite 220-yard sides finds that the straight-across distance is 110 yards. Find the area of the field in acres, where an acre is 4,840 square yards.

6. A parallelogram has sides 20 inches and 15 inches long. The straight-across distance between the 20-inch sides is 12 inches. Calculate the straight-across distance between the 15-inch sides. (*Hint*: use the fact that the area can be calculated in two ways.)

7. Find the areas of the following triangles.
   (a) Base 11 inches, height 16 inches
   (b) Base 31 inches, height 43 inches
   (c) Base 27 inches, height 37 inches

8. Two sides of a triangle are 39 inches and 52 inches. When the 39-inch side is used as the base, the vertical height is 48 inches. What is the vertical height when the 52-inch side is used as the base?

Figure 6-14

Illustration for problems 11 and 12.

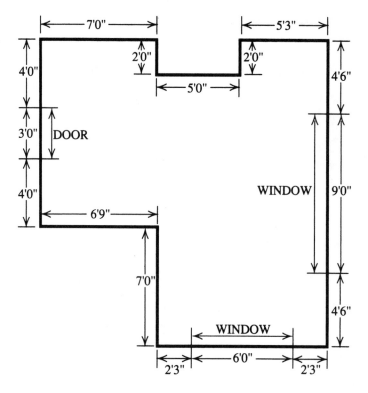

9. A piece of property has two square corners. The side that joins these square corners is 200 yards long. Measuring from each square corner, the two adjoining sides are 106.5 yards and 256.5 yards. The fourth side, joining the ends of these sides, is 250 yards long. What is the area of this property in acres? (*Hint*: treat this area in two parts, an oblong and a square-cornered triangle.)

10. A piece of property measures 300 yards by 440 yards. The owner wants to keep a smaller piece inside that piece that measures 110 yards by 44 yards, and sell the rest. What is the area he wants to sell?

11. Find the area of the floor in the room illustrated by Fig. 6-14.

12. The walls of the room in question 11 are 7 feet high. Doors and windows, at the positions shown, run all the way from the floor to the ceiling. What is the total surface area of the walls?

7
Time, Percentages, and Graphs

You can use various linear units to measure height, width, or depth. That is the first dimension. Area is the second dimension, and volume is the third. And then you have time! It, too, is a dimension. In this chapter, you'll see how time relates to the three so-called *spatial dimensions*. You'll also learn how changing quantities, in any dimension, can be considered as percentages and portrayed as graphs.

TIME AS A DIMENSION

In mathematics, any measurable quantity can be called a dimension. We use a variety of schemes and devices to measure time. The ancients used sand in an *hourglass*. Until recently, we used spring-wound or weight-driven *clocks* and *watches*. Quartz-crystal electronic clocks use alternating-current (ac) signals at precise frequencies to count "ticks" of time.

According to the above definition, we can call temperature, weight, brightness of light, loudness of sound, and intensity of taste "dimensions," although people rarely think of them that way. Even taste has at least four "dimensions": sweet, sour, bitter, and salty. For now, let's confine ourselves to the traditional four dimensions commonly called height, width, depth, and time. This gives us a *four-dimensional continuum*. Albert Einstein coined a term for this in the early 1900s, and it has been used ever since: *space-time*. Some folks turn this around and call it *time-space*.

TIME AND SPEED

Combining length one way with length another way defines an area. Similarly, combining time with length or distance defines *speed*. Suppose you walk at a steady rate to travel a distance of 1 mile in 15 minutes. If you don't change that pace, you will go another mile in the next 15 minutes. In an hour (60 minutes) you'll go 4 miles. Your speed is therefore 4 miles per hour (mph or mi/h).

Keep it up, and you'll do 4 more miles in the next hour, and so on. To find the total distance traveled, you multiply speed by time. The same is true of driving. If you drive at a speed of 60 mph, you go a mile every minute. In 60 minutes (1 hour), you will go 60 miles. In 1 second, you'll go 1/60 mile, or 88 feet, based on the fact that there are 5,280 feet in a mile.

To calculate speed, you must know how far you go and how long it takes you to get there. If you travel 300 miles in 6 hours at a steady speed, you must be doing 50 miles every hour. Six hours at 50 miles each hour is 300 miles. Speed is distance (a length dimension) divided by time.

AVERAGE SPEED

In the last section, it was assumed that speed was steady. This is almost never the case, of course! Suppose that, on a 300-mile journey, you go 30 miles in the first hour, 45 miles in the second hour, 60 miles in the third hour, and 55 miles in each of the fourth, fifth, and sixth hours. A total of 300 miles are traveled in 6 hours (Fig. 7-1). Your *average speed* for the trip is therefore 300 miles divided by 6 hours: 300/6 = 50 mph. You don't travel at a steady speed all the way. You stop now and then to get fuel or to have lunch, or because a traffic light in a town turns red. Your speed at any given time is called the *instantaneous speed*.

Watch your car's speedometer. You will find that you can seldom go at a steady speed even for a few minutes, unless you are on a freeway in the Great Plains on a calm day with the cruise control set! Measuring the distance covered every hour shows the average speed during that hour. Instead of noting the distance traveled every hour, you might notice your speed on the speedometer every few minutes. The speedometer will show how your speed varies during each hour. Perhaps you have to brake to a stop during the hour. What does that do to the average? Later in this course, you will see how calculus can help you answer that question.

Figure 7-1

To find the average speed over the course of a trip, we divide the total distance by the total time. In this example, we go 300 miles in 6 hours, so the average speed is 300/6 = 50 mph.

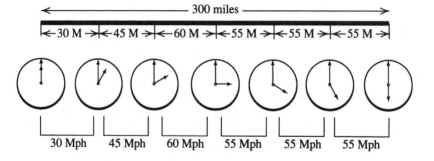

Figure 7-2

A racecar driver can calculate her average speed over the course of a 1-mile lap by measuring the time it takes to complete the lap.

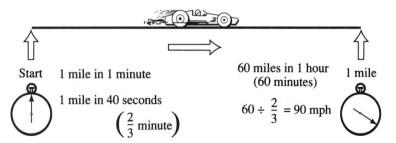

Start   1 mile in 1 minute   60 miles in 1 hour   1 mile
(60 minutes)

1 mile in 40 seconds

$\left(\dfrac{2}{3} \text{ minute}\right)$ $60 \div \dfrac{2}{3} = 90 \text{ mph}$

Now imagine a race car driver who watches her speed closely. Suppose that each lap of track is exactly 1 mile long. The average speed for a lap can be figured by timing that lap. If a lap is made in exactly 1 minute, the average speed for that lap is 60 mph. If a mile requires 40 seconds (2/3 of a minute), her average speed for that mile is 90 mph, as shown in Fig. 7-2.

Suppose you time a 50-mile race that takes place by driving 10 laps on a 5-mile course. The times made by one driver for the 10 laps are precisely 6, 5, 4, 3.5, 4, 3.5, 5.5, 4.5, 4, and 4 minutes. The total time for 50 miles is therefore exactly 44 minutes. If you want to determine the average speed for the entire race, divide the total distance in miles (50) by the total time in hours (44/60). You can use your calculator to figure out that the average speed for the race is 50 / (44/60) = 68.181818 . . . mph. Using your knowledge of decimals and fractions, you can see that this is 68-18/99 mph, which can be simplified to 68-2/11 mph. All this assumes, of course, that your measurements are accurate and exact enough to justify the number of significant figures you include in your final answer.

MAKING UP FOR LOST TIME

Suppose you travel at 30 mph for exactly 1 hour. When you have completed the trip, you will have gone 30 miles, and the average speed for the hour will have been 30 mph. Now suppose you keep going for another hour. You go 30 mph for the first 50 minutes and then 60 mph for the last 10 minutes, so you travel 25 miles at 30 mph and 10 miles at 60 mph for a total of 35 miles. The average speed for the second hour is 35 mph. Over the whole 2-hour journey, you have gone 65 miles, so the average speed for the 2 hours is 32.5 mph. From all this, you can see that the average speed for a given length of time traveled increases as more of the time is spent moving at a higher speed.

Now suppose you have 45 miles to go and an hour to do it in. If you travel at a steady speed, you must go 45 mph. If you go 30 mph for the first 10 minutes, that's 5 miles traveled, leaving you with 40 miles to go in 50 minutes, or 48 mph. If you go 30 mph for the first 30 minutes, you'll have gone 15 miles, leaving yourself with 30 miles to cover in 30 minutes. So for the rest of the trip, you'll have to go 60 mph. Now figure this out: What if you go 30 mph for the first 50 minutes? How fast will you have to go for the rest of the trip, to finish in the allotted hour?

FRACTIONAL GROWTH

Speed and rate of growth are similar ideas, rather like comparing the hare and the tortoise methods of travel. Similar—but different too! You measure time in seconds, minutes, hours, days, or weeks. You measure distance in miles, hundreds of feet, yards, feet, inches, or fractions of an inch. It all depends on how time and distance are related. Watching an auto race is fun. Watching a tree grow is dull.

If a seedling is 2 inches high today, and tomorrow it's 3 inches high, that sounds like fast growth. However, if a 30-foot tree grows to 30 feet 1 inch by tomorrow, you'll have to measure it to see if it grew at all. When you add 1 inch to 2 inches, that's a big growth. But added to 30 feet, 1 inch seems like nothing. Absolute growth and relative growth are vastly different concepts!

The seedling grew from 2 inches to 3 inches, increasing by one-half of yesterday's height. The 30-foot tree was $30 \times 12 = 360$ inches tall at first, and then it added 1 inch. That is only 1/360 as a fraction of the height of the tree. Considered as a fractional increase, the seedling grew 180 times as much as the tree overnight, even though both actual increases were 1 inch. As the seedling keeps growing taller day by day, say at the rate of 1 inch daily for weeks on end, the fractional daily increase will get smaller and smaller. The same thing will also happen with the tree, but the change in the fractional rate of growth will take place more slowly because it was tiny to begin with.

PERCENTAGES

The *percentage* method is a standard way to express fractional increases or decreases. Percentages were developed before decimals to make working in fractions simpler. Decimals might be easier to understand directly, but percentages were used for so many years that they became a sort of mathematical habit, especially in economics and statistics.

Percentages began because fractions are clumsy. If you were asked which is the bigger fraction, 2/5 or 3/8, could you answer just by looking at them? Converting to decimals makes the comparison easy: 2/5 = 0.4 and 3/8 = 0.375.

Percentages always use 100 as the denominator. The numerator is written as a number followed by the word "percent" or the symbol %. You should get used to seeing either the word or the symbol. In this book we'll use both. Percentages, like decimals, make it easy to compare fractions at a glance. You can easily see that 40 percent (or 40/100) is more than 37.5 percent (or 37.5/100).

When a percentage expresses a change in a value, it always refers to the starting size or number. If you say something grew by 1 inch, you don't know whether to think that's fast or slow, unless you know how big it was at the beginning. The seedling in the above example grew 1 inch from a starting height of 2 inches, so the percentage increase was $1/2 = 50/100 = 50$ percent. The tree grew 1 inch from a starting height of 360 inches, so the percentage increase was $1/360 = 0.2777 \ldots / 100 = 0.2777 \ldots$ percent. You might want to round that up to 0.28 percent or even 0.3 percent.

When you have any fractional expression—something with a numerator and a denominator—you can easily find the percentage with a calculator. Just divide the fraction out so the calculator shows it as a decimal. Then multiply that number by 100 to get the figure as a percentage.

PERCENTAGES WITH MONEY

When dealing with money, you often hear about percentages. If railway or airline fares increase, it's usually figured as a percentage. Dividends on stocks and other investments are paid as a percentage, so that everything can be divided fairly.

Suppose that a commuter railway company has to raise fares because its operating costs have increased. It would not be fair to charge everyone $10 more. That would raise a one-time $1 fare to $11, an increase of $10 / $1 = 1,000 / 100 = 1,000\%$. But it would raise a $100 full-year pass to $110, a much smaller percentage increase of $10 / $100 = 10/100 = 10\%$.

Here is another example. Suppose the profits for a company are $10,000 in a given year, and there are 200 stockholders. If $50 is given to each of the stockholders, is that fair? Most likely not! One stockholder might have invested $100 in the company, another $100,000. It would be inequitable for them both to receive $50 of the profit for the year.

Such things are worked out on a percentage basis. If the railway's costs rise from $100,000,000 to $110,000,000, that is a 10% increase. To get this money back in fares charged, each fare should increase by 10%. Thus, the $1 fare would increase to $1.10, and the $100 full-year pass would increase to $110. Similarly, if profits are $10,000 on a total investment of $2,000,000 in a given year and the dividend rate is 0.5%, then the stockholder who invested $100 gets $0.50, and the one who invested $100,000 gets $500.

PERCENTAGES UP AND DOWN

Here is an important thing to keep in mind when you are calculating percentages. We've already mentioned it, but it bears repeating, with a "bullet" for emphasis!

- **Always use the starting figure of a transaction or calculation as the 100% point.**

Suppose a woman buys property for $200,000 and its value increases, so she sells it for $250,000. She's made 25% profit on the deal. It cost her $200,000; she recovered that and made $50,000 more. The profit is $25 for every $100 of the starting price.

But the second owner is not so lucky. The value decreases, so he sells it for its original price of $200,000. Since it is back to its original price, after having gone up 25%, you might think it dropped 25%. But it hasn't! The second owner paid $250,000. Of his original investment, he gets $200,000. The loss is $50,000, the same as the profit the first owner made. But now it's a fraction or percentage of $250,000 (the starting price), not of $200,000. The loss is 20% because he lost $20 for every $100 of his purchase price.

Does this seem odd to you? A 25% increase is the same fractional change, reversed, as 20% decrease. Some things in mathematics are counterintuitive. The "discrepancy" here depends on the extent of the change. Smaller percentages come nearer to being the same, either way. For larger percentages, the difference is larger. A doubling of the starting value represents an increase of 100%. But the reverse, which is halving, is a decrease of only 50%. If the initial value increases by an amount equal to 10 times itself (say from 8 to 88), that is an increase of 1,000%. But if it goes down to only one-tenth as much (say from 100 to 10, which is 100 − 90), it is a decrease of 90%. Make up and solve a few of these problems on your own, and pretty soon you'll develop a good sense of how it works.

Here's an interesting twist. There is no limit to the percentage by which something can increase. You can, at least in theory, have something increase by a trillion percent! But a decrease can never be more than 100%, which represents a decrease from some defined starting value all the way down to zero.

THE BAR GRAPH

Visual presentation of statistics is common these days. Television and Internet advertisements use it all the time (even if their facts are questionable!). Visual comparison conveys an impression more quickly and effectively than plain numbers.

The *bar graph* is one of the most common methods of graphical representation. Numbers are replaced by lines or bars having various lengths. Suppose, for example, you come across old records for the membership in your local golf and country club. You see data for years from 1953, when the club was founded, through 1961. The membership figures for the successive years are 47, 52, 65, 73, 76, 77, 85, 96, and 110. To show this growth, you can draw bars to represent the number of members for each year. Figure 7-3 shows two ways this can be done, called the *horizontal bar graph* (at A) and the *vertical bar graph* (at B).

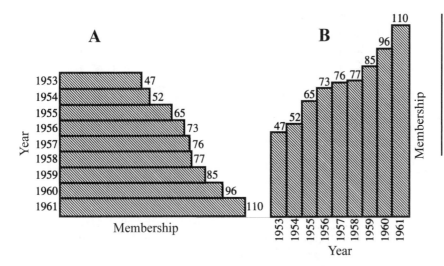

In a basic bar graph, the lengths of the bars are directly proportional to the quantities rendered. If you draw each member at 1/16 of an inch, for example, then a bar representing 10 members measures 10/16 (or 5/8) of an inch long, and a bar representing 100 members measures 100/16 (or 6-1/4) inches long. The thicknesses of the bars are all the same in most bar graphs, although in more complicated ones called *histograms* the thicknesses of the bars can vary in order to portray a second dimension.

OTHER GRAPHS

Graphical presentations can clarify things in many ways. A college might want to show how many of its alumni belong to certain groups such as engineers, doctors, lawyers, salesmen, factory workers, shop assistants, truck drivers, and musicians. A bar graph can work well for this purpose, but there are other ways, too.

If you want to see how much of the whole is represented by each particular group, you can use a *box graph*. A rectangle of suitable height and width is marked off in 100 units, and each group is given an area that represents its percentage of the total. Because the total is 100%, all the widths together must exactly fill the box, as shown in Fig. 7-4A.

Another scheme to show this situation involves dividing or "slicing" up a circle, just like you would slice up a pie. In fact, because it looks like a sliced-up pie (Fig. 7-4B), it is often called a *pie graph*. You divide up the circumference into 100 equal angular parts (10 are shown here to make it clearer). Then you divide the circumference according to the percentages in the groups, and you draw lines from each marker to the center (Fig. 7-4C).

Figure 7-4

At A, a box graph can help show proportions of various groups. At B, a pie graph can do the same thing. At C, a pie graph is constructed by dividing a circle into 100 equal angular parts.

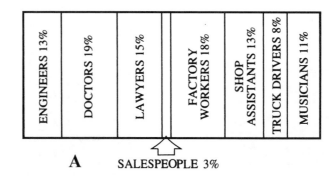

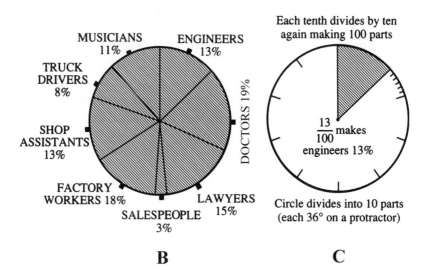

A — SALESPEOPLE 3%

B

C

INTERPOLATION

Graphs are used to help make calculations and to figure out how much of something will be needed to do a certain task. Suppose tests on an electric motor relate the electrical wattage input to the mechanical horsepower output. The results are tabulated, from electric power running "light" with no mechanical load to electric power inputs for various mechanical horsepower outputs.

Suppose you have a job that requires a certain amount of horsepower not shown in the table at the bottom of Fig. 7-5. How do you determine the electric power it needs? A graph, as shown above the table, makes the job easier. You place points on squared paper to show all the figures in the table, and then you join the dots you have made. This gives you a line or curve from which you can visually guess at intermediate values.

You need not know anything about electricity to make this graph. But if you do have some electrical expertise, you will notice that this motor does not convert all the

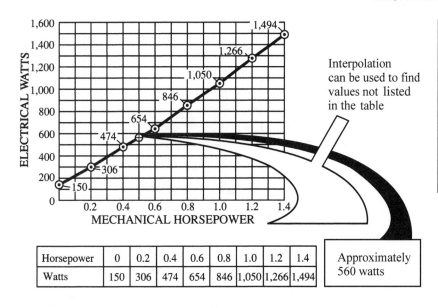

Figure 7-5

Graphs can be used for interpolation. Here, we determine the amount of electrical power that a certain motor will need to produce 0.5 horsepower of mechanical work.

Interpolation can be used to find values not listed in the table

Approximately 560 watts

Horsepower	0	0.2	0.4	0.6	0.8	1.0	1.2	1.4
Watts	150	306	474	654	846	1,050	1,266	1,494

electricity it receives to mechanical work. If it did, it would get 1 horsepower from 746 electrical watts, not 1,050 watts! The extra electricity is wasted as heat in the motor. An electrical engineer would say that the motor has an *efficiency* of less than 100%—another use of percentages! If you have to put in 1,050 watts of electricity to get 746 watts of mechanical power out of the motor, then the motor efficiency is 746 / 1,050 = 0.71 = 71%, at least for a load that demands exactly 1 horsepower from the motor.

In this situation, you can find the amount of electric power that is needed for any particular job by reading it on your graph. This process is called *interpolation*. Figure 7-5 shows an example. If you need 0.5 horsepower from the motor, it will consume about 560 watts of electricity.

GRAPHS CAN HELP YOU EVALUATE DATA AND FIND ERRORS

Graphs can show things that are not obvious from the data used to make them. Graphs also are useful to check and interpret data, and to find and correct mistakes.

Consider a meter that has numbers on its scale at 20 and 30, with unnumbered "hash marks" in between to represent the single units as shown in Fig. 7-6. The reading in this case should be 23 units. But suppose you write down 27 by mistake when tabulating the data, and that is the only mistake you make. You plot your graph, using your figures. All the points line up nicely, except the one that you wrongly copied as 27. This "off-the-track" point immediately shows you that something is wrong, and it also shows you where the error occurred. So you take that particular meter reading again.

Specific information can often be presented in multiple ways. Figure 7-7 shows three ways to graphically show a race car driver's performance as he makes 10 laps around a

Figure 7-6

Graphs can be used to detect and locate errors when reading meters or collecting data.

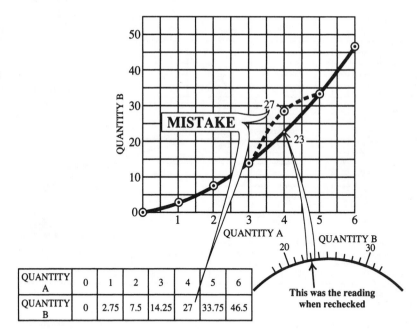

QUANTITY A	0	1	2	3	4	5	6
QUANTITY B	0	2.75	7.5	14.25	27	33.75	46.5

This was the reading when rechecked

Figure 7-7

At A, a graph showing time vs. distance for a racer who drives 10 laps around a 5-mile track. At B, a bar graph showing average speed for each lap. At C, a bar graph showing the elapsed time for each lap.

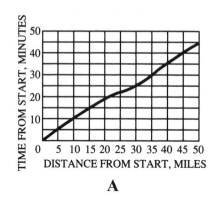

A

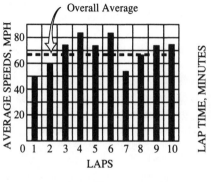

B

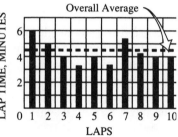

C

5-mile track. Each graph gives you a different "picture" of the driver's performance. The first graph (A) is based on the cumulative distance traveled, the second (B) is based on the lap-by-lap average speed, and the third (C) is based on the time per lap.

QUESTIONS AND PROBLEMS

This is an open-book quiz. You may refer to the text in this chapter (and earlier ones, too, if you want) when figuring out the answers. Take your time! Consider all values given as exact, so you don't have to worry about significant figures. The correct answers are in the back of the book.

1. How far will a car go at 35 mph for 36 minutes? (*Hint*: find what fraction of an hour is represented by 36 minutes.)

2. A riverboat makes a water speed (at which its motor propels it through the water) of 10 mph. The river has a downstream current flowing at 2 mph. How fast does the boat go (a) upstream and (b) downstream?

3. How long will the riverboat in question 2 take to make a journey of 96 miles (a) upstream and (b) downstream?

4. Traveling at a water speed of 10 mph, the riverboat in question 2 burns one-half ton (1 ton is 2,000 pounds) of fuel per hour. How much fuel does the boat use on a 96-mile journey (a) upstream and (b) downstream?

5. If the boat reduces its downstream water speed to make the downstream journey in the same length of time as the upstream journey, and if that reduces fuel burned in proportion to the water-speed reduction, will reducing the downstream water speed save fuel? If so, by how much? What would be the percentage saving (on the round trip)?

6. If the boat reduces its water speed on the upstream run, will it save fuel or use more? How much? What percentage?

7. A man invests $50,000 in stock. For the first year, it pays a dividend of 5% on his investment. At the same time, the value of his stock rises to $60,000. If he sells the stock, how much profit will he make (a) in cash and (b) as a percentage?

8. During early weeks of growth, a tree's height is recorded every week. For 5 successive weeks, the heights are 16 inches, 24 inches, 36 inches, 54 inches, and 81 inches. What is the percentage growth per week for each of the 4 weeks? What is the percentage for the whole month?

9. Make a graph of the tree's growth (from question 8) for the month. From the graph, estimate the height of the tree in the middle of the second week.

10. A race track has an 8-mile lap that consists of 5 miles with many hairpins, corners, and grades. The remaining 3 miles have straightaways and banked curves. The best time any car can make over the 5-mile part is 6 minutes, but the 3-mile section lets drivers "open up." Two drivers tie for the best on the 5-mile part, but one averages 90 mph on the 3-mile part, and the other averages 120 mph. Find the average speed for each on the whole lap.

11. A car is checked for mileage per gallon and is found to get 32 miles per gallon (mpg) for straight turnpike driving. How far will it go on a tankful if the tank holds 18 gallons?

12. A company needs printed-circuit boards for which two processes are available. The first needs a tool that costs $2,000, then makes boards for 15 cents apiece. The other uses a procedure that initially costs $200, then makes boards for 65 cents each. Find the cost per board by each process, assuming the total quantity ordered is 100, 500, 1,000, 2,000, 5,000, and 10,000 units.

13. Plot a graph of the cost per board by the two processes (question 12) for quantities from 1,000 to 10,000 boards. For how many boards would the costs of both processes be the same?

14. Driving a car at a steady 40 mph produces a mileage of 28 mpg. Driving the same car at 60 mph reduces the mileage to 24 mpg. On a journey of 594 miles, how much gas will be saved by driving at the slower speed? How much longer will the journey take?

15. A man pays $200,000 for some property. After a year its value rises 25%, but he does not sell. During the next year its value drops 10%, after which he sells. What profit did he make on his original investment (a) in cash and (b) in percent? Why was it not $25 - 10 = 15\%$ profit?

16. After all allowances and deductions have been made, a man's taxable income for a given year is $120,000. How much tax will he pay at 20% on the first $30,000 and 22% on the rest?

17. A square-cornered triangle with 12-foot and 16-foot adjacent sides (against the square corner) has the same area as a parallelogram with opposite pairs of sides that are 10 feet and 16 feet long. What is the perpendicular distance between the 16-foot sides of the parallelogram?

Part 2

ALGEBRA, GEOMETRY, AND TRIGONOMETRY

8

First Notions in Algebra

As problems get more complicated, so does the arithmetic used to solve them. People invented multiplication and division to shortcut the repeating of addition and subtraction. *Algebra* was developed to handle more involved problems, where the older shortcuts didn't help much.

SHORTCUTS FOR LONG PROBLEMS

Figure 8-1 shows a fairly simple problem that algebra can help you solve: fencing for a double-fenced enclosure. You are told two facts at the beginning of this problem. First, the length of the inner enclosure must always be triple its width. Second, the spacing between the two fences must always be 3 feet.

If you know the inner enclosure size, how much fencing is needed? Given a certain total length of fencing, how big an inner enclosure will it make? The first question can be answered by arithmetic if you are willing to make a table. The information in Fig. 8-2A shows the arithmetic needed for this part. The second part isn't so simple.

GRAPHS HELP YOU FIND ERRORS AND SOLUTIONS

With all the steps in the calculations, you might make a mistake. By graphing your results (Fig. 8-2B), you can find mistakes. The results should be on a straight line or on a smooth curve, depending on the type of problem. A point that appears "off the track" alerts you to the likelihood that you have made a mistake.

Make the graph with squared paper. Choose a scale that uses as much paper as possible without making the values awkward to read. Here, each small space represents 20 feet of fencing. In this example, you can see by interpolation that if you have a total of 456 feet of fencing, the width of the inner enclosure must be very close to, or perhaps exactly, 27 feet.

Figure 8-1

A problem involving two rectangular fence enclosures, one inside the other. The length of the inner enclosure is three times its width. The fences are 3 feet apart all the way around.

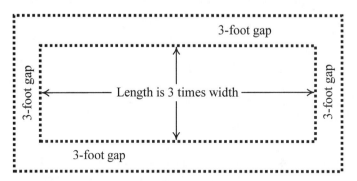

MANIPULATING EQUATIONS

The above-described schemes involve making the same series of calculations several times with different numbers. A pocket calculator makes these problems easier than they used to be. However, it's still a long way around, especially if you want only one answer. An example is shown in Fig. 8-3.

Without actually using algebra, first write all the pieces, in terms of either width or feet, and add them. The total length of fencing, which you can determine by looking at Fig. 8-1 and counting, is 16 widths plus 24 feet. You calculate each answer in the table with this formula, using only one multiplication ($\times$ 16) and one addition (+ 24). Or, turn everything around with a process you will become more familiar with in algebraic terms. You can use an *equation*:

$$\text{Total length of fencing} = 16 \text{ widths} + 24 \text{ feet}$$

The words on either side of the equals sign have the same value; they are different ways of naming the same amount. Next, if you subtract 24 feet from the total that appears on either side of the equals sign, the statement or equation will still be true. Each side will just be 24 feet less than it was before:

$$\text{Total length of fencing} - 24 \text{ feet} = 16 \text{ widths}$$

Divide both sides by 16, and it will still be true. Each side will be 1/16 of what it was:

$$\text{Length of fencing} - 24 \text{ feet, all divided by } 16 = \text{width}$$

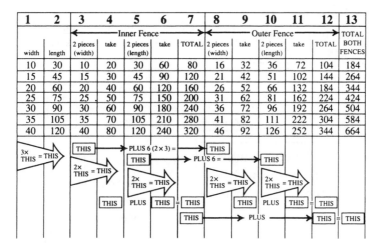

1	2	3	4	5	6	7	8	9	10	11	12	13
		Inner Fence					Outer Fence					TOTAL BOTH FENCES
width	length	2 pieces (width)	take	2 pieces (length)	take	TOTAL	2 pieces (width)	take	2 pieces (length)	take	TOTAL	
10	30	10	20	30	60	80	16	32	36	72	104	184
15	45	15	30	45	90	120	21	42	51	102	144	264
20	60	20	40	60	120	160	26	52	66	132	184	344
25	75	25	50	75	150	200	31	62	81	162	224	424
30	90	30	60	90	180	240	36	72	96	192	264	504
35	105	35	70	105	210	280	41	82	111	222	304	584
40	120	40	80	120	240	320	46	92	126	252	344	664

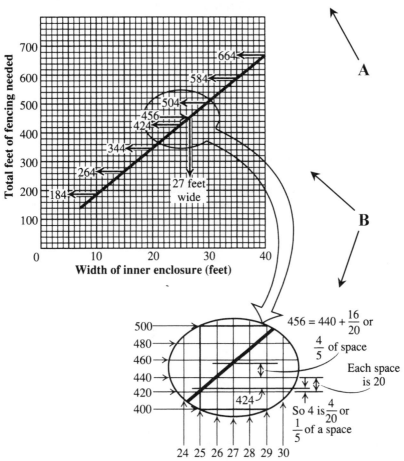

Figure 8-2
At A, a table showing some data for the fencing problem. At B, a graphical presentation of some of that data.

Figure 8-3

At A, finding the total length of fencing when the width of the inner enclosure is 25 feet. At B, finding the dimensions of the inner enclosure when the total length of fencing is 456 feet.

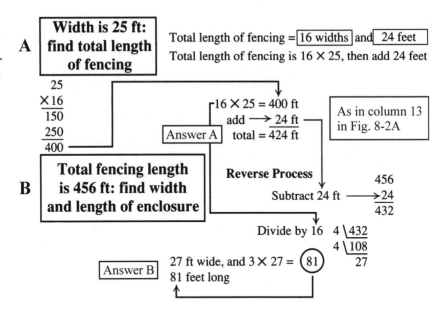

WRITING IT AS ALGEBRA

Writing problems of this kind can be shortened by using a letter of the alphabet to stand for each original measurement or quantity in the problem. Let's say you write w for width. Length is specified as 3 times the width, so you can write length as $3w$, meaning 3 times w. (In algebra, there is no multiplication symbol when one or both of the quantities being multiplied are letters instead of numbers.) Finally, write f for the total length of fencing. Then the above three equations, replacing the words with the letters (called *variables*), look like this:

$$f = 16w + 24$$

and

$$f - 24 = 16w$$

and

$$(f - 24) / 16 = w$$

In this last expression, you might wonder, "What do the *parentheses* mean?" You can probably figure this out from the context, but it will be explained in greater detail shortly.

When using algebra, as in arithmetic, always be sure that the units are consistent. Here they are all in feet. Don't mix different units for the same quantity! Don't specify some

ARITHMETIC			ALGEBRA		
163	means	100	16w	means	$\overline{10}$
	plus	60		plus	6
	plus	3	all multiplied by		w

To multiply 16 by 3 :			To multiply 16 by x :		
multiply 6 by 3		18	multiply 6 by x		6x
multiply 10 by 3		30	multiply 10 by x		10x
Add together		48	A total of 16 times x		16x

16w + 24 means:
16 is multiplied by w,
but 24 is not

Figure 8-4

Comparisons in notation between arithmetic and algebra. Note the differences!

lengths in inches or yards, and others in feet. Avoid expressing area in square feet in one part of a problem, and in square yards someplace else in the same problem.

In arithmetic, we write numbers using figures in a row. Thus, 23 is 2 tens plus 3 ones. In algebra with variables, the expression xy does not mean x tens and y ones, but x multiplied by y. Get used to this, because it's the way everybody does it! As you look at such new things, ask yourself, "What does this new way of writing mean?" That's one difference between arithmetic and algebra, of which you'll see more as you go on. Figure 8-4 shows some more comparisons in notation between arithmetic and algebra.

In some computers, algebra is a little different from "regular" algebra. Multiplication is sometimes shown by an asterisk between the quantities to be multiplied, whether they are both numbers, one number and one variable, or both variables. Not only that, but the space is often missing on either side of the symbol. Instead of 3×4 you might see 3*4; instead of $5x$ you might see 5*x; instead of xy you might see x*y. We won't be doing much with computer algebra, but you should be aware of this notation in case you ever come across it.

ITALICS—OR NOT?

From now on, you'll see that variables are italicized in the text but not in the illustrations. In text, italics are used to avoid confusion. For example, "Suppose a rocket ship accelerates at a rate of a meters per second." Even if you read this two or three times, you might have trouble figuring out what it means! You should write, "Suppose a rocket ship accelerates at a rate of a meters per second." But in a drawing or graph, you can label an axis or object "a" (without italics) and, if your readers see the text that goes with the graph, they'll know that either "*a*" or "a" can stand for "acceleration." You should get used to both ways of denoting variables. In this book, they're italicized in the text but not in illustrations.

GROUPING SYMBOLS

When algebra was taught in the olden days, the use of parentheses and other *grouping symbols* was standardized. Some uses that you will see today were considered unnecessary. Parentheses are used a lot more now because we often want to know which to do first, add or multiply, for instance. It's better to put in a couple of unnecessary grouping symbols and get things right, than to be obsessed with using the absolute minimum number of grouping symbols (in the name of "elegance" or some other noble cause) and wind up making a mistake!

Here is a good rule to use when you see an expression with parentheses and other grouping symbols: Do the calculations within the inside parentheses first. What is inside the parentheses is regarded as a single quantity. For example, when you see $2(w + 6)$, it means that the quantity $w + 6$ is multiplied by the 2 outside the parentheses. The addition is done first, and then the multiplication. But if you see the expression $2w + 6$ without parentheses, you should do first the multiplication and then the addition. That is, multiply w by 2, and then add 6.

Three kinds of grouping symbols are commonly used. They have different names, depending on whom you ask. The most common terms are parentheses for (and), *brackets* for [and], and *braces* for { and }. Brackets are used outside of parentheses, and braces are used outside of brackets. Here are some examples of expressions that build up in complexity, requiring more and more of these symbols:

$$x + 2y = 3$$

$$z(x + 2y) = 3z$$

$$w[z(x + 2y)] = 3zw$$

$$(a + b)w[z\,(x + 2y)] = (a + b)\,(3zw)$$

$$c\{(a + b)w[z(x + 2y)]\} = c[(a + b)\,(3zw)]$$

Sometimes, only plain parentheses are used, and the expressions can still be interpreted although you are more likely to make a mistake. Here is how the above progression of equations looks when only parentheses are used:

$$x + 2y = 3$$

$$z(x + 2y) = 3z$$

$$w(z(x + 2y)) = 3zw$$

$$(a + b)w(z(x + 2y)) = (a + b) (3zw)$$

$$c((a + b)w(z(x + 2y))) = c((a + b) (3zw))$$

These sequences of equations aren't necessarily all in the simplest possible form. They're presented only to show you how the grouping symbols work.

Here are two rules to remember when you see (or write) a complicated equation with parentheses, brackets, and braces:

- Do the calculations in the innermost groups first, and work outward, on both sides of the equals sign separately.
- Count the number of opening and closing symbols. The number of opening symbols should always be the same as the number of closing symbols in the whole expression. If they aren't, something is wrong with the way the expression is written!

SOLVING PROBLEMS WITH GROUPING SYMBOLS

Figure 8-5 shows a rather messy equation in algebra, using a single variable x. You can solve this equation for x—that is, determine what number x must be—by removing the parentheses and brackets, following the rule as explained above, starting from the inside ones. Then multiply things out as shown in Fig. 8-6.

4 times the quantity $x + 5$ is $4x + 20$

and

2 times the quantity $x - 3$ is $2x - 6$

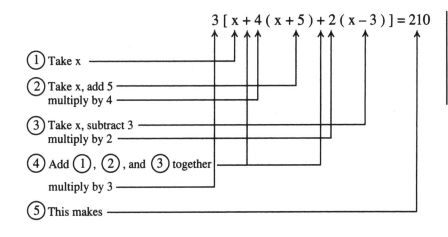

$$3 [x + 4 (x + 5) + 2 (x - 3)] = 210$$

① Take x
② Take x, add 5 multiply by 4
③ Take x, subtract 3 multiply by 2
④ Add ①, ②, and ③ together multiply by 3
⑤ This makes

Figure 8-5
Interpreting an expression that contains multiple parentheses and brackets.

Figure 8-6

Solution to the
equation shown
in Fig. 8-5.

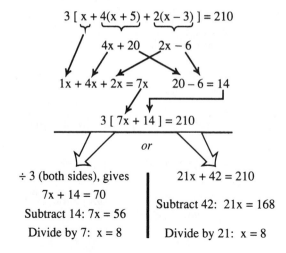

Multiply each part of what's inside the parentheses by what's outside them. Then collect the pieces inside the brackets. Add the three x terms: x (don't write the 1 in algebra), $4x$, and $2x$. That makes $7x$. Add two plain numbers, 20 and -6, getting $20 - 6 = 14$. Now you can multiply the whole thing by 3:

3 times the quantity $7x + 14$ is $21x + 42$

So you get this equation:

$$21x + 42 = 210$$

This is a lot simpler than what you began with! Now to solve it: If 3 times what's in the brackets is 210, then what's in the brackets must be 1/3 of 210, or 70. That means $7x + 14 = 70$. If you subtract 14 from both sides of this, you get $7x = 56$. Now divide both sides by 7, so you get $x = 8$. That's the number you wanted to find.

Now check your solution to be sure it is right! You should always check in algebra, just as you do in arithmetic. You have 8 as the value for x. "Plug it in" to the left-hand side of the original equation (at the top of Fig. 8-6) and grind it out. If there are no parentheses or brackets in a part of an expression, do the multiplications first, and then the additions. You should get 210, like this:

$$3[x + 4(x + 5) + 2(x - 3)]$$
$$3[8 + 4(8 + 5) + 2(8 - 3)]$$
$$3[8 + 4 \times 13 + 2 \times 5]$$
$$3[8 + 52 + 10]$$
$$3 \times 70$$
$$210$$

To recap, here are the rules if you need to simplify a complicated expression with grouping symbols, or figure out what quantity it represents. Perform these steps in this order:

* Group everything within parentheses, brackets, and braces from the inside out.
* Do all the products and quotients.
* Do all the sums and differences.
* Simplify the final expression as much as you can.

PUTTING A PROBLEM INTO ALGEBRA

Here's a problem that is "made to order" for algebra. Eleven young people (some boys, some girls) go out to eat together at a bargain fast-food establishment. Each boy orders something that costs $1.25. Each girl orders something that costs $1.60. The total check comes to $16.20. How many boys are in the group? How many girls?

One way of expressing the problem is shown in Fig. 8-7. Each price is written as a certain number of nickels. (You can also write it in pennies or dollars, although dollars would require that you use decimals. Whatever units you use, stick to them throughout the problem.) Figure 8-8A shows how the problem can be solved. Once you've done that, you should check your work (Fig. 8-8B).

MAGIC BY ALGEBRA

Here's a trick you might use at a party. Ask several people to think of a number (but not say what it is) and write it on a piece of paper. Then tell them to add 5, multiply by 2, subtract 4, multiply by 3, add 24, divide by 6, and finally subtract the number they first

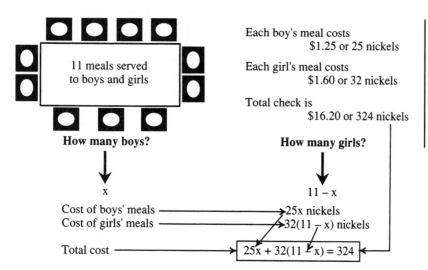

Each boy's meal costs
$1.25 or 25 nickels

Each girl's meal costs
$1.60 or 32 nickels

Total check is
$16.20 or 324 nickels

11 meals served
to boys and girls

How many boys?

How many girls?

x

$11 - x$

Cost of boys' meals ⟶ 25x nickels
Cost of girls' meals ⟶ $32(11 - x)$ nickels

Total cost ⟶ $25x + 32(11 - x) = 324$

Figure 8-7

A group of boys and girls go out to eat and share the cost. How many boys are in the group? How many girls?

Figure 8-8

Here is one way to solve the problem shown in Fig. 8-7 and described in the text. At A, a solution is found. At B, the answers are checked to be sure they are right.

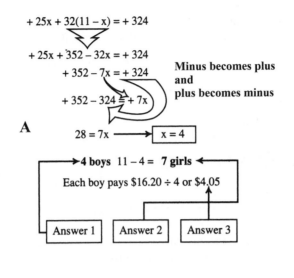

$$+ 25x + 32(11 - x) = + 324$$

$$+ 25x + 352 - 32x = + 324$$

$$+ 352 - 7x = + 324 \qquad \text{**Minus becomes plus and plus becomes minus**}$$

$$+ 352 - 324 = + 7x$$

A $\qquad 28 = 7x \longrightarrow \boxed{x = 4}$

4 boys $\quad 11 - 4 = \quad$ **7 girls**

Each boy pays $16.20 ÷ 4 or $4.05

| Answer 1 | Answer 2 | Answer 3 |

4 boys at $1.25 = $5.00
7 girls at $1.60 = $11.20

B $\qquad$ Total check $\qquad$ $16.20

Now check each statement!

$$25x + 352 - 32x = 324$$

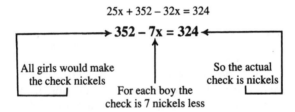

$$352 - 7x = 324$$

All girls would make the check nickels

For each boy the check is 7 nickels less

So the actual check is nickels

wrote down. Give them time to do these calculations, and then declare, "The final answer is 7."

The fact that all the people think of different numbers, but all get the same answer (assuming no one makes a mistake in his or her arithmetic), seems almost like magic. But you can see how it works if you let x stand for the starting number. Figure 8-9 shows what happens when you do the operations in the order described above, starting with x. The variable x "vanishes" at the end because you subtract it from itself.

A BOAT-IN-THE-RIVER PROBLEM

Suppose a man rents a motorboat that will go 30 miles per hour in still water. First he drives it downstream for 10 minutes. Then he drifts with the current for 30 minutes. Finally, he heads upstream for the remaining 20 minutes. Where does he finish, relative to his starting point?

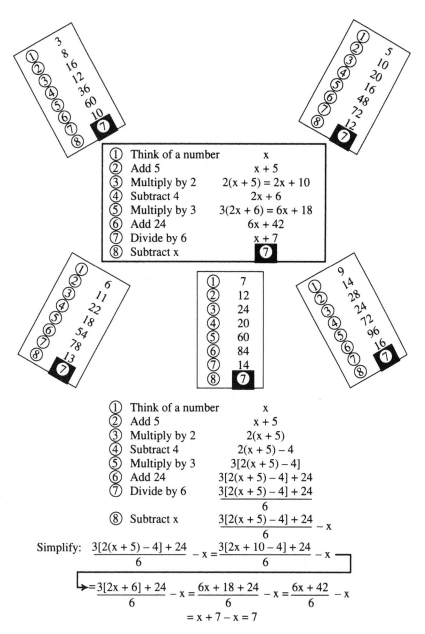

Figure 8-9

A popular "think-of-a-number" trick. No matter what number we start with, the final answer is 7.

Figure 8-10 illustrates this problem. You don't know how fast the river is flowing, so write an x for the river's rate of flow in miles per hour. This variable gives us an expression for his final position, in miles downstream:

$$5 + (1/6)x + (1/2)x - 10 - [(-1/3)x]$$

The minus is multiplied by a minus because upstream is exactly the opposite direction from downstream. If the river were not flowing, he'd go 10 miles upstream (the -10).

Figure 8-10

A "boat-in-the-river" problem, showing how two minuses can make a plus.

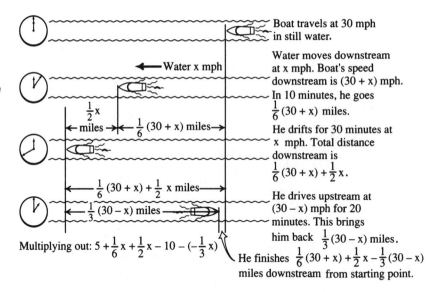

Boat travels at 30 mph in still water.

Water moves downstream at x mph. Boat's speed downstream is (30 + x) mph. In 10 minutes, he goes $\frac{1}{6}(30 + x)$ miles.

He drifts for 30 minutes at x mph. Total distance downstream is $\frac{1}{6}(30 + x) + \frac{1}{2}x$.

He drives upstream at (30 – x) mph for 20 minutes. This brings him back $\frac{1}{3}(30 - x)$ miles. He finishes $\frac{1}{6}(30 + x) + \frac{1}{2}x - \frac{1}{3}(30 - x)$ miles downstream from starting point.

Multiplying out: $5 + \frac{1}{6}x + \frac{1}{2}x - 10 - (-\frac{1}{3}x)$

Because the river is flowing at x mph, he floats less than 10 miles upstream by $(1/3)x$ miles. He's heading upstream so it slows him down: minus times minus.

You can use an imaginary floating log to help solve this. The boat's position, relative to the log, is given by numbers without x. In one hour, the log floats x miles downstream. You can substitute values for x to get various "answers" (Fig. 8-11). The exact answer depends on what you choose for x, the speed of the river current. Mathematicians would say that the answer is a *function* of x. Collecting the number terms, $+ 5$ and $- 10$, the boat finishes 5 miles upstream from wherever the log finishes.

Figure 8-11

Working towards a solution to the "boat-in-the-river" problem. We now know the position of the boat as a function of the water flow speed.

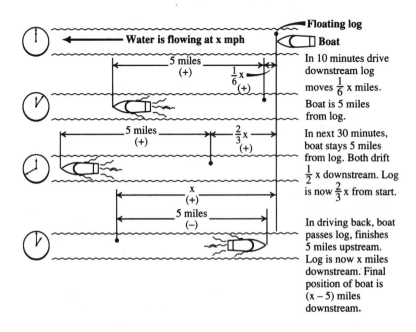

Floating log
Boat

In 10 minutes drive downstream log moves $\frac{1}{6}$ x miles. Boat is 5 miles from log.

In next 30 minutes, boat stays 5 miles from log. Both drift $\frac{1}{2}$ x downstream. Log is now $\frac{2}{3}$ x from start.

In driving back, boat passes log, finishes 5 miles upstream. Log is now x miles downstream. Final position of boat is (x – 5) miles downstream.

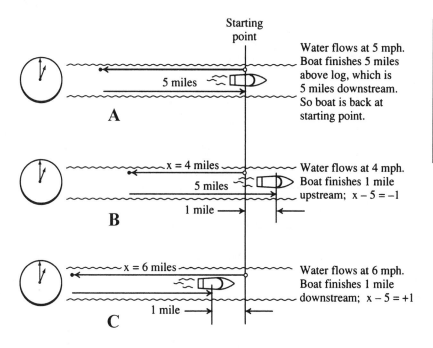

Figure 8-12

Solutions to the "boat-in-the-river" problem for water flow (river current) speeds of 5 mph (at A), 4 mph (at B), and 6 mph (at C).

Figure 8-12 shows three examples for river current speeds of 5 mph, 4 mph, and 6 mph. If the stream is flowing at 5 mph (at A), the log travels 5 miles downstream, so the boat driver finishes where he started. If the stream flows at 4 mph (at B), he finishes 1 mile upstream from his starting point. If the stream flows at 6 mph (at C), he finishes 1 mile downstream from his starting point.

NUMBER PROBLEMS

Suppose someone tells you that a certain number has a ones digit that is twice the tens digit, but adding 18 reverses its digits. What is the number? You could try a few numbers until you find the one that works. But algebra gives you a more direct route.

Assume that the tens digit is x. That means its actual numerical value is $10x$. You are told that the ones digit is twice the tens digit, so it must be $2x$. Therefore, the whole number has a value of $10x + 2x$, or $12x$. Now, add 18. That gives you $12x + 18$. What was the ones digit is now the tens digit. The tens digit in the new number must therefore be $20x$ instead of $10x$. The units digit is now x instead of $2x$. The new number is therefore $20x + x$, or $21x$.

You can now write an equation putting these two descriptions together:

$$12x + 18 = 21x$$

Subtract $12x$ from both sides, and you get

$$18 = 9x$$

Figure 8-13

Solution to a number problem in which adding 18 reverses the digits.

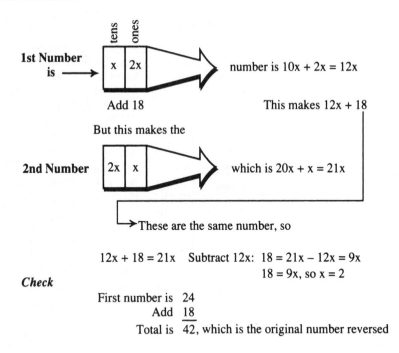

1st Number is → number is $10x + 2x = 12x$

Add 18   This makes $12x + 18$

But this makes the

2nd Number   which is $20x + x = 21x$

These are the same number, so

$12x + 18 = 21x$ Subtract 12x: $18 = 21x - 12x = 9x$
$18 = 9x$, so $x = 2$

Check

First number is 24
Add 18
Total is 42, which is the original number reversed

Divide each side by 9, getting $2 = x$. You originally decided that x was to be the tens digit. That means the number you're seeking has a tens digit of 2. The ones digit is twice that, or 4. The "mystery number" is therefore 24. Checking it out, if you add 18 to 24, you get 42, which is what you get when you reverse the digits in 24. The whole solution process for this problem is illustrated in Fig. 8-13.

QUESTIONS AND PROBLEMS

This is an open-book quiz. You may refer to the text in this chapter (and earlier ones, too, if you want) when figuring out the answers. Take your time! Consider all values given as exact, so you don't have to worry about significant figures. The correct answers are in the back of the book.

1. When a certain number x is subtracted from 12, the result is the same as when the number is multiplied by 2. Write, and solve, an equation to find the unknown number x.

2. When a certain number y is multiplied by 19, the result is the same as when the number is added to 36. Write, and solve, an equation to find the unknown number y.

3. Solve the following "word equations" for the unknown number z.
(a) z divided by 2, and then added to 10, equals 20.
(b) 4 times the quantity (z minus 7) equals 0.
(c) 10 divided by z, then added to 7, equals 12.

4. Suppose $x = 5$ and $y = 7$. Then what is the value of xy? Why isn't it 57? How can 57 be expressed in terms of x and y?

5. Simplify the following expressions.
(a) $5[3x - 2(5x + 7)] - 9$
(b) $14 + 2[x + 5(2x + 3)]$

6. Write down and simplify an expression for the following sequence of statements: A number has 5 added, then is multiplied by 3. The same original number has 6 added, then is multiplied by 4. Next, the same original number has 7 added, then is multiplied by 50. Finally, all three of these three results are added together.

7. A number has 3 added, then is multiplied by 4. The same number is multiplied by 4, then 3 is added. These two results are added, multiplied by 5, and 6 is added to the total. Write down this expression and simplify it. If the total is 361, what was the original number?

8. A professional society's membership costs $20 per year for full members and $8 per year for student members. Membership totals 2000 with annual dues receipts of $35,200. How many full members and how many student members does this society have?

9. The ones digit of a number is 2 more than its tens digit. When multiplied by 3, the tens digit is the same as the ones digit. What is the number? (*Hint*: use x for the tens digit.)

10. A two-figure number has a ones digit that is 1 more than the tens digit. When the number is multiplied by 4, the ones digit is what the tens digit was before, and the tens digit is 3 times what the ones digit was before. What was the original number? (*Hint*: use x for the original tens digit.)

11. Use algebra to show that in any number where the ones digit is 1 more than the tens digit, adding 9 will reverse the digits.

12. Use algebra to show that in any number where the ones digit is greater than the tens digit, adding 9 times the difference between the digits reverses them. (*Hint*: use a for the tens digit, and use $a + x$ for the ones digit.)

13. By substituting various values of x into the following two expressions, say what is different about them. Show why the second expression is unique.
(a) $(3x + 7) / (x + 1)$
(b) $(3x + 9) / (x + 3)$

14. In each of the following equations, y is on one side of the equals sign, and an expression containing x is on the other. Mathematicians would say that these equations express y in terms of x. In each case, make a transposition that will put x by itself on one side of the equals sign, with the correct expression containing y on the other. That is, change them so they express x in terms of y.

    (a) $y = x + 5$
    (b) $y = (x/3) - 2$
    (c) $y = 6(x + 2)$
    (d) $y = 1 / (x - 1)$
    (e) $y = 3x - 7$
    (f) $y = (5x + 4) / 3$

15. Show that the following equations cannot be solved for x.

    (a) $4(x - 5) = 4x - 18$
    (b) $5x + 3 = 5x - 7$
    (c) $(2x + 16) / 4 = (x + 12) / 2$

16. Draw graphs showing several x and y values for each of the following equations. By doing this, you will see that equations (a) and (c) are fundamentally different from equations (b) and (d). What is the difference?

    (a) $y = x + 2$
    (b) $y + 1 = x(3x - 1)$
    (c) $5 + x = 2 - 2y$
    (d) $x = 2y(y - 1)$

17. Choose any number. Add 6 to this number. Multiply the resulting sum by 3. Then from this, subtract 12. Divide the result by 3. Finally, subtract the number you chose to begin with. The answer will always be 2. Write down a series of expressions showing how this "magic number" problem works.

9
Some "School" Algebra

In this chapter, you'll learn some basic techniques that have always been taught in first-year "school" algebra courses. You'll also learn how to solve sets of equations that have more than one variable.

ORDERLY WRITING IN ALGEBRA

Although parentheses can clarify problems with "long multiplication," removing the parentheses to perform the multiplication needs understanding, so you can see how algebra is like arithmetic. Your pocket calculator does long multiplication in arithmetic. It's not easy to get your calculator to do it in algebra.

Figure 9-1 shows four examples of long multiplication in algebra. Each one is a little more complicated than the one above it. At A, you multiply a parenthetical expression by a number (outside the parentheses). At B, you multiply a parenthetical expression by x times a number ($3x$). At C, you multiply two parenthetical expressions together, each of which contains a *term* in x (that is, a number times x). At one point here, you get x times itself. This is x *squared*, written x^2. At D, again, you multiply two parenthetical expressions by each other. But this time, the two expressions contain different variables, one called x and the other called y.

SHOWING "PLACE" IN ALGEBRA

You can compare how generations of students once wrote long multiplication in arithmetic (before pocket calculators did it for them) with something similar in algebra. Successive places in arithmetic, moving left in columns, stand for successively bigger powers of 10. Farthest to the right are the ones, which are technically 10 to the zeroth power, written 10^0. Next comes a number of tens (10^1), then a number of hundreds (10^2), then a number of thousands (10^3), and so on.

Figure 9-1

"Long multiplication" in algebra. At A, multiplication by a number. At B, multiplication by x times a number. At C, the product of two expressions containing the same variable. At D, the product of two expressions containing different variables.

A $4(x + 3)$

$$\begin{array}{r} x + 3 \\ 4 \\ \hline 4x + 12 \end{array}$$

B $3x(5x + 7)$

$$\begin{array}{r} 5x + 7 \\ 3x \\ \hline 15x^2 + 21x \end{array}$$

C $(3x + 4)\,(5x + 6)$

$$\begin{array}{r} 5x + 6 \\ 3x + 4 \\ \hline 20x + 24 \\ 15x^2 + 18x \\ \hline 15x^2 + 38x + 24 \end{array}$$

D $(7x + 6)\,(5y + 4)$

$$\begin{array}{r} 5y + 4 \\ 7x + 6 \\ \hline 30y + 24 \\ 35xy + 28x \\ \hline 35xy + 28x + 30y + 24 \end{array}$$

In algebra, we have expressions called *polynomials* ("polynomial" means "multiple terms") in which the places, called *terms*, are separated by plus or minus signs. Successive terms, moving from right to left, contain plain numbers farthest to the right. Next comes a number called a *coefficient* times x, then a coefficient times x^2, then a coefficient times x^3, and so on. If the coefficient happens to be 0 for a particular power of x, then that part of the polynomial does not have to be written down because it doesn't contribute anything. Here is an example of a big polynomial:

$$-15x^8 + 5x^7 + 2x^6 - 12x^4 - x^3 + 7x^2 + 5x$$

Note that the term for x^5 is missing; that means its coefficient is 0. Also, the plain number at the end happens to be 0, so it is not written down either.

In arithmetic, you know the relationship between figures in successive places. They always are in a ratio of 10:1. In algebra, there is no fixed relationship between successive terms in a polynomial. But it is consistent—x always has the same value in the same problem—even if you don't know what the value is. If x happens to be equal to 3, then x^2 is 3×3, or 9. After that, you go by successively higher powers of 3. These are 27, 81, 243, and so on. If x is 5, then powers move up through 5, 25, 125, 625, and so on.

DIMENSION IN ALGEBRA

Increasing powers of x correspond to successive spatial dimensions. When you multiply a length by a length, the result is an area. Multiply the area by another length, and the result is a volume. That is why x times x is called x squared, and x times x times x is called x cubed. A cube is the three-dimensional equivalent of a square.

Figure 9-2 illustrates how two polynomials in algebra can be multiplied. One of them has two dimensions because it has x^2 in it; the other has only one dimension because the highest power of x is 1. The product of these polynomials produces an expression containing x^3, representing three dimensions.

When you multiply powers of a variable, you add the total number of dimensions. So if you multiply x^2 by x (two dimensions times one dimension), you get x^3 (2 + 1, or three, dimensions). If you multiply y^3 by y^2 (three dimensions times two dimensions), you get y^5 (3 + 2, or five, dimensions). This fact comes out over and over in mathematics!

On the left-hand side of Fig. 9-2, the expression contains actual numbers as the coefficients. On the right-hand side, letters from the beginning of the alphabet (a through e) replace the numbers. Mathematicians would say that the expressions on the left are *specific*, while the ones on the right are *general*. If you substitute $a = 3$, $b = 5$, $c = 4$, $d = 7$, and $e = 6$, the general expression on the right becomes the specific example on the left. The letters allow you to fill in any other numbers for the coefficients, and the general form still gives you the answer in terms of powers of x. When such letters are used, they are called *constants*. Constants can have different values in different problems, but their values always remain the same within a particular problem.

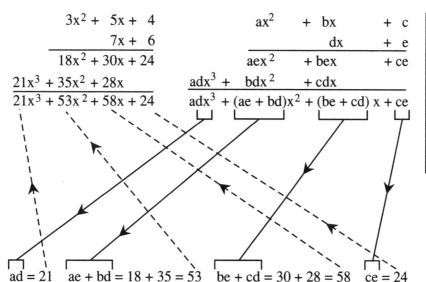

Figure 9-2

"Long multiplication" of polynomials, one of dimension 2 and the other of dimension 1. The expression on the left is specific, and the one on the right is general.

By convention, letters from the beginning of the alphabet (*a*, *b*, *c*, and so on) are often used to represent constants, and letters from the end of the alphabet (*z*, *y*, *x*, and so on backward) are commonly used to represent variables. But there's no clear-cut dividing line in the middle of the alphabet, so you might see the letter *n* used as a constant in one problem and as a variable in another.

EQUATIONS AND INEQUALITIES

An *equation* is a mathematical statement with an equals sign ($=$) in the middle and numbers or expressions on either side. For any equation being worked out, the quantities on either side of the equals sign are always the same—*always*—whether this fact is obvious or not. Figure 9-3 shows how expressions and equations relate, how an equation can often be portrayed as a graph, and how solutions to an equation can appear as points on a graph.

In "school" algebra, equations were taught once the students knew how expressions could be written. Then came *inequations*, more often called *inequalities*. In a strict inequality, the quantities on either side of the central symbol *cannot* be equal. The standard "not equals" sign looks like an equals sign with a forward slash through it ($\neq$). In "nonstrict" inequalities, the quantities on the either side of the central symbol are *not necessarily* equal, but they sometimes can be.

There are various ways in which two quantities can fail to be strictly equal. Here are examples of the four most common forms of inequality, using the standard symbols:

$$3 > 2 \text{ means "3 is greater than 2"}$$

$$y \geq x \text{ means "}y\text{ is greater than or equal to }x\text{"}$$

Figure 9-3

Here, $3x + 5$ is an expression, and $y = 3x + 5$ is an equation that appears as a straight line on a graph. Specific points on the line are shown, one where $x = 20$ and $y = 65$, and the other where $x = 17$ and $y = 56$.

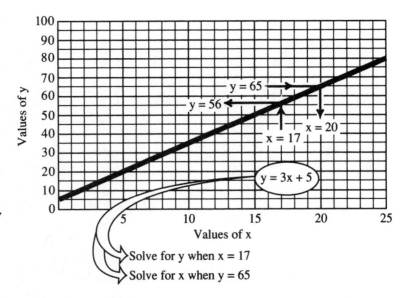

$$4 < u \text{ means "4 is less than } u\text{"}$$

$$w \leqslant z \text{ means "} w \text{ is less than or equal to } z\text{"}$$

The first and third examples above are strict inequalities, because the quantities on either side of the symbol cannot be equal. The second and fourth inequalities are not strict, because the two quantities can be equal (but are more likely not).

A CONSECUTIVE-NUMBER PROBLEM

Suppose you have three consecutive numbers. The first is divided by 2, added to the next and divided by 3, added to the third and divided by 4, which yields a fourth consecutive number. You need to find these numbers. That would be difficult using plain arithmetic! But algebra makes it easy, as you can see by following along in Fig. 9-4. To begin, write the first three numbers as x, $x + 1$, and $x + 2$. Then do what the problem says to produce the fourth number, which will be $x + 3$. Algebra derives the "mystery numbers" directly. In arithmetic, you can only guess at various numbers until you hit the right ones.

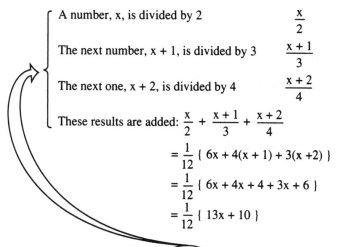

A number, x, is divided by 2 $\qquad \dfrac{x}{2}$

The next number, x + 1, is divided by 3 $\qquad \dfrac{x+1}{3}$

The next one, x + 2, is divided by 4 $\qquad \dfrac{x+2}{4}$

These results are added: $\dfrac{x}{2} + \dfrac{x+1}{3} + \dfrac{x+2}{4}$

$$= \frac{1}{12} \{ 6x + 4(x + 1) + 3(x + 2) \}$$

$$= \frac{1}{12} \{ 6x + 4x + 4 + 3x + 6 \}$$

$$= \frac{1}{12} \{ 13x + 10 \}$$

Figure 9-4
An equation can be "built up" and then solved to find a sequence of unknowns.

This gives the result of doing $\boxed{\text{THIS}}$ to any three consecutive numbers.

But the three we want give a fourth consecutive number, x + 3; so:

$$\frac{1}{12} \{ 13x + 10 \} = x + 3$$

Multiply by 12: $13x + 10 = 12(x + 3) = 12x + 36$

Subtract 12x + 10: $\qquad x = 26$

As required

CHECK $\qquad \dfrac{26}{2} + \dfrac{27}{3} + \dfrac{28}{4} = 13 + 9 + 7 = 29$

SIMULTANEOUS EQUATIONS

Often a problem can be solved with only one variable. Sometimes it is easier to use two or more variables. If you know that 4 times one number plus 5 times another number adds up to 47, and 5 times the first number plus 4 times the second number adds up to 43, how would you find the answers?

The numbers can be found in several ways. Don't think that one method is the only correct one. Sometimes you can figure out a way to solve a problem that's easier and more direct than the method shown in a textbook.

Here is one way to solve a pair of *simultaneous* equations representing the above problem. In this context, "simultaneous" means "occurring together" or "being solved at the same time." Let the unknown numbers be x and y. First, write down the equations, according to the "word problem" above, right over each other like this:

$$4x + 5y = 47$$
$$5x + 4y = 43$$

Now multiply both sides of the top equation by 5, and both sides of the bottom equation by 4. This will give you

$$20x + 25y = 235$$
$$20x + 16y = 172$$

Now subtract the bottom equation from the top one. When you do this, the terms containing x subtract from each other to give you 0. The result is an equation in only one variable:

$$25y - 16y = 235 - 172$$

which can be simplified to

$$9y = 63$$

Dividing each side of this by 9 tells you that $y = 7$. Now you can "plug in" this value for y in either of the original equations. Let's use the top one and solve it for x. The process goes like this:

$$20x + 25y = 235$$
$$20x + (25 \times 7) = 235$$
$$20x + 175 = 235$$
$$20x = 235 - 175$$

$$20x = 60$$

$$x = 60/20$$

$$x = 3$$

The solution to this pair of equations is therefore $x = 3$ and $y = 7$. You can write this as an *ordered pair*:

$$(x,y) = (3,7)$$

In this example of ordered-pair notation, x is first and y is second. But you can just as well write them the other way around:

$$(y,x) = (7,3)$$

Note that there is no space after the comma in an ordered pair, as there is in an ordinary list or sequence.

This same pair of simultaneous equations can also be solved by using a graph. Both of the equations produce straight lines when you draw them on graph paper. The two lines intersect at a point that shows the solution (Fig. 9-5).

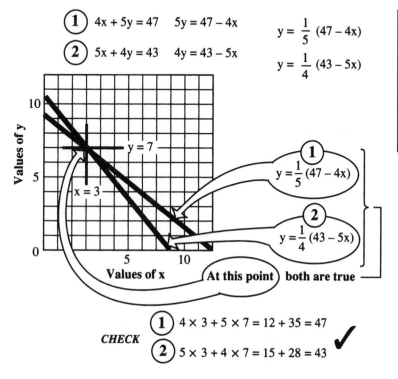

Figure 9-5

A graphical representation of two simultaneous equations. The solution corresponds to the point where the lines intersect.

Figure 9-6

At A, setting up simultaneous equations to solve a "fraction problem." At B, solving the equations to find the numerator and denominator of the fraction.

A

Fraction is $\dfrac{x}{y}$

First fact: $\dfrac{x-1}{y-1} = \dfrac{1}{2}$

Second fact: $\dfrac{x+1}{y+1} = \dfrac{3}{5}$

(1) Multiply by $2(y-1)$:
$$2(x-1) = y - 1$$
$$2x - 2 = y - 1$$
$$2x - y = 1$$

(2) Multiply by $5(y+1)$:
$$5(x+1) = 3(y+1)$$
$$5x + 5 = 3y + 3$$
$$5x - 3y = -2$$

B

$$2x - y = 1 \quad \text{(1)}$$
$$5x - 3y = -2 \quad \text{(2)}$$

To eliminate y

Multiply (1) by 3

Signs are the SAME, so SUBTRACT (2)

$$6x \ominus 3y = 3$$
$$5x \ominus 3y = -2$$
$$\overline{x = 5}$$

Substitute in (1) $2 \times 5 - y = 1$ $\quad y = 10 - 1 = 9$

CHECK Fraction is $\dfrac{5}{9}$:

$$\dfrac{5-1}{9-1} = \dfrac{4}{8} = \dfrac{1}{2}$$

$$\dfrac{5+1}{9+1} = \dfrac{6}{10} = \dfrac{3}{5}$$

A FRACTION PROBLEM

Suppose you have a "mystery fraction." You don't know what its numerator and denominator are, but someone tells you two things. First, subtracting 1 from both the numerator and the denominator gives you the fraction 1/2. Second, adding 1 to both the numerator and the denominator gives you the fraction 3/5.

Let x be the numerator and let y be the denominator of the fraction you want to find. Figure 9-6 shows how you can express this problem as a pair of simultaneous equations, and then how you can solve these equations to find the mystery fraction.

A LITTLE MORE PRACTICE

Figure 9-7 shows two more examples of how you can eliminate one variable in a pair of simultaneous equations. The steps in the process are always the same. First, multiply one or both equations by numbers that will make the variable to be eliminated have the

A

$$x + y = 30 \quad ①$$
$$4x - 3y = 1 \quad ②$$

To eliminate y

Multiply ① by 3

Signs are the OPPOSITE, so ADD ②

Substitute in ① $13 + y = 30$

$$3x + 3y = 90$$
$$4x - 3y = 1$$
$$7x \qquad = 91$$
$$x \qquad = 91 \div 7 = 13$$
$$y = 30 - 13 = 17$$

CHECK

$$13 + 17 = 30$$
$$4 \times 13 - 3 \times 17 = 52 - 51 = 1$$

B

$$x + 3y = 40 \quad ①$$
$$12x - 7y = 7 \quad ②$$

To eliminate y

Multiply ① by 7

Multiply ② by 3

Signs are OPPOSITE, so ADD

$$7x + 21y = 280$$
$$36x - 21y = 21$$
$$43x \qquad = 301$$

$$x = 301 \div 43 = 7$$

Substitute in ① $7 + 3y = 40 \qquad 3y = 40 - 7 = 33$

$$y = 33 \div 3 = 11$$

CHECK $\quad 7 + 3 \times 11 = 7 + 33 = 40$

$$12 \times 7 - 7 \times 11 = 84 - 77 = 7$$

Figure 9-7

Two examples showing how to eliminate a variable to solve a pair of simultaneous equations.

same coefficient in both equations. Then either add or subtract the two equations in that form. If the signs are the same, subtract. If they are opposite, add.

SOLVING BY SUBSTITUTION

In the examples of simultaneous equations you have worked with so far, you eliminated one variable, leaving an equation in only one variable that was easy to solve. Then you went back and "plugged in" the value for the solved variable to find the value of the remaining unknown.

Figure 9-8 shows an alternative method that can sometimes be simpler. You rearrange one of the equations to get one variable in terms of the other, and then you substitute the equation for the first variable in the other equation. This, again, gives

Figure 9-8

An alternative method of solving simultaneous equations, in which one equation is rearranged and then substituted into the other.

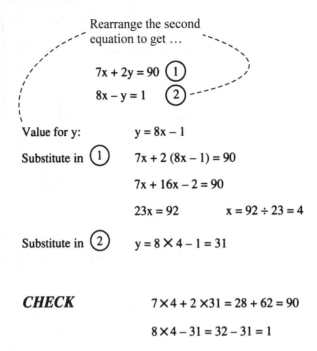

Rearrange the second equation to get ...

$$7x + 2y = 90 \quad ①$$

$$8x - y = 1 \quad ②$$

Value for y: $y = 8x - 1$

Substitute in ① $7x + 2(8x - 1) = 90$

                     $7x + 16x - 2 = 90$

                     $23x = 92 \qquad x = 92 \div 23 = 4$

Substitute in ② $y = 8 \times 4 - 1 = 31$

CHECK $7 \times 4 + 2 \times 31 = 28 + 62 = 90$

                    $8 \times 4 - 31 = 32 - 31 = 1$

you a one-variable equation that you can easily solve. The rest of the process is the same as before. Interestingly, the student who "discovered" this scheme was having problems with algebra! As always, remember to check your result.

SOLVING FOR RECIPROCALS

Figure 9-9 shows a simultaneous equation problem in which the variables appear as *reciprocals*, or in which it is easier to solve for reciprocals than for the variables themselves. The reciprocal of a quantity, also called the *multiplicative inverse*, is equal to 1 divided by that quantity. Here, we solve for $1/x$ and $1/y$, instead of for x and y.

One thing you should learn from studying simultaneous equations is that you can often save time by using common sense to find the easiest way, rather than trying to apply a single rigid routine to every problem.

LONG DIVISION AND FACTORING

Long division can help you understand how algebra works, and it is particularly helpful in developing an understanding of dimension. Compare the two ways of doing long division shown in Fig. 9-10, one for arithmetic and the other for algebra. Note the similarities. Study this awhile. You should be able to "figure it out" well enough to do problems 2 and 3 in the quiz that follows.

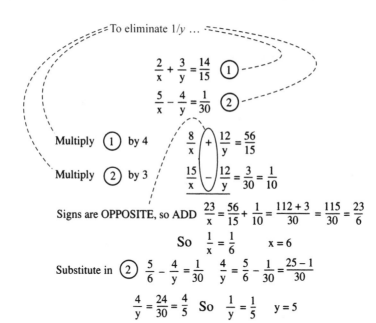

Figure 9-9
Sometimes it's easier to solve for reciprocals of variables, rather than for the variables themselves.

To eliminate $1/y$ ...

$$\frac{2}{x} + \frac{3}{y} = \frac{14}{15} \quad \text{①}$$

$$\frac{5}{x} - \frac{4}{y} = \frac{1}{30} \quad \text{②}$$

Multiply ① by 4 $\qquad \frac{8}{x} + \frac{12}{y} = \frac{56}{15}$

Multiply ② by 3 $\qquad \frac{15}{x} - \frac{12}{y} = \frac{3}{30} = \frac{1}{10}$

Signs are OPPOSITE, so ADD $\quad \frac{23}{x} = \frac{56}{15} + \frac{1}{10} = \frac{112 + 3}{30} = \frac{115}{30} = \frac{23}{6}$

So $\quad \frac{1}{x} = \frac{1}{6} \qquad x = 6$

Substitute in ② $\quad \frac{5}{6} - \frac{4}{y} = \frac{1}{30} \qquad \frac{4}{y} = \frac{5}{6} - \frac{1}{30} = \frac{25 - 1}{30}$

$$\frac{4}{y} = \frac{24}{30} = \frac{4}{5} \quad \text{So} \quad \frac{1}{y} = \frac{1}{5} \qquad y = 5$$

CHECK $\quad \frac{2}{6} + \frac{3}{5} = \frac{1}{3} + \frac{3}{5} = \frac{5 + 9}{15} = \frac{14}{15}$

$$\frac{5}{6} - \frac{4}{5} = \frac{25 - 24}{30} = \frac{1}{30}$$

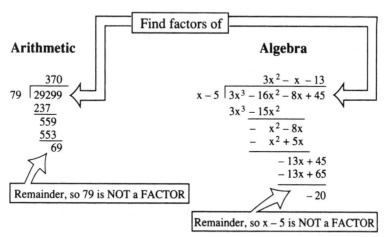

Figure 9-10
Long division in algebra is similar to long division in arithmetic. It can be useful when finding the factors of a polynomial.

Find factors of

Arithmetic

```
        370
79 | 29299
     237
     559
     553
      69
```

Remainder, so 79 is NOT a FACTOR

Algebra

$$
\begin{array}{r}
3x^2 - x - 13 \\
x - 5 \overline{\big)\, 3x^3 - 16x^2 - 8x + 45} \\
3x^3 - 15x^2 \\
\hline
-\ x^2 - 8x \\
-\ x^2 + 5x \\
\hline
-13x + 45 \\
-13x + 65 \\
\hline
-20
\end{array}
$$

Remainder, so x – 5 is NOT a FACTOR

```
        353
83 | 29299
     249
     439
     415
     249
     249
     •••
```

83 and 353 ARE FACTORS

$$
\begin{array}{r}
x^2 - 7x + 9 \\
3x + 5 \overline{\big)\, 3x^3 - 16x^2 - 8x + 45} \\
3x^3 + 5x^2 \\
\hline
-21x^2 - 8x \\
-21x^2 - 35x \\
\hline
+27x + 45 \\
27x + 45 \\
\hline
\end{array}
$$

3x + 5 and $x^2 - 7x + 9$ ARE FACTORS

QUESTIONS AND PROBLEMS

This is an open-book quiz. You may refer to the text in this chapter (and earlier ones, too, if you want) when figuring out the answers. Take your time! Consider all values given as exact, so you don't have to worry about significant figures. The correct answers are in the back of the book.

1. Perform the following multiplications.
   (a) $(x + 1)(x - 3)$
   (b) $(x - 3)(x - 5)$
   (c) $(x + 1)(x - 1)$
   (d) $(x + 1)(x^2 - x + 1)$

2. Perform the following divisions. In each case, check your result by multiplication.
   (a) $(x^2 - 2x - 3) / (x + 1)$
   (b) $(x^2 - 8x + 15) / (x - 3)$
   (c) $(x^2 - y^2) / (x + y)$
   (d) $(x^4 + 2x^3 - x^2 + 4) / (x + 2)$

3. Use long division to find the two factors of each of the following.
   (a) $x^2 + 2x - 35$
   (b) $x^3 + x^2 - 5x - 5$
   (c) $x^3 + x^2 - 7x - 3$

4. Solve the following pairs of simultaneous equations. If there is no solution, explain.
   (a) $4x - 8y = 2$ and $-3x + 5y = -5$
   (b) $2x + 3y = 0$ and $7x - 2y = 2$
   (c) $y = 2x + 4$ and $y = -2x - 6$
   (d) $y = 4x + 2$ and $y = 4x - 2$

5. A rectangle has certain dimensions. Making it 2 feet wider and 5 feet longer increases its area by 133 square feet. Making it 3 feet wider and 8 feet longer increases its area by 217 square feet. What were its original dimensions? (*Hint*: use simultaneous equations.)

6. Separate a number c into two parts, such that a times one part is equal to b times the other part. If $c = 28$, $a = 3$, and $b = 4$, what are the parts? (*Hint*: it might be easiest to find the second part first.)

7. A fraction is somewhere between 3/4 and 4/5. Adding 3 to both the numerator and the denominator makes the fraction equal to 4/5, but 4 subtracted from each makes the fraction equal to 3/4. What is the fraction?

8. In another fraction, adding 1 to both the numerator and the denominator makes the fraction equal to 4/7, but 1 subtracted from each makes the fraction equal to 5/9. What is the fraction?

9. Adding 1 to the numerator and denominator of another fraction makes it equal 7/12, but 1 subtracted from each makes it equal to 9/16. What is the fraction?

10. The highest two of four consecutive numbers, multiplied together, produce a product that is 90 more than the lowest two multiplied together. What are the numbers?

11. An "investigation" of five consecutive numbers yields the fact that if the middle three numbers are multiplied together, they are 15 more than the first, middle, and last numbers multiplied. What are the numbers?

12. A man has an option on a piece of land. He was told that the parcel is 50 feet longer than it is wide. A survey shows that it is 10 feet less in width than he was told. The seller offers him an extra 10 feet in length. Does the man get the same total area? If not, how much does he lose on the deal? Does it depend on the actual dimensions?

10 Quadratic Equations

Simultaneous equations come from problems that have two or more answers "at the same time." *Quadratic equations* can represent another kind of problem where different answers exist, not "at the same time" but as alternatives. Quadratic equations are also called *second-order equations* because they involve raising a variable to the second power (squaring it).

A QUADRATIC EXAMPLE

Suppose you have 80 feet of fencing available to make a rectangular enclosure such as a pen for your dog. How much area will the fence enclose? That, of course, depends on the width and the length of the rectangle, both of which can vary. If the rectangle is W feet wide by L feet long, then $2W + 2L = 80$. Dividing through by 2 tells you that $W + L = 40$. The area A of the enclosure is the product WL. By substituting $40 - W$ for L in that equation, the area can be written as $W(40 - W)$ square feet, which multiplies out to $40W - W^2$. So you know that $A = 40W - W^2$. If you substitute the more conventional variable notations x for the width dimension and y for the area, you get $y = 40x - x^2$.

The nature of the problem is shown in Fig. 10-1. Two different ways of solving it using tables are shown in Fig. 10-2. In the first method, you multiply the width x by the length $(40 - x)$ to get the area. In the second method, you multiply the width x by 40 to get $40x$, and then subtract the width squared (x^2) to get the area. The results are always the same, no matter what the width happens to be.

LINEAR VS. QUADRATIC FORMS

The equations you worked with in Chap. 9 are called *linear equations*, referring to a form that can be written like this:

$$y = ax + b$$

Figure 10-1

There is 80 feet of fencing available to form a rectangular enclosure. What is the area of the enclosure?

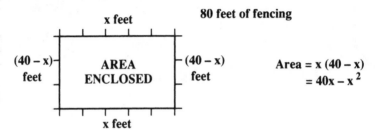

where a and b are constants. The term "linear" arises from the fact that the graph of such an equation is always a straight line when drawn on conventional graph paper with the two axes at right angles to each other.

Quadratic equations are a little more complicated. When a quadratic is graphed, the result is always a curve. This type of equation can always be reduced to the following *standard form for a quadratic*

$$y = ax^2 + bx + c$$

where a, b, and c are constants, and $a \neq 0$. (If a were equal to 0, it would be a linear equation.) Writing in letters for the coefficients gives you a general form for an equation in which many different problems can be expressed.

A QUADRATIC GRAPH IS SYMMETRIC

If you plot the values for the area $40x - x^2$ against the width x in the foregoing "fence problem" on graph paper, you get a curve, as shown in Fig. 10-3. If x is either 0 or 40, then $40x - x^2$ has a value of zero because $40x = x^2$. When $x = 20$, the value of $40x - x^2$ reaches a maximum of 400.

You can draw a vertical line where $x = 20$ and see that the curve appears as two "mirror images," one on either side of the line. Such a curve is said to have *bilateral symmetry*.

Figure 10-2

Two different ways of calculating the area of a rectangular enclosure whose perimeter is 80 feet.

x	0	5	10	15	20	25	30	35	40	
40 − x	40	35	30	25	20	15	10	5	0	1st Method
Area x(40 − x)	0	175	300	375	400	375	300	175	0	

Same Answers

x	0	5	10	15	20	25	30	35	40	
40x	0	200	400	600	800	1000	1200	1400	1600	
x^2	0	25	100	225	400	625	900	1225	1600	2nd Method
Area 40x − x^2	0	175	300	375	400	375	300	175	0	

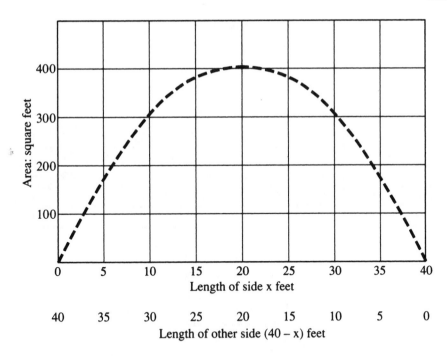

All quadratic equations, when plotted on standard rectangular graph paper, produce curves that have bilateral symmetry. The curve in Fig. 10-3 (and the curve for every quadratic) is called a *parabola*. Here, the parabola opens downward. Some quadratics produce graphs in which the parabola opens upward, to the right, or to the left.

SOLVING A QUADRATIC EQUATION

Suppose you know that, as well as using 80 feet of fencing, the enclosure has an area of 300 square feet. What are the length and the width? One way to solve this problem is to write this:

$$40x - x^2 = 300$$

Then you can rearrange the equation to get it into standard quadratic form:

$$x^2 - 40x + 300 = 0$$

Now, notice what happens if you tabulate the values of the left-hand side of this equation and plot the results on a graph as shown in Fig. 10-4. Here, the curve is inverted from the one in Fig. 10-3, and the positions of the values are changed so they straddle the horizontal zero line. The zero line becomes the locator of the solutions because the equation is expressed so that a particular quantity is equal to 0.

This "changing around" of an equation is called *transposition*. It reproduces the same curve in a different position.

Figure 10-4

At A, transposition of $40x - x^2 = 300$, and a table of values that reveals two solutions where $x^2 - 40x + 300 = 0$. At B, a graph of the transposed equation that shows the solutions.

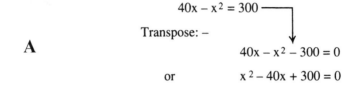

A

$$40x - x^2 = 300$$

Transpose: −

$$40x - x^2 - 300 = 0$$

or

$$x^2 - 40x + 300 = 0$$

x	0	5	10	15	20	25	30	35	40
x^2	0	25	100	225	400	625	900	1225	1600
$40x$	0	200	400	600	800	1000	1200	1400	1600
$x^2 - 40x$	0	−175	−300	−375	−400	−375	−300	−175	0
$x^2 - 40x + 300$	300	125	0	−75	−100	−75	0	125	300

B

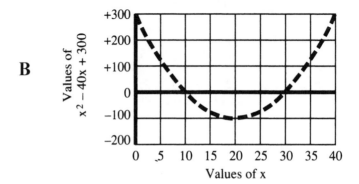

FACTORS OF A QUADRATIC

The expression $x^2 - 40x + 300$ can be broken down into a product of two factors: $(x - 10)$ and $(x - 30)$. Mathematically:

$$x^2 - 40x + 300 = (x - 10)(x - 30)$$

Now tabulate the values of $x - 10$ and $x - 30$ for various values of x as shown in Fig. 10-5A, and plot lines that represent these factors (Fig. 10-5B). The "factor lines" cross the "zero line" when $x = 10$ and when $x = 30$, respectively. The quadratic curve passes through the zero line at the same points as the lines for $x - 10$ and $x - 30$.

Finding the factors of a quadratic equation, formed by transposing it into standard form, gives the solutions of that equation. If the factors have a minus sign (as they do here), the solutions are the corresponding positive quantities. Later, you'll see that quadratics with factors containing plus signs produce solutions that are negative numbers.

A $x^2 - 40x + 300 = (x - 10)(x - 30)$

x	0	5	10	15	20	25	30	35	40
x – 10	–10	–5	0	+5	+10	+15	+20	+25	+30
x – 30	–30	–25	–20	–15	–10	–5	0	+5	+10
(x – 10)(x – 30)	+300	+125	0	–75	–100	–75	0	+125	+300

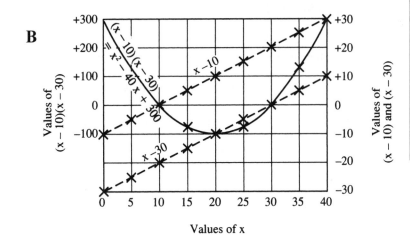

Figure 10-5

At A, factors of $x^2 - 40x + 300$, and a table of values that reveals two solutions where $(x - 10)(x - 30) = 0$. At B, a graph of the whole quadratic (curve) along with graphs of the two factors (straight lines). The combined value is 0 when either factor is 0.

FINDING FACTORS

To find factors, you often need some guidelines, rules, or tricks to help. You must examine the coefficients of the terms in a particular quadratic to see how it is constructed. Each factor of an expression having the form $ax^2 + bx + c = 0$, where a, b and c might be either positive or negative, will be of the form $dx + e$, where d and e might be either positive or negative.

Now look at Fig. 10-6. In each expression A through D, the coefficient of x^2 is 3. If you think about this for a little while, you will realize that if any of these expressions

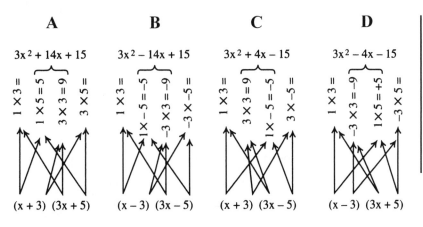

Figure 10-6

Finding factors to solve quadratic equations. Note that all four equations A, B, C, and D are different, even though their factors look a lot alike!

has factors that contain only whole numbers, one of the factors must be x plus or minus a whole number, and the other must be $3x$ plus or minus a whole number. This will be true because 3 is a prime number, whose only possible factors are 1 and itself. In all these expressions, the *numeric term* (the one without any x attached to it) is 15. So, in the two factors, one must have plus or minus 3 and the other must have plus or minus 5. Or else one must have plus or minus 1 and the other must have plus or minus 15. Try all the possibilities! And keep in mind that the above rules only apply to factors that contain whole numbers. We'll deal with factors containing fractions or square roots later.

Study this system until you can find the factors without too much trouble. You can even make up some practice problems of your own. There is a certain technique to this method of factoring, and you can get good at it only by doing a lot of practice.

Now make a "line and curve" graph for the following equation and its factors:

$$3x^2 - 14x + 15 = 0$$

You will get a graph in which the lines for the factors are not parallel (Fig. 10-7). But just as before, the solutions come at the two places where the straight lines cross the zero line.

Figure 10-7

The quadratic equation $3x^2 - 14x + 15 = 0$, and a graph of the curve along with graphs of the two factors (straight lines). The roots appear when either factor is 0.

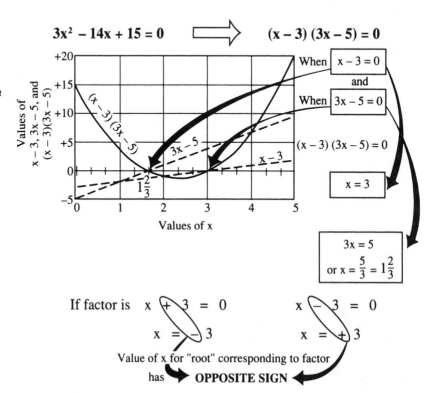

$$x^2 - 5x + 6 = 0 \implies (x - 2)(x - 3) = 0$$

A Easy

$$x^2 - 6x + 5 = 0 \implies (x - 1)(x - 5) = 0$$

B Not so easy

$$x^2 - 6x + 6 = 0 \implies$$

$$x^2 - 6\tfrac{1}{3}x + 10 = 0 \implies$$

?

Figure 10-8

At A, two quadratics that are easy to factor. At B, two quadratics that are not so easy to factor.

COMPLETING THE SQUARE

Figure 10-8 shows two examples of equations that are easy to factor by trial and error (at A) and two examples of equations that are hard to factor that way (at B). By means of a scheme called *completing the square*, you can often find the factors of stubborn equations such as those in Fig. 10-8B.

The process of completing the square can be understood by using a geometric figure such as the one in Fig. 10-9. Your expression starts in the standard quadratic form:

$$ax^2 + bx + c = 0$$

Now look at the diagram closely. Notice that if $N = (x + n)$, then $N^2 = (x + n)^2$ which, when multiplied out, becomes

$$N^2 = x^2 + 2nx + n^2$$

Study this for awhile, and you will begin to see the correspondence between the algebraic expression and the geometric figure.

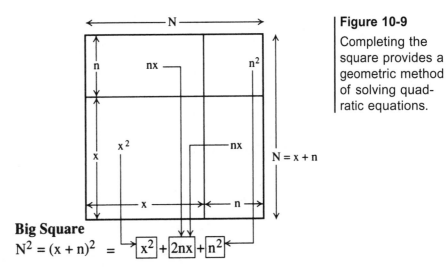

Figure 10-9

Completing the square provides a geometric method of solving quadratic equations.

Big Square

$$N^2 = (x + n)^2 = \boxed{x^2} + \boxed{2nx} + \boxed{n^2}$$

Figure 10-10

Using the technique of completing the square to simplify the quadratic equations shown in Fig. 10-8B.

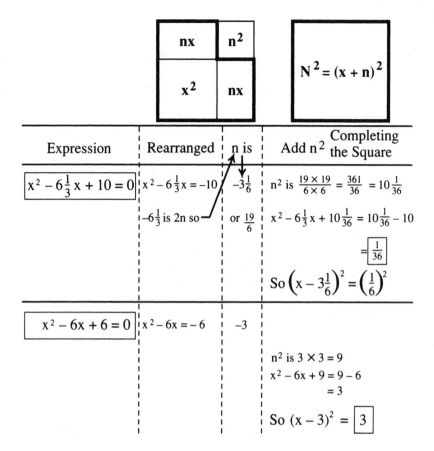

Expression	Rearranged	n is	Add n^2	Completing the Square
$x^2 - 6\frac{1}{3}x + 10 = 0$	$x^2 - 6\frac{1}{3}x = -10$	$-3\frac{1}{6}$	n^2 is $\frac{19 \times 19}{6 \times 6} = \frac{361}{36} = 10\frac{1}{36}$	
	$-6\frac{1}{3}$ is 2n so	or $\frac{19}{6}$	$x^2 - 6\frac{1}{3}x + 10\frac{1}{36} = 10\frac{1}{36} - 10$	
			$= \boxed{\frac{1}{36}}$	
			So $\left(x - 3\frac{1}{6}\right)^2 = \left(\frac{1}{6}\right)^2$	
$x^2 - 6x + 6 = 0$	$x^2 - 6x = -6$	-3		
			n^2 is $3 \times 3 = 9$	
			$x^2 - 6x + 9 = 9 - 6$	
			$\qquad = 3$	
			So $(x - 3)^2 = \boxed{3}$	

Figure 10-11

Finding the solutions of the factored forms of the equations derived in Fig. 10-10.

$$\boxed{(x - 3\tfrac{1}{6})^2 = (\tfrac{1}{6})^2}$$

(FIRST EQUATION)

So $(x - 3\frac{1}{6}) = +\frac{1}{6}$ or $-\frac{1}{6}$

If $x - 3\frac{1}{6} = +\frac{1}{6}$ $\quad x = 3\frac{1}{6} + \frac{1}{6} = \boxed{3\frac{1}{3}}$

If $x - 3\frac{1}{6} = -\frac{1}{6}$ $\quad x = 3\frac{1}{6} - \frac{1}{6} = \boxed{3}$

$\left.\right\}$ **Two Answers**

$$\boxed{(x - 3)^2 = 3}$$

(SECOND EQUATION)

So $(x - 3) = +\sqrt{3}$ or $-\sqrt{3}$

If $x - 3 = +\sqrt{3}$ $\quad x = \boxed{3 + \sqrt{3}}$

If $x - 3 = -\sqrt{3}$ $\quad x = \boxed{3 - \sqrt{3}}$

$\left.\right\}$ **Two Answers**

(FIRST EQUATION)

$$x = 3\tfrac{1}{3} \qquad x^2 = \frac{10 \times 10}{3 \times 3} = \frac{100}{9} = 11\tfrac{1}{9}$$

$$6\tfrac{1}{3}x = \frac{10}{3} \times \frac{19}{3} = \frac{190}{9} = 21\tfrac{1}{9}$$

$$\boxed{x^2 - 6\tfrac{1}{3}x = 11\tfrac{1}{9} - 21\tfrac{1}{9} = -10}$$

Both answers check

or $x = 3 \qquad x^2 = 9$

$$6\tfrac{1}{3}x = 3 \times \frac{19}{3} = 19$$

$$\boxed{x^2 - 6\tfrac{1}{3}x = 9 - 19 = -10}$$

Figure 10-12

Figure 10-12
To check solutions, "plug them into" the original equations. The resulting arithmetic should turn out correct. If it doesn't, a mistake was made somewhere!

(SECOND EQUATION)

$$x = 3 + \sqrt{3} \qquad x^2 = 3^2 + 2 \times 3 \times \sqrt{3} + \sqrt{3}^2 = 9 + 6\sqrt{3} + 3 = 12 + 6\sqrt{3}$$

$$6x = 6(3 + \sqrt{3}) = 18 + 6\sqrt{3}$$

$$\boxed{\begin{aligned} x^2 - 6x &= 12 + 6\sqrt{3} - (18 + 6\sqrt{3}) \\ &= 12 + 6\sqrt{3} - 18 - 6\sqrt{3} = -6 \end{aligned}}$$

$$x = 3 - \sqrt{3} \qquad x^2 = 3^2 - 2 \times 3 \times \sqrt{3} + \sqrt{3}^2 = 9 - 6\sqrt{3} + 3 = 12 - 6\sqrt{3}$$

$$6x = 6(3 - \sqrt{3}) = 18 - 6\sqrt{3}$$

Both answers check

$$\boxed{\begin{aligned} x^2 - 6x &= 12 - 6\sqrt{3} - (18 - 6\sqrt{3}) \\ &= 12 - 6\sqrt{3} - 18 + 6\sqrt{3} = -6 \end{aligned}}$$

Now let's get back to those tough equations in Fig. 10-8B. We'll take them in reverse order, doing the bottom equation first and then the top one. The factors are found and solutions derived as shown in Figs. 10-10 and 10-11. The work is checked according to the process shown in Fig. 10-12. This can be a little "messy," but if you're careful, it will work out. Pay attention to the signs! Remember these "sign facts":

- Plus times plus equals plus
- Plus times minus equals minus
- Minus times plus equals minus
- Minus times minus equals plus

WHAT THE ANSWERS MEAN

You might wonder how completing the square can produce two solutions to an equation. Only one solution seems obvious because only one area will fill the space. Figure 10-13A can help you visualize the two answers in terms of geometry.

Figure 10-13

At A, geometric portrayal of quadratic solutions derived from completing the square. At B, the geometric portrayal of the solutions to the first equation in Fig. 10-11.

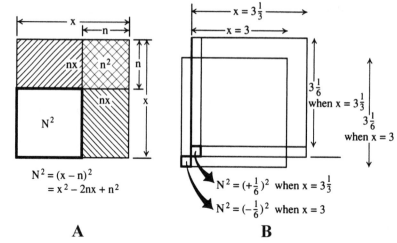

$N^2 = (x - n)^2$
$= x^2 - 2nx + n^2$

$N^2 = (+\frac{1}{6})^2$ when $x = 3\frac{1}{3}$

$N^2 = (-\frac{1}{6})^2$ when $x = 3$

A   **B**

Look at the significance of $N^2 = (x - n)^2$. Here, x^2 is the big square. The final square that represents N^2 is smaller, so x^2 is diminished by two quantities' areas nx. The two rectangles that represent nx overlap by another square that represents n^2. So, as you found in algebra, the geometric construction supports it:

$$N^2 = x^2 - 2nx + n^2$$

Figure 10-13B shows the geometric way of looking at the first equation in Fig. 10-11, where the solutions are 3 or 3-1/3.

THE QUADRATIC FORMULA

Look at the following two quadratic equations:

$$-3x^2 - 4x - 2 = 0$$
$$4x^2 + 3x - 5 = 0$$

These equations are difficult to put into factored form. But there is a method known as the *quadratic formula* that can be used to solve quadratics that are not "factor-friendly." Remember the general quadratic equation

$$ax^2 + bx + c = 0$$

where $a \neq 0$. The solution(s) to this equation can be found by using this formula:

$$x = [-b \pm (b^2 - 4ac)^{1/2}] / 2a$$

The symbol $\pm$ is read "plus or minus" and is a way of compacting two mathematical expressions into one. Written separately, the equations are

$$x = [-b + (b^2 - 4ac)^{1/2}] / 2a$$
$$x = [-b - (b^2 - 4ac)^{1/2}] / 2a$$

Figure 10-14

Four examples in which the quadratic formula is used to solve equations that aren't "factor-friendly."

Important!

$+ \boxed{\begin{matrix} + \text{Watch} + \\ \text{the} \\ - \text{Signs} - \\ + \end{matrix}}$

For $ax^2 + bx + c = 0$

$$x = -\frac{b}{2a} \pm \sqrt{\frac{b^2}{4a^2} - \frac{c}{a}}$$

A $3x^2 + 14x + 15 = 0$

$$x = -\frac{14}{6} \pm \sqrt{\frac{196}{36} - \frac{15}{3}}$$

$a = 3$
$b = 14$
$c = 15$

$$= -\frac{7}{3} \pm \sqrt{\frac{49}{9} - \frac{45}{9}} = -\frac{7}{3} \pm \sqrt{\frac{4}{9}} = -\frac{7}{3} \pm \frac{2}{3} = \boxed{-\frac{5}{3}} \text{ or } \boxed{-3}$$

B $3x^2 - 14x + 15 = 0$

$$x = \frac{14}{6} \pm \sqrt{\frac{196}{36} - \frac{15}{3}}$$

$a = 3$
$b = -14$
$c = 15$

$$= \frac{7}{3} \pm \sqrt{\frac{4}{9}} = \frac{7}{3} \pm \frac{2}{3} = \boxed{\frac{5}{3}} \text{ or } \boxed{3}$$

C $3x^2 + 4x - 15 = 0$

$$x = -\frac{2}{3} \pm \sqrt{\frac{4}{9} + \frac{15}{3}}$$

$a = 3$
$b = 4$
$c = -15$

$$= -\frac{2}{3} \pm \sqrt{\frac{4}{9} + \frac{45}{9}} = -\frac{2}{3} \pm \sqrt{\frac{49}{9}} = -\frac{2}{3} \pm \frac{7}{3} = \boxed{-3} \text{ or } \boxed{+\frac{5}{3}}$$

D $3x^2 - 4x - 15 = 0$

$$x = \frac{2}{3} \pm \sqrt{\frac{4}{9} + \frac{15}{3}}$$

$a = 3$
$b = -4$
$c = -15$

$$= \frac{2}{3} \pm \sqrt{\frac{49}{9}} = \frac{2}{3} \pm \frac{7}{3} = \boxed{-\frac{5}{3}} \text{ or } \boxed{+3}$$

The fractional exponent means the 1/2 power, which is the positive square root.

Figure 10-14 demonstrates how four quadratic equations can be solved by using the quadratic formula (in a slightly different form than the one shown above, but still exactly the same formula). The equations to be solved are shown in the figure as A, B, C, and D in this order:

$$3x^2 + 14x + 15 = 0$$
$$3x^2 - 14x + 15 = 0$$

$$3x^2 + 4x - 15 = 0$$

$$3x^2 - 4x - 15 = 0$$

As an exercise, you might want to prove that the formula at the top right in Fig. 10-14 is the same as the one shown in this section of the text.

A QUADRATIC PROBLEM

Sometimes, a problem leading to a quadratic equation seems to have only one true answer. Being realistic people, we naturally ask what the other answer means. Here is an example. Suppose you are told that a picture to be framed is twice as wide as it is high. The frame provides a 3-inch margin around all sides. The total area—frame and picture—is 260 square inches. What are the dimensions of the picture and the frame?

If we make x the height and $2x$ the width of the picture, the dimensions of the frame will be $(x + 6)$ inches high by $(2x + 6)$ inches wide. These dimensions multiply to get the area of the frame in square inches:

$$(x + 6)(2x + 6) = 2x^2 + 18x + 36$$

If you want, you can draw a diagram to make this easier to visualize. But first try to work it out without using a diagram.

We are told that the total area of the frame with the picture in it is 260 square inches. That means

$$2x^2 + 18x + 36 = 260$$

We can reduce this to standard form and lowest terms as follows:

$$2x^2 + 18x + 36 = 260$$

$$2x^2 + 18x - 224 = 0$$

$$x^2 + 9x - 112 = 0$$

This can be factored to

$$(x - 7)(x + 16) = 0$$

so we have two answers that are characteristic of quadratics, 7 or -16.

The positive answer is easy. Remember that x is the height of the picture, and $2x$ is its width. So we know that the picture is 7 inches high by 14 inches wide. The frame must therefore be $7 + 6 = 13$ inches high by $14 + 6 = 20$ inches wide. When you multiply these two numbers, you get 260 square inches. It checks out!

What about the "negative solution"? A picture with negative dimensions has no obvious real-world meaning. But let's investigate anyway. The picture dimensions, according to the negative solution, are -16 inches high by -32 inches wide. By adding twice 3 inches each way, the frame dimensions are $(-16 + 6)$ by $(-32 + 6)$ inches, or -10 inches high by -26 inches wide. This also multiplies to an area of 260 square inches, because a minus times a minus yields a plus. Mathematically, it checks out!

QUESTIONS AND PROBLEMS

This is an open-book quiz. You may refer to the text in this chapter (and earlier ones, too, if you want) when figuring out the answers. Take your time! Consider all values given as exact, so you don't have to worry about significant figures. The correct answers are in the back of the book.

1. Find the factors and say what values of x make the following expressions equal to zero.

(a) $x^2 + 7x - 8$
(b) $3x^2 - 16x + 13$
(c) $7x^2 - 48x - 7$
(d) $30x^2 - 73x + 40$

2. Solve the following by completing the square.

(a) $x^2 - 4x = 45$
(b) $x^2 - 6 = x$
(c) $x^2 - 7x + 7 = 0$
(d) $x^2 - 12x = 4$

3. Solve the following quadratics by formula.

(a) $5x^2 - 2x - 7 = 0$
(b) $7x^2 - 4x - 3 = 0$
(c) $x - (1/x) = 24/5$
(d) $x + (1/x) = 10$

4. Solve the following quadratics and explain anything unusual you observe about the solutions.

(a) $5x^2 - 3x - 2 = 0$
(b) $5x^2 - 3x = 0$
(c) $5x^2 - 3x + 9/20 = 0$

5. A quantity is required, such that adding it to twice its reciprocal will produce a sum of 4. What is the quantity? Check both results. (Remember that the reciprocal of a quantity is equal to 1 divided by that quantity.)

6. The length of a rectangular enclosure is 10 feet less than twice its width. Its area is 2800 square feet. Find its dimensions. Explain the negative answer and check the positive answer.

7. Extending each side of a square enclosure by 6 feet makes its area 4 times as big. Find the original side length and explain the negative answer.

8. Find three successive numbers whose sum is 3/8 the product of the smaller two numbers. (*Hint*: take x as the middle number.) Explain the less obvious solution.

9. In mowing a rectangular lawn that measures 60 feet by 80 feet, how wide a strip around the edge must be mowed for one-half the grass to be cut? Explain the second answer.

10. At a party, a man tried to run a "think of a number" game and gave these instructions: Think of a number, double it, subtract 22, multiply it by the number you first thought of, divide by 2, add 70, and subtract the number you first thought of. He said that the answer was 35. Only two people, who had used different numbers, had that answer. What two numbers did those two use?

11. The height of a small box is 1 inch less than its width, and the length is 2 inches more than its width. If the total area of its sides is 108 square inches, what are its dimensions? Assume that all the sides are rectangles.

12. The "negative solution" to problem 11 leads to another set of dimensions whose total surface area is also 108 square inches. What are these dimensions?

Have you noticed that when two factors have the same terms but with the sign between them different, the product is the difference between the terms, each squared? For example, you can multiply the two terms $(a + b)(a - b)$ to yield $a^2 - b^2$. You can also factorize $a^2 - b^2$ into $(a + b)(a - b)$. This is a shortcut that you can use in working out some algebra problems. Let's look at a few more.

SUM AND DIFFERENCE IN GEOMETRY

The sum and difference principle can be seen in geometry as well as in algebra. If you take the smaller square from the bigger square in Fig. 11-1, you are left with the shaded area. This is like taking b^2 from a^2. If you cut the upright piece off and lay it end-on to the other piece, the resulting rectangle has the dimensions $a + b$ and $a - b$. This is the geometric way of saying

$$a^2 - b^2 = (a + b)(a - b)$$

Here's another little shortcut that uses subtraction and squares to help you do certain long multiplication problems in your head. Suppose you want to multiply 37 by 43. If you happen to notice that $37 = 40 - 3$ and $43 = 40 + 3$, and you also notice that $40^2 = 1600$ and $3^2 = 9$, then it is not much of a mental leap to see that 37 times 43 must be $1600 - 9$, which you can straightaway tell is equal to 1591. Try making up some other problems similar to this so you get the "hang" of it!

DIFFERENCE OF SQUARES FINDS FACTORS

You can also find shortcuts in algebra. The polynomial expression $x^4 + x^2 + 1$ has no odd powers of x, so you can think of x^2 as the variable and thereby rewrite the whole polynomial as $(x^2)^2 + (x^2) + 1$. Another way to do this is to let y represent x^2, so it

Figure 11-1

The difference of two squares is equal to the sum times the difference of the original numbers.

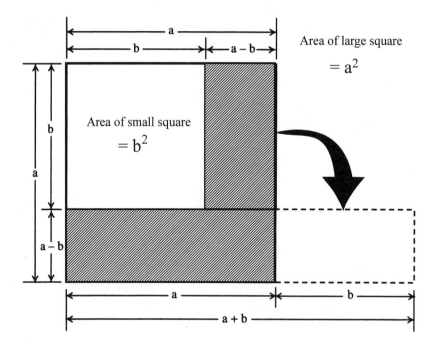

Area of large square

$$= a^2$$

Area of small square

$$= b^2$$

simplifies to $y^2 + y + 1$. This polynomial in y doesn't have obvious factors, so it is natural to suppose that the original polynomial in x does not factor easily either. Not so fast!

As you work with polynomials, you'll get used to seeing certain facts. One of the more well-known facts is $(y + 1)^2 = y^2 + 2y + 1$. (You can substitute anything else for y, of course, and it will still work.) From this it is not hard to see the following:

$$[(x^2) + 1]^2 = x^4 + 2x^2 + 1$$

This happens to be x^2 larger than the expression you want to factorize, and x^2 is a square. So the expression $x^4 + x^2 + 1$ is the difference between two squares:

$$x^4 + x^2 + 1 = [(x^2) + 1]^2 - x^2$$

Written that way, the factors are $[(x^2 + 1) + x]$ and $[(x^2 + 1) - x]$, which you can rearrange into a more conventional form: $(x^2 + x + 1)(x^2 - x + 1)$. So the original polynomial does indeed have two fairly simple factors:

$$x^4 + x^2 + 1 = (x^2 + x + 1)(x^2 - x + 1)$$

You can check this equation by multiplying out the right-hand side, crossing out all the terms that are identical but with opposite sign (thus canceling out to 0), and making sure you end up with the left-hand side. The process is a bit tedious, and the signs can be tricky, but doing this can help you see how factors behave. It is also a good habit to get into, because it forces you to check your work and make sure it is right.

FINDING SQUARE ROOTS

Years ago, students learned a routine to find square roots without a calculator or even its venerable ancestor the *slide rule*. This scheme, called *extracting the square root*, is more of a game now than anything useful or necessary, but it can help you understand not only square roots, but also a lot of other things that computers and calculators do.

Start by looking at squares and the square roots of easier numbers in arithmetic, algebra, and geometry, using numbers that "work out." See how this works for a square root that is a two-digit number, so the square has three or four digits. In Fig. 11-2, the

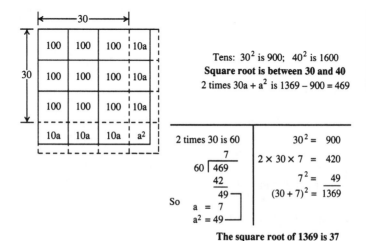

Figure 11-2

A procedure for "manually" finding the square root of 1369.

Tens: 30^2 is 900; 40^2 is 1600
Square root is between 30 and 40
2 times $30a + a^2$ is $1369 - 900 = 469$

2 times 30 is 60

$$30^2 = 900$$
$$2 \times 30 \times 7 = 420$$
$$7^2 = 49$$
$$(30 + 7)^2 = 1369$$

$$\begin{array}{r} 7 \\ 60 \overline{\smash{)}469} \\ 42 \\ \hline 49 \\ \end{array}$$

So $a = 7$
$a^2 = 49$

The square root of 1369 is 37

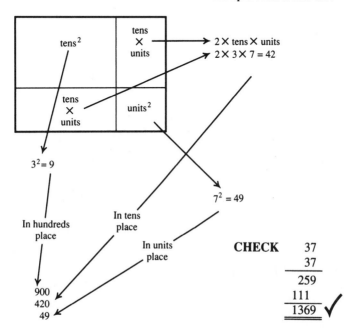

$2 \times$ tens $\times$ units
$2 \times 3 \times 7 = 42$

tens^2

tens
$\times$
units

$\begin{array}{r} \text{tens} \\ \times \\ \text{units} \end{array}$

units^2

$3^2 = 9$

$7^2 = 49$

In hundreds place

In tens place

In units place

CHECK

$$\begin{array}{r} 37 \\ 37 \\ \hline 259 \\ 111 \\ \hline 1369 \end{array} \checkmark$$

900
420
49

Figure 11-3

A "longhand" way of finding the first few digits of the square root of 2.

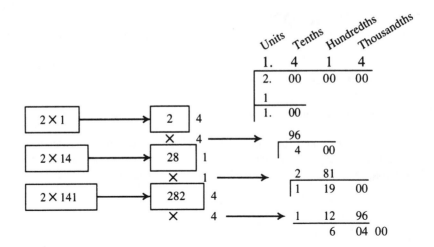

problem $1369^{1/2} = 37$, which can also be expressed as $37^2 = 1369$, is worked through both forward and backward.

Where the square or the root has a decimal fraction part, mark places in the square from the decimal point each way. Where you have no decimal point, the right end of the number is where the decimal point would be. Where the root continues, perhaps indefinitely, you have a decimal part to the root. The first four digits of the square root of 2, a well-known number in mathematics, are found by this method in Fig. 11-3.

Figure 11-4 shows two more examples. At A, the square root of 37.94 is found to the first four figures. At B, the same thing is done with 379.4. This is how math students

Figure 11-4

At A, finding the square root of 37.94. At B, finding the square root of 379.4. Be sure the decimal point stays in the right place!

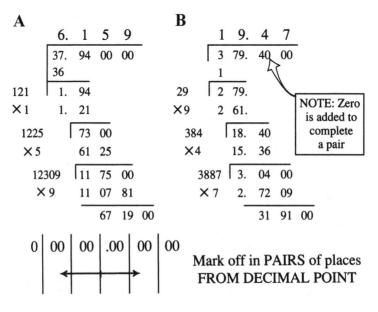

Mark off in PAIRS of places
FROM DECIMAL POINT

used to derive square roots to several digits before they had calculators. (Even the slide rule could work out problems to only two or three significant figures.) The first thing they did was to mark off the digits in pairs from the right. Why do that? For example, the square root of 9, which has just one digit, is 3; but the square root of 90, which has two digits, is more than 9 but less than 10. So, a one-digit root can become a square that has one or two digits. That is why you mark the digits in pairs: to know whether the first digit of the root is taken from a number between 1 and 10 or one between 10 and 100.

If you check the results of Fig. 11-4 with a calculator, you'll see that these methods don't round off to the nearest significant digit. They only generate digits one by one. So the answer is not rounded off, but truncated.

SIGNS IN SUCCESSIVE ROOTS

Looking at powers and roots leads to a whole new branch of mathematics. Note that the square root of any positive number can be positive or negative. For example, both 2 and -2, when squared, will give you 4; both 10 and -10, when squared, will give you 100. If you pursue a cube root, you'll find that the cube root of a positive number must be positive and the cube root of a negative number must be negative. For example, 2 cubed is 8, -2 cubed is -8, 10 cubed is 1000, and -10 cubed is -1000. Does this mean that the cube root of any number can only have one answer? For the numbers most laypeople work with, the answer is yes. But for the mathematician, the answer is no! You'll see why shortly.

Now take the fourth root, which is the square root of the square root. You should not have much trouble figuring out that the fourth root of a positive number can be either positive or negative, just like the square root. But something seems to be missing. The first square root can be either positive or negative. You can take the root of the positive root again, but what about the negative root? Should you just forget about it? Not if you're a mathematician!

REAL VS. IMAGINARY NUMBERS

Studying these kinds of problems led to the concept of *imaginary numbers*. Centuries ago, negative numbers weren't defined because they didn't represent real things, or so it seemed. You've seen some examples already when working with pictures and frames that turned out to have "negative dimensions." These didn't represent real objects. But in some situations, negative numbers can be used in calculation to get valid answers, even if they strike you as a little bit strange. You might say that driving -20 mph straight west is the same as going 20 mph straight east, for example.

The roots we have worked with so far have been *real number* roots. A real number is one that you can put on a conventional *number line*. All the real numbers can be denoted as points on a continuous straight line with a "zero point" defined and then extending off indefinitely in either direction. The operations of addition, subtraction, multiplication, and exponentiation (raising to a whole-number power) can all be defined over the set of real numbers. If # represents any one of these operations and x and y are real numbers, then $x \# y$ is also a real number.

There are situations that don't fit into this definition, however. For example, if you try to divide a real number by 0, you don't get a real number. To this day, expressions such as 5/0 or 0/0 remain undefined in conventional mathematics. Another exception arises when you raise a negative real number to certain fractional powers such as 1/2. Taking the square root of -1, for example, does not give you a real number. But in this case, the operation has been defined. The positive square root of -1, also written as $-1^{1/2}$, is defined as the *unit imaginary number*. Mathematicians denote it as i, and engineers more often call it j or the j *operator*.

Just as a negative times a negative is a positive for real numbers, i times i (or i^2) is a negative real number. When you multiply out $i \times i \times i$ to get i^3, the result is $-i$. If you multiply again by i to get i^4, you get 1. Continuing on causes you to go in a complete circle with every four iterations, like this:

$$i^2 = -1$$
$$i^3 = -i$$
$$i^4 = 1$$
$$i^5 = i$$
$$i^6 = -1$$
$$i^7 = -i$$
$$i^8 = 1$$
$$i^9 = i$$
$$i^{10} = -1$$

etc.

Figure 11-5 shows some of the properties of imaginary numbers and how they interact with real numbers.

COMPLEX NUMBERS

A *complex number* is the sum of a real number and an imaginary number. The general form for a complex number c is

$$c = a + ib$$

Individual complex numbers can be portrayed as points on a coordinate plane. One axis is a real number line, and the other axis is an imaginary number line. The intersection

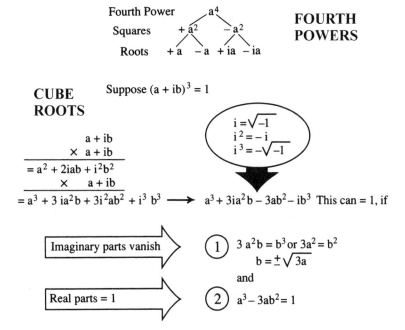

Figure 11-5
Some properties of imaginary and complex numbers.

SQUARE ROOTS

Real roots $\sqrt{+1} = +1 \text{ or } -1$

Imaginary roots $\sqrt{-1} = +i \text{ or } -i$
$i^2 = -1$

Fourth Power a^4

Squares $+a^2$ $-a^2$

Roots $+a$ $-a$ $+ia$ $-ia$

FOURTH POWERS

CUBE ROOTS

Suppose $(a + ib)^3 = 1$

$i = \sqrt{-1}$
$i^2 = -i$
$i^3 = -\sqrt{-1}$

$a + ib$
$\times \; a + ib$
$= a^2 + 2iab + i^2b^2$
$\times \qquad a + ib$
$= a^3 + 3ia^2b + 3i^2ab^2 + i^3 b^3 \longrightarrow a^3 + 3ia^2b - 3ab^2 - ib^3$ This can $= 1$, if

Imaginary parts vanish

① $3a^2b = b^3 \text{ or } 3a^2 = b^2$
$b = \pm\sqrt{3a}$

and

Real parts $= 1$

② $a^3 - 3ab^2 = 1$

point between the real and imaginary number lines corresponds to the value 0 on the real number line and the value $i0$ on the imaginary number line. The real and imaginary zeros are identical; that is, $0 = i0$. Therefore, they are both represented by the same point.

COMPLEX NUMBERS FIND NEW ROOTS

Now that you are "armed" with an entirely new set of numbers, you can look for less obvious cube roots of 1. As things turn out, there are three cube roots of 1! Sometimes you can't even imagine that something might exist in mathematics until you decide to look for it—and then, there it is!

Start by writing down $(a + ib)^3 = 1$. The ultimate goal here is to find all the possible values of a and b for which this equation holds true. This looks like a formidable task, because you will have to multiply out a sum that is cubed. But it's not really too hard because the presence of the i causes some of the terms to cancel out. You get

$$(a + ib)^3 = a^3 + i3a^2b - 3ab^2 - ib^3$$

This expression must be equal to 1 which is a *pure real number*, meaning it has no imaginary part. It's technically the same as the complex number $1 + i0$. The real part therefore must be

$$a^3 - 3ab^2 = 1$$

because the equation can only be true if the imaginary terms in the polynomial "disappear." The imaginary part, which is 0, must satisfy another equation:

$$i3a^2b - ib^3 = 0$$

Divide this through by ib to get

$$3a^2 - b^2 = 0$$

Then rearrange, to obtain

$$b^2 = 3a^2$$

Now substitute $3a^2$ into the real number equation in place of b^2 and solve:

$$a^3 - 3ab^2 = 1$$
$$a^3 - 3a(3a^2) = 1$$
$$a^3 - 9a^3 = 1$$
$$-8a^3 = 1$$
$$a^3 = -1/8$$
$$a = -1/2$$

Remember that $b^2 = 3a^2$. So plug in the solution for a and solve:

$$b^2 = 3 \times (-1/2)^2$$
$$b^2 = 3 \times 1/4$$
$$b^2 = 3/4$$
$$b = \pm(3/4)^{1/2}$$
$$b = \pm(3^{1/2}) / 2$$

Remember that a square root of a positive real number can be either positive or negative! So you actually have two different solutions:

$$a = -1/2 \quad \text{and} \quad b = +(3^{1/2}) / 2$$

or

$$a = -1/2 \quad \text{and} \quad b = -(3^{1/2}) / 2$$

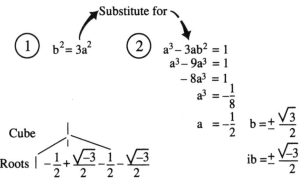

Figure 11-6

Summary of how the nonreal-number cube roots of 1 are found.

Check $\left(-\frac{1}{2}+\frac{\sqrt{-3}}{2}\right)^3 = -\frac{1}{8}+\frac{3}{8}\sqrt{-3}+\frac{9}{8}-\frac{3}{8}\sqrt{-3}$

$$= \frac{9}{8}-\frac{1}{8}=1$$

$\left(-\frac{1}{2}-\frac{\sqrt{-3}}{2}\right)^3 = -\frac{1}{8}-\frac{3}{8}\sqrt{-3}+\frac{9}{8}+\frac{3}{8}\sqrt{-3}$

$$= \frac{9}{8}-\frac{1}{8}=1$$

Now you should recall that we wanted to find the possible values if $a + ib$, any of which could be a cube root of -1. They are

$$1 + i0$$
$$-1/2 + i(3^{1/2}) / 2$$
$$-1/2 - i(3^{1/2}) / 2$$

Figure 11-6 is a summary of how the non-real-number cube roots of 1 are found, and—you guessed it!—a check of the solutions to be sure that they work.

A PROBLEM USING SIMULTANEOUS EQUATIONS

Sometimes a problem with two variables has two possible solutions that are given by quadratic equations. Suppose a rectangular holding pen is fenced on only three sides, while the fourth side is left entirely open. Those three sides require 20 feet of fencing to enclose an area of 48 square feet. What are the dimensions of the pen? Refer to the upper left-hand part of Fig. 11-7.

Figure 11-7

Summary of how the solutions to the "partially enclosed area" problem are found.

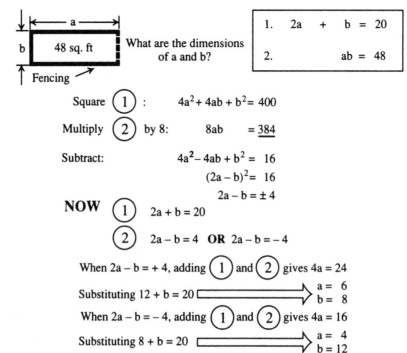

What are the dimensions of a and b?

1. $2a + b = 20$
2. $ab = 48$

Square ① : $4a^2 + 4ab + b^2 = 400$

Multiply ② by 8: $8ab = \underline{384}$

Subtract: $4a^2 - 4ab + b^2 = 16$

$(2a - b)^2 = 16$

$2a - b = \pm 4$

NOW ① $2a + b = 20$

② $2a - b = 4$ **OR** $2a - b = -4$

When $2a - b = +4$, adding ① and ② gives $4a = 24$

Substituting $12 + b = 20$ → $a = 6$, $b = 8$

When $2a - b = -4$, adding ① and ② gives $4a = 16$

Substituting $8 + b = 20$ → $a = 4$, $b = 12$

From this problem, you get the following equations, using *a* for the lengths of the two opposite fenced sides and *b* for the lengths of the other opposite sides, only one of which is fenced:

$$2a + b = 20$$

$$ab = 48$$

One way to minimize the work here is to square each side of the first equation. When you do that, you get

$$(2a + b)^2 = 400$$

This multiplies out to

$$4a^2 + 4ab + b^2 = 400$$

Now multiply each side of the second original equation by 8. This gives you

$$8ab = 384$$

Now subtract this whole equation from the previous equation: left side from left side, and right side from right side. This gives you

$$4a^2 + 4ab + b^2 - 8ab = 400 - 384$$

which simplifies to

$$4a^2 - 4ab + b^2 = 16$$

Note that the left-hand side of this equation is a perfect square. You can change it to

$$(2a - b)^2 = 16$$

Then taking the square root of each side, you get two equations because the square root of 16 can be either 4 or −4:

$$2a - b = 4$$

or

$$2a - b = -4$$

Now remember that $2a + b = 20$. There are two solutions that satisfy this and the above two equations:

$$a = 6 \quad \text{and} \quad b = 8$$

or

$$a = 4 \quad \text{and} \quad b = 12$$

All this is summarized in Fig. 11-7.

QUESTIONS AND PROBLEMS

This is an open-book quiz. You may refer to the text in this chapter (and earlier ones, too, if you want) when figuring out the answers. Take your time! Consider all values given as exact, so you don't have to worry about significant figures. The correct answers are in the back of the book.

1. Using the difference-of-squares method, find the following products. Use long multiplication (or a calculator) to check your results.
   (a) 63×77 (b) 85×95 (c) 117×123
   (d) 193×207 (e) 49×51

2. Find the square root of each of the following numbers.
   (a) 179,776   (b) 20,164   (c) 456,976
   (d) 9,920.16   (e) 12,769

3. Find the square root of each of the following numbers to three decimal places.
(a) 30   (b) 50   (c) 60
(d) 70   (e) 80

4. Solve the following pairs of simultaneous equations.
(a) $x + y = 20$ and $xy = 96$
(b) $x - y = 5$ and $x^2 + y^2 = 53$
(c) $3x + y = 34$ and $xy = 63$
(d) $x - y = 6$ and $x^2 + y^2 = 26$

5. The force exerted by wind is proportional to the square of the wind speed. Suppose a wind blows at 30 mph and exerts x amount of relative force on objects in the physical environment. Then how much relative force is exerted by a wind blowing at the following speeds? Calculate your answers to two significant digits.
(a) 10 mph   (b) 20 mph   (c) 40 mph
(d) 60 mph   (e) 100 mph

6. The intensity of a light beam varies according to the inverse of the square of the distance. Suppose that, at a distance of 10.00 feet (ft), a light beam has a relative brilliance of x "light units." Then how bright will the beam be at the following distances? Calculate your answers to four significant digits.
(a) 2.000 ft   (b) 5.000 ft   (c) 15.00 ft
(d) 25.00 ft   (e) 100.0 ft

7. A box has a volume of 480 cubic inches and a surface area (all six rectangular sides added together) of 376 square inches. It is 6 inches high. Find the other two dimensions.

8. A rectangular lot of 10 acres (435,600 square feet) must be enclosed with a fence. Assuming that the lot is square with 660 feet on each side, 2460 feet of fencing is bought. But because the area is actually rectangular and not square, the fencing covers three sides and exactly one-half of the fourth side. What are the dimensions of the lot?

9. Suppose, in Prob. 8, that the fencing was 110 feet short of completing the enclosure. What are the dimensions of the lot in this case? Why do the alternative answers in Prob. 8 differ, but here the dimensions are the same either way?

10. Two numbers multiplied together are 432. One divided by the other leaves a quotient of 3. What are the numbers?

11. Add the following pairs of complex numbers. Note that in this problem and the next two, the square root of -1 is represented by j.
(a) $2 + j2$ and $1 - j4$
(b) $-2 - j6$ and $9 - j2$
(c) $1 + j2$ and $1 - j2$
(d) $-2 - j3$ and $2 + j3$

12. In each of the following pairs of complex numbers, subtract the second from the first. (*Hint*: multiply the second number by -1 and then add.)

    (*a*) $7 + j7$ and $1 + j4$

    (*b*) $-2 + j10$ and $-3 - j2$

    (*c*) $6 + j6$ and $6 - j6$

    (*d*) $-4 - j5$ and $-4 + j3$

13. Find the products of the following pairs of complex numbers.

    (*a*) $8 + j2$ and $1 - j2$

    (*b*) $-3 - j0$ and $3 - j2$

    (*c*) $0 + j2$ and $7 - j2$

    (*d*) $6 - j3$ and $6 + j3$

12 Mechanical Mathematics

In this chapter, you'll learn how mathematics is applied to basic problems in the branch of physics known as *classical mechanics*. This involves the relationship among *force*, *mass*, *displacement* (or distance), *velocity* (or speed), and *acceleration*. Ultimately, mechanical phenomena can take the form of *work*, *energy*, or *power*.

WHAT IS FORCE?

Force is a measure of "push." A cart with a heavy load begins to move steadily when it is pushed (Fig. 12-1A). If the cart has a lighter load, the same "push" will make it go much faster (Fig. 12-1B). Alternatively, greater push can move the heavier load more quickly, too (Fig. 12-1C).

Force is needed to start and stop movement. However, in the absence of friction, movement would continue unchanged indefinitely, unless or until some force changed it or stopped it (Fig. 12-2). For simplicity, assume that no friction exists; then force is needed only to start, stop, or change motion. From observation, force is directly proportional to the weight to be moved. Force is also directly proportional to the acceleration (rate at which motion increases) or the deceleration (rate at which motion decreases).

UNITS OF FORCE

The unit for measuring force in the foot-pound-second system is the *poundal* (pdl). It is the force needed to accelerate (or decelerate) a weight (usually called mass) of one pound (1 lb) so that it changes its speed by one foot per second per second (1 ft/s^2).

The metric unit of force is the *newton* (N). A force of 1 N accelerates or decelerates a mass of one kilogram (1 kg) by one meter per second per second (1 m/s^2). Another rarely used metric unit is the *dyne* (dyn), which accelerates a mass of one gram (1 g) by one centimeter per second per second (1 cm/s^2).

Figure 12-1

At A, a heavy load moves slowly when pushed. At B, a light load is moved more quickly by the same effort. At C, a heavy load moves faster when pushed harder.

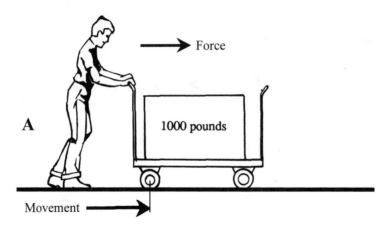

A

Force

1000 pounds

Movement

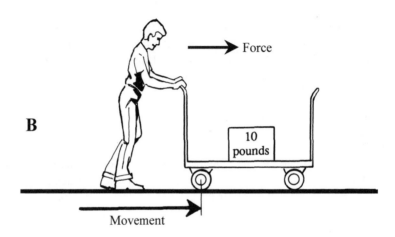

B

Force

10 pounds

Movement

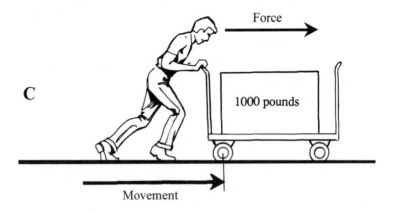

C

Force

1000 pounds

Movement

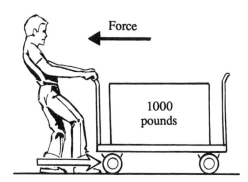

Force

1000
pounds

Figure 12-2
Force is needed to slow down or stop movement, as well as to start it or increase it.

Whatever units you measure force with, they are equivalent to mass times acceleration. That is proportional to mass multiplied by distance, divided by time squared. That is what "per second per second" means. It isn't a misprint! It can be translated to "per second, every second" or "per second squared." You will better understand this concept as you continue through this chapter.

Notation is not always consistent in classical mechanics. For example, some texts will abbreviate second(s) as sec rather than s. Miles per hour is sometimes abbreviated as mph and at other times as mi/h. You'll see all four of these notational variants in this book. That will help you get used to them all; you'll know what is meant by the context.

ACCELERATION, SPEED, AND DISTANCE

The "per second per second" part sounds confusing at first, as if I repeated myself by mistake. Look at it in a way that avoids this repetition.

Assume that you travel by car at 40 miles per hour (mph or mi/h). For the next minute, you accelerate steadily. Then one minute (1 min) later, you are moving at 60 mi/h. This acceleration is 20 miles per hour per minute [(mi/h)/min]. Although exact repetition is avoided in this case, "per hour per minute" is still used. This is not a standard unit of acceleration, but it's a perfectly valid one, and it should help you understand the principle by not using repetitive units.

If acceleration is steady, the average speed during that minute will be midway between the start and finish speeds, that is, 50 mi/h. The distance traveled during that minute will be the same as if this average speed had been used for the whole minute, that is, 5/6 mi.

A standard unit of acceleration is *feet per second per second* (ft/s^2). Suppose that you undergo a steady, straight-line acceleration of 10 ft/s^2 from a standstill. At the start, you are not moving. After 1 s, you will have accelerated by 10 ft/s^2 to a speed of 10 ft/s. The average speed for the first second will be 5 ft/s, halfway between 0 ft/s and 10 ft/s. So you will travel 5 ft during this first second.

During the next second, the speed will change from 10 ft/s to 20 ft/s, for an average speed of 15 ft/s. Thus, you will travel 15 ft in that second, a total of 20 ft from your starting position. The average speed over 2 s is 10 ft/s.

During the third second, the speed increases from 20 ft/s to 30 ft/s, an average of 25 ft/s, to travel 25 ft, making a total of 45 ft from the starting point. The average speed over 3 s is 15 ft/s.

You can tabulate distances traveled for any number of seconds from the start, as shown in Fig. 12-3A. If you plot the result as a graph, you get a quadratic curve, the familiar parabola (Fig. 12-3B).

If t is the total time from the start in seconds (when the speed is 0, a standstill) and a is the acceleration in feet per second per second, then the speed v at the end of time t

Figure 12-3

An example of how acceleration affects distance traveled as time goes by, tabulated (A) and plotted as a graph (B). In this case, the acceleration rate is a constant 10 ft/s².

Number of second from start	Speed			Total distance from start
	At beginning of second	At end of second	Average for second	
	Feet per second			
			Also distance in feet	
1st	0	10	5	5
2nd	10	20	15	20
3rd	20	30	25	45
4th	30	40	35	80
5th	40	50	45	125

A

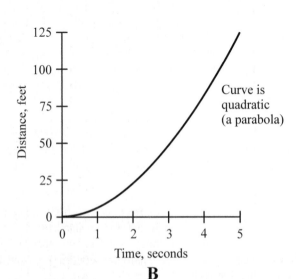

Curve is quadratic (a parabola)

B

is *at* (*a* times *t*) feet per second. (Speed is symbolized *v* for "velocity" here, because speed is actually an expression of straight-line velocity.) So the average speed between a starting speed of 0 ft/s and a finishing speed of *at* ft/s is *at*/2 ft/s. Because the time is *t* seconds, the distance *d* in feet traveled from the start is (*at*/2) × *t*, or *at*²/2. So you can say these things:

$$v = at$$

$$d = at^2/2$$

Notice that acceleration uses units of length (distance) per time per time, such as feet per second per second (ft/s²) or miles per hour per minute [(mi/h)/min]. Speed results from multiplying acceleration by time, and thus it is length per time, such as feet per second (ft/s) or miles per hour (mph). Distance covered is obtained by multiplying speed by time. For example, feet per second (ft/s) times seconds (s) yields feet because the seconds "cancel out."

You can also look at it the other way. Speed is a measure of distance covered per unit time. So you would divide feet by seconds to get feet per second (ft/s). Acceleration is a measure of speed change per unit time. You would divide feet per second by seconds to get feet per second per second, or feet per second squared (ft/s²). You can manipulate the above two formulas to get

$$a = v/t = 2d/t^2$$

These formulas apply only when the starting speed is 0 and the starting time is also 0. If either of these things is not the case, or if the acceleration is not constant, things get more complicated.

FORCE AND WORK

Force is a measure of "push" or "pull," as stated earlier. Work (as a mathematical term) is a measure of what is done by a force. If nothing moves, such as when you lean against something, the force is there, but no work is performed. Work results when an applied force causes movement.

Force is a mass (*m*) times an acceleration (*a*). Work (*w*) is proportional to applied force (*F*) and distance moved (*d*). Work is equal to force times distance (*Fd*), so it must be proportional to mass times acceleration times distance (*mad*). Distance moved, using constant acceleration from a standstill, is acceleration multiplied by time squared, divided by 2 (*at*²/2). So work is mass times acceleration times *at*²/2, or *ma*²*t*²/2. Because speed (*v*) is acceleration multiplied by time (*at*), this expression for work can be simplified to

$$w = mv^2/2$$

Units of work are the *foot-poundal* (ft · pdl), the *joule* (J), and the *erg*. The foot-poundal represents the work done by a force of 1 pdl moving through 1 ft. The joule, which is the standard unit of work used by most scientists and engineers these days, represents the work done by a force of 1 N moving through 1 m (1 N · m). The erg, occasionally used when the amount of work done is very small, represents 1 dyn moving through 1 cm (1 dyn · cm). The small elevated dot in the *composite unit* expressions technically represents multiplication.

The formula $w = mv^2/2$ represents the work of bringing a specific mass to a certain speed, regardless of the acceleration. If the acceleration is 10 ft/s², then 80 ft and 4 s are required to reach a speed of 40 ft/s. If the acceleration is 5 ft/s², it takes 8 s and 160 ft. Either way, 800,000 ft · pdl will move 1000 lb from standstill to 40 ft/s, although time and distance differ.

WORK AND ENERGY

Work and energy are expressed in the same units. For example, suppose an archer pulls back the string of a bow. The string is pulled back by a force that is equal to (or a little greater than) the tension of the string. The amount of work that is needed to pull the bow string back is stored in the bow as energy. When the archer releases the arrow, the string's thrust accelerates it to flight speed. This work transfers the energy of the drawn bow to the energy of the arrow in flight.

Energy is a capacity for doing work and, conversely, work is the transfer of energy from one form or place to another. So both use the same units—foot-poundals, joules, or ergs—according to the system of units employed.

When the energy E is in the form of a mass in motion, the appropriate formula is

$$E = mv^2/2$$

This formula can be used to determine the work needed to attain this motion, or to determine the energy actually manifest in it.

ENERGY AND POWER

Power is an expression of the rate at which work is done, or the rate at which energy is transferred from one form to another. As work is force applied over a distance, power is force applied over a distance per unit time.

You already know the units used, but an example will illustrate the relations between power and the other quantities. Assume the question relates to power vs. the weight of a car. Suppose that one motor unit develops a power of 290,000 foot-poundals per second (ft · pdl/s), and another has twice the power, or 580,000 ft · pdl/s. Coupled with these power levels, different weights must be moved. One car weighs 1500 lb, and the other weighs 3000 lb.

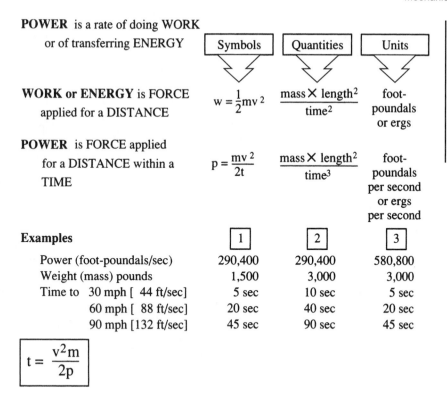

POWER is a rate of doing WORK or of transferring ENERGY

WORK or ENERGY is FORCE applied for a DISTANCE

POWER is FORCE applied for a DISTANCE within a TIME

	Symbols	Quantities	Units
WORK or ENERGY	$w = \frac{1}{2}mv^2$	$\dfrac{\text{mass} \times \text{length}^2}{\text{time}^2}$	foot-poundals or ergs
POWER	$p = \dfrac{mv^2}{2t}$	$\dfrac{\text{mass} \times \text{length}^2}{\text{time}^3}$	foot-poundals per second or ergs per second

Examples

	1	2	3
Power (foot-poundals/sec)	290,400	290,400	580,800
Weight (mass) pounds	1,500	3,000	3,000
Time to 30 mph [44 ft/sec]	5 sec	10 sec	5 sec
60 mph [88 ft/sec]	20 sec	40 sec	20 sec
90 mph [132 ft/sec]	45 sec	90 sec	45 sec

$$t = \frac{v^2 m}{2p}$$

Figure 12-4

An example of how power and weight relate to the time required for two different cars to reach speeds of 30, 60, and 90 mi/h.

Energy is expressed in the form $mv^2/2$. So, power must be in the form $mv^2/2t$. Transposing this, using the symbol p for power, we find that the time for a mass to reach a given speed is $mv^2/2p$. From this formula you can find the time taken by the smaller power unit with the smaller weight, the smaller power unit with the greater weight, and the greater power unit with the greater weight. If you want, you can complete the set by taking the greater power with the smaller weight, but we won't deal with that situation here.

Figure 12-4 tabulates the time needed in each case to reach 44 ft/s, 88 ft/s, and 132 ft/s, which are the speeds that correspond to 30 mi/h, 60 mi/h, and 90 mi/h. Notice that the time needed is related to the square of the speed to be reached. At constant acceleration, the speed is directly proportional to the elapsed time. At constant power, acceleration must diminish as speed increases.

GRAVITY AS A SOURCE OF ENERGY

Until now, to keep the units basic (such as the pound, the foot per second, and so on), gravity has been left out of our calculations involving force, work, and power. The constant vertical force of gravity that acts around us, however, provides a convenient means of storing and concentrating energy.

Figure 12-5

Principle of the pile driver. At A and B, work lifting a weight stores energy. At C, energy is released when the weight is allowed to fall back down.

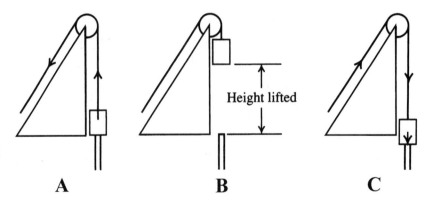

A   **B**   **C**

Height lifted

A pile driver illustrates this principle. First, work is done by lifting a weight against the force of gravity (Fig. 12-5A). The weight is not accelerated upward, but it is lifted steadily against a constant force, which is gravity pulling downward. Just as energy or work is force times distance, this energy takes the form of distance lifted times weight (Fig. 12-5B). Moving twice the distance requires twice the work, and therefore stores twice the energy. Speed is not involved—yet. Less power requires more time to do the same amount of work.

When the weight reaches the top, it is released to drop on the pile. Gravity is a mutual pull between earth and any mass. Doubling the mass doubles the pull (weight). So, in free fall, any object will drop at the same acceleration, which is approximately 32 ft/s² or 9.8 m/s² at the earth's surface.

As the object accelerates downward, it stores energy at the rate of $mv^2/2$ as a result of its motion (Fig. 12-5C). When it hits the pile, this stored energy is concentrated for a very short time, thrusting the pile downward.

WEIGHT AS FORCE

Weight provides a steady downward force F on any object with mass. The force is found from the mass m on which gravity acts, multiplied by the acceleration a of gravity, which produces 32 ft/s² or 9.8 m/s². That is,

$$F = ma$$

The force needed to prevent an object from falling is equal to the pull of gravity on the weight. As the pull of gravity accelerates the weight downward, the force of gravity on a mass of 1 lb is equal to 32 pdl. On a 2-lb mass, the force is 64 pdl.

If the mass is 1 kg, the force is 9.8 N. If the mass is 2 kg, the force is 19.6 N, and so on.

Otherwise stated, when you use gravity on a *mass* of 1 lb, it becomes a 1-lb *weight*, exerting a force of 1 lb (32 pdl in absolute nongravitational units). The same thing happens with kilograms and newtons. For this reason, in basic or absolute force units, pounds and kilograms express mass, which is a scientific way of saying how much "stuff" an object has. However, in gravitational units, pounds and kilograms refer to weight. That is the way the layperson usually thinks of mass, anyway.

GRAVITATIONAL MEASURE OF WORK

In gravitational measure of work, force does not have to accelerate a mass. Gravity exerts a force continuously on everything. If something doesn't fall, it's so because an equal force supports it, pushing it up.

If a 10-lb object (gravity acting on a 10-lb mass) rests on the floor, the object presses on the floor with a weight of 10 lb, which is a force of 320 pdl. Correspondingly, the floor pushes upward against the object with a force of 320 pdl to prevent it from falling. If a 10-kg object (gravity acting on a 10-kg mass) rests on the floor, the object presses on the floor with a weight of 10 kg, which is a force of 98 N.

You might ask, "Does the floor change its upward force according to what is on it?" The answer, surprisingly enough, is yes. If you hold the 10-lb object, your feet press on the floor, and the floor presses on your feet with a force that is 10 lb more than it would be with only you standing on the floor.

All the time that these forces operate, they balance. They are said to be *forces in equilibrium*. If the floor cannot provide enough upward push, it collapses, and work—most likely destructive—is done!

ENERGY FOR CONSTANT ACCELERATION

It is simplest to assume that acceleration is constant, which means that speed increases at a uniform rate, such as 1 ft/s^2. Then, as well as representing a steady growth in speed, acceleration represents a steady force. However, constant acceleration does not correspond with a constant rate of work.

The faster an object goes, the more power is produced by a given applied force. Remember that work is equal to force times distance. Therefore, exerting or maintaining a constant force at higher and higher speed produces (or requires, depending on the situation) more and more work, energy, or power as time goes by. Figure 12-6 illustrates an example of this.

Figure 12-6

How much energy is needed to maintain constant acceleration?

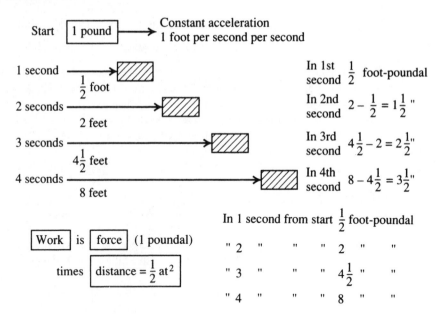

Start | 1 pound | → Constant acceleration
1 foot per second per second

1 second → $\frac{1}{2}$ foot

In 1st second $\frac{1}{2}$ foot-poundal

2 seconds → 2 feet

In 2nd second $2 - \frac{1}{2} = 1\frac{1}{2}$ "

3 seconds → $4\frac{1}{2}$ feet

In 3rd second $4\frac{1}{2} - 2 = 2\frac{1}{2}$ "

4 seconds → 8 feet

In 4th second $8 - 4\frac{1}{2} = 3\frac{1}{2}$ "

In 1 second from start $\frac{1}{2}$ foot-poundal

" 2 " " " 2 " "

" 3 " " " $4\frac{1}{2}$ " "

" 4 " " " 8 " "

| Work | is | force | (1 poundal)

times | distance $= \frac{1}{2} at^2$

POTENTIAL AND KINETIC ENERGY

The two forms of mechanical energy are called *potential energy* and *kinetic energy*. They can be described as follows.

- Potential energy is a function of the position of an object in the presence of a force. It is proportional to force and distance.

- Kinetic energy exists as a result of an object in motion. It is proportional to mass and speed squared.

Potential energy can be built from movement. But as long as start and finish are both the same or at some constant speed, movement is not involved in the calculation, as it is with kinetic energy. In the pile driver, for instance, a little more force is needed to start its upward movement. While it ascends at a constant rate, force and movement are both constant. A little less force is used to reach the top, if it stops before being released. The overall work needed to lift it is weight times height lifted.

KINETIC ENERGY AND SPEED

At constant acceleration, such as when a weight falls by the pull of gravity, energy builds in proportion to time squared. This happens because energy is proportional to speed squared, as shown in the example of Fig. 12-7.

Viewed another way, energy is proportional to force times distance. But because distance, at constant acceleration, increases in proportion to time squared, energy also

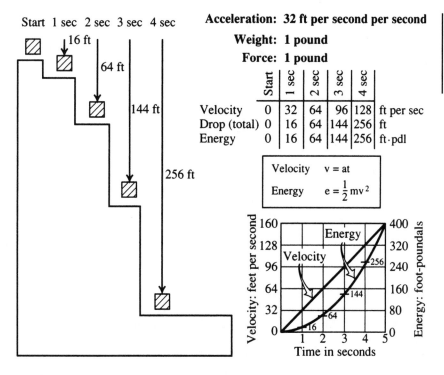

Figure 12-7
Kinetic energy is proportional to the square of the speed.

increases in proportion to time squared. Constant acceleration means that the speed grows in direct proportion to the elapsed time.

ACCELERATION AT CONSTANT POWER

The rate of work (power) at constant acceleration increases with speed, requiring progressively greater power during acceleration. By rearranging the formula that relates kinetic energy and power, you get

$$mv^2/2 = pt$$

So if power p and mass m are both constant, the speed v must increase in proportion to the square root (or 1/2 power) of the time t. You can see this by rearranging the above equation as follows:

$$mv^2/2 = pt$$

$$v^2/2 = pt/m$$

$$v^2 = 2pt/m$$

$$v = (2pt/m)^{1/2} = t^{1/2}(2p/m)^{1/2}$$

Figure 12-8

Constant power enables an accelerated mass to reach a speed of 100 ft/s in 20 s from a standstill start. Note that the acceleration (slope of the curve) decreases with the passage of time.

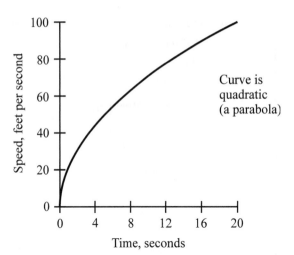

Curve is quadratic (a parabola)

To illustrate, suppose that constant power enables an accelerated object to reach a speed of 100 ft/s in 20 s from a standstill start. You can calculate the speed at any time during this 20-s period. You can make this calculation without knowing the mass or power involved. After the full 20 s, the speed reaches 100 ft/s. Note that 100 squared is 10,000. Divide this number in proportion to time: 2,000 for 4 s, 4,000 for 8 s, 6,000 for 12 s, and 8,000 for 16 s. Then take the square roots to find the speed at each of these times in feet per second. When you plot the points on a graph and connect them with a smooth curve, you get a graph like Fig. 12-8.

In this situation, the acceleration is greatest at the beginning. Then as the speed builds, the acceleration drops. The faster an object goes, the slower is the rate at which its speed increases. Because of this, one-half the final speed is reached in only 1/4 of the time.

HOW A SPRING STORES ENERGY

Another way to store potential energy is with a spring, similar to the principle of the archer's bow from earlier in this chapter. Assume that a spring supports only the weight that is attached to it, as shown in Fig. 12-9. Now imagine that you progressively apply greater force, in addition to that caused by the 1-lb weight itself. This will compress the spring. Suppose that 3-in compression requires 2 lb of extra force, 6-in compression requires 4 lb of extra force, 9-in compression requires 6 lb of extra force, and 12-in compression requires 8 lb of extra force, as shown.

The force that is applied to compress the spring by 12 in (or 1 ft) uniformly grows from 0 at the start to 8 lb at the finish. The average force over the 1 ft of compression must therefore be 4 lb. As a result, the energy stored in the spring, when it is compressed by the 8-lb force, is 4 ft · lb. This is equivalent to $4 \times 32 = 128$ ft · pdl. All this energy

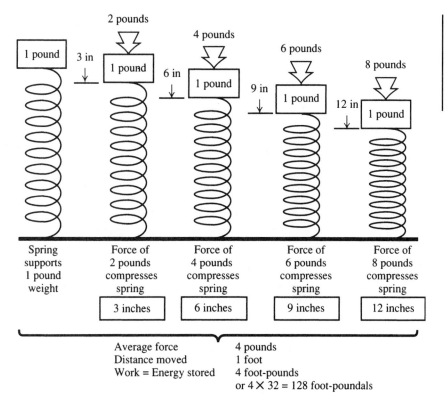

Figure 12-9
A stressed spring stores energy. In this case the stress takes the form of compression.

Average force | 4 pounds
Distance moved | 1 foot
Work = Energy stored | 4 foot-pounds
or 4 × 32 = 128 foot-poundals

will remain stored in the form of potential energy as long as the 8-lb force holds the spring compressed.

HOW A SPRING TRANSFERS ENERGY

Suppose that the 8-lb force holding the spring compressed is suddenly released. When this is done, the spring starts to propel the 1-lb weight upward. The initial force—right at the instant the spring is released—is 8 lb. As the weight moves upward, the accelerating force diminishes. But the speed continues to increase, because as long as stress remains on the spring, it exerts an upward force. The total energy—potential plus kinetic—remains constant in this type of system.

At the instant when the spring is half decompressed (to 6 in), the force has dropped to 4 lb, or 128 pdl. The average force represented in this compression is 2 lb, or 64 pdl, and the distance over which this average force is applied is 6 inches (in), or 0.5 ft. So the remaining potential energy is $64 \times 0.5 = 32$ ft · pdl. Of the original 128 ft · pdl, 96 ft · pdl has been turned into kinetic energy. This equation for the kinetic energy must be $E = mv^2/2$; m is 1 lb, so v^2 must be $2 \times 96 = 192$. That means that the instantaneous speed v is the square root of 192, or 13.856 ft/s. When the spring is fully decompressed, all the energy is kinetic, so now $v^2 = 256$ and $v = 16$ ft/s. This is shown in Fig. 12-10.

Figure 12-10

When a compressed spring is released, energy is converted from potential to kinetic.

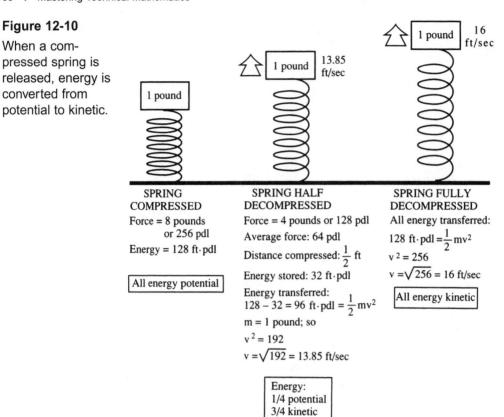

SPRING COMPRESSED

Force = 8 pounds or 256 pdl

Energy = 128 ft·pdl

All energy potential

SPRING HALF DECOMPRESSED

Force = 4 pounds or 128 pdl

Average force: 64 pdl

Distance compressed: $\frac{1}{2}$ ft

Energy stored: 32 ft·pdl

Energy transferred:
128 − 32 = 96 ft·pdl $= \frac{1}{2}mv^2$

m = 1 pound; so

$v^2 = 192$

$v = \sqrt{192} = 13.85$ ft/sec

Energy:
1/4 potential
3/4 kinetic

SPRING FULLY DECOMPRESSED

All energy transferred:

128 ft·pdl $= \frac{1}{2}mv^2$

$v^2 = 256$

$v = \sqrt{256} = 16$ ft/sec

All energy kinetic

RESONANCE CYCLE

The transfer of energy from potential to kinetic in the spring-and-weight arrangement, as shown in Fig. 12-10, forms the first part of a *resonance cycle*. At the instant the spring has become fully decompressed, the weight is moving upward at 16 ft/s. The spring now starts to decelerate the weight, because the spring is going into tension (pulling down, instead of pushing up). For each 3-in movement upward past the neutral position, the spring applies a tension of 2 lb until the upward displacement reaches 1 ft, where the tension becomes 8 lb. This is based on the assumption that the *spring constant*, or force-vs.-displacement ratio, is the same throughout the entire ±1-ft range of compression and tension discussed here. (This is true of most springs in the real world, unless they are compressed or stretched so far that they are permanently deformed.)

As with the compression, the average force of tension over the range of movement is 4 lb, so the potential energy is once again 4 lb, or 128 ft · pdl. All the energy is again potential, and the weight is momentarily stationary.

Since the weight has reached the upper extreme, an equal acceleration downward starts the second half of the cycle. Another interchange of energy, from potential to kinetic,

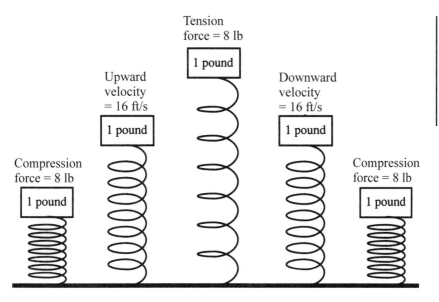

Tension
force = 8 lb

1 pound

Upward
velocity
= 16 ft/s

1 pound

Downward
velocity
= 16 ft/s

1 pound

Compression
force = 8 lb

1 pound

Compression
force = 8 lb

1 pound

continues until the neutral position is again reached. At this point, all the energy is kinetic, and the speed is 16 ft/s downward, steady (for the moment), neither accelerating nor decelerating. Then as the weight continues downward, compression starts again, until the weight comes to rest with the spring fully compressed, 1 ft down, with an 8-lb force pushing it back up (Fig. 12-11). And again, as before when the spring was fully compressed, all the energy is potential.

This up-and-down process, called *oscillation*, would go on forever in an ideal system. But in practice, the energy gradually *dissipates*; it transfers to other forms, notably thermal energy ("heat"). Friction absorbs some of the kinetic energy. The nature of the metal from which the spring is made causes some inefficiency in the system. Even air resistance plays a role. The *excursion* (maximum amount of up-and-down displacement) and speed therefore slowly diminish, and the weight eventually comes to rest.

TRAVEL AND SPEED IN RESONANT SYSTEM

You started with an assumed compression of 1 ft, which led to a maximum speed of 16 ft/s. Suppose the initial compression is only 6 in, or that friction has decreased the excursion to this magnitude. The maximum force is now 4 lb rather than 8 lb (128 pdl instead of 256 pdl). The distance over which the average force was 2 lb (64 pdl) is now compressed to 6 in rather than 12 in. So the maximum potential energy is 32 ft · pdl. When the weight passes through the neutral position, all this energy will become kinetic; v^2 will now be 64, so v is 8 ft/s.

Notice that halving the maximum upward or downward extent of travel also halves the maximum speed reached. The object travels one-half the distance at one-half the speed, so it performs the entire cycle in the same time. Interestingly, regardless of the magnitude of the oscillation, resonance still requires the same time to complete a full cycle. This length of time is called the *period of oscillation*. The number of oscillations per unit time is called the *frequency of oscillation*. So as a weight-and-spring system "dies down" after having been set in motion, the period and frequency stay the same, even though the excursion gets smaller and smaller.

The principle of resonance is used in the balance wheels of old-fashioned clocks, the pendulums of grandfather clocks, and many similar devices—not just mechanical, but also electrical, electronic, and atomic.

QUESTIONS AND PROBLEMS

This is an open-book quiz. You may refer to the text in this chapter (and earlier ones, too, if you want) when figuring out the answers. Take your time! Consider all values given as exact, so you don't have to worry about significant figures. The correct answers are in the back of the book.

1. Suppose a car accelerates uniformly from standstill to 40 mi/h in 3 min. How far will it travel in those 3 min?

2. In the next 6 min, the car increases its speed at a steady rate, from 40 mi/h to 60 mi/h. How far will it travel in these 6 min?

3. The same car brakes to a stop in 30 s. If the deceleration is uniform during these 30 s, how far will the car travel before it stops?

4. From the fact that 1 mi = 5280 ft and 1 h = 3600 s, find the speed in miles per hour that corresponds to 88 ft/s.

5. During takeoff, an aircraft builds up a thrust that accelerates it at 16 ft/s^2. Its take-off speed is 240 mi/h. Find the time from releasing the brakes until the plane lifts into the air. How much runway is required?

6. A gun can use cartridges with pellets of two sizes, one that is twice the weight of the other. If the heavier pellet leaves with a muzzle speed of 150 ft/s, find the muzzle speed of the lighter pellet, assuming that the explosive charge develops the same energy in each case.

7. An electric car's motor and transmission develop constant power during maximum acceleration. This particular car (it's no drag-racing machine!) can reach 60 mi/h in 20 seconds. In how long will it reach 30 mi/h? 45 mi/h?

8. If the weight of car and driver (in question 7) is 3000 lb, what time is necessary to reach 30 mi/h, 45 mi/h, and 60 mi/h when an additional load of 1000 lb is carried?

9. Find the power developed by the motor and transmission of the same car in foot-poundals per second.

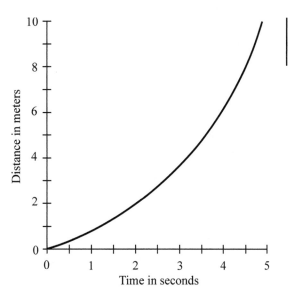

Figure 12-12
Illustration for problem 11.

10. A spring-and-weight resonance system can be changed, by altering either the weight or the spring. By figuring the effect of such change on maximum speed reached from a given starting deflection, deduce the effect of (1) doubling the weight and (2) halving it.

11. Using the accompanying graph of distance vs. time (Fig. 12-12), make rough estimates of the speed of the object, in meters per second, for the instants of time corresponding to 1 s, 2 s, 3 s, and 4 s. Explain how you deduced these results.

12. Using the accompanying graph of speed vs. time (Fig. 12-13), determine the approximate acceleration of the object, in meters per second per second, for the

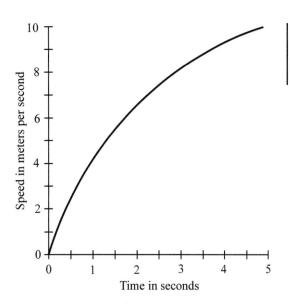

Figure 12-13
Illustration for problems 12 and 13.

Figure 12-14

Illustration for
problem 14.

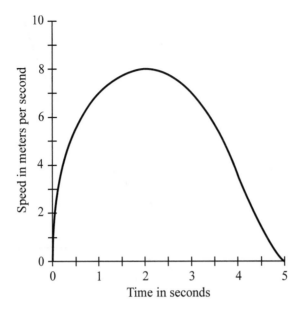

instants of time corresponding to 1 s, 2 s, 3 s, and 4 s. Explain how you deduced these values.

13. Assuming the object described by Fig. 12-13 has a constant mass and there is no friction, tension, or gravitational influence, how does the applied force vary qualitatively with time? Explain how you know this. (Think of a mass in outer space, propelled by a small rocket.)

14. Figure 12-14 shows the speed vs. time for an object that moves faster and faster for a while and then slows down. Thus, the curve appears somewhat "bell-shaped." Qualitatively, how does the acceleration vary with time? When is the acceleration greatest? When is it smallest? What is the approximate acceleration when the elapsed time is 2 s?

13
Ratio and Proportion

Thus far, the concepts of *ratio* and *proportion* have been covered with respect to the simple idea of similarity. In this chapter, you'll learn how ratios and proportions can be used to solve practical problems. You'll also get a first taste of *trigonometry*, the branch of mathematics that relates lengths to angles.

FRACTIONS, PROJECTIONS, EXTREMES, AND MEANS

A fraction expresses a ratio or proportion in terms of the relationship between the numerator and denominator. Different numbers can be used to write the same value as a fraction—the same relationship between the numerator and denominator. For example, 1/2 can also be written as 2/4, 3/6, 4/8, or, in general, $x/(2x)$, where x is any positive real number. Proportions can also be denoted by using colons. The foregoing would then be written as 1:2, 2:4, 3:6, 4:8, or, in general, $x:2x$.

One application that sets the scene for later concepts is the use of ratio or proportion relative to projected images, as shown in Fig. 13-1. This concept is sometimes called the *aspect ratio*. The old-fashioned standard television (TV) screen aspect ratio is 4:3, meaning 4 units horizontally to 3 units vertically. This aspect ratio is still used in many computers today, although some of the newer computer displays have larger aspect ratios. Movie screens have various aspect ratios, particularly now that wide screens have become popular.

Figure 13-2 shows several ways that ratios and proportions can be expressed, along with some of the properties of the variables in these expressions. Note especially how *cross products* work. In an equation containing two ratios, the left-hand numerator and the right-hand denominator are called the *extremes*, while the left-hand denominator and the right-hand numerator are called the *means*. In any equation of this form, the product of the extremes is equal to the product of the means.

Figure 13-1

An example of ratio or proportion in a projected image. In this case the aspect ratio is 4:3, regardless of the distance from the projector.

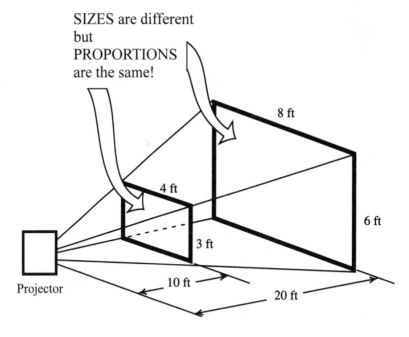

SIZES are different but PROPORTIONS are the same!

8 ft

4 ft

6 ft

3 ft

Projector

10 ft

20 ft

Figure 13-2

Several ways in which ratios and proportions can be written. The product of the extremes always equals the product of the means.

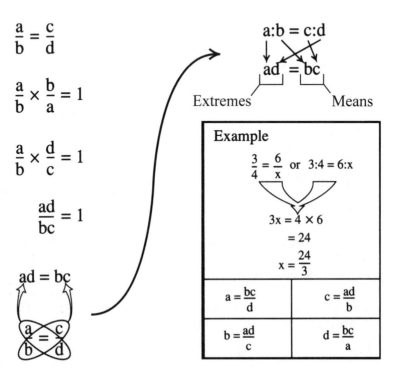

$$\frac{a}{b} = \frac{c}{d}$$

$$\frac{a}{b} \times \frac{b}{a} = 1$$

$$\frac{a}{b} \times \frac{d}{c} = 1$$

$$\frac{ad}{bc} = 1$$

$$ad = bc$$

$$\frac{a}{b} = \frac{c}{d}$$

a:b = c:d

ad = bc

Extremes   Means

Example

$\frac{3}{4} = \frac{6}{x}$ or 3:4 = 6:x

$3x = 4 \times 6$

$= 24$

$x = \frac{24}{3}$

$a = \frac{bc}{d}$	$c = \frac{ad}{b}$
$b = \frac{ad}{c}$	$d = \frac{bc}{a}$

PROPORTIONS IN PRACTICE

Suppose a woman's will explicitly says that her estate is to be divided among her three daughters in proportion to their ages. The amount of the estate is $78,000, and their ages are 53, 47, and 30. How much does each get?

First, assume that the basis of the proportion is $x per year of age to each person. This means the daughters get $53x, $47x, and $30x, respectively. That adds up to $130x. You know that the total is $78,000. So the equation is

$$130x = 78,000$$

This can be easily solved if you divide through by 130 on each side, getting $x = 600$. After you substitute into the expressions $53x$, $47x$, and $30x$, the daughters receive $31,800, $28,200, and $18,000 respectively. To check, add these amounts. They sum up to $78,000.

SHAPE AND SIZE

Ratio and proportion form a good basis for showing the distinction between shape and size. This principle can be demonstrated clearly with triangles, the simplest geometric figures having straight edges.

If a triangle is expanded in proportion, the lengths of its respective sides maintain the same ratio (Fig. 13-3). The triangle has the same shape, but it differs in size. Because it has the same shape, it also has the same interior angles. All angles marked (1) are equal, those marked (2) are also equal, as are those marked (3).

The sides with one crossmark have the same proportion to the sides with two crossmarks in each triangle. The proportion between sides with two and three crossmarks, or

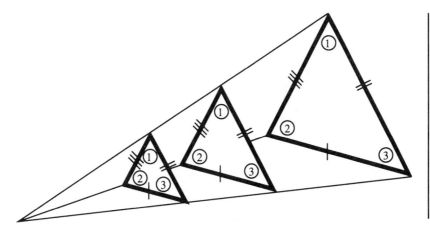

Figure 13-3

These three triangles have the same shape even though their sizes differ, because the corresponding angles (1), (2), and (3) are the same. The proportion among the lengths of the three sides is the same in each triangle.

between sides with one and three crossmarks, are the same in each triangle, or with any others that have the same shape.

Triangles that have the same shape but not necessarily the same size are called *similar triangles*. Although their sides might be longer or shorter, they are always in the same proportion. They are "magnified" or "minimized" copies of one another.

ABOUT ANGLES IN TRIANGLES

When two straight lines cross (mathematicians use the word "intersect"), the opposite angles are equal. Figure 13-4A shows how this statement can be demonstrated. Draw square (right-angle) corners at the intersection, based on each of the heavy lines. You can easily see that any pair of angles numbered (1) and (2) has a total angular measure equivalent to two square corners. (This is called a *straight angle* and has a measure of 180°.) One pair of square corners consists of top angle (2) and left side angle (1). Another pair consists of top angle (2) and right-side angle (1). The total is the same, two square corners, and both use the same angle (2), so the two angles (1) must be the same. Similarly, you can show that the two angles marked (2) are equal.

Next, if two parallel lines intersect a third line (Fig. 13-4B), the angles at the intersections are equal if taken in correct pairs. To show this, complete a rectangle with square corners. As with the single intersection, you can now see that angles numbered (1) are all equal, as are all angles numbered (2).

Figure 13-4

At A, opposite angles have equal measure when two lines intersect. At B, all the angles marked (1) have equal measure, as do all the angles marked (2). At C, the measures of the interior angles in a triangle add up to two square corners (a straight angle).

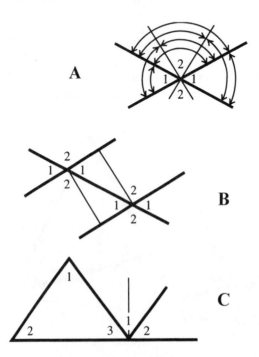

In any triangle, its three interior angles always add up to two square corners. To show this statement (Fig. 13-4C), extend one side at one corner and draw a line parallel to the opposite side. Because of their positions, relative to parallel lines, corresponding angles marked (1) and (2) are equal. Where the side is extended are three angles numbered (1), (2), and (3) that add up to two square corners. Therefore, the corresponding angles inside the triangle, also numbered (1), (2), and (3), must also add up to two square corners.

USE OF RIGHT TRIANGLES

A right triangle, also called a square-cornered triangle, becomes a very important building block in other shapes and sizes, whether in triangles or in more-complicated shapes. Any acute triangle (one in which all the interior angles are smaller than right angles) can be divided into two right triangles in three different ways, as shown in Fig. 13-5A. In the triangles shown, one such division is done with a thin line, the other two with dashed lines.

If all the angles are acute, all three ways of dividing into two square-cornered triangles are additive, so the original triangle consists of the two square-cornered ones added together. However, if one angle of a triangle is obtuse (wider than a square corner), two of the possible divisions require a difference, rather than a sum. The original triangle is the larger square-cornered triangle minus the smaller one.

An interesting fact about these divisions, which we will not prove here, is that the three dividing lines from the corners of the original triangle, formed by making them perpendicular to the opposite side, always intersect at a single point. "Perpendicular" means the two lines create two square-cornered angles. In an acute triangle, the point of intersection is inside the triangle (Fig. 13-5B). In an obtuse triangle, the point is found only by extending all three perpendiculars (dashed lines). This exercise begins to show the importance of square-cornered (right) triangles as building blocks.

Figure 13-5
At A, any acute triangle can be divided into two square-cornered ones in three different ways. At B, the perpendiculars in an acute triangle always intersect at a single point inside the triangle.

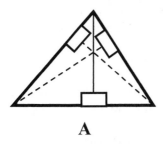

A

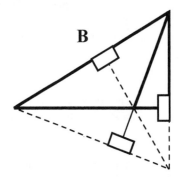

B

Figure 13-6

In a right triangle with one acute angle fixed, the ratio of the length of the opposite side to the length of the longest side is constant, regardless of the size of the triangle.

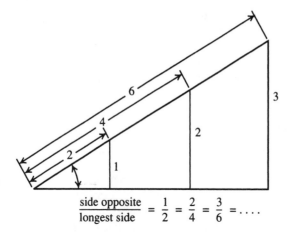

$$\frac{\text{side opposite}}{\text{longest side}} = \frac{1}{2} = \frac{2}{4} = \frac{3}{6} = \dots$$

ANGLES IDENTIFIED BY RATIOS

Angles determine the shape of a triangle and also the ratios between the lengths of its sides. Any of the three sides can be changed to alter all three angles, so the relationship between the side ratio and the angle becomes rather involved.

Consider the specific case of a right triangle. Because the measures of its three interior angles add up to two right angles, the other two angles always add up to one right angle. A right triangle that has a fixed measure for one of its acute angles can therefore have only one shape. Regardless of how big (or small) you draw the triangle, the ratio between the lengths of any two sides always is the same for the particular angle of interest. An example of this is shown in Fig. 13-6. The ratio 1/2 identifies the angle, shown by the arc with arrows at each end, uniquely. No other angle can produce a right triangle whose opposite and longest sides have lengths in this same ratio. In the example shown here, the length of the opposite side is always exactly one-half the length of the *hypotenuse* (longest side).

THEOREM OF PYTHAGORAS

Suppose we have a right triangle defined by points A, B, and C whose sides have lengths a, b, and c, respectively. Let c be the length of the side opposite the right angle, which is the hypotenuse (Fig. 13-7). Then the following equation holds:

$$a^2 + b^2 = c^2$$

The converse of this is also true: If there is a triangle whose sides have lengths a, b, and c, and the above equation holds, then that triangle is a right triangle.

This formula can be stated verbally as follows: "The square of the length of the hypotenuse of a right triangle is equal to the sum of the squares of the lengths of the other two sides."

Figure 13-7

The Theorem of Pythagoras for right triangles.

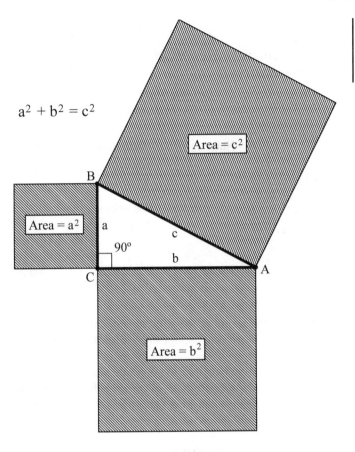

$$a^2 + b^2 = c^2$$

This is called the *theorem of Pythagoras* or the *Pythagorean theorem*. It is named after the Greek mathematician *Pythagoras of Samos* who lived in the 4th century B.C.

NAMES FOR ANGLE RATIOS

As you proceed in studying mathematics, the use of ratios to identify angles assumes an important role. You need names to identify them. These names are the terminology of trigonometry.

The right triangle used to identify the ratios has three sides. The ratio of any two of its three sides uniquely identifies a certain acute angle. In picking those two sides, you have six possible choices. Three of these choices result in the *basic trigonometric functions*.

The *sine* is defined as the length of the opposite side divided by the length of the hypotenuse. Each quantity is written with "sin" followed by a letter or symbol to identify that particular angle. In Fig. 13-8, the angle is shown by A, so we write "sin A" to represent the sine of angle A.

Figure 13-8

The trigonometric functions can be defined as ratios of the lengths of the sides of a right triangle.

The three basic functions in trigonometry

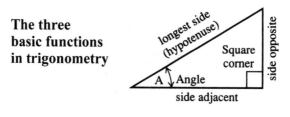

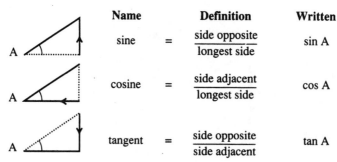

	Name		Definition	Written
	sine	=	side opposite / longest side	sin A
	cosine	=	side adjacent / longest side	cos A
	tangent	=	side opposite / side adjacent	tan A

The *cosine* is the length of the adjacent side divided by the length of the hypotenuse. Cosine is abbreviated "cos" so for the cosine of angle A, we write "cos A."

The *tangent* uses the length of the opposite side divided by the length of the adjacent side. Tangent is abbreviated "tan," so we write the tangent of angle A as "tan A." (If you have previously learned that a tangent is a line that touches the circumference of a circle, this usage is entirely different.)

The little diagrams at the left in Fig. 13-8 show how to remember the relationship of the ratio to the angle. The line with the arrow is the numerator. It leads to the side that forms the denominator. Remember that these names (sine, cosine, and tangent) represent ratios of lengths. A ratio identifies the angles, regardless of how large or how small the triangle happens to be.

SPOTTING THE TRIANGLE

Remembering the ratios in trigonometry requires as much time and practice as learning the multiplication tables in arithmetic. What can be more confusing, needing more care, is spotting the right sides for the ratio when the angle is not in the position used in the previous section. Regardless of where the angle is, you must construct a right triangle (or use one that's already there) one way or another. Then the ratio follows the definition. A sine, for example, is always the opposite side over the hypotenuse. The other ratios follow similar layouts.

Figure 13-9 shows four possible positions that you might encounter. At A, the triangle is in the "standard orientation" from the previous section. At B, the right angle is on top. At C, the triangle is in the same orientation as at A, but the angle in question is different. At D, the triangle has the angle in question at the bottom, with the right angle at the upper left.

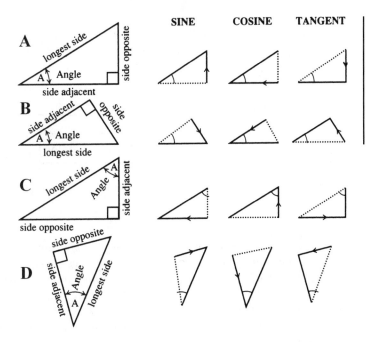

Figure 13-9

It takes some practice to remember trigonometric ratios when the triangles are not oriented in a familiar-looking way.

The little diagrams at the right in Fig. 13-9 show the relationship of each ratio to the angle for the sine, cosine, and tangent. The line with the arrow is the numerator. It leads to the side that forms the denominator.

DEGREE MEASURE OF ANGLES

Degree measure for angles requires a complete circle (rotation) in a flat surface, which is 360 degrees (symbolized 360°). A half rotation is 180°. A quarter rotation, which is a right angle, is 90°. Acute angles are less than 90°, and obtuse angles are more than 90° but less than 180° (Fig. 13-10).

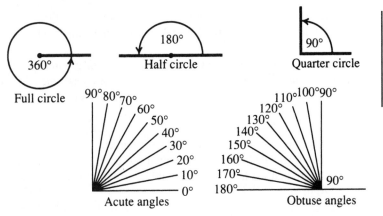

Figure 13-10

Degree measures for angles are based on a full circle of 360 equal increments.

Figure 13-11

Equilateral (A) and right isosceles (B) triangles give rise to two special angles, 60° and 45°.

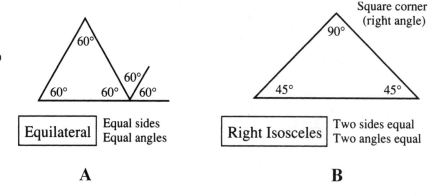

A

B

Two other angles of particular interest, 60° and 45°, are based on special triangles. The first is an *equilateral triangle*, in which all three sides are equally long. It also has all equal angles, each 60°. The other special angle comes from a right triangle in which the two shorter sides have equal lengths. Because the three angles must add up to 180° and one of them is 90°, the other two must each be 45°. This polygon is called a *right isosceles triangle*. "Isosceles" means "having two equal sides." These two noteworthy situations are shown in Fig. 13-11.

THREE APPLICATIONS OF TRIGONOMETRY

Figure 13-12 illustrates three examples of the use of trigonometry to solve practical problems. Use your calculator to find the sines, cosines, and tangents of angles.

Problem 1. Suppose that an observation point at sea level is exactly 20 mi from a mountain peak (Fig. 13-12A). The elevation of the peak above sea level is exactly 5°, viewed from this point. How high is the peak?

Solution 1. The relationship involves the adjacent side (a distance of 20 mi), the height, the opposite side, and an angle of 5°. This is the tangent ratio. A calculator tells you that tan 5° = 0.0875, rounded off to four decimal places. The height is therefore 0.0875 times the 20-mi distance. Convert the figure to feet by multiplying by 5280, getting 105,600 ft. Multiply this by 0.0875 and you have the peak height, 9,240 ft.

Problem 2. A skyscraper is viewed from a point on the ground exactly 50 ft away from its vertical wall (Fig. 13-12B). The angle to the horizontal line of sight is exactly 84°. How high is the building? Assume your eyes are at ground level and the ground near the building is perfectly horizontal.

Solution 2. This problem again involves the tangent. A calculator will tell you that tan 84° = 9.5144, rounded off to four decimal places. By multiplying this figure by 50 ft, you can figure that the height of the building is 475.7 ft. You might want to round this off to 476 ft.

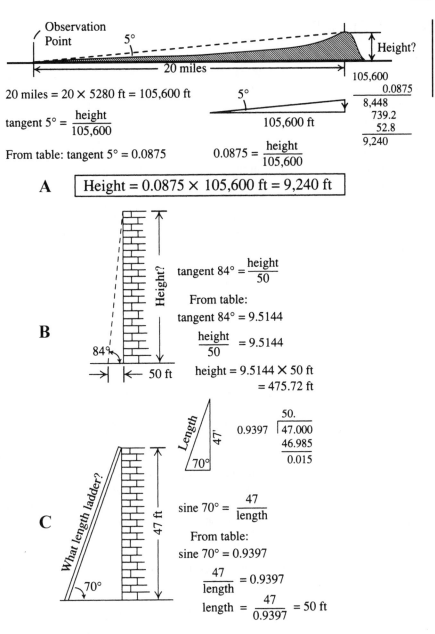

Figure 13-12

Three practical problems using trigonometry. See text for discussion.

20 miles = 20 × 5280 ft = 105,600 ft

tangent 5° = $\dfrac{\text{height}}{105,600}$

From table: tangent 5° = 0.0875 $0.0875 = \dfrac{\text{height}}{105,600}$

A   Height = 0.0875 × 105,600 ft = 9,240 ft

tangent 84° = $\dfrac{\text{height}}{50}$

From table:
tangent 84° = 9.5144

$\dfrac{\text{height}}{50} = 9.5144$

height = 9.5144 × 50 ft
= 475.72 ft

B

sine 70° = $\dfrac{47}{\text{length}}$

From table:
sine 70° = 0.9397

$\dfrac{47}{\text{length}} = 0.9397$

length = $\dfrac{47}{0.9397}$ = 50 ft

C

Problem 3. A ladder must reach the roof of a building exactly 47 ft high. The slope of the ladder, when resting against the building, should be exactly 70° (Fig. 13-12C). What ladder length is necessary?

Solution 3. The solution here involves the ratio of the opposite side to the hypotenuse, which is the sine, but it's the inverse. Using a calculator, you can find sin 70° = 0.9397, rounded off to four decimal places. Dividing 47 by 0.9397 gives the needed ladder length as a tiny bit more than 50 ft.

QUESTIONS AND PROBLEMS

This is an open-book quiz. You may refer to the text in this chapter (and earlier ones, too, if you want) when figuring out the answers. Take your time! Consider all values given as exact, so you don't have to worry about significant figures. The correct answers are in the back of the book.

1. The aspect ratio for a TV picture is 4:3. A wide-screen movie is transmitted so that the picture fills the full height of the TV screen. What proportion of the width must be lost at the sides, if the aspect ratio of the movie picture is 2:1?

2. Another way of transmitting the picture in Prob. 1 is to include the full picture width and mask off an area (top and bottom) where the picture does not fill the TV screen. What proportion of the TV screen will be masked off (top and bottom)?

3. A man wills his estate to his 5 children: 3 boys and 2 girls. It calls for each to get an amount proportional to his or her age at the time of his death. This man is old-fashioned, and he has decided that the boys should get twice the rate for their ages that the girls do. When the will is made, the boys' ages are 40, 34, and 26, while the girls are aged 37 and 23. If the father dies in the same year, what will each receive from an estate of $22,100?

4. If the father in Prob. 3 lives 10 years after making the will, the estate has not changed in value, and all five children are still living, how much will each get?

5. The two sides of a right triangle that adjoin the right angle are 5 in and 12 in long. What is the length of the hypotenuse?

6. A highway gradient is measured as the rise in altitude divided by the distance along the pavement surface. Suppose that an 8,000-ft length of straight highway maintains a gradient of 1:8 (1 to 8). Find the altitude gained in this distance, and the amount by which the distance measured horizontally falls short of 8,000 ft. Use the Pythagorean theorem.

7. At a distance of 8 mi, the elevation of a mountain peak, viewed from sea level, is 9°. Some distance farther away, still at sea level, the elevation is 5°. What is the height of the peak, and the distance of the second viewpoint?

8. A railroad track stretches for 3 mi at a gradient of 1:42 up, then 5 mi of 1:100 up, then 2 mi level, then 6 mi of 1:250 down, then 4 mi level, and finally 5 mi at 1:125 up. How much higher is the finish point than the starting point?

9. A house is to have a roof slope of 30°, gabled in the middle. The width of the house is 40 ft, and the roof is to extend 2 ft horizontally beyond the wall. What distance from the ridge of the gable to the guttering is required for rafters?

10. By how much could the rafter length be reduced in the house (Prob. 9) by making the roof slope 20° rather than 30°?

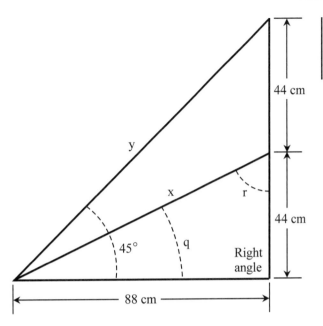

Figure 13-13
Illustration for problems 12 through 16.

11. A house wall is 50 ft high. A ladder used to scale it is 60 ft long. How far from the base of the wall must the ladder be placed for its top to just reach the top of the house wall? Solve this in two ways. First, use trigonometry to derive the answer to four significant figures. Second, use the Pythagorean theorem.

12. In Fig. 13-13, all length units are in centimeters (cm). Assume the lengths and angle measures to be exact. From the data given, use the Pythagorean theorem to derive the length of side x to four significant digits.

13. Use trigonometry, not the Pythagorean theorem, to derive the length of side y in Fig. 13-13 to four significant digits.

14. Use trigonometry to derive the measure, in degrees, of angle q in Fig. 13-13 to the nearest one-tenth of a degree.

15. Use trigonometry to derive the measure, in degrees, of angle r in Fig. 13-13 to the nearest one-tenth of a degree.

16. Find the measure of angle r in Fig. 13-13 to the nearest one-tenth of a degree by using a method that does not involve trigonometry.

14
Trigonometric and Geometric Calculations

In this chapter you will learn how to find the sine, cosine, and tangent of the sum of two angles. It doesn't work like removing the parentheses in algebra. You'll also learn how to find the trigonometric functions of the difference of two angles, and of angle multiples.

RATIOS FOR SUM ANGLES

Suppose angle A measures 30° and angle B measures 45°. With your calculator, you can find that $\sin 30° = 0.5000$ and $\sin 45° = 0.7071$, both to four significant figures. Adding the two gives you something bigger than 1.

It is apparent that $\sin A + \sin B$ cannot be the same as $\sin (A + B)$ in this case. No sine (or cosine, for that matter) can be more than 1, because the ratio has the hypotenuse of a right triangle, which is the longest side, as its denominator. The numerator can never be larger than the denominator because neither the adjacent side nor the opposite side can be longer than the hypotenuse. A sine or cosine can never be greater than 1, so such an answer must be wrong.

The sine of a sum is not the same as the sum of the sines. The same is true for the cosine and tangent. We need new formulas for these.

The sine of the sum of two angles A and B can be found by using this formula:

$$\sin (A + B) = \sin A \cos B + \cos A \sin B$$

The cosine of the sum of two angles A and B can be found as follows:

$$\cos (A + B) = \cos A \cos B - \sin A \sin B$$

FINDING TAN $(A + B)$

The formula for the tangent of the sum of two angles can be derived from the formulas for the sine and cosine. Remember that, for any angle, the tangent is equal to the sine divided by the cosine. Using that fact, you can write this:

$$\tan (A + B) = \sin (A + B) / \cos (A + B)$$

In a way that does it! But, using the formulas for the sine and cosine of the sum of two angles, you can expand this to

$$\tan (A + B) = (\sin A \cos B + \cos A \sin B) / (\cos A \cos B - \sin A \sin B)$$

Divide both the numerator and denominator of the fraction on the right-hand side of this equation by $\cos A \cos B$, which turns all the terms into tangents, giving

$$\tan (A + B) = (\tan A + \tan B) / (1 - \tan A \tan B)$$

Figure 14-1 shows geometrically how this process works with right triangles, canceling out the lengths of the sides to simplify the result.

Figure 14-1

Geometric demonstration of how the tangent of the sum of two angles can be figured out.

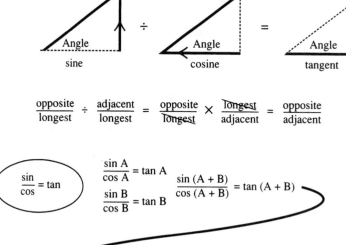

A   **sin 75°**

$$= \sin(30° + 45°) = \sin 30° \cos 45° + \cos 30° \sin 45°$$

$$\frac{1 + \sqrt{3}}{2\sqrt{2}} = \frac{\sqrt{2} + \sqrt{6}}{4}$$

$\sqrt{2} = 1.414$
$\sqrt{6} = 2.449$
3.863

$$= \frac{1}{2} \times \frac{1}{\sqrt{2}} + \frac{3}{2} \times \frac{1}{\sqrt{2}}$$

$$\frac{0.\textcircled{966}}{4 \overline{)3.863}}$$

$$= \frac{1 + \sqrt{3}}{2\sqrt{2}} = \boxed{0.966}$$

B   **cos 75°**

$$= \cos(30° + 45°) = \cos 30° \cos 45° - \sin 30° \sin 45°$$

$$\frac{\sqrt{3} - 1}{2\sqrt{2}} = \frac{\sqrt{6} - \sqrt{2}}{4}$$

$\sqrt{6} = 2.449$
$\sqrt{2} = 1.414$
1.035

$$= \frac{\sqrt{3}}{2} \times \frac{1}{\sqrt{2}} - \frac{1}{2} \times \frac{1}{\sqrt{2}}$$

$$\frac{0.\textcircled{259}}{4 \overline{)1.035}}$$

$$= \frac{\sqrt{3} - 1}{2\sqrt{2}} = \boxed{0.259}$$

C   **tan 75°**

$$= \tan(30° + 45°) = \frac{\tan 30° + \tan 45°}{1 - \tan 30° \tan 45°}$$

$$\frac{\sqrt{3} + 1}{\sqrt{3} - 1} = \frac{(\sqrt{3} + 1)^2}{(\sqrt{3} - 1)(\sqrt{3} + 1)} = \frac{4 + 2\sqrt{3}}{3 - 1}$$

$$= \frac{\frac{1}{\sqrt{3}} + 1}{1 - \frac{1}{\sqrt{3}}} = \frac{\sqrt{3} + 1}{\sqrt{3} - 1}$$

$$= \frac{4 + 2\sqrt{3}}{2} = \boxed{2 + \sqrt{3}}$$

$$= \boxed{3.732}$$

Figure 14-2

The sine, cosine, and tangent of a 75° angle can be derived from the values for 30° and 45° angles by using the angle-sum formulas. At A, the sine; at B, the cosine; at C, the tangent.

AN EXAMPLE: RATIOS FOR 75°

You can show the ratios for sine, cosine, and tangent of 75° by considering this angle as the sum of a 30° angle and a 45° angle. Substitute into the sum formula and then reduce the result to its simplest form before you evaluate the square roots in each case. Figure 14-2 shows these processes for sin 75° (at A), cos 75° (at B), and tan 75° (at C). If you use your calculator for evaluation, it will make little difference whether you simplify the expressions first or just plow through them!

RATIOS FOR ANGLES GREATER THAN 90°

So far, only the trigonometric ratios for acute angles (between 0° and 90°) have been considered. But angles can go over 90°, and even over 180°. To simplify classification of angles according to size, they are divided into *quadrants*.

A quadrant is one-quarter of a circle. Because the circle is commonly divided into 360 angular degrees, the quadrants are named by 90° segments. The first quadrant has angles measuring 0° to 90°. The second quadrant goes from 90° to 180°, the third goes from 180° to 270°, and the fourth goes from 270° to 360°.

Figure 14-3

At A, quadrants in a coordinate system, along with a rotating vector, can define trigonometric functions. At B, the signs of the variables in the four quadrants.

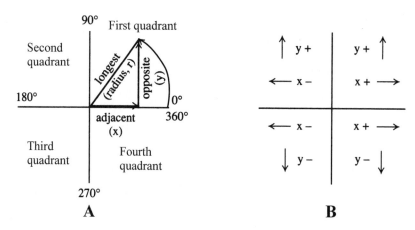

A

B

You can draw lines to represent the quadrant boundaries, with 0° or 360° going horizontally to the right, 90° going vertically upward, 180° going horizontally to the left, and 270° going vertically downward. You get a coordinate graph system like that shown in Fig. 14-3A. Progressively larger angles are defined by a rotating arrow of constant length called a *vector*, starting from 0° and rotating counterclockwise. The back-end point of the vector is always at the center of the coordinate system where the axes intersect. Horizontal elements are x: positive to the right, negative to the left. Vertical elements are y: positive up, negative down (Fig. 14-3B). The length of the rotating vector is r. So in this system, the sine of an angle is y/r, the cosine is x/r, and the tangent is y/x. The vector of length r is always positive. That way, the sign of the ratios can be easily figured out for the various quadrants.

In Fig. 14-4, the signs of the three ratios have been tabulated for the four quadrants. The drawing also shows how the equivalent angle in the first quadrant "switches" as the vector passes from one quadrant to the next. In the first quadrant, the sides can be defined in the standard triangle format for the sine, cosine, and tangent. As you move into bigger angles in the remaining quadrants, the opposite side is always the vertical (y). What was called the adjacent side is always the horizontal (x). The hypotenuse is always the rotating vector (r). Can you see a pattern in the way these trigonometric ratios for angles vary?

RATIOS FOR DIFFERENCE ANGLES

In a format similar to that used for the sum angles, here are the sine and cosine formulas for difference angles:

$$\sin (A - B) = \sin A \cos B - \cos A \sin B$$

and

$$\cos (A - B) = \cos A \cos B + \sin A \sin B$$

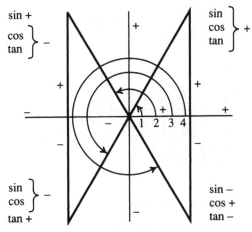

Figure 14-4

Trigonometric ratios in the four quadrants. Note how the signs vary.

Quadrant	Angle	Sine	Cosine	Tangent
1st	A	$\frac{+}{+} = +$	$\frac{+}{+} = +$	$\frac{+}{+} = +$
2nd	$180° - A$	$\frac{+}{+} = +$	$\frac{-}{+} = -$	$\frac{+}{-} = -$
3rd	$180° + A$	$\frac{-}{+} = -$	$\frac{-}{+} = -$	$\frac{-}{-} = +$
4th	$-A$	$\frac{-}{+} = -$	$\frac{+}{+} = +$	$\frac{-}{+} = -$

An interesting method of finding the formula for difference angles uses the sum formula already obtained, but makes B negative. From our investigation of the signs for various quadrants, negative angles from the first quadrant will be in the fourth quadrant. Making this substitution produces the same results arrived at in the section "Ratios for sum angles" at the beginning of this chapter.

Finding the tangent formula follows the same method, either going through substitution into the sine and cosine formulas or, more directly, by making tan $(-B)$ equal to $-\tan B$. Either way you get the following formula:

$$\tan (A - B) = (\tan A - \tan B) / (1 + \tan A \tan B)$$

RATIOS THROUGH THE FOUR QUADRANTS

You can deduce a few more ratios with the sum and difference formulas. You already did ratios for 75°. Now, do those for 15°. Note that $45° - 30° = 15°$. These formulas can give you values for angles at 15° intervals through the four quadrants, all the way from 0° to 360°. Plotting them out for the full circle, you can see how the three ratios change as the vector sweeps through the four quadrants (Fig. 14-5). If you want to plot a lot more points, you can use a calculator. This can be especially interesting for values of the tangent near 90° and 270°.

Figure 14-5

Graphs of the sine, cosine, and tangent values for angles between 0° and 360°.

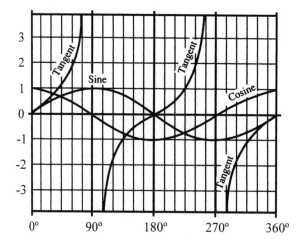

Both the sine and cosine "wave" up and down between +1 and −1. Notice that the "waves" are displaced by 90°, one from the other. The tangent starts out like the sine curve but quickly sweeps up, and at 90° it "blows up." Informally you can say that it goes off to "positive infinity." But then it "appears" from "negative infinity" on the other side of 90°. Going through the 180° point, the tangent curve duplicates what it does going through 0° or 360° (whichever way you view it). At 270°, it repeats what it did at 90°.

PYTHAGORAS IN TRIGONOMETRY

A formula can often be simplified, as was found by deriving the tangent formulas from the sine and cosine formulas and changing it from terms using one ratio to terms using another ratio. In doing this, the Pythagorean theorem, expressed in trigonometry ratios, is handy.

Assume that a right triangle has a hypotenuse 1 unit long and an angle of measure A, as shown in Fig. 14-6. Then the opposite side has a length of sin A, and the adjacent side has a length of cos A. From that, the Pythagorean theorem shows that

$$(\cos A)^2 + (\sin A)^2 = 1$$

This statement is true for any value of A. The above formula is more often written this way:

$$\cos^2 A + \sin^2 A = 1$$

In this context, $\cos^2 A$ means $(\cos A)^2$, and $\sin^2 A$ means $(\sin A)^2$. The angle's cosine or sine, not the angle itself, is to be squared. If you wrote them as $\cos A^2$ or $\sin A^2$, the equations would mean something else.

The
formulas ...

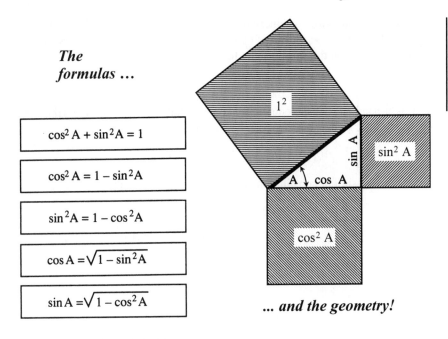

Figure 14-6
The Pythagorean Theorem implies some useful formulas relating the sine and cosine.

$$\cos^2 A + \sin^2 A = 1$$

$$\cos^2 A = 1 - \sin^2 A$$

$$\sin^2 A = 1 - \cos^2 A$$

$$\cos A = \sqrt{1 - \sin^2 A}$$

$$\sin A = \sqrt{1 - \cos^2 A}$$

... and the geometry!

The above formula can be transposed. Two other forms are

$$\cos^2 A = 1 - \sin^2 A$$

and

$$\sin^2 A = 1 - \cos^2 A$$

From these, taking the positive square roots of both sides, you can derive the following:

$$\cos A = (1 - \sin^2 A)^{1/2}$$

and

$$\sin A = (1 - \cos^2 A)^{1/2}$$

where the 1/2 power of a quantity represents the positive square root of that quantity.

DOUBLE ANGLES

The sum formulas, along with the Pythagorean theorem, are used for angles that are twice, three times, or a greater whole-number multiple of any original angle. Let's derive the formulas for the sine and cosine of 2A.

The sum formula works whether both angles are the same or different. You can have $\sin (A + B)$ or $\sin (A + A)$. But of course, $\sin (A + A)$ is really $\sin 2A$. So you have the following, based on the sum formula for the sine:

$$\sin 2A = \sin A \cos A + \cos A \sin A$$

The right-hand side of the equation is a sum whose addends are the same product in opposite order, so the statement can be simplified to

$$\sin 2A = 2 \sin A \cos A$$

You can make similar substitutions for the cosine and tangent:

$$\cos 2A = \cos (A + A)$$
$$= \cos A \cos A - \sin A \sin A$$
$$= \cos^2 A - \sin^2 A$$

and

$$\tan 2A = \tan (A + A)$$
$$= (\tan A + \tan A) / (1 - \tan A \tan A)$$
$$= 2 \tan A / (1 - \tan^2 A)$$

HIGHER MULTIPLE ANGLES

The formulas for the triple angle ($3A$) follow straightaway from the formulas for the double angle. It's as simple as writing $3A = 2A + A$, reapplying the sum formulas, and substituting $2A$ in place of B to get everything into terms of ratios for the simple angle A. This process can be carried on indefinitely, building upward to get formulas for the angle $4A$, the angle $5A$, and so on.

Here are the formulas for the sine, cosine, and tangent of $3A$. If you want, you can derive these (that is, grind them out manually!) from the double-angle formulas as an optional exercise:

$$\sin 3A = 3 \sin A - 4 \sin^3 A$$
$$\cos 3A = 4 \cos^3 A - 3 \cos A$$
$$\tan 3A = (3 \tan A - \tan^3 A) / (1 - 3 \tan^2 A)$$

The notation rules for cubes (third powers) of trigonometric ratios are in the same format as the rules for squares. Remember this for any positive whole number n:

$$\sin^n A = (\sin A)^n$$
$$\cos^n A = (\cos A)^n$$
$$\tan^n A = (\tan A)^n$$

PROPERTIES OF THE ISOSCELES TRIANGLE

You have already seen that a right triangle is a useful building block for other shapes. An isosceles triangle has slightly different uses. An isosceles triangle has two equal sides and two equal angles opposite those two sides. A perpendicular from the third angle (not one of the equal angles) to the third side (not one of the equal sides) bisects that third side. That is, it divides it into two equal parts, making the whole triangle into mirror-image right triangles, as shown in Fig. 14-7.

Any triangle except a right triangle can be divided into three adjoining isosceles triangles by dividing each side into two equal parts and erecting perpendiculars from the points of bisection. Where any two of these bisecting perpendiculars meet, if lines are drawn to the corners of the original triangle, the three lines have equal lengths. That is so because two of them form the sides of an isosceles triangle. So the perpendicular from the third side of the original triangle must also meet in the same point.

This statement is true whether the original triangle is acute or obtuse. Figure 14-8 shows an example with an acute triangle, where the meeting point is inside. In the case of an obtuse triangle, the meeting point still exists, but it is outside the triangle rather than inside it.

If you try to apply this rule to a right triangle, perpendiculars from the midpoint of the hypotenuse to the other two sides bisect those two sides. But the meeting point of those perpendiculars lies on the hypotenuse, so the third isosceles triangle disappears.

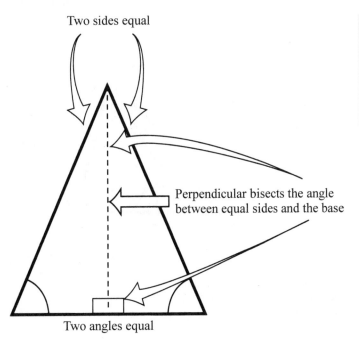

Two sides equal

Perpendicular bisects the angle between equal sides and the base

Two angles equal

Figure 14-7

Properties of an isosceles triangle. The two equal angles are opposite the two equal sides.

Figure 14-8

Perpendiculars from the centers of the three sides of an acute triangle meet at a single point, which forms the top vertex for three isosceles triangles.

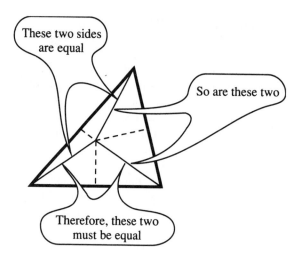

These two sides are equal

So are these two

Therefore, these two must be equal

ANGLES IN A CIRCLE

The center point of a circle is at an equal distance from every point on its circumference. This equal distance is the *radius* of the circle. A line segment from the center of a circle to any point on its circumference is called a *radial*, but it can also be called a radius.

If you draw a triangle such that its three *vertices* (corner points) all lie on the circumference of a circle, the circle is said to *circumscribe* the triangle, and the triangle is said to be *inscribed* within the circle. In such a triangle, the perpendiculars from the midpoints of its sides all meet at the circle's center, and radii from the corners of the triangle divide it into three isosceles triangles. Figure 14-9 shows an example.

Figure 14-9

A triangle inscribed within a circle has some interesting properties.

Circumscribing circle

$180° - 2A$

$180° - 2B$

$180° - 2C$

B

A

A

C

C

B

Base

$2A + 2B + 2C = 180°$ (main triangle)
Angle at center on base: $180° - 2C = 2A + 2B$
Angle at apex (on circle) on $\left.\begin{array}{c} \\ \text{same base} \end{array}\right\}$ $A + B$

Suppose you name the equal pairs of angles in each isosceles triangle as follows: A, A, B, B, C, and C. You can see clearly in Fig. 14-9 that the original large triangle has one angle $A + B$, one angle $B + C$, and one angle $A + C$. The three angles therefore total $2A + 2B + 2C$. This, you know, adds up to 180°, because the sum of the measures of the angles in any triangle must add up to 180°. Expressed as an equation, this is

$$2A + 2B + 2C = 180°$$

In any isosceles triangle, the angle at the apex is 180° minus twice the measure of either of the other two angles. From the above equation, it follows that $180 - 2A$ must be the same as $2B + 2C$. Consider the angles opposite from the part of the circle against which the top left side of the triangle sits. The angle at the center is $2B + 2C$, as just deduced. The angle at the circumference is $B + C$.

You will find that, for any segment of a circle, the angle at the circumference is always one-half the angle at the center. Figure 14-10 shows how this works for a specific segment of a circle (determined by the horizontal line).

The situation described in the previous section leads to an interesting fact about angles in circles containing inscribed triangles. Instead of identifying the angles with a side

Each of these
angles ...

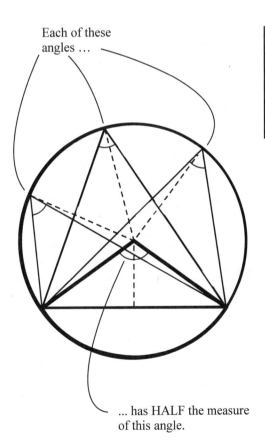

... has HALF the measure
of this angle.

Figure 14-10

For any segment of a circle, the angle at the circumference is always half the angle at the center.

of a triangle, you can use an arc (portion of the circumference) of the circle. The important thing is the angle that corresponds to the arc at the center. A part of the circumference of a circle that is identified by the angle at the center is called a *chord* of the circle.

ANGLES IN A SEMICIRCLE

Any angle drawn touching the circumference, using a chord as termination for the lines bounding the angle, must be just one-half the angle at the center. Thus, all the angles in a circle, based on the same chord, must be equal. Suppose that the chord has an angle of 120°. The angles at the circumference will all be exactly 60°. A special case is the *semicircle* (an exact half circle). The angle at the center is a straight line (180°). Every angle at the circumference of a semicircle is exactly 90° (a right angle). Any triangle inscribed within a semicircle is therefore a right triangle, as shown in Fig. 14-11.

DEFINITIONS

The previous sections have often discussed angles that add up to either a right angle (90°) or two right angles (180°). When the measures of two angles add up to 180°, they are called *supplementary angles*. When the measures of two angles add up to 90°, they are called *complementary angles*.

Figure 14-11

A triangle inscribed within a semicircle is always a right triangle.

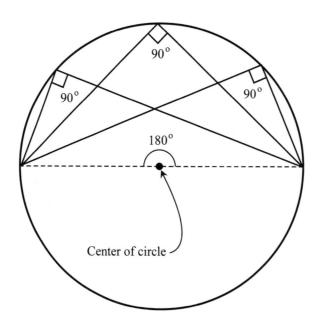

QUESTIONS AND PROBLEMS

This is an open-book quiz. You may refer to the text in this chapter (and earlier ones, too, if you want) when figuring out the answers. Take your time! Consider all values given as exact, so you don't have to worry about significant figures. The correct answers are in the back of the book.

1. The sine of angle A is 0.8 and the sine of angle B is 0.6. From this information, and without using a calculator's trig buttons, find the following.
  (a) $\tan A$
  (b) $\tan B$
  (c) $\sin (A + B)$
  (d) $\cos (A + B)$
  (e) $\sin (A - B)$
  (f) $\cos (A - B)$
  (g) $\tan (A + B)$
  (h) $\tan (A - B)$

2. At the equator, the earth has a radius of 4,000 mi. (This is an approximate figure, but for this problem, consider it exact.) Angles around the equator are measured in meridians of longitude, with a north-to-south line through Greenwich, England, as the 0° reference. Suppose that a point on Mt. Kenya is on the equator at 37.5° east of Greenwich, and a point on the island of Sumatra is on the equator at 100.5° east of Greenwich. How far apart are these two places, measured along an imaginary straight line through the earth?

3. If sights were made horizontally from the observation points in Prob. 2 (due east from Mt. Kenya and due west from Sumatra), at what angle would the lines of sight cross in space above the earth?

4. At a certain time, exactly synchronized at both places in the above scenario, a satellite is observed. On Mt. Kenya, the elevation of a line of sight, centered on the satellite, is 58° above horizontal, precisely eastward. On Sumatra, the elevation of a line of sight, centered on the satellite, is 58° above horizontal, precisely westward. How far away is the satellite along a line of sight from Sumatra? From Kenya?

5. The cosine of a certain angle is exactly twice the sine of the same angle. What is the tangent of this angle? You don't need either tables or a calculator to solve this.

6. The sine of a certain angle is exactly 0.28. Find the cosine and tangent without the trig functions on your calculator.

7. The sine of a certain angle is exactly 0.6. Find the sine of twice this angle and the sine of three times this angle.

8. Find the sine and cosine of an angle exactly twice that of question 7.

9. Using 15° as a unit angle and the formulas for ratios of 2*A* and 3*A*, find the values of sin 30° and sin 45°.

10. Using 30° as a unit angle and the formulas for ratios of 2*A* and 3*A*, find the values for sin 60° and sin 90°.

11. Using 45° as a unit angle and the formulas for ratios of 2*A* and 3*A*, find values for tan 90° and tan 135°. Confirm your results from the quadrant information in Fig. 14-4.

12. Using 60° as a unit angle and the formulas for ratios of 2*A* and 3*A*, find values for cos 120° and cos 180°. Confirm your results from the quadrant information in Fig. 14-4.

13. Using 90° as a unit angle and the formulas for ratios of 2*A* and 3*A*, find values for cos 180° and cos 270°. Check your results from the quadrant boundary information in Fig. 14-4.

14. Using the tangent formulas for multiple angles and your calculator, find the tangents for three times 29°, three times 31°, three times 59°, and three times 61°. Account for the changes in sign between three times 29° and three times 31°, and between three times 59° and three times 61°.

15. The sine of an angle is 0.96. Find the sine and cosine of twice the angle.

16. A problem leads to the following algebraic expression containing the cosine function:

$$8 \cos^2 A + \cos A = 3$$

Solve for cos *A*, and state in which quadrant the angle representing each solution will come. Give approximate values obtained with your calculator.

Part 3

ANALYSIS AND CALCULUS

15

Systems of Counting

The *decimal* (base-10) counting system is used almost exclusively by laypeople throughout the world. This system is also used in general science. However, in some disciplines—especially computer programming and engineering—other counting systems are used. In this chapter, you'll learn about these systems and how they relate to the decimal method of counting. You'll also learn how powers and roots can be expressed as fractional exponents.

MECHANICAL COUNTERS

Before electronic calculators were invented, people used counters with little wheels that carried numbers. The numbers that showed through the front window were like those that electronic digital devices display nowadays. If you took the cover off the wheel assembly, you could see how it worked.

The rightmost wheel counted from 0 to 9 in a decimal system. When it came to 9, it would move from 9 to 0, and the next wheel to the left would move from 0 to 1. Every time the first wheel passed from 9 to 0, the next wheel would advance 1 more, until it got up to 9. Then two wheels would read 99. As the first wheel moved from 9 to 0 this time, the next one would also move from 9 to 0, and the third wheel would move from 0 to 1, making the whole display read 100. When the third wheel reached 9 and then returned to 0, the fourth wheel would change from 0 to 1, and the display would indicate 1000. This could go on up to many thousands or millions, depending on the number of wheels in the assembly. Figure 15-1 shows the general idea of this system.

Even today, some mechanical meters work in a manner similar to "ancient" mechanical wheel counters. If you have an old car, its odometer is mechanical. Some utility meters still use mechanical dials or rotating drums.

Figure 15-1

A mechanical wheel counting machine that works in base 10.

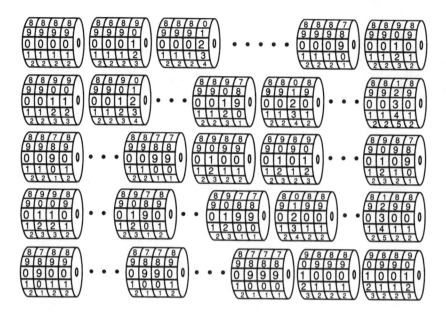

DUODECIMAL NUMBERS

The decimal system is not the only system that people can use. Years ago, some cultures used the *duodecimal system*, counting by 12s instead of by 10s. This system is also known as *base 12, radix 12,* or *modulo 12*. To use a wheel-counter system in base 12, you need two more numbers on each wheel. In the wheels shown in Fig. 15-2, the extra symbols are "t" and "e" for ten and eleven. You count like this: 0, 1, 2, 3, 4, 5, 6, 7, 8, 9, t, e, 10, 11, 12, . . ., 18, 19, 1t, 1e, 20, 21, 22, and so on. Here, 10 represents what you normally think of as 12 or a *dozen*; 100 represents what you would usually call 144 or a *gross*; 1000 represents what you would usually call 12 gross, and so on.

CONVERSION FROM DECIMAL TO DUODECIMAL

Why, you may ask, should anybody ever want to work in the duodecimal system when it's almost never used in the real world anymore? There's a good reason: Something unfamiliar makes you think! As a result, after you get over the initial "confusion curve," you'll have a better understanding of how numbering systems work in general.

To find how many times a number counts up to 12, you divide the number by 12 in the familiar decimal system. An example is shown in Fig. 15-3, where a decimal number is converted to duodecimal form.

First, you divide the original number by 12. The remainder at the bottom is the number of 1s left over after a number of complete 12s in the quotient have been

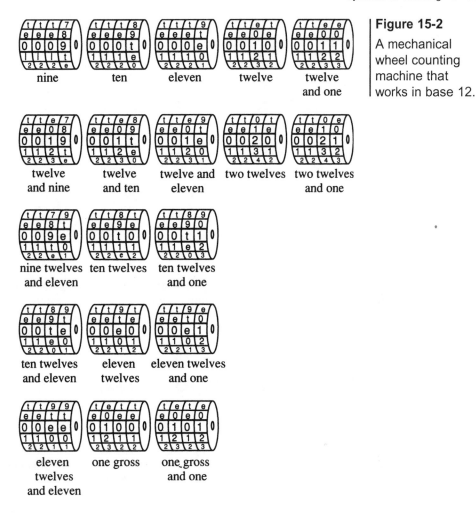

Figure 15-2

A mechanical wheel counting machine that works in base 12.

nine   ten   eleven   twelve   twelve and one

twelve and nine   twelve and ten   twelve and eleven   two twelves   two twelves and one

nine twelves and eleven   ten twelves   ten twelves and one

ten twelves and eleven   eleven twelves   eleven twelves and one

eleven twelves and eleven   one gross   one gross and one

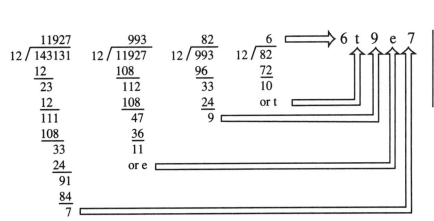

Figure 15-3

Conversion of decimal 143,131 to duodecimal form.

Table 15-1

Duodecimal multiplication table.

x	2	3	4	5	6	7	8	9	t	e
2	4	6	8	t	10	12	14	16	18	1t
3	6	9	10	13	16	19	20	23	26	29
4	8	10	14	18	20	24	28	30	34	38
5	t	13	18	21	26	2e	34	39	42	47
6	10	16	20	26	30	36	40	46	50	56
7	12	19	24	2e	36	41	48	53	5t	65
8	14	20	28	34	40	48	54	60	68	74
9	16	23	30	39	46	53	60	69	76	83
t	18	16	34	42	50	5t	68	76	84	92
e	1t	29	38	47	56	65	74	83	92	t1

passed on to the 12s counter. Then you divide by 12 again. This time the remainder is 11. In duodecimal, all numbers up to 11 use a single digit, so you call this e. You can follow through the rest of the conversion by examining the "flow" process in Fig. 15-3.

CONVERSION FROM DUODECIMAL TO DECIMAL

If you want to convert a number from duodecimal to decimal, simply reverse the process. Using duodecimal, you must divide the duodecimal number by 10 however many times is necessary. You can use the *duodecimal multiplication table* (Table 15-1) to help you find products in this unfamiliar system.

Suppose you want to convert duodecimal 6t9e7 to decimal form. You divide repeatedly by 10. Remember that in the duodecimal system, 10 is expressed as t, and 11 is represented as e. Follow the "flow" of Fig. 15-4 to see how the process works.

Figure 15-4

Conversion of duodecimal 6t9e7 to decimal form. Refer to the duodecimal multiplication table for products to complete the "long division" process.

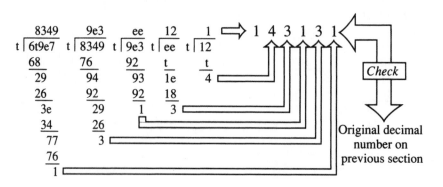

OCTAL NUMBERS

Another numbering scheme, called the *octal number system*, has eight, or 2 cubed (2^3), symbols. It is also called *base 8*, *radix 8*, or *modulo 8*. Every digit is an element of the following set:

$$\{0, 1, 2, 3, 4, 5, 6, 7\}$$

Counting proceeds from 7 directly to 10, from 77 directly to 100, from 777 directly to 1000, and so on. There are no numerals 8 or 9. In octal notation, decimal 8 is expressed as 10, and decimal 9 is expressed as 11.

HEXADECIMAL NUMBERS

Still another scheme, commonly used in computer practice, is the *hexadecimal number system*. It is so named because it has 16, or 2 to the fourth power (2^4), symbols. These digits are the usual 0 through 9 plus six more, represented by A through F, the first six letters of the alphabet. The digit set is

$$\{0, 1, 2, 3, 4, 5, 6, 7, 8, 9, A, B, C, D, E, F\}$$

In this number system, A is the equivalent of decimal 10, B is the equivalent of decimal 11, C is the equivalent of decimal 12, D is the equivalent of decimal 13, E is the equivalent of decimal 14, and F is the equivalent of decimal 15. This scheme is also called *base 16*, *radix 16*, or *modulo 16*.

BINARY NUMBERS

The *binary number system* is a method of expressing numbers using only the digits 0 and 1. It is sometimes called *base 2*, *radix 2*, or *modulo 2*. The digit immediately to the left of the radix or "decimal" point is the ones digit. The next digit to the left is the twos digit; after that comes the fours digit. Moving farther to the left, the digits represent 8, 16, 32, 64, and so on, doubling every time. To the right of the radix point, the value of each digit is cut in half again and again—that is, 1/2, 1/4, 1/8, 1/16, 1/32, 1/64, and so on.

Consider an example using the decimal number 94:

$$94 = (4 \times 10^0) + (9 \times 10^1)$$

In the binary number system the breakdown is

$$1011110 = (0 \times 2^0) + (1 \times 2^1) + (1 \times 2^2)$$
$$+ (1 \times 2^3) + (1 \times 2^4) + (0 \times 2^5) + (1 \times 2^6)$$

When you work with a computer or calculator, you give it a decimal number that is converted into binary form. The computer or calculator does its operations with zeros and ones, which are represented by different voltages or signals in electronic circuits. When the process is complete, the machine converts the result back into decimal form for display.

The difficulty about working directly in binary is that each place has only two "states," which are 0 and 1. You don't count "up to" something and then move to the next place. If you already have 1, the next 1 puts it back to 0 and passes a 1 to the next place. If you have a row of ones, then adding another 1 shifts them all back to 0 and passes a 1 to the next place (from right to left).

COMPARING THE VALUES

Table 15-2 compares numerical values in the decimal, binary, octal, and hexadecimal systems for decimal values 0 through 64. In general, as the number base increases, the numeral representing a given value becomes "smaller."

Table 15-2

Comparison of numerical values for decimal numbers 0 through 64.

DECIMAL	BINARY	OCTAL	HEXADECIMAL
0	0	0	0
1	1	1	1
2	10	2	2
3	11	3	3
4	100	4	4
5	101	5	5
6	110	6	6
7	111	7	7
8	1000	10	8
9	1001	11	9
10	1010	12	A
11	1011	13	B
12	1100	14	C
13	1101	15	D
14	1110	16	E
15	1111	17	F
16	10000	20	10
17	10001	21	11
18	10010	22	12
19	10011	23	13
20	10100	24	14
21	10101	25	15
22	10110	26	16
23	10111	27	17

DECIMAL	BINARY	OCTAL	HEXADECIMAL
24	11000	30	18
25	11001	31	19
26	11010	32	1A
27	11011	33	1B
28	11100	34	1C
29	11101	35	1D
30	11110	36	1E
31	11111	37	1F
32	100000	40	20
33	100001	41	21
34	100010	42	22
35	100011	43	23
36	100100	44	24
37	100101	45	25
38	100110	46	26
39	100111	47	27
40	101000	50	28
41	101001	51	29
42	101010	52	2A
43	101011	53	2B
44	101100	54	2C
45	101101	55	2D
46	101110	56	2E
47	101111	57	2F
48	110000	60	30
49	110001	61	31
50	110010	62	32
51	110011	63	33
52	110100	64	34
53	110101	65	35
54	110110	66	36
55	110111	67	37
56	111000	70	38
57	111001	71	39
58	111010	72	3A
59	111011	73	3B
60	111100	74	3C
61	111101	75	3D
62	111110	76	3E
63	111111	77	3F
64	1000000	100	40

Table 15-2

Comparison of numerical values for decimal numbers 0 through 64. (*Continued*)

Figure 15-5

At A, decimal-equivalent values for columns in a binary numeral. At B, conversion of decimal 1546 to binary form.

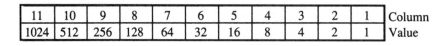

11	10	9	8	7	6	5	4	3	2	1	Column
1024	512	256	128	64	32	16	8	4	2	1	Value

A

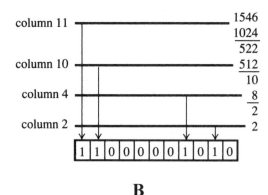

B

CONVERTING DECIMAL TO BINARY

Figure 15-5A shows the decimal values of the places in a binary numeral that have a 1 instead of a 0 for the first 11 columns, working from right to left. This comparison can be useful when you want to convert a decimal number to binary.

As an example, suppose you want to convert decimal 1546 to binary. First, the 11th column of binary is the equivalent of decimal 1024. That puts a 1 in the 11th column of binary. Subtract 1024 from 1546, leaving 522. Next, the 10th column in binary is 512, so subtract 512 from 522, leaving 10, and put a 1 in the 10th column of binary. With 10 left, the next binary digit that you can use is the fourth column, which is 8. So we pass over the columns from ninth to fifth, put a 1 in the fourth column, and subtract 8 from 10, leaving 2. The 2 puts a 1 in the second column of binary, finishing the conversion. You can follow this procedure visually by looking at the "flow" of Fig. 15-5B.

BINARY MULTIPLICATION

Although you enter data into your calculator or computer in the familiar decimal notation, the device uses binary to perform the actual operations.

As an exercise, try "playing calculator" to multiply 37 by 27. First, convert each number to binary. Figure 15-6 shows the conversions of decimal 37 and decimal 27 to binary. Figure 15-7 shows multiplication in binary. It works just as you would do it in ordinary long multiplication, but in a system where no numbers above 1 are allowed. Every digit must be either 1 or 0.

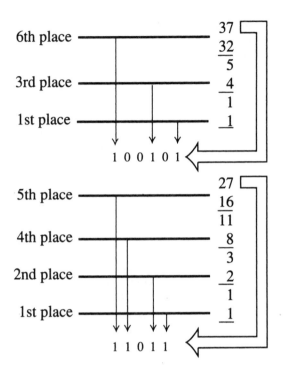

6th place ——————— 37
32
—
5
3rd place ——————— 4
—
1
1st place ——————— 1

1 0 0 1 0 1

5th place ——————— 27
16
—
11
4th place ——————— 8
—
3
2nd place ——————— 2
—
1
1st place ——————— 1

1 1 0 1 1

Place	Decimal Number
1	1
2	2
3	4
4	8
5	16
6	32
7	64
8	128
9	256
10	512

Figure 15-6

Conversion of decimal 37 and 27 to binary form in preparation for binary multiplication.

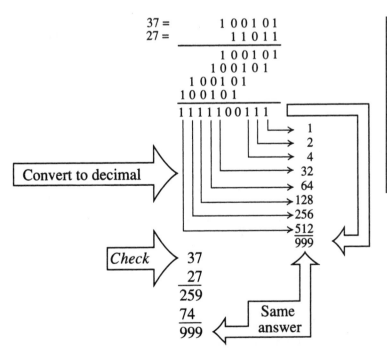

37 =   1 0 0 1 0 1
27 =   1 1 0 1 1
_____
         1 0 0 1 0 1
       1 0 0 1 0 1
     1 0 0 1 0 1
   1 0 0 1 0 1
   1 1 1 1 1 0 0 1 1 1

→ 1
→ 2
→ 4
→ 32
→ 64
→ 128
→ 256
→ 512
———
999

Convert to decimal

Check   37
         27
        ———
        259
         74
        ———
        999

Same answer

Figure 15-7

Multiplication of binary 100101 (the equivalent of decimal 37) by binary 11011 (the equivalent of decimal 27), and conversion back to decimal form for checking.

Working from the right, the first three places each have only one 1, which appears in the sum. The fourth place has two ones, which make a 0 in that place and pass 1 to the fifth place, which already has a 1 of its own; so it becomes 0 and passes a 1 to the sixth place. This place already has two ones, so that place goes to 1 again and passes a 1 to the seventh place, where there are again two ones. This place now has a 1, and it passes a 1 to the eighth place. The eighth place has no ones, so the 1 passed is entered and that's the end of the "passing left." The remaining two places each have a single 1, which gets "brought down." The product, in binary, is therefore 1111100111—a 10-place numeral! Follow this visually by looking at the "flow" in Fig. 15-7.

You can convert the binary number back to decimal by putting the decimal equivalent of each binary place where a 1 exists. Adding the decimal equivalents comes to 999. To check, multiply 37 by 27, the old-fashioned long way. Or if you're in a hurry, use your calculator!

The binary method of "long multiplication" seems longer to you than the decimal method, doesn't it? Technically, in terms of the actual number of steps required, it really is longer. But the calculator doesn't "know" the difference because it performs a huge number of operations every second. It goes the "long way" around and calculates more quickly than you can by means of the familiar "short way," just as a light beam can get to the moon faster than you can drive your car across town!

BINARY DIVISION

Here's another way to convert decimals to binary. Figure 15-8A shows binary equivalents for numbers from 1 to 9 in each decimal place in the base-10 system. To illustrate its use, two decimal numbers, 4773 and 37, are converted to binary by adding up thousands, hundreds, tens, and ones. You can follow the "flow" of this conversion process by examining Fig. 15-8B.

Figure 15-8C shows the "flow" for division of the binary equivalent of 4773 by the binary equivalent of 37. Subtracting the binary for 37, which is 100101, in the top places of the dividend is exact with no remainder. What is left is the binary for 37 in the last place. So the quotient in binary form is 10000001.

To convert the binary number back to decimal, use some more subtraction in binary and refer again to Fig. 15-8A. The first subtraction is the binary for 100, which leaves 11101. For the binary of 20, which leaves 1001, subtract the binary for 9. So working through binary, dividing 4773 by 37 leaves 129 as the quotient.

INDICES

In any system of numbers, the position of a particular digit indicates a power of the number on which the system is based. In the binary system, according to where the 1 appears, it represents some power of 2. In the fourth place, for example, it is the third

Decimal	Binary	Decimal	Binary	Decimal	Binary	Decimal	Binary
1000	1111101000	100	1100100	10	1010	1	1
2000	11111010000	200	11001000	20	10100	2	10
3000	101110111000	300	100101100	30	11110	3	11
4000	111110100000	400	110010000	40	101000	4	100
5000	1001110001000	500	111110100	50	110010	5	101
6000	1011101110000	600	1001011000	60	111100	6	110
7000	1101101011000	700	1010111100	70	1000110	7	111
8000	1111101000000	800	1100100000	80	1010000	8	1000
9000	10001100101000	900	1110000100	90	1011010	9	1001

A

4773:
4000	111110100000
700	1010111100
70	1000110
3	11
4773	1001010100101

B

37:
30	11110
7	111
37	100101

```
            10000001
100101 | 1001010100101
        100101
        _____
          100101
          100101
          _____
```

Check

```
            10000001
100 ——— 1100100
        11101
20  ——— 10100
        1001
9   ——— 1001
        1001
```

```
     129
37 | 4773
     37
     ___
     107
     74
     ___
     333
     333
```

    9
   ___
   129

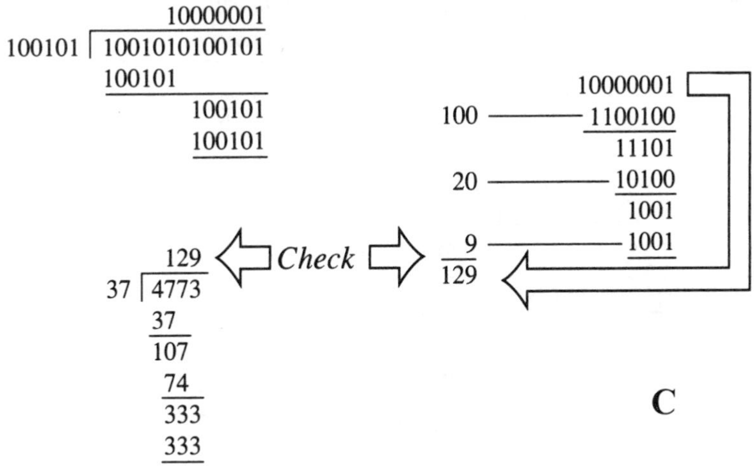

C

Figure 15-8

At A, an alternative binary conversion scheme. At B, conversion of decimal 4773 and 37 to binary form using this alternative method. At C, division of binary 1001010100101 (the equivalent of decimal 4773) by binary 100101 (the equivalent of decimal 37), and conversion back to decimal form for checking.

power of 2, which is 8. Figure 15-9A is a comparison between powers of 2 (for the binary system) and powers of 10 (for the decimal system).

From this list, you can infer some rules for using *indices* that help you take further shortcuts in multiplication and division. First, remember that multiplication and division

Figure 15-9

At A, powers of 2 and 10 for exponents ranging from 10 to −10. At B, examples of how exponents add for multiplication and subtract for division.

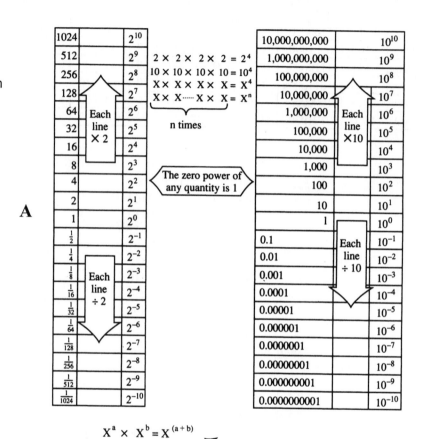

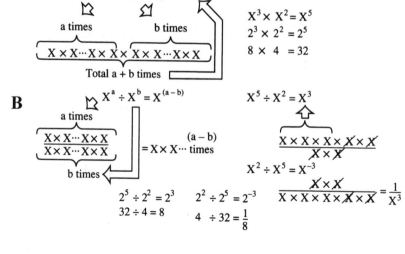

are shortcut methods for performing repeated addition and subtraction. Indices are shortcut methods for repeated multiplication and division.

Suppose you have to multiply X^a by X^b (X to the ath power times X to the bth power), where X is any number, but a and b are whole numbers that can be positive, negative, or zero. In this case, the product is equal to $X^{(a+b)}$. When you multiply powers, you add exponents. You can easily see this if you write X multiplied by itself a times, then multiply the product by X multiplied by itself b times. The total number of times that you multiply X by itself is $a + b$. An example is shown in the top part of Fig. 15-9B, where $a = 3$ and $b = 2$.

Now try division. Dividing X^a by X^b gives a quotient of $X^{(a-b)}$. You can check this answer by multiplying X by itself a times for the numerator of a fraction, and using X times itself b times for the denominator. You can cancel b times the number of X's in the numerator and leave a remainder of X's in the numerator that is $a - b$ times. To illustrate, make $a = 5$ and $b = 2$. Then X^5 divided by X^2 equals X^3. If you use 2 for X, then $X^5 = 32$, $X^2 = 4$, and $X^3 = 8$. In this case, 32 divided by 4 is equal to 8. This is shown in the bottom part of Fig. 15-9B.

NEGATIVE AND FRACTIONAL EXPONENTS

It's important to distinguish between the inverse of a number and the inverse of a power. A negative exponent denotes the multiplicative inverse or reciprocal of the number raised to the power, indicated by the index. For example:

$$2^{-3} = 1/2^3 = 1/8$$

$$4^{-2} = 1/4^2 = 1/16$$

$$(-5)^{-2} = 1/(-5)^2 = 1/25$$

$$-(5^{1/2}) = -1/5^2 = -1/25$$

Fractional exponents indicate positive roots. For example:

$$2^2 = 4, \text{ therefore } 4^{1/2} = 2$$

$$2^3 = 8, \text{ therefore } 8^{1/3} = 2$$

$$2^4 = 16, \text{ therefore } 16^{1/4} = 2$$

$$2^5 = 32, \text{ therefore } 32^{1/5} = 2$$

Fractional exponents can also represent powers and positive roots combined. For example:

$$4^{3/2} = (4^3)^{1/2} = 8$$

$$8^{2/3} = (8^2)^{1/3} = 4$$

$$2^{5/3} = (2^5)^{1/3} = 3.1748\ldots$$

$$5^{5/2} = (5^5)^{1/2} = 55.901\ldots$$

Note that when you take a power of a power, you multiply the exponents. When you take the root of a power or the power of a root, the resulting exponent has the power in its numerator and the root in its denominator.

SURDS

Surds are an old-fashioned way of writing roots, but you will still see them once in awhile. A surd looks something like a check mark ($\sqrt{}$). Before the fractional index notation introduced in the previous section came into vogue, it was customary to use a surd in front of a number to indicate its square root.

As an example, the symbol $\sqrt{4}$ represents the square root of 4. Putting a small, elevated numeral 3 in front of the surd ($\sqrt[3]{}$) indicates the cube root of the number after the surd. So $\sqrt[3]{27}$ means the cube root of 27. Putting a small n or any other letter or number in front of the surd ($\sqrt[n]{}$) represents the nth root or the $(1/n)$th power, where n can be any positive whole number.

If the number X after the surd has a power p with a q in front of the surd, the expression can be written as $X^{p/q}$ (the qth root of X to the pth power). For example:

$$\sqrt[2]{4^3} = 4^{3/2} = 8$$

$$\sqrt[3]{8^2} = 8^{2/3} = 4$$

$$\sqrt[3]{2^5} = 2^{5/3} = 3.1748\ldots$$

$$\sqrt[2]{5^5} = 5^{5/2} = 55.901\ldots$$

A surd followed by a *vinculum* (horizontal line) over the top of an expression indicates the square root of the whole expression. If there is an elevated numeral n before the surd, it indicates the nth root of the whole expression.

QUESTIONS AND PROBLEMS

This is an open-book quiz. You may refer to the text in this chapter (and earlier ones, too, if you want) when figuring out the answers. Take your time! Consider all values given as exact, so you don't have to worry about significant figures. The correct answers are in the back of the book.

The questions and problems on these pages assume a knowledge of earlier parts of this book. If you have difficulty with a problem, try others first and then return to the one that is difficult. These questions are designed so that you have to exercise some initiative in applying the principles that have been introduced up to this point.

1. Find the decimal equivalent of the fraction 1/37. Determine the error that occurs when this decimal equivalent is found to three significant digits.

2. Using the binary system, multiply 15 by 63 and convert back to decimal. Check your result by directly multiplying the decimal numbers.

3. Using the binary system, divide 1,922 by 31 and convert back to decimal. Check your result by directly dividing the decimal numbers.

4. Find the values of the following expressions.
    (a) $16^{3/4}$ (b) $243^{0.8}$ (c) $25^{1.5}$
    (d) $64^{2/3}$ (e) $343^{4/3}$

5. Convert the following numbers from decimal to binary. As a check, convert them back.
    (a) 62   (b) 81   (c) 111
    (d) 49   (e) 98   (f) 222
    (g) 650   (h) 999   (i) 2000

6. Convert the following numbers from binary to decimal. As a check, convert them back.
    (a) 101   (b) 1111   (c) 10101
    (d) 111100   (e) 110111000110

7. Multiply 129 by 31 in the decimal system. Multiply the binary equivalents of these numbers. Now, suppose that an error is made in the second digit from the right in the second number in the decimal product, so 129 is multiplied by 41 instead of by 31. Also suppose that a similar error occurs in the binary system, so the second digit from the right in the second number is reversed. Compare the relative error in the decimal system with the error in the binary system.

8. Evaluate the expression $(a^2 + b^2)^{1/2}$ for the following values.
    (a) $a = 4$ and $b = 3$
    (b) $a = 12$ and $b = 5$
    (c) $a = 24$ and $b = 7$
    (d) $a = 40$ and $b = 9$
    (e) $a = 60$ and $b = 11$
    (f) $a = 84$ and $b = 13$
    (g) $a = 112$ and $b = 15$

What does each pair have in common?

9. Evaluate the expression $(a^2 + b^2)^{1/2}$ for the following values.
    (a) $a = 8$ and $b = 6$
    (b) $a = 15$ and $b = 8$
    (c) $a = 24$ and $b = 10$
    (d) $a = 35$ and $b = 12$

(e) $a = 48$ and $b = 14$
(f) $a = 63$ and $b = 16$

What does each pair have in common?

10. Write as simple decimal numbers, without fractions, the following expressions.
(a) 100^2 (b) $100^{1/2}$
(c) 100^{-2} (d) $100^{-1/2}$

On the basis of the four results you derived for (a) through (d) above, find the values of the following expressions by the method of adding and subtracting indices.
(e) $100^{3/2}$ (f) $100^{5/2}$
(g) $100^{-3/2}$ (h) $100^{-5/2}$

11. Using a calculator's square root function button only, evaluate the following to at least three decimal places.
(a) $100^{1/4}$ (b) $100^{1/8}$
(c) $100^{1/16}$ (d) $100^{1/32}$

12. As the exponent in Prob. 11 is repeatedly cut in half, that is, 1/64, 1/128, 1/256, 1/512, and so on, what will the whole expression ultimately approach? Why?

13. Find the values, correct to three decimal places, of the following expressions.
(a) $32^{0.1}$ (b) $32^{0.2}$ (c) $32^{0.3}$
(d) $32^{0.4}$ (e) $32^{0.5}$ (f) $32^{0.6}$
(g) $32^{0.7}$ (h) $32^{0.8}$ (i) $32^{0.9}$

14. Evaluate the following expressions, using a calculator if you wish. Where applicable, render expressions to at least three decimal places.
(a) $(10^2 - 2^6)^{1/2}$ (b) $(36^2 - 8^3)^{1/2}$ (c) $(28^2 - 21^2)^{1/3}$
(d) $(52 - 32)^{1/4}$ (e) $(172 - 152)^{1/6}$ (f) $6{,}561^{1/2}$
(g) $6{,}561^{-1/2}$ (h) $6{,}561^{1/4}$ (i) $6{,}561^{-1/4}$
(j) $6{,}561^{1/8}$ (k) $6{,}561^{-1/8}$

16

Progressions, Permutations, and Combinations

Whatever system you use in arithmetic or algebra, a sequence of numbers can reveal a pattern that will help you to develop and check that sequence in various ways. Such a pattern is unlike the ones in multiplication tables. They depend on the system or notation that was used: decimal, octal, hexadecimal, binary, or whatever. Sequence patterns exist independently of the numbering base.

ARITHMETIC PROGRESSION

Five *arithmetic progressions* are listed below. Four are specific, and the fifth is general. First is the counting number sequence itself. Second is the even numbers. (Odd numbers would be similar, but starting with 1 and adding 2 for each successive number.) Third is the numbers divisible by 3. Fourth is the numbers that add 3 to each term, but the sequence begins with 1 instead of 3.

$$1, 2, 3, 4, 5, \ldots$$

$$2, 4, 6, 8, 10, \ldots$$

$$3, 6, 9, 12, 15, \ldots$$

$$1, 4, 7, 10, 13, \ldots$$

$$a, a + d, a + 2d, a + 3d, a + 4d, \ldots$$

The last sequence gives the general form of an arithmetic progression where a is the first term and each succeeding term is d more than the previous one. (The d stands for *difference between successive terms*.) If the numbers diminish instead of increase, d is a negative number.

Figure 16-1 portrays an arithmetic progression as a sequence of equally spaced vertical lines. Each line is longer than its neighbor to the left by the same amount. The lower endpoints of all the vertical lines lie on a single horizontal line. As a result, the upper endpoints lie on a slanted, but straight, line.

Figure 16-1

The terms in an arithmetic progression increase in a constant, linear manner.

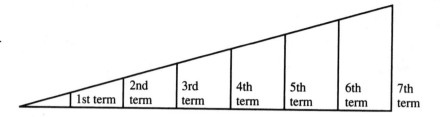

GEOMETRIC PROGRESSION

In an arithmetic progression, each term differs from the previous term by the same amount, added or subtracted. In a *geometric progression*, each term is multiplied by the same amount to get the next term. Just as the difference in arithmetic progression can either add or subtract, so the ratio of one term to the next in geometric progression can either expand or contract successive terms.

Here are some examples of geometric progressions. In the first case, each term is twice the previous one. In the second case, each term is 3 times the previous one. In the third, each term is 1.5 times the previous one. In the fourth, each term is one-half of the previous one. In the fifth case, each term is 2/3 of the previous one.

$$1, 2, 4, 8, 16, \ldots$$

$$1, 3, 9, 27, 81, \ldots$$

$$16, 24, 36, 54, 81, \ldots$$

$$1, 1/2, 1/4, 1/8, 1/16, \ldots$$

$$9, 6, 4, 2, 8/3, 16/9, \ldots$$

$$a, ar, ar^2, ar^3, ar^4, \ldots$$

The last sequence gives the general form for all geometric progressions where a is the first term and r is the *ratio of successive terms*, a constant by which each term is multiplied to get the next one.

Figure 16-2 shows the two types of geometric progression. At A is an expanding sequence. Read this diagram from left to right. At B is a contracting sequence—or, as mathematicians call it, a *converging sequence*. You should read this diagram from right to left. Notice that the expanding sequence quickly "runs away" (goes off scale), while the converging sequence gets smaller and smaller, indefinitely. The construction in each case uses similar right triangles to represent the changing ratio.

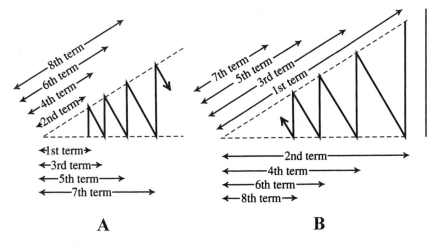

Figure 16-2
Pictorial renditions
of geometric pro-
gressions. At A,
an increasing
progression; at B,
a decreasing
progression.

HARMONIC PROGRESSION

A third kind of progression has been taught in schools. It is not used as often as arithmetic and geometric progressions. You can learn something about progression in general by studying these three types.

The easy way to understand the nature of a *harmonic progression* is to think of it as a sort of "inverse arithmetic progression." Instead of each term increasing by d (a constant difference), each term diminishes by being divided by an arithmetic series. If the first term is a, the second term is a divided by $1 + d$, the third term is divided by $1 + 2d$, and so on, like this:

$$a, a/(1 + d), a/(1 + 2d), a/(1 + 3d), a/(1 + 4d), \ldots$$

Three numerical examples of harmonic progressions are shown below. In the first sequence, $d = 1$. In the second sequence, $d = 1/2$. In the third sequence, $d = 1$ again, but the first term is 1 instead of 60.

$$60, 30, 20, 15, 12, \ldots$$

$$60, 40, 30, 24, 20, \ldots$$

$$1, 1/2, 1/3, 1/4, 1/5, \ldots$$

These patterns in numbers serve a variety of purposes that will be developed as you study them. For example, they form the basis for series that can define the trigonometric ratios for angles.

Now let's see what happens when we add the terms in a sequence. The sum of the terms in a sequence, either up to a certain point or for the entire sequence, is called a *series*. This distinction is important: a series is the sum of terms in a sequence.

ARITHMETIC SERIES

Often, you must sum a sequence by adding its terms. The long way would be to write down all the terms and add them. When there are a lot of terms in an arithemetic sequence, you have two ways of getting the sum.

If you know the first term a, the last term l, and the number of terms n, then the sum of the terms in an arithmetic sequence is n times the average term. Because the terms increase (or decrease) uniformly, the average term is midway between the first one and the last one. The midpoint can be found by adding together the first and last terms and then dividing this sum by 2. So the sum S of the whole sequence is

$$S = n(a + l) / 2$$

If you don't know the last term but you do know the first term a, the difference d, and the number of terms n, then the last term is $a + (n - 1)d$. So the sum of the first and last terms is $a + a + (n - 1)d$, or $2a + (n - 1)d$. Now you can find the sum S of the whole sequence using the average method derived just a moment ago and simplify, getting

$$S = n[2a + (n - 1)d] / 2$$
$$= na + n(n - 1)d / 2$$

The symbol universally used for the sum of a sequence is an oversized uppercase Greek letter sigma (Σ) with the lower and upper limits of the sum shown as small numbers or variables at the lower right and upper right, respectively. Figure 16-3 illustrates the

Figure 16-3

Principle of summing the terms in an arithmetic sequence.

First term: a
Second   "   a + d
Third   "   a + 2d
Fourth   "   a + 3d
Nth   "   a + (n − 1)d

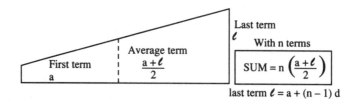

a, a + d, a + 2d, a + 3d,..., a + (n − 1)d

First term
a

Average term
$\frac{a + \ell}{2}$

Last term
ℓ
With n terms

$$\text{SUM} = n \left(\frac{a + \ell}{2} \right)$$

last term $\ell = a + (n - 1)\,d$

$\sum_{1}^{n}$ $\begin{pmatrix} \text{Greek capital} \\ \text{sigma} \end{pmatrix}$
means

Sum of terms 1 through n

$$\text{SUM} = n \left(\frac{2a + (n - 1)\,d}{2} \right)$$

$$\sum_{1}^{n} = na + \frac{n(n - 1)\,d}{2}$$

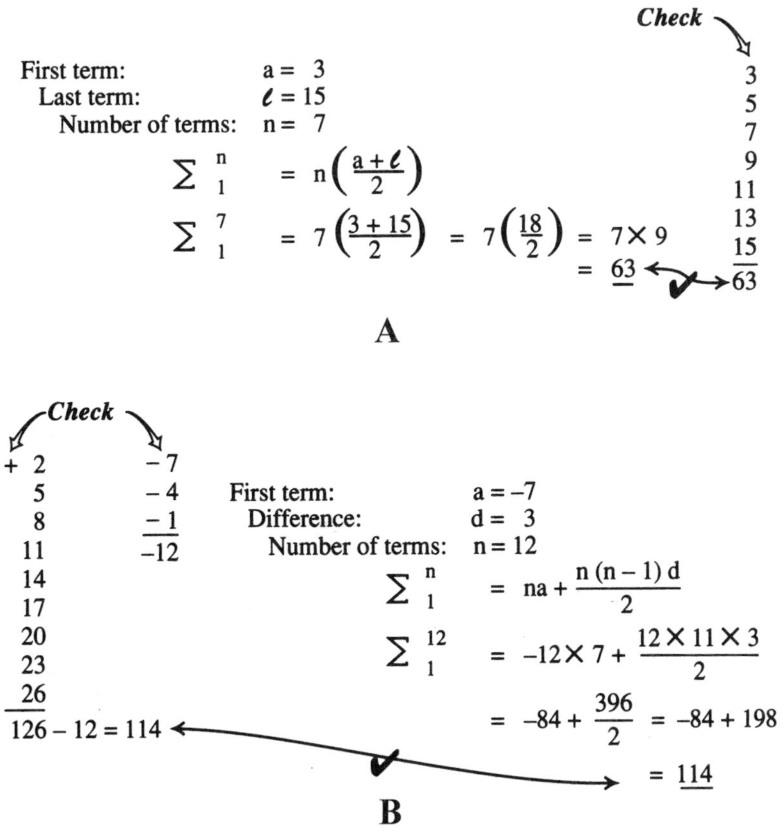

$$\begin{aligned}
&\text{First term:} &a &= 3\\
&\text{Last term:} &\ell &= 15\\
&\text{Number of terms:} &n &= 7
\end{aligned}$$

$$\sum_{1}^{n} = n\left(\frac{a+\ell}{2}\right)$$

$$\sum_{1}^{7} = 7\left(\frac{3+15}{2}\right) = 7\left(\frac{18}{2}\right) = 7\times 9$$
$$= 63$$

Check

3
5
7
9
11
13
15
63

A

Figure 16-4

Sums of arithmetic sequences using the formulas derived in the text. At A, the series 3 + 5 + 7 + 9 + 11 + 13 + 15. At B, the series (−7) + (−4) + (−1) + 2 + 5 + ⋯ to 12 terms.

Check

+ 2	− 7
5	− 4
8	− 1
11	−12
14	
17	
20	
23	
26	

126 − 12 = 114

$$\begin{aligned}
&\text{First term:} &a &= -7\\
&\text{Difference:} &d &= 3\\
&\text{Number of terms:} &n &= 12
\end{aligned}$$

$$\sum_{1}^{n} = na + \frac{n(n-1)d}{2}$$

$$\sum_{1}^{12} = -12\times 7 + \frac{12\times 11 \times 3}{2}$$

$$= -84 + \frac{396}{2} = -84 + 198$$

$$= 114$$

B

principles described in this section, using the summation sign Σ with lower and upper limits of the sums. Figure 16-4 shows two examples of how arithmetic sequences can be summed using the principles in this section.

GEOMETRIC SERIES

Adding the terms in a geometric sequence is not so simple, but the trick that you learn here can be helpful when you encounter series in the future.

Because a is a factor common to all the terms in a geometric sequence, it can be put outside parentheses with a list of all the multipliers inside. Now, if you multiply every term by ratio r, they all move one to the right, as shown in Fig. 16-5. In one line, you have r times the sum. However, it begins with r times a and ends with r^n times a. The original sum has, inside the brackets, terms that begin with 1 and finish with r^{n-1}.

Keep following along in Fig. 16-5. Assuming that r is greater than 1, subtract the original sum from r times the sum. All the middle terms disappear because they are all the same. Inside the brackets, then, only -1 and r^n remain. The sum on the other side of

Figure 16-5

Principle of summing the terms in a geometric sequence.

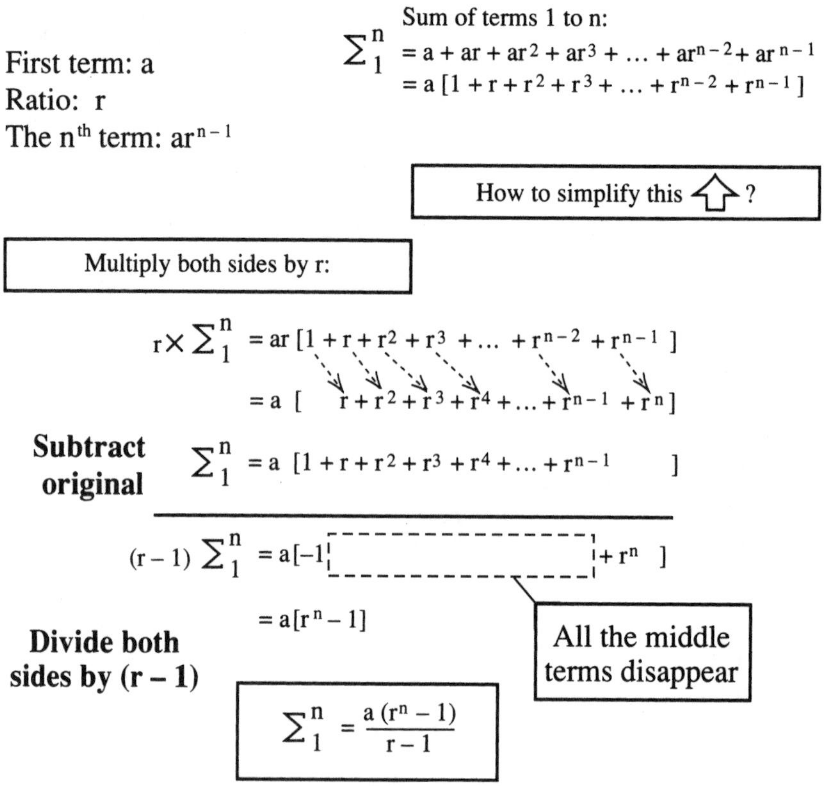

First term: a
Ratio: r
The nth term: ar^{n-1}

Sum of terms 1 to n:

$$\sum_1^n = a + ar + ar^2 + ar^3 + \dots + ar^{n-2} + ar^{n-1}$$
$$= a\,[1 + r + r^2 + r^3 + \dots + r^{n-2} + r^{n-1}]$$

How to simplify this ⇧ ?

Multiply both sides by r:

$$r \times \sum_1^n = ar\,[1 + r + r^2 + r^3 + \dots + r^{n-2} + r^{n-1}]$$
$$= a\,[\quad r + r^2 + r^3 + r^4 + \dots + r^{n-1} + r^n]$$

Subtract original

$$\sum_1^n = a\,[1 + r + r^2 + r^3 + r^4 + \dots + r^{n-1} \qquad]$$

$$(r-1)\sum_1^n = a[-1 \qquad\qquad + r^n\ ]$$

$$= a[r^n - 1]$$

All the middle terms disappear

Divide both sides by (r – 1)

$$\sum_1^n = \frac{a\,(r^n - 1)}{r - 1}$$

the equation is multiplied by $r - 1$. The sum S can now be found by dividing both sides by $r - 1$. You get

$$S = a(r^n - 1)/(r - 1)$$

Figure 16-6 illustrates how you find a couple of geometric series, one to the first 10 terms and the other to the first seven terms. In each case, you can verify the result by doing it the long way (that is, by adding all the terms up one by one).

You can regard the checking as a sort of reversible process. If you haven't used the formula for some time, you might not feel sure about it. Do it the long way, and then by using the formula. That will "prove" two things: first, that you didn't make a mistake in doing it the long way; second, that the formula really works!

CONVERGENCE

The geometric sequences you have considered so far have been expanding or *diverging* because each term is larger than the one before it. This occurs when r is greater than 1. In a *converging* geometric sequence, each term is smaller than the one before it.

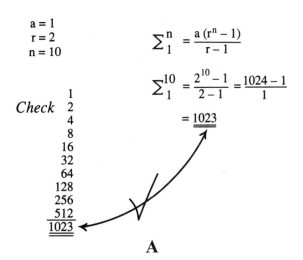

$$a = 1$$
$$r = 2$$
$$n = 10$$

$$\sum_1^n = \frac{a\,(r^n - 1)}{r - 1}$$

$$\sum_1^{10} = \frac{2^{10} - 1}{2 - 1} = \frac{1024 - 1}{1}$$

$$= \underline{1023}$$

$$\begin{array}{r} 1 \\ Check \quad 2 \\ 4 \\ 8 \\ 16 \\ 32 \\ 64 \\ 128 \\ 256 \\ \underline{512} \\ \underline{\underline{1023}} \end{array}$$

A

Figure 16-6

Sums of geometric sequences using the formulas derived in the text. At A, the series $1 + 2 + 4 + 8 + 16 + \cdots$ to 10 terms. At B, the series $4 + 12 + 36 + 108 + \cdots$ to seven terms.

$$a = 4$$
$$r = 4$$
$$n = 7$$

$$\begin{array}{r} 4 \\ Check \quad 12 \\ 36 \\ 108 \\ 324 \\ 972 \\ \underline{2916} \\ \underline{\underline{4372}} \end{array}$$

$$\sum_1^n = \frac{a\,(r^n - 1)}{r - 1}$$

$$\sum_1^7 = \frac{4 \times (3^7 - 1)}{3 - 1}$$

$$= \frac{4 \times (2187 - 1)}{3 - 1} = \frac{4 \times 2186}{2}$$

$$= 4372$$

$$3^3 = 27$$
$$3^4 = 81$$
$$ \overline{27}$$
$$ 216$$
$$3^7 = \overline{2187}$$

B

A story exists about an Eastern potentate who offered a philosopher a reward for some work. He offered him a chess board with grains of wheat on each of its 64 squares. He would put one grain on the first, two grains on the second, four on the third, eight on the fourth, and so on, until he got to the 64th square. It didn't sound like much, until he figured it. The grand total is $2^{64} - 1$, which is 18,446,744,073,709,551,615 grains. That was more wheat than he expected, or could ever hope to provide! That series not only diverges; it "explodes"!

When r is positive but less than 1, then r^n is also less than 1. If n is large, such as happens when you determine a series for a large number of terms, r^n approaches zero when r is between (but not including) 0 and 1. For this reason, the sum of a converging geometric sequence always approaches a defined value. An interesting thing happens if you sum "to infinity." You continue until the sequence disappears into nothingness. When

Figure 16-7

Principle of summing the terms in a converging infinite geometric sequence.

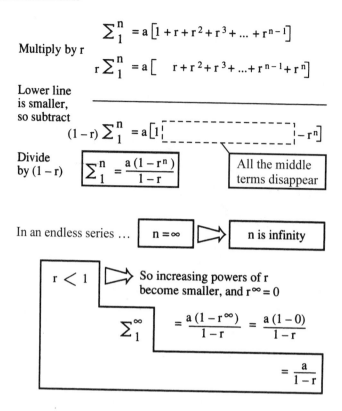

$$\sum_1^n = a\left[1 + r + r^2 + r^3 + \dots + r^{n-1}\right]$$

Multiply by r

$$r\sum_1^n = a\left[\quad r + r^2 + r^3 + \dots + r^{n-1} + r^n\right]$$

Lower line is smaller, so subtract

$$(1-r)\sum_1^n = a\left[1 \begin{array}{c} \boxed{} \end{array} - r^n\right]$$

Divide by $(1-r)$

$$\boxed{\sum_1^n = \dfrac{a(1-r^n)}{1-r}}$$

All the middle terms disappear

In an endless series … $\boxed{n = \infty}$ ▷ $\boxed{n \text{ is infinity}}$

$r < 1$ ▷ So increasing powers of r become smaller, and $r^\infty = 0$

$$\sum_1^\infty = \frac{a(1-r^\infty)}{1-r} = \frac{a(1-0)}{1-r}$$

$$= \frac{a}{1-r}$$

r is larger than 0 but less than 1, you can imagine that r^n becomes equal to 0 when n is "equal to infinity." So, the formula for the sum S becomes very simple:

$$S = a(0 - 1)/(r - 1)$$

$$= -a/(r - 1)$$

$$= a/(1 - r)$$

Figure 16-7 shows how this method works when we treat "infinity" as a number. In that case, when r is less than 1 but larger than 0, we get "r to the infinitieth power equals 0."

RATE OF CONVERGENCE

Here is an example that can help you see how the use of different series for the same calculation can make the work easier. The sum of three different converging sequences produces a value of 4 "at infinity." But the series converge toward 4 at different rates.

The first sequence, shown in Fig. 16-8A, is 1, 3/4, 9/16, 27/64, ... Notice that the sum to six terms is 3-295/1024. That is still quite far from 4. The second sequence

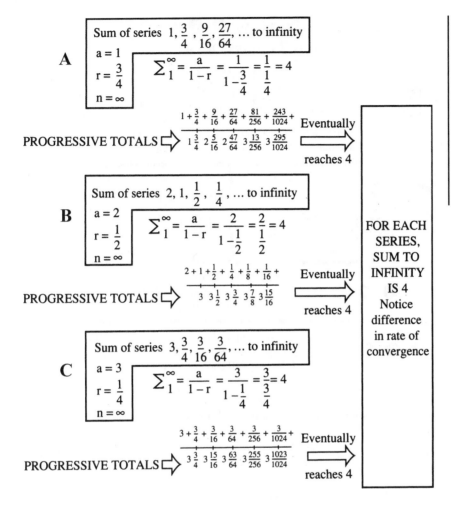

Figure 16-8

Converging infinite geometric series using the formula derived in the text. At A, the series converges rather slowly. At B, the series converges faster. At C, the series converges faster still.

(Fig. 16-8B) is 2, 1, 1/2, 1/4, ... Notice that the sum to six terms is 3-15/16. That is only 1/16 short of the ultimate value of 4. The third sequence (Fig. 16-8C) is 3, 3/4, 3/16, 3/64, ... Notice that the sum to six terms is now 3-1023/1024, which is only 1/1024 away from its ultimate value. Each of these series has a successively greater rate of convergence than the one before it.

PERMUTATIONS

Imagine that seven horses are entered in a race. Any one of the seven can come in first. For each winner, six are left that can come in second. The possibilities for first and second place are 7 × 6, or 42. For each possible pair of horses coming in first and second, five choices are left for third place. This makes 7 × 6 × 5, or 210 possibilities for the first three places.

Figure 16-9

Seven articles can be arranged in 5,040 different orders or permutations. When six places are filled, only one choice exists for the seventh place; there's no need to multiply by it.

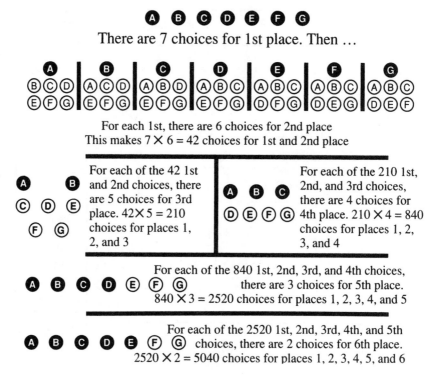

A B C D E F G

There are 7 choices for 1st place. Then …

For each 1st, there are 6 choices for 2nd place
This makes 7 × 6 = 42 choices for 1st and 2nd place

For each of the 42 1st and 2nd choices, there are 5 choices for 3rd place. 42 × 5 = 210 choices for places 1, 2, and 3

For each of the 210 1st, 2nd, and 3rd choices, there are 4 choices for 4th place. 210 × 4 = 840 choices for places 1, 2, 3, and 4

For each of the 840 1st, 2nd, 3rd, and 4th choices, there are 3 choices for 5th place. 840 × 3 = 2520 choices for places 1, 2, 3, 4, and 5

For each of the 2520 1st, 2nd, 3rd, 4th, and 5th choices, there are 2 choices for 6th place. 2520 × 2 = 5040 choices for places 1, 2, 3, 4, 5, and 6

To find the order of the rest, the possibilities are $7 \times 6 \times 5 \times 4 \times 3 \times 2 \times 1$. The factor of 1 at the end is superfluous, of course; it represents the only horse left! The total product here tells us that the race can end up in any one of 5,040 different ways, as shown in Fig. 16-9. These different ways are called *permutations*.

Now let's consider a more general problem. In how many ways can a certain number of items be taken from a set? The usual symbols for expressing this are k articles chosen from an available n. Examining the figures used so far, you could write

$$n(n - 1)(n - 2)(n - 3) \cdots (n - n + 1)$$

Mathematicians use a shortened way of writing these equations known as *factorial notation*. The accepted symbol is an exclamation point after the number. Factorial notation is defined as the product of every positive whole number from 1 up to the number before the exclamation point. Therefore,

$$n! = n(n - 1)(n - 2)(n - 3) \cdots (2)(1)$$

Again, the factor of 1 is superfluous, because when you multiply any number by 1, you just get that same number again.

Note that $1! = 1$; that is trivial. For completeness, and to eliminate confusion in certain calculations, the value of $0!$ is defined as being equal to 1. Factorial values are defined only for the *natural numbers*, that is, the nonnegative whole numbers. So there's no such thing as 2.84! or $(-7)!$ or $(2/3)!$ or $(31/2)!$ But don't get fooled here. There can, in fact, be such things as $-(17!)$ or $-(258!)$ or $6!/2$.

Most scientific calculators provide an $x!$ or $n!$ button that will display the factorial expansion of whatever number was displayed just before that button was pressed. It won't work if you enter anything other than a natural number, or if you enter a number whose factorial expansion is too big for the calculator to handle.

The number of possible permutations of k articles taken from an available n is equal to the product of all the numbers from n down to $n - k + 1$. That seems to be the simplest form. But if you divide $n!$ by $(n - k)!$, you can cancel the shorter string of numbers from the longer, and then you are left with a product of all the numbers from n down to $(n - k + 1)$. You get

$$n(n - 1)(n - 2)(n - 3) \cdots (n - k + 1) = n! / (n - k)!$$

The symbol for permutations is an uppercase letter P with two subscripts, one in front of the P and one after it. The one before it says from how many the choice is made. The one after it says how many are chosen. So the formula for permutations of k taken from n looks like this:

$$_nP_k = n! / (n - k)!$$

COMBINATIONS

Permutations consider the order in which the articles are chosen. In a *combination*, the order is not important.

Suppose you are asked to pick three letters from the first 10 letters of the alphabet, A though J. You have 10 choices for your first pick, nine choices for second, and eight for third. That multiplies out to $10 \times 9 \times 8$, or 720 possible choices. Suppose your choice picks the first three letters, A, B, and C. These could have been picked in any order. Six different orders exist for which you could pick the first three letters.

Permutations take order into account. Combinations deal only with which items are picked, in any order. Pursuing the 10-letter choice from which you picked three letters, 720 permutations exist (taking order into account). However, if you only want to know which three letters are picked and order does not matter, you must divide 720 by 6. That gives you 120 different combinations.

The symbol for combinations is a capital letter C, written in the same style as the P for permutations. The formula for combinations of k taken from n is

$$_nC_k = n! / [k!(n - k)!]$$

Figure 16-10

The number of combinations of three articles chosen from 10 is the same as the number of combinations of 10 − 3, or seven, articles chosen from the same 10.

PERMUTATIONS of 3 taken from 10 includes

$$\left.\begin{array}{l} \text{ABC} \\ \text{ACB} \\ \text{BAC} \\ \text{BCA} \\ \text{CAB} \\ \text{CBA} \end{array}\right\} \begin{array}{l} \text{6 different orders of} \\ \text{each combination} \end{array}$$

How many COMBINATIONS of 3 taken from 10?

$$\frac{10 \times 9 \times 8}{3 \times 2 \times 1} = \frac{720}{6} = 120$$

How many COMBINATIONS of 7 taken from 10?

Each 3 taken leaves 7, so it should be the same answer

$$\frac{10 \times 9 \times 8 \times 7 \times 6 \times 5 \times 4}{7 \times 6 \times 5 \times 4 \times 3 \times 2 \times 1} = \frac{10 \times 9 \times 8}{3 \times 2 \times 1} = 120$$

Notice that the same formula also gives the number of combinations of $(k - n)$ items taken from n. This tells you that the number of ways a certain number k can be taken from a bigger number n is equal to the combinations of those left in the same case. An example is shown in Fig. 16-10. Here, you can see that

$$_{10}C_3 = {_{10}C_{(10 - 3)}}$$

POWERS OF A BINOMIAL

A *binomial* is any expression that consists of two terms with either a plus or a minus sign between them. For a general form of binomial, write $a + b$. (The value of b can be negative so you can stick with the plus sign in the general notation.) When such an expression is raised to successive powers (squared, cubed, fourth, and so on), it generates a successively more complicated series of terms, each of which consists of a power of a, a power of b, or a product of powers of a and b.

You can multiply successive expressions of the general binomial $a + b$ to get the pattern of terms that forms. In any of the powers of the binomial, start with a raised to that power, followed by terms that consist of successively lower powers of a multiplied by successively higher powers of b, until you get to the last term, which is b raised to the highest power. Each of the product terms, with powers of both a and b, has a numerical *coefficient*. That is a constant by which the product of powers of a and b is multiplied.

$$a + b \longleftarrow \text{is a binomial}$$

multiply by .... $a + b$
$$\overline{a^2 + 2ab + b^2} \qquad \text{is } (a + b)^2$$

multiply by .... $a + b$ again
$$\overline{a^3 + 3a^2b + 3ab^2 + b^3} \qquad \text{is } (a + b)^3$$
$$a + b$$
$$\overline{a^4 + 4a^3b + 6a^2b^2 + 4ab^3 + b^4}$$

A

Figure 16-11

At A, the binomial expansions for $(a + b)^2$, $(a + b)^3$, and $(a + b)^4$. At B, the arrangement of coefficients for binomial expansions up to $(a + b)^{10}$, illustrated in the form of a pyramid.

Power Index	Coefficients
1	1 1
2	1 2 1
3	1 3 3 1
4	1 4 6 4 1
5	1 5 10 10 5 1
6	1 6 15 20 15 6 1
7	1 7 21 35 35 21 7 1
8	1 8 28 56 70 56 28 8 1
9	1 9 36 84 126 126 84 36 9 1
10	1 10 45 120 210 252 210 120 45 10 1

B

Figure 16-11A shows the binomial expansions for $(a + b)$ up to the fourth power. Figure 16-11B shows the pattern of coefficients up to the 10th power by forming a pyramid.

Now write out the expansion: $(a + b)^n$. It is $(a + b)(a + b)(a + b). . ., n$ times. Next, multiply all those terms together. The first term of each is a^n. The next one is almost as easy: take $(n - 1)$ a's and one b. You have n of those. Now, it begins to get involved: the combinations in which you take $(n - 2)$ a's and two b's.

After going through this step, write the general term in the whole expansion. It has three parts: the coefficient, which is the combination of k [you are at the $(k + 1)$st term] a's taken with $(n - k)$ b's; then a^{n-k}, and finally b^k, all multiplied together.

ALTERNATIVE NOTATION FOR MULTIPLYING NUMBERS

Until now, to indicate multiplication of numbers, the "times" sign or tilted cross ($\times$) has been used. You can also use a raised dot, like an elevated period. To avoid making it look like a decimal point, keep the dot above the base line ($\cdot$). This notation saves space when you have a lot to write. You'll see this notation often from now on. Figure 16-12 shows two *binomial expansions*, $(a + b)^7$ and $(10 + 3)^5$, using this notation.

Figure 16-12

Powers of binomials can be expanded into series. Here, we see how to derive the binomial expansions of $(a + b)^7$, shown at A, and $(10 + 3)^5$, shown at B. Note the elevated dots to represent multiplication of numbers.

$$(a + b)^n = a^n + na^{n-1}b + {}_nC_2 \, a^{n-2}b^2 + \ldots + {}_nC_k \, a^{n-k}b^k + \ldots$$

$$= a^n + na^{n-1}b + \frac{n(n-1)}{2} a^{n-2}b^2 + \ldots + \frac{n(n-1)\cdots(n-k+1)}{k!} a^{n-k}b^k + \ldots$$

$$(a + b)^7 = a^7 + 7a^6b + \frac{7 \cdot 6}{2} a^5b^2 + \frac{7 \cdot 6 \cdot 5}{3 \cdot 2} a^4b^3 + \frac{7 \cdot 6 \cdot 5 \cdot 4}{4 \cdot 3 \cdot 2} a^3b^4 + \ldots$$

These two the same; from here on repeat ⟶

$$= a^7 + 7a^6b + 21a^5b^2 + 35a^4b^3 + 35a^3b^4 + 21a^2b^5 + 7ab^6 + b^7$$

A

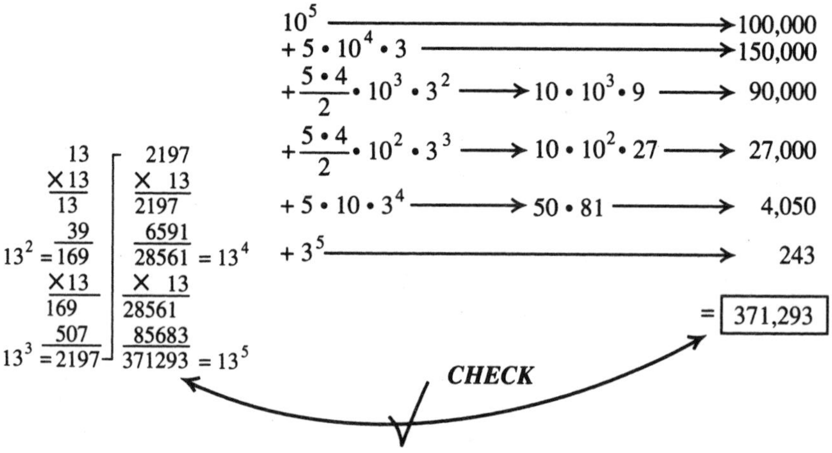

$$
\begin{array}{ll}
13 & 2197 \\
\times 13 & \times \ 13 \\
\hline
13 & 2197 \\
39 & 6591 \\
\hline
13^2 = 169 & 28561 = 13^4 \\
\times 13 & \times \ 13 \\
\hline
169 & 28561 \\
507 & 85683 \\
\hline
13^3 = 2197 & 371293 = 13^5
\end{array}
$$

$$10^5 \longrightarrow 100{,}000$$
$$+ \, 5 \cdot 10^4 \cdot 3 \longrightarrow 150{,}000$$
$$+ \frac{5 \cdot 4}{2} \cdot 10^3 \cdot 3^2 \longrightarrow 10 \cdot 10^3 \cdot 9 \longrightarrow 90{,}000$$
$$+ \frac{5 \cdot 4}{2} \cdot 10^2 \cdot 3^3 \longrightarrow 10 \cdot 10^2 \cdot 27 \longrightarrow 27{,}000$$
$$+ \, 5 \cdot 10 \cdot 3^4 \longrightarrow 50 \cdot 81 \longrightarrow 4{,}050$$
$$+ \, 3^5 \longrightarrow 243$$
$$= \boxed{371{,}293}$$

CHECK

B

QUESTIONS AND PROBLEMS

This is an open-book quiz. You may refer to the text in this chapter (and earlier ones, too, if you want) when figuring out the answers. Take your time! Consider all values given as exact, so you don't have to worry about significant figures. The correct answers are in the back of the book.

1. Identify the following progressions according to type: arithmetic, geometric, or harmonic. State the first term and the value of d or r.
   (a) 1, 5, 9, 13, 17
   (b) −9, −3, 3, 9, 15
   (c) 16, 12, 9, 6-3/4, 5-1/16
   (d) −81, 54, −36, 24, −16
   (e) 1512, 1260, 1080, 945, 840, 756

(f) 1, 1/2, 1/4, 1/8, 1/16, . . .

(g) 1, 2, 4, 8, 16, . . .

2. Sum the following series.

(a) $1 + 3 + 5 + 7 + 9 + \cdots$ (through the 20th term)

(b) $23 + 25 + 27 + \cdots + 69 + 71 + 73$

(c) $5 + 10 + 20 + 40 + 80 + \cdots$ (through the 11th term)

(d) $5 - 10 + 20 - 40 + 80 - \cdots$ (through the 10th term)

(e) $100 + 50 + 25 + 12.5 + \cdots$ (through the sixth term)

(f) $100 - 50 + 25 - 12.5 + \cdots$ (through the sixth term)

(g) $100 - 50 + 25 - 12.5 + \cdots$ (through the seventh term)

(h) $6 + 17 + 28 + 39 + 50 + \cdots$ (through the 19th term)

3. For each of the following sequences, determine whether or not it converges. If a given sequence converges, find the sum of the entire infinite series.

(a) $1 + 1/2 + 1/4 + 1/8 + 1/16 + \cdots$

(b) $1 + 1/2 + 1/3 + 1/4 + 1/5 + \cdots$

(c) $900 + 90 + 9 + 0.9 + 0.09 + \cdots$

(d) $10 - 6 + 3.6 - 2.16 + 1.296 - \cdots$

(e) $700 + 210 + 63 + 18.9 + 5.67 + \cdots$

(f) $10 + 9 + 8.1 + 7.29 + 6.561 + \cdots$

(g) $180 - 144 + 115.2 - 92.16 + \cdots$

(h) $256 + 128 + 64 + 32 + 16 + \cdots$

4. Evaluate the following permutations.

(a) $_{50}P_3$ (b) $_{10}P_5$ (c) $_{12}P_6$

(d) $_{10}P_4$ (e) $_7P_6$

5. Evaluate the following combinations.

(a) $_{50}C_3$ (b) $_{10}C_5$ (c) $_{12}C_6$

(d) $_{10}C_4$ (e) $_7C_6$

6. A new telephone area code is created. This opens up a new block of seven-digit telephone numbers. The only numbers that cannot be assigned are those whose first digit is 0. How many different seven-digit telephone numbers can be assigned in the new area code?

7. Suppose that an additional restriction is placed on the telephone numbers in the previous problem: The first digit cannot be 0, 1, or 9. How many different seven-digit telephone numbers can be assigned in this system?

8. Twelve horses run in a race. If all the horses have equal ability:

(a) What are the odds that you can name the correct winner?

(b) What are the odds that you can name the horses that come in first and second, but not necessarily in order?

(c) What are the odds that you can name the horses that come in first and second in the correct order?

9. In the same race as the one in the previous problem, what are the odds that a particular horse will come in among the first three?

10. Assume the following:
- There are 200 billion (2×10^{11}) stars in our galaxy besides our sun.
- To support life as we know it, a star must resemble our sun and have at least one earthlike planet.
- The chance of an earthlike planet evolving life as we know it is 1 in 5.
- Exactly 6 percent of the stars in our galaxy resemble our sun.
- Of these, exactly 3 percent have at least one earthlike planet.

Given these premises, how many planets in our galaxy can be expected to evolve life as we know it?

11. Imagine that you have read the following statistics:
- Exactly 1 out of every 5 males over the age of 18 has had a speeding ticket within the past 12 months.
- Exactly 1 out of every 10 females over the age of 18 has had a speeding ticket within the past 12 months.
- Men and women frequent singles bars in equal numbers.

You choose a person at random in a singles bar where, presumably, everyone is over 18 years of age. What is the probability that you will choose a person who has had a speeding ticket within the past 12 months?

12. Suppose you choose one man and one woman at random, in the same bar as in the previous problem. What is the probability that you will choose a man and a woman who have both had speeding tickets in the past 12 months?

13. A tossed coin is equally likely to land heads or tails. For two tosses, the probability of the coin landing heads both times is 1 in 2^2, or 1/4. For three tosses, the probability of the coin landing heads in every case is 1 in 2^3, or 1/8. For four tosses, the probability of the coin landing heads in every case is 1 in 2^4, or 1/16. In general, if the coin is tossed n times, the chance of it landing heads every time is $1/2^n$. Based on these facts, suppose you toss a coin 14 times and it comes up heads every single time. What is the probability that it will come up heads on the 15th toss?

17
Introduction to Derivatives

The relationship between a quantity and its rate of change can provide shortcuts in calculating. These rates, called *derivatives*, are sometimes the only way to find an accurate result. In this chapter, you'll learn how rates of change can be determined by a technique called *differentiation*.

RATES OF CHANGE

To start understanding the basic principle behind differentiation, consider a car traveling along a highway. The rate at which it moves along the highway is its *speed*. If the direction is also specified, or if the car moves in a straight line, speed can be called *velocity*. In contrast, *acceleration* is the rate at which velocity changes. When the speed increases, the acceleration is positive; when the speed decreases, the acceleration is negative (this is also called *deceleration*).

Before radar speed guns were invented, police timed the movement of a car between two points. If the time was less than what traveling at the legal speed would require, the driver got a ticket. A smart driver would see the first cop and slow down to a speed well below the limit before she got to the second cop. The time check would show that she wasn't speeding, even though she was when she saw the first cop. Radar speed guns stopped this practice by reading speed at an instant, instead of averaging it over a distance.

The weight-and-spring system, described earlier in this book, shows how velocity and acceleration change during its movement. How do you check those facts? *Differential calculus* (sometimes called *infinitesimal calculus*), which is straightforward despite its imposing name, helps you study these types of problems.

INFINITESIMAL CHANGES

Any relationship can be plotted as a graph. Any graph can have an algebraic equation to express the function that is plotted. Some equations are simple, and others are complicated. Those that relate to the real world have two types of variables: *independent* and *dependent*. In an equation such as $y = x^n$, x is the independent variable on which the value of y depends. That means x is the independent variable, and y is the dependent variable.

A tiny change in y, divided by the tiny change in x that causes it, gives you the *slope* of a graph at a point. Use the symbols dx and dy to represent *infinitesimals* (vanishingly small changes) of x and y. Compared to x and y, these infinitesimals are too small to be measurable. You can make them as minuscule as you want, either positively or negatively, as long as you don't make them equal to zero. This makes interesting things happen! When the technique of infinitesimals was first worked out by Isaac Newton and Gottfried Leibniz in the 17th century, the results, which we now call calculus, seemed like mathematical magic!

Figure 17-1 shows the general solution for the slope of a graph for $y = x^n$. The binomial expansion is used on $(x + dx)^n$. Then $y = x^n$ is subtracted from the expression for $y + dy = (x + dx)^n$, leaving you with an equation for dy in terms of dx. The smaller you make a change, the smaller the higher powers become. As dx/x approaches zero, so does the value of $(dx)^2/x^2$, but faster! Using your imagination, you can make these infinitesimals so small that you can treat them as zeros, even though they really aren't. They converge toward zero, but never quite get all the way there.

Now apply these principles to a few cases and see how they work. Start with the simplest possible equation, $y = x$, as shown in Fig. 17-2. Dividing dy by dx, after subtracting $y = x$, we have $dy/dx = 1$. This statement is easily seen to be correct, because for every change of 1 in x, the value of y also changes by 1.

Now take $y = x^2$, as shown in Fig. 17-3. From the formula already derived for the derivative of a variable raised to a power, $dy/dx = 2x$. Plotting the curve for $y = x^2$, for integer values of x from -5 to 5, draw a smooth curve through the points. If your curve represents the true curve for $y = x^2$, you can draw right triangles under the curve with a base

Figure 17-1

Infinitesimal changes and the principle of binomial expansion can be used to find the derivative dy/dx of the function $y = x^n$.

Infinitesimal increase in x: $x + dx$

Corresponding increase in y: $y + dy$

$$y = x^n$$

$$y + dy = (x + dx)^n$$

$$(x + dx)^n = x^n + nx^{n-1}\,dx + \frac{n(n-1)}{2}x^{n-2}(dx)^2 + \ldots$$

Because $\dfrac{dx}{x} = 0$ $\dfrac{(dx)^2}{x^2} = 0^2$ – doubly infinitesimal infinite convergence

$$(x + dx)^n = x^n + nx^{n-1}\,dx$$
$$y + dy \; = x^n + nx^{n-1}\,dx$$
$$y \qquad\; = x^n$$

$$\rule{6cm}{0.4pt}$$

$$dy \; = \; nx^{n-1}\,dx \quad \text{or} \quad \boxed{\dfrac{dy}{dx} = nx^{n-1}}$$

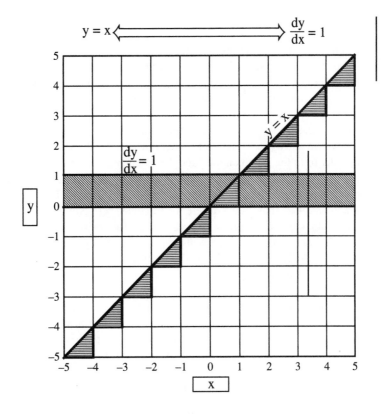

Figure 17-2

The derivative of the function $y = x$ is $dy/dx = 1$.

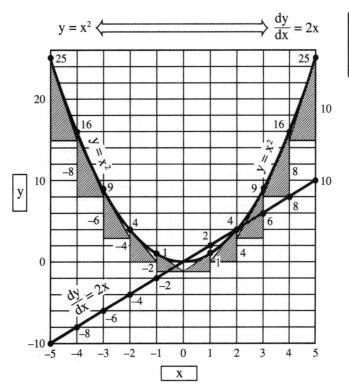

Figure 17-3

The derivative of the function $y = x^2$ is $dy/dx = 2x$.

length of 1, and a slanting hypotenuse whose top just touches the curve at the same slope as the curve. In each case, the vertical height of the triangle is $2x$ for the particular point. Values of dy/dx can be plotted as another graph (this time a straight line), which shows the slope of the first curve at every point.

Now try the graph of $y = x^3$, as shown in Fig. 17-4. From the general formula, you can easily find that $dy/dx = 3x^2$. For higher powers of x, the scale for y has to be compressed in order to fit the graph on the page. Draw tangent triangles for the unit horizontal base, just as you did in the previous graph, and measure the height to verify the formula. Then plot a graph for dy/dx. You'll get a parabola for this, with a shape similar to that of the graph of $y = x^2$ (but with a different vertical aspect). Notice that the cubic curve, unlike the square curve or parabola, curves upward for positive values of x and downward for negative values of x. Its slope is always upward, except when $x = 0$, where it is momentarily horizontal, the zero slope. The slope is equally positive for the same values of x, positive and negative.

Take one more power of x for the time being: $y = x^4$, as shown in Fig. 17-5. Compressing the y scale again to get the graph on the page, plot from $x = -5$ to $x = 5$. This curve is similar to that for $y = x^2$, in that values of y are positive for both positive and negative values of x. But the curvature is different. Correspondingly, the curve for dy/dx

Figure 17-4

The derivative of the function $y = x^3$ is $dy/dx = 3x^2$.

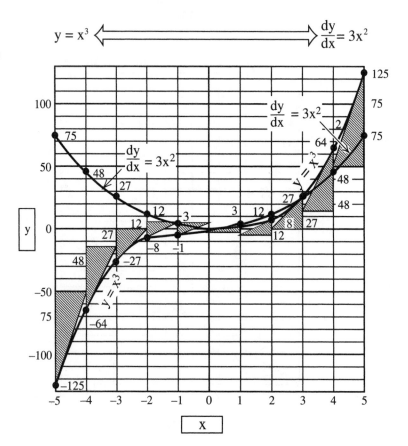

$$y = x^3 \xleftrightarrow{\hspace{4cm}} \frac{dy}{dx} = 3x^2$$

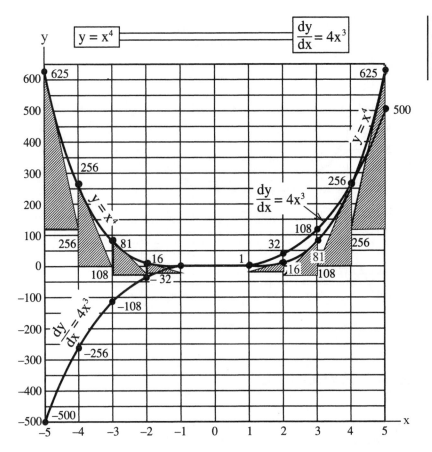

is negative when x is negative, because the $y = x^4$ slope is downward (from left to right). Once again, the triangle constructions verify the formula result.

SUCCESSIVE DIFFERENTIATION

In the preceding examples, the slopes of the curves were found by a single differentiation of y with respect to x, denoted dy/dx. Some problems require *successive differentiation* (as mathematicians call it). For example, velocity or speed is a rate of change of position, also called displacement or distance. Acceleration is a rate of change of velocity or speed. You can even talk about the rate of change of acceleration, sometimes called *jerk*. Velocity is the *first derivative* of position, acceleration is the *second derivative* of position, and jerk is the *third derivative* of position.

Starting with $y = x^4$, you can repeatedly differentiate to find the *successive derivatives*. Previously the simplest equation, $y = x^n$, was used. As things work out, a constant will transfer to the derivative. The work in Fig. 17-6 is an example of this. The general statement is

If $y = ax^n$, then $dy/dx = anx^{n-1}$

Figure 17-6

The first, second, third, and fourth derivatives of $y = x^4$.

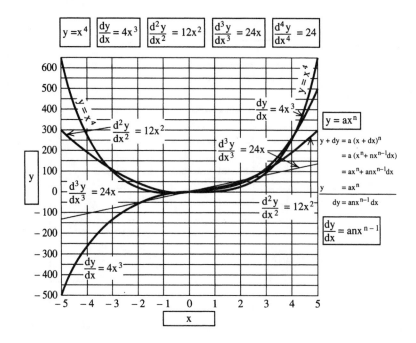

DIFFERENTIATING A POLYNOMIAL

Suppose an equation for y in terms of x takes the form of a *polynomial* with several terms that involve different powers of x, all added up. In general form, this might be

$$y = ax^m + bx^n + \cdots$$

It could continue to any number of terms, although only two are taken here.

Increasing x by dx, substitute $x + dx$ for x each time it occurs in the original equation. Similarly, increase y by dy to make $y + dy$. Using the same method as that shown in Fig. 17-1, you get

$$y + dy = a(x + dx)^m + b(x + dx)^n + \cdots$$
$$= ax^m + amx^{m-1}\, dx + bx^n + bnx^{n-1}\, dx + \cdots$$

If you subtract the parts that correspond to y, the parts that correspond with dy are left:

$$dy = amx^{m-1}\, dx + bnx^{n-1}\, dx + \cdots$$

Dividing through by dx gives an expression showing that dy/dx is equal to the sum of the derivatives of the individual terms of the polynomial:

$$dy/dx = amx^{m-1} + bnx^{n-1} + \cdots$$

This rule holds true for all expressions made up of terms added together. An easy way to remember it is this: "The derivative of a sum is equal to the sum of the derivatives." Here is an example:

$$y = x^5 - 50x^3 + 520x$$

The derivative is

$$dy/dx = 5x^4 - 150x^2 + 520$$

Notice that where the original term has a minus sign (as does $50x^3$), the term in the derivative also has a minus sign, as long as the original power is positive.

Now work through this with tabulated values of y and dy/dx, term by term, for integer values of x from $x = -6$ to $x = 6$. Study the curves as they are shown in Fig. 17-7. Where the curve for y reaches a maximum or minimum, the curve for dy/dx passes

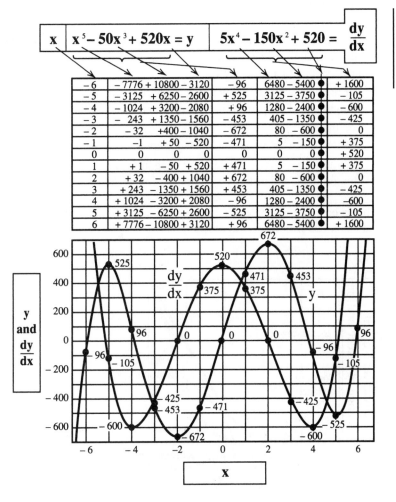

Figure 17-7

Table and graphs for $y = x^5 - 50x^3 + 520x$ and its derivative.

through a zero value. Momentarily, y is neither increasing nor decreasing. Where the curve for y crosses the zero line (except at the ends where both curves go off scale), the dy/dx curve is at a maximum positive or negative. At those points, the slope of y reaches a maximum, either up or down.

SUCCESSIVE DIFFERENTIATION OF MOVEMENT

Successive differentiation can be applied to movement (for example, a traveling car). Time is the independent variable because whatever happens, time continues. Distance is the dependent variable because it depends on time. The first derivative of distance is velocity, which is expressed in terms of distance traveled per unit time. The second derivative of distance is acceleration, the rate of change of velocity, measured as distance per unit time squared. Jerk, or rate of change of acceleration, is expressed as distance per unit time cubed. These relationships can be demonstrated during five successive time intervals for a hypothetical moving car, as shown in Fig. 17-8.

- In the first interval, the car is stationary. The distance (from any other point) is fixed. Velocity and acceleration are both zero.

- For the second interval, acceleration increases. Assume a steady rate, shown by the straight slanting line. Velocity increases with a quadratic curve. Distance begins increasing with a cubic curve.

- For the third interval, acceleration holds constant, so velocity increases steadily in a straight line. That part of the distance curve is quadratic.

- For the fourth interval, acceleration decreases back to zero. Velocity follows an inverted quadratic curve, and distance approaches a straight line.

- For the fifth interval, acceleration is again zero. Velocity is constant (a horizontal straight line), and distance is a steeply sloping straight line.

This is only one example. Of course, there are infinitely many other ways in which a car can move with respect to time.

Figure 17-8

Here is how distance, velocity, acceleration, and rate of acceleration change (jerk) might vary with time for a moving car.

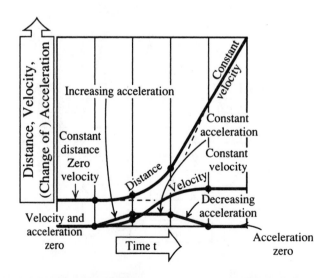

CIRCULAR MEASURE OF ANGLES

Angles can be measured in a variety of ways. Degree measure, shown at A in Fig. 17-9, divides a complete circle (rotation or revolution) into 360 equal *degrees*, symbolized deg or °. The number of degrees tells you how much of a complete circle the angle represents. Advocates of the metric system originally wanted to divide the circle into 100 equal parts, which they called *grades*. This measure, shown in Fig. 17-9B, is rarely used nowadays, but the principle is the same as for the others. Dividing a circle into *quadrants* is quite basic, as shown at C.

The last method of measuring angles considers a ratio of distances. In many fields of mathematics, physics, and engineering, an angle is defined as the ratio of the arc length around the circumference of a circle, divided by the radius of that same circle. This unit, called the *radian* and symbolized rad, is illustrated at D in Fig. 17-9.

The circular measure of angles works well with the definitions of trigonometric ratios. Sine, cosine, and tangent are all ratios that identify an angle. None of them is conveniently proportional to all angles. Sine and tangent begin at small angles, in direct proportion to the angle, but this proportionality breaks down long before the first right angle. The cosine begins at 1 and decreases, slowly at first, reaching zero at the first right angle.

Arc length, measured along the circumference of a circle and divided by the radius, is always proportional to the angle that the ratio identifies. Measuring distance around the

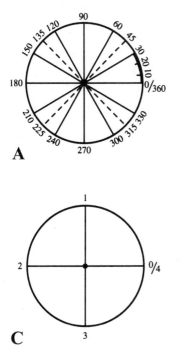

A

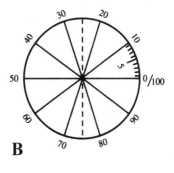

B

C

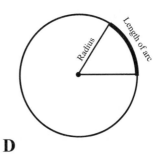

D

Figure 17-9

Measures of angles. At A, by degrees. At B, by equal parts with 100 to a complete circle. At C, by quadrants, with four parts to a circle. At D, by radians, defined as arc length divided by radius.

Figure 17-10

Comparison of circle and triangle definitions for the basic trig ratios, along with tabulations for several angles.

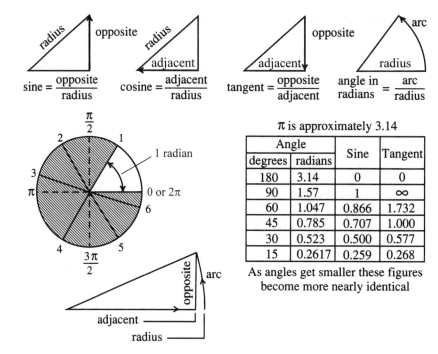

$$\text{sine} = \frac{\text{opposite}}{\text{radius}}$$

$$\text{cosine} = \frac{\text{adjacent}}{\text{radius}}$$

$$\text{tangent} = \frac{\text{opposite}}{\text{adjacent}}$$

$$\frac{\text{angle in}}{\text{radians}} = \frac{\text{arc}}{\text{radius}}$$

π is approximately 3.14

Angle		Sine	Tangent
degrees	radians		
180	3.14	0	0
90	1.57	1	∞
60	1.047	0.866	1.732
45	0.785	0.707	1.000
30	0.523	0.500	0.577
15	0.2617	0.259	0.268

As angles get smaller these figures become more nearly identical

circumference, the first semicircle (180°) accommodates a little over 3 radii. A whole circle accommodates twice as many, a little over 6 radii. The ratio of the length of a semicircular arc to its radius has the universal symbol π (the Greek lowercase letter pi). For the present, you can use its approximate value of 3.14.

On this basis, a right angle is exactly $\pi/2$ rad. From this, you can derive radian values for angles of 60°, 45°, 30°, and 15° as shown in Fig. 17-10. As you tabulate the radian equivalents, sines, and tangents, notice how close the angle in radians, the sine of the angle, and the tangent of the angle are at 15°. For very small angles, the figures are even closer. As the angle approaches zero, the radian measure, the sine, and the tangent all come together. You can verify this with a calculator. Likewise, as the angle increases toward $\pi/2$ rad, or 90°, the differences between the values becomes greater.

DERIVATIVES OF SINE AND COSINE

Circular measure allows you to apply the principles of differential calculus concerning infinitesimal changes. As already pointed out, for very small angles, the ratio for the sine is almost the same as its circular measure in radians. So for a nearly zero angle, you can consider that $\sin dx = dx$. Because the adjacent equals the hypotenuse at zero angle and $1/1 = 1$, it follows that $\cos dx = 1$. These values are true regardless of the value of x, because they concern only a vanishingly small angle dx.

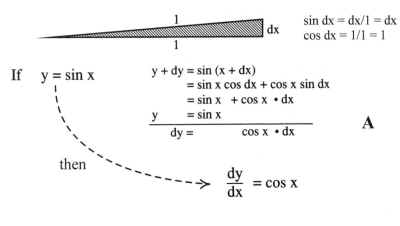

Figure 17-11

At A, differentiation of the sine function. At B, differentiation of the cosine function. Note the use of the infinitesimal angle *dx*, where sin *dx* = *dx* and cos *dx* = 1.

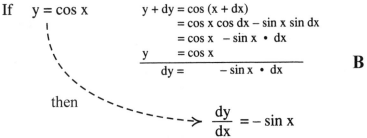

Figure 17-11 shows two examples of how the sine and cosine functions can be differentiated. At A in the figure, $y = \sin x$. Apply the sum formula to the right-hand part of this equation:

$$y + dy = \sin(x + dx)$$

Using the formula you learned earlier for the sine of a sum, and substituting $\sin dx = dx$ and $\cos dx = 1$, you get

$$y + dy = \sin x \cos dx + \cos x \sin dx$$
$$= \sin x + \cos x \, dx$$

Now take away the original part, $y = \sin x$. This leaves

$$dy = \cos x \, dx$$

Dividing both sides by dx, you obtain the derivative you're looking for:

$$dy/dx = \cos x$$

Similarly, at B in the figure, if $y = \cos x$, you can again use the sum formula and substitute for $\cos dx$ and $\sin dx$, so you get $dy/dx = -\sin x$.

SUCCESSIVE DERIVATIVES OF SINE

These facts about trigonometric ratios can help you draw curves representing the sine and the cosine. The curves provide an excellent way of learning how these functions are related. In Fig. 17-12, the first, second, third, and fourth derivatives of the sine function are tabulated for values of x between 0 rad and 2π rad ($0°$ and $360°$). Note that

$$d/dx\ \sin x = \cos x$$

$$d^2/dx^2\ \sin x = d/dx\ \cos x = -\sin x$$

$$d^3/dx^3\ \sin x = d/dx\ (-\sin x) = -\cos x$$

$$d^4/dx^4\ \sin x = d/dx\ (-\cos x) = \sin x$$

The notation d^n/dx^n means that you should take the nth derivative, with respect to x, of whatever comes after it.

Differentiating the sine yields the cosine, differentiating the cosine yields minus the sine, differentiating minus the sine yields minus the cosine, and differentiating minus the cosine yields the sine. If you keep on taking derivatives after that, you will cycle through these same four functions over and over. Figure 17-13 shows graphs of these functions based on the tabulated values, with both degree measure and radian measure. Note that all the curves have the same basic shape. This shape is called a *sine wave* or *sinusoid*. Also note that the waves all have identical heights and identical lengths. The only difference is their horizontal position, called the *phase*. Each time you take a derivative of one of these functions, you shift the phase to the left by $90°$, or $\pi/2$ rad.

Figure 17-12

Tabulated values for successive derivatives of the sine function for selected angles in radians.

When y is ...		Values of x								
		0	$\frac{\pi}{3}$	$\frac{\pi}{2}$	$\frac{2\pi}{3}$	π	$\frac{4\pi}{3}$	$\frac{3\pi}{2}$	$\frac{5\pi}{3}$	2π
sin x	$y =$	0	0.866	1	0.866	0	−0.866	−1	−0.866	0
	$\frac{dy}{dx} =$	1	0.5	0	−0.5	−1	−0.5	0	0.5	1
cos x	$y =$	1	0.5	0	−0.5	−1	−0.5	0	0.5	1
	$\frac{dy}{dx} =$	0		−1		0		+1		0
− sin x	$y =$	0	−0.866	−1	−0.866	0	0.866	1	0.866	0
	$\frac{dy}{dx} =$	− 1	−0.5	0	0.5	1	0.5	0	−0.5	−1
− cos x	$y =$	− 1	−0.5	0	0.5	1	0.5	0	−0.5	−1
	$\frac{dy}{dx} =$	0		1		0		−1		0
sin x	$y =$	0	0.866	1	0.866	0	−0.866	−1	−0.866	0
	$\frac{dy}{dx} =$	1	0.5	0	−0.5	−1	0.5	0	0.5	1

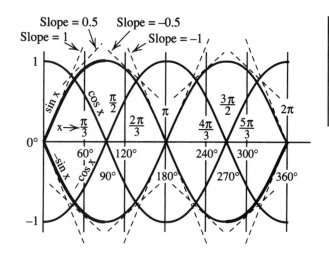

Figure 17-13

Curves showing successive derivatives of the sine function for angles in degrees and radians.

QUESTIONS AND PROBLEMS

This is an open-book quiz. You may refer to the text in this chapter (and earlier ones, too, if you want) when figuring out the answers. Take your time! Consider all values given as exact, so you don't have to worry about significant figures. The correct answers are in the back of the book.

1. Find dy/dx for the following functions:
   (a) $y = 5x^2$
   (b) $y = x^2 + 3x - 5$
   (c) $y = (x + 5)(x - 2)$
   (d) $y = x^2 + \cos x$
   (e) $y = 4x^3 - 4x^2 - 4x - 4$
   (f) $y = \sin x + 2 \cos x$
   (g) $y = 5x^4 + 2x^2$
   (h) $y = x^2 + 4x^4$

2. Suppose a car starts from a stationary position and accelerates at a rate of 5 miles per hour per second [5 (mi/h)/s]. How far will the car travel in the first 10 s after it starts?

3. In the situation of the previous problem, how fast will the car be moving, in miles per hour, after the first 10 s has elapsed?

4. Suppose the car in the scenario of Probs. 2 and 3 stops accelerating (the acceleration becomes zero) at the end of the first 10 s, and then the car continues traveling indefinitely at a zero acceleration rate. How far will the car have gone from its initial position 20 s after it started moving?

5. An object in free fall, in the gravitational field of the earth, accelerates downward at a rate of 32 feet per second per second (32 ft/s²), assuming there is no effect from air

resistance. Suppose you drop a lead ball (heavy enough to overcome all air resistance) off the top of a building 20 stories tall, where each story is 10 ft high. How long will it be, in seconds, from the time the ball is dropped until it hits the ground?

6. In the situation of Prob. 5, how fast will the ball be traveling, in feet per second, when it hits the ground?

7. Refer to the graph of acceleration versus time shown by Fig. 17-14. Suppose a car starts out traveling at a speed of 44 ft/s. Write an equation for the speed v of the car in feet per second as a function of the elapsed time t in seconds, based on the information given in this graph.

8. Write an equation, based on the information in Fig. 17-14, for the displacement s of the car in feet as a function of the elapsed time t in seconds. Assume that when $t = 0$, $s = 0$.

9. Based on the information in Fig. 17-14 and the derived equations, how fast will the car be traveling, in feet per second, after the following lengths of time? Express your answers in decimal form to four significant digits.
    (*a*) 1.000 s   (*b*) 2.000 s
    (*c*) 5.000 s   (*d*) 10.00 s

10. Based on the information in Fig. 17-14 and the derived equations, how far will the car have traveled, in feet, from its starting point after the following lengths of time? Express your answers in decimal form to four significant digits.
    (*a*) 1.000 s   (*b*) 2.000 s
    (*c*) 5.000 s   (*d*) 10.00 s

11. Refer to the graph of voltage versus time in Fig. 17-15. The curve is a sine wave. The maximum voltage is +8.00, and the minimum voltage is −8.00. Write down an equation for the voltage V (at any given instant) in volts as a function of time t in seconds.

Figure 17-14

Illustration for problems 7 through 10.

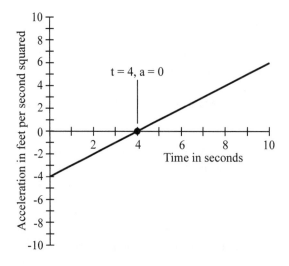

t = 4, a = 0

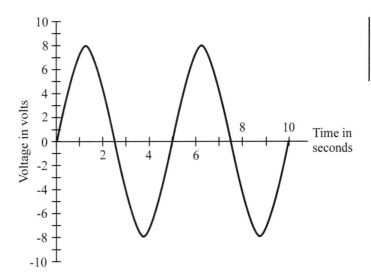

Figure 17-15

Illustration for problems 11 and 12.

12. How rapidly, in volts per second (V/s), is the voltage in Fig. 17-15 changing at the following times t? Use a calculator if you need one, and express your answers to four significant digits.

(a) $t = 0.000$ s   (b) $t = 2.000$ s
(c) $t = 2.500$ s   (d) $t = 5.000$ s
(e) $t = 6.000$ s   (f) $t = 7.500$ s

13. What are the measures, in radians, of the following angles? Use a calculator and specify to four significant digits.

(a) $10.00°$ (b) $30.00°$
(c) $75.00°$ (d) $145.0°$
(e) $220.0°$ (f) $300.0°$

14. What are the measures, in degrees, of the following angles? Use a calculator and specify to four significant digits.

(a) 0.2000 rad   (b) 0.5000 rad
(c) 1.000 rad   (d) 1.700 rad
(e) 2.200 rad   (f) 3.500 rad

15. Assume that the earth makes one complete revolution around the sun, relative to the distant stars, in exactly 365.25 days. Also assume that the earth revolves at a constant speed at all times in its orbit, and that the orbit of the earth is a perfect circle with the sun at the center. (These assumptions are not quite true in reality, but assuming them simplifies this problem tremendously!) Given this information:

(a) How many degrees of arc does the earth advance around the sun in one day? Express your answer to five significant digits.
(b) How many radians of arc does the earth advance around the sun during the month of April, which has exactly 30 days? Express your answer to five significant digits.

18
More about Differentiation

The previous chapter showed that successive differentiation of sine waves results in the same equation with its starting point shifted. Starting at sin x, successive derivatives are cos x, $-\sin x$, $-\cos x$, sin x, cos x, $-\sin x$, ..., cycling through these four over and over. It is particularly easy with x in radians. If the angle is in degrees, it has to be converted into radians for this method to work. Sometimes the quantity is not itself an angle, but is something that trigonometry can represent as an angle.

FREQUENCY AND PERIOD OF A SINE WAVE

Suppose you are analyzing waves and you come across the equation $y = \sin 2x$ instead of $y = \sin x$. Now consider this:

$$y + dy = \sin 2(x + dx)$$
$$= \sin 2x \cos 2\, dx + \cos 2x \sin 2\, dx$$
$$= \sin 2x + 2\, dx \cos 2x$$

If you let $y = \sin 2x$ in the last expression above, then you can substitute, subtract y from each side, and finally divide each side by dx, getting

$$y + dy = y + 2\, dx \cos 2x$$

$$dy = 2\, dx \cos 2x$$

$$dy/dx = 2 \cos 2x$$

Next, take $y = \sin 3x$. The derivative can be found in the same way. You get

$$dy/dx = 3 \cos 3x$$

Using a general multiplier a, the derivative is

$$dy/dx = a \cos ax$$

There are two shortened or alternative forms of writing this:

$$d/dx \, (\sin ax) = a \cos ax$$

$$(\sin ax)' = a \cos ax$$

Using the simple equation $y = \sin x$, the slope of the curve at the zero (starting) point is equal to its magnitude at maximum, which is why the cosine has the same amplitude as the sine wave. When the multiplier a is introduced, then a times as many waves occur in the same amount of time, so the slope of the original wave is a times as steep at every point. The graph of the wave is compressed horizontally (along the time axis) by a factor of a. An engineer or scientist would say that the *frequency* of the wave (the number of complete cycles per unit time) is multiplied by a. A mathematician might say that the *period* of the wave (the length of time required for one complete cycle to occur) is divided by a. Both mean the same thing. The *amplitude*, or top-to-bottom difference, of the derivative is therefore multiplied by a.

SINUSOIDAL MOTION

Now assume the equation $y = A \sin bt$ represents the motion of some object with passage of time t. The constant A represents the maximum movement from the reference (or zero) position, and the variable y represents the instantaneous distance from this reference position at time t. The constant b expresses how fast the object moves every time it passes through the reference position, and therefore how many times the object will make one complete excursion back and forth in a given time.

Refer to the diagrams in Fig. 18-1. The graph at A shows the vertical displacement as a function of time for some object moving up and down in a sine wave motion. In illustration B, you see that the instantaneous vertical velocity is the first derivative of the instantaneous displacement, given by dy/dt. It figures to

$$dy/dt = Ab \cos bt$$

As shown in drawing C, the instantaneous vertical acceleration is the second derivative of the instantaneous displacement, given by the equation

$$d^2y/dt^2 = -Ab^2 \sin bt$$

Notice that the maximum velocity occurs every time the object passes through the reference position, and zero velocity occurs at each extreme. Zero acceleration occurs when velocity is a maximum, and is a maximum at each extreme of the displacement.

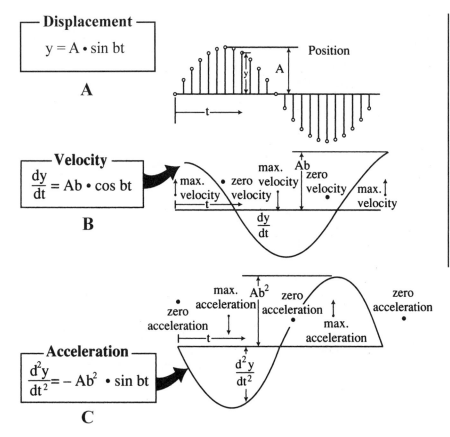

Displacement

$$y = A \cdot \sin bt$$

A

Velocity

$$\frac{dy}{dt} = Ab \cdot \cos bt$$

B

Acceleration

$$\frac{d^2y}{dt^2} = -Ab^2 \cdot \sin bt$$

C

Figure 18-1

An object oscillating with sine-wave motion. At A, vertical displacement as a function of time. At B, velocity is the first derivative of displacement with respect to time. At C, acceleration is the second derivative of displacement with respect to time.

Notice also that the reference position and points of zero acceleration coincide. Maximum excursion and maximum acceleration also coincide.

The instantaneous acceleration is b^2 times the instantaneous displacement (in whatever units are used), and it is of the opposite sign. When the displacement is at its maximum upward, acceleration is at its maximum downward, and vice versa. Here, positive signs mean upward displacement, velocity, or acceleration; negative signs mean downward displacement, velocity, or acceleration.

HARMONIC MOTION

The existence of sine wave motion in mechanical, electrical, acoustical, and other systems is a manifestation of the cyclic interchange of energy. Such a system has a characteristic period and a characteristic frequency, regardless of the amplitude of the movement. When the same consistent units of distance and time are used (feet and seconds, for example), the period is always equal to the reciprocal of the frequency.

Harmonic motion is the name given to the movement an object or variable quantity makes during a cyclic period that is *sinusoidal* (representable as a sine wave). In the example shown by Fig. 18-2, the displacement and movement are sinusoidal. In an electrical system, the voltage and current would be sinusoidal. In an acoustical system, the airflow and pressure variation would be sinusoidal.

The natural relationship is fixed by the quantity b. Only at one frequency, which makes $b = 2\pi/t$, does this natural relationship hold, where energy interchanges with no external force applied. In a spring-loaded system such as the one shown in Fig. 18-2, the value of b^2 is determined by the stiffness of the spring, called the *spring constant*, and

Figure 18-2

Aspects of harmonic motion. At A, graphs of instantaneous displacement and velocity for comparison. At B, the instantaneous acceleration increases with increasing distance from the reference or zero position.

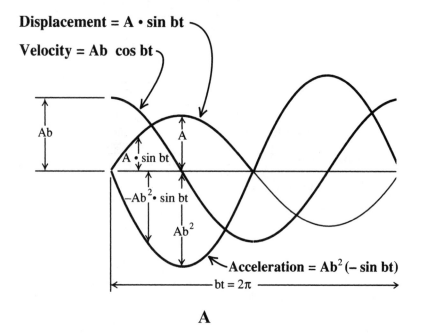

Displacement = **A** • **sin** *bt*

Velocity = **Ab** **cos** *bt*

Ab

$A \cdot \sin bt$

A

$-Ab^2 \cdot \sin bt$

Ab^2

Acceleration = **Ab**^{2} **(– sin** *bt***)**

$bt = 2\pi$

A

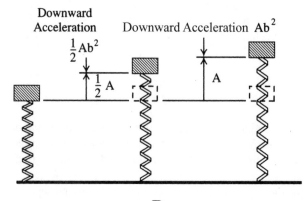

Downward Acceleration

$\frac{1}{2} Ab^2$

$\frac{1}{2} A$

Downward Acceleration **Ab**^{2}

A

B

by the mass of the moving object. If you change either one of these factors, then b^2 changes, and therefore b changes too. The result is a change in the *natural frequency*, more often called the *resonant frequency*, of the system.

LINEAR AND NONLINEAR RELATIONSHIPS

In a spring-loaded system, *pure harmonic motion*, also called *simple harmonic motion*, at a specific resonant frequency occurs only if the spring is linear. That means the relationship between the amount of compression or tension and the force produced by the spring is a straight-line graph. In a linear spring, if every inch results in a force of 15 poundals (Fig. 18-3A), the force for successive inches will increase in uniform steps: 15, 30, 45, 60, 75, and so on, for both compression and tension (stretching).

In the real world, springs are not always linear. The extra force might increase as the spring becomes more fully compressed (Fig. 18-3B). Instead of increasing to 15, 30, 45, 60, and 75 poundals for successive inches, the figures might run as 16, 34, 54, 76, and

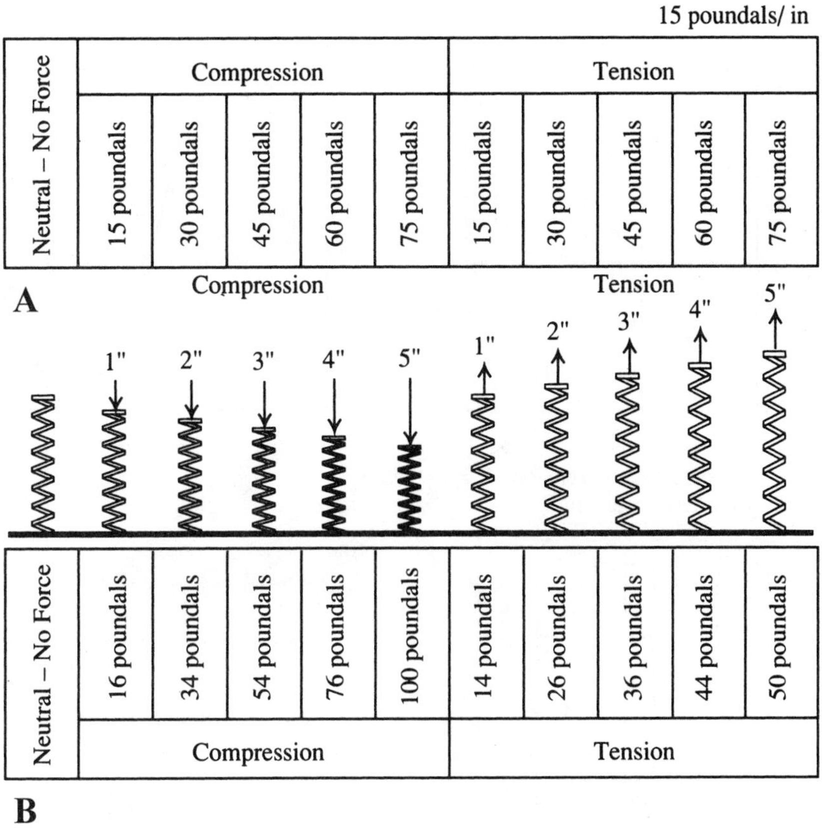

Figure 18-3
At A, behavior of a hypothetical linear spring. At B, behavior of a hypothetical nonlinear spring.

100 poundals. In tension, the effect might be reversed, so successive forces of tension, at inch intervals, would read 14, 26, 36, 44, and 50 poundals. Such a spring is nonlinear, because the force is not proportional to displacement.

Plotting the force/displacement relationship of the nonlinear spring shown in Fig. 18-3B as a graph, you find that it can be resolved in two components, as shown in Fig. 18-4A. The linear part (straight dashed line) is the same as the linear spring: 15 poundals for every inch. Then the nonlinear part, in this case called a *square-law* component and shown as a dashed parabolic curve, is proportional to the square of displacement. Because displacement is in opposite directions, one is considered positive, the other negative. In one direction, the square-law component will add to the linear force. The other way, it will take away from the linear force. The overall or composite force is shown in Fig. 18-4A as a solid curve.

Now imagine that the movement is somehow made sinusoidal so the force produced is determined by the sinusoidal variation in position. You can show this position by plotting motion and force separately, each against time, as shown in Fig. 18-4B and C. At B, the dashed curve is a true sinusoid, and the solid-line curve represents the force produced by the actual nonlinear spring.

Figure 18-4

At A, components and overall force produced by a nonlinear spring. At B and C, force and motion are plotted separately against time.

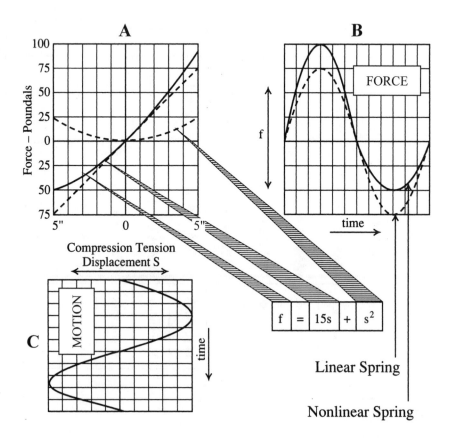

MULTIPLES AND POWERS

Finding expressions for multiple angle functions in terms of powers of unit angle functions can be pursued systematically, as shown in Table 18-1. The first step is to extend the multiple angle formulas as far as you need, one multiple at a time. Using sum formulas, first find expressions for sin 2A and cos 2A; regard 2A as A + A. Next, regarding 3A as 2A + A, you can find expressions for sin 3A and cos 3A. Regarding the expression (n + 1)A as A + nA, substitute previously found expressions for sin nA and cos nA. Table 18-1 takes this as far as 6A. The working is not shown. You might wish to put yourself methodically through each step.

For the even powers, both $\sin^n A$ and $\cos^n A$ use a form of the expression for cos nA, and substitute it for the lower powers. For the odd powers, take the expression for sin nA to get $\sin^n A$ and the expression for cos nA to get $\cos^n A$ with similar substitutions. Here again, only the results are listed in Table 18-1. Try a few to see how to do it.

The substitutions used to derive these expressions can become involved in detail working. Nothing is really difficult; it's just "messy." The routine becomes familiar with practice, but the sheer number of substitutions means that a mistake can creep in at any point. You need a simple means to check the results.

Table 18-1
Formulas for multiple angles and powers of the sine and cosine functions.

$\sin 2A = 2 \sin A \cos A$	$\cos 2A = \cos 2A - \sin^2 A$
	$= 2 \cos^2 A - 1$
	$= 1 - 2 \sin^2 A$
$\sin 3A = 3 \sin A - 4 \sin^3 A$	$\cos 3A = 4 \cos^3 A - 3 \cos A$
$\sin 4A = 4 \sin A (2 \cos^3 A - \cos A)$	$\cos 4A = 1 - 8 \cos^2 A + 8 \cos^4 A$
	$= 1 - 8 \sin^2 A + 8 \sin^4 A$
$\sin 5A = 5 \sin A - 20 \sin^3 A + 16 \sin^5 A$	$\cos 5A = 5 \cos A - 20 \cos^3 A + 16 \cos^5 A$
$\sin 6A = \cos A (6 \sin A - 32 \sin^3 A + 32 \sin^5 A)$	$\cos 6A = 32 \cos^6 A - 48 \cos^4 A + 18 \cos^2 A - 1$
	$= 1 - 18 \sin^2 A + 48 \sin^4 A - 32 \sin^6 A$

$\sin^2 A = \frac{1}{2}(1 - \cos 2A)$	$\cos^2 A = \frac{1}{2}(1 + \cos 2A)$
$\sin^3 A = \frac{1}{4}(3 \sin A - \sin 3A)$	$\cos^3 A = \frac{1}{4}(3 \cos A + \cos 3A)$
$\sin^4 A = \frac{1}{8}(3 - 4 \cos 2A + \cos 4A)$	$\cos^4 A = \frac{1}{8}(3 + 4 \cos 2A + \cos 4A)$
$\sin^5 A = \frac{1}{16}(10 \sin A - 5 \sin 3A + \sin 5A)$	$\cos^5 A = \frac{1}{16}(10 \cos A + 5 \cos 3A + \cos 5A)$
$\sin^6 A = \frac{1}{32}(10 - 15 \cos 2A + 6 \cos 4A - \cos 6A)$	$\cos^6 A = \frac{1}{32}(10 + 15 \cos 2A + 6 \cos 4A + \cos 6A)$

In Table 18-2, two angles are used for each check: $A = 0$ and $A = \pi/2$ rad, which is 90°. Whatever the multiple, nA is always 0 when A is 0, so sin nA should always be 0 and cos nA always 1 in this column. For the $\pi/2$ column, nA should always be equal to n right angles. So for sin nA starting with $n = 2$, the sequence S will be

$$S = 0, -1, 0, 1, 0, -1, 0, 1, \ldots$$

For cos nA starting with $n = 2$, you will have the sequence C as follows:

$$C = -1, 0, 1, 0, -1, 0, 1, 0, \ldots$$

Table 18-2
Some numerical examples involving multiple angles and powers of the sine and cosine functions.

Quantity	Values for A = 0	Values for $A = \frac{\pi}{2}$
$\sin 2A = 2 \sin A \cos A$	$2 \cdot 0 \cdot 1 = 0$	$2 \cdot 1 \cdot 0 = 0$
$\cos 2A = \cos^2 A - \sin^2 A$	$1 - 0 = 1$	$0 - 1 = -1$
$\sin 3A = 3 \sin A - 4 \sin^3 A$	$3 \cdot 0 - 4 \cdot 0 = 0$	$3 \cdot 1 - 4 \cdot 1 = -1$
$\cos 3A = 4 \cos^3 A - 3 \cos A$	$4 \cdot 1 - 3 \cdot 1 = 1$	$4 \cdot 0 - 3 \cdot 0 = 0$
$\sin 4A = 4 \sin A (2 \cos^3 A - \cos A)$	$4 \cdot 0 (2 - 1) = 0$	$4 \cdot 1 (2 \cdot 0 - 1 \cdot 0) = 0$
$\cos 4A = 1 - 8 \cos^2 A + 8 \cos^4 A$	$1 - 8 + 8 = 1$	$1 - 8 \cdot 0 + 8 \cdot 0 = 1$
$\sin 5A = 5 \sin A - 20 \sin^3 A + 16 \sin^5 A$	$5 \cdot 0 - 20 \cdot 0 + 16 \cdot 0 = 0$	$5 \cdot 1 - 20 \cdot 1 + 16 \cdot 1 = 1$
$\cos 5A = 5 \cos A - 20 \cos^3 A + 16 \cos^5 A$	$5 \cdot 1 - 20 \cdot 1 + 16 \cdot 1 = 1$	$5 \cdot 0 - 20 \cdot 0 + 16 \cdot 0 = 0$
$\sin 6A = \cos A (6 \sin A - 32 \sin^3 A + 32 \sin^5 A)$	$1(6 \cdot 0 - 32 \cdot 0 + 32 \cdot 0) = 0$	$0(6 \cdot 1 - 32 \cdot 1 + 32 \cdot 1) = 0$
$\cos 6A = 32 \cos^6 A - 48 \cos^4 A = 18 \cos^2 A - 1$	$32 - 48 + 18 - 1 = 1$	$0 - 0 + 0 - 1 = -1$
$\sin^2 A = \frac{1}{2}(1 - \cos 2A)$	$\frac{1}{2}(1 - 1) = 0$	$\frac{1}{2}(1 + 1) = 1$
$\cos^2 A = \frac{1}{2}(1 - \cos 2A)$	$\frac{1}{2}(1 + 1) = 1$	$\frac{1}{2}(1 - 1) = 0$
$\sin^3 A = \frac{1}{4}(3 \sin A - \sin 3A)$	$\frac{1}{4}(0 - 0) = 0$	$\frac{1}{4}(3 + 1) = 1$
$\cos^3 A = \frac{1}{4}(3 \cos A - \cos 3A)$	$\frac{1}{4}(3 + 1) = 1$	$\frac{1}{4}(0 + 0) = 0$
$\sin^4 A = \frac{1}{8}(3 - 4 \cos 2A + \cos 4A)$	$\frac{1}{8}(3 - 4 + 1) = 0$	$\frac{1}{8}(3 + 4 + 1) = 1$
$\cos^4 A = \frac{1}{8}(3 + 4 \cos 2A + \cos 4A)$	$\frac{1}{8}(3 + 4 + 1) = 1$	$\frac{1}{8}(3 - 4 + 1) = 0$
$\sin^5 A = \frac{1}{16}(10 \sin A - 5 \sin 3A + \sin 5A)$	$\frac{1}{16}(0 - 0 + 0) = 0$	$\frac{1}{16}(10 + 5 + 1) = 1$
$\cos^5 A = \frac{1}{16}(10 \cos A - 5 \cos 3A + \cos 5A)$	$\frac{1}{16}(10 + 5 + 1) = 1$	$\frac{1}{16}(0 + 0 + 0) = 0$
$\sin^6 A = \frac{1}{32}(10 - 15 \cos 2A + 6 \cos 4A - \cos 6A)$	$\frac{1}{32}(10 - 15 + 6 - 1) = 0$	$\frac{1}{32}(10 + 15 + 6 + 1) = 1$
$\cos^6 A = \frac{1}{32}(10 + 15 \cos 2A + 6 \cos 4A - \cos 6A)$	$\frac{1}{32}(10 + 15 + 6 + 1) = 1$	$\frac{1}{32}(10 - 15 + 6 - 1) = 0$

In the powers table, 0^n is always equal to 0, and 1^n is always equal to 1. So $\sin^n A$ will always be 0 for $A = 0$ and 1 for $A = \pi/2$ rad. Similarly, $\cos^n A$ will always be 1 for $A = 0$ and 0 for $A = \pi/2$ rad. All these check columns should have either 0 or 1 in the appropriate pattern. If one of the coefficients in the detail working has gone wrong, it will almost always cause a different result in one or both of these checks.

FUNCTIONS

An expression that "performs some action" on a variable is a *function* of that variable. You have already considered several functions without calling them that. Powers, roots, trig ratios, and many other relations are all functions.

To work with functions in a broad sense, because they can take various forms, a general form is $y = f(x)$, which is read "y is a function of x" or "y equals f of x." This expression can mean any function of x, with the implication that it is a function for which you can figure out successive derivatives.

Of course, you can use letters other than f to denote functions, letters other than x to denote independent variables, and letters other than y to denote dependent variables. You might see expressions like this:

$$q = g(z)$$
$$z = H(r)$$

or even symbolic functions and variables like this:

$$\phi = \psi(\omega)$$

The derivative of a function can be denoted in various ways. Several ways to denote the first derivative where $y = f(x)$ include these:

$$dy/dx$$
$$df(x)/dx$$
$$df/dx$$
$$f'(x)$$
$$df$$
$$f'$$

All these refer to "the derivative of the function f." The last two do not specify any particular variable, but usually this variable will be evident from context. Second derivatives would be written like this:

$$d^2y/dx^2$$
$$d^2f(x)/dx^2$$

$$d^2f/dx^2$$
$$f''(x)$$
$$d^2f$$
$$f''$$

DERIVATIVE OF CONSTANT

You've learned some general rules about differentiation. Let's review these and then look at some more. First, note that the derivative of a constant without any variable attached to it is always equal to 0. Stating this the way a mathematician would do it, we let f be a function of x such that $f(x) = c$, where c is a real number constant. Then

$$dc/dx = 0$$

DERIVATIVE OF SUM OF TWO FUNCTIONS

Now suppose that f and g are two functions of a variable x. Maybe these two functions are identical, but more likely they are different. Consider the sum of these two functions $f + g$, which is $f(x) + g(x)$ for every possible value of x. We can define this sum another way:

If $y = f(x)$ and $z = g(x)$, then $(f + g)(x) = y + z$

In this sort of situation, as you've already learned, the derivative of the sum is equal to the sum of the derivatives:

$$d(f + g)/dx = df/dx + dg/dx$$

You can also write it like this:

$$(f + g)' = f' + g'$$

DERIVATIVE OF DIFFERENCE OF TWO FUNCTIONS

The rule for differentiating the difference of two functions follows directly from the rule for the sum, because subtraction is simply the addition of a negative. Suppose that f and g are functions of a variable x, and let $f - g = f(x) - g(x)$ for all possible values of x. Then

$$d(f - g)/dx = df/dx - dg/dx$$

You can also write it like this:

$$(f - g)' = f' - g'$$

DERIVATIVE OF FUNCTION MULTIPLIED BY A CONSTANT

Here's another rule that you have already learned. Let f be a function of a variable x. Suppose c is a constant by which $f(x)$ is multiplied to get $cf(x)$, which can be shortened to cf. Then you can move the constant outside the derivative, like this:

$$d(cf)/dx = c(df/dx)$$

You can also write

$$(cf)' = cf'$$

DERIVATIVE OF PRODUCT OF TWO FUNCTIONS

Suppose f and g are two functions of a variable x. Define the product of f and g as follows:

$$f \cdot g = f(x) \cdot g(x)$$

for any specific value of the variable x. This is just a simplified notation calling $f(x)$ by the shorter name f, and calling $g(x)$ by the shorter name g. Then you can find the derivative of the product function by using this formula:

$$d(f \cdot g)/dx = f \cdot (dg/dx) + g \cdot (df/dx)$$

You might find this expression easier to understand:

$$(f \cdot g)' = f \cdot g' + g \cdot f'$$

DERIVATIVE OF PRODUCT OF THREE FUNCTIONS

The rule for multiplying three functions follows from the rule for multiplying two functions. (In fact, if you are patient, you can derive a general rule for multiplying any number of functions!) Let f, g, and h be three functions of a variable x. Define the product of f, g, and h as follows:

$$f \cdot g \cdot h = f(x) \cdot g(x) \cdot h(x)$$

Then you can find the derivative of the product function this way:

$$d(f \cdot g \cdot h)/dx = f \cdot g \cdot (dh/dx) + f \cdot h \cdot (dg/dx) + g \cdot h \cdot (df/dx)$$

You might find this expression simpler:

$$(f \cdot g \cdot h)' = f \cdot g \cdot h' + f \cdot h \cdot g' + g \cdot h \cdot f'$$

DERIVATIVE OF QUOTIENT OF TWO FUNCTIONS

Let f and g be two functions of a variable x, and define $f/g = [f(x)]/[g(x)]$. Then

$$d(f/g)/dx = [g \cdot (df/dx) - f \cdot (dg/dx)]/g^2$$

where $g^2 = g(x) \cdot g(x)$, not to be confused with d^2g/dg^2. You might find it easier to write it like this:

$$(f/g)' = (g \cdot f' - f \cdot g')/g^2$$

RECIPROCAL DERIVATIVES

Suppose that f is a function, and x and y are variables such that $y = f(x)$. The following formulas are both true as long as the derivatives exist for all values of the variables:

$$dy/dx = 1/(dx/dy)$$

$$dx/dy = 1/(dy/dx)$$

DERIVATIVE OF FUNCTION RAISED TO A POWER

Let f be a function of a variable x. Suppose n is a positive whole number. Then you can find the derivative of $f(x)$ raised to the nth power by using this formula:

$$d(f^n)/dx = n(f^{n-1}) \cdot df/dx$$

where f^n denotes f multiplied by itself n times, not to be confused with the nth derivative. You might find it easier to write the formula like this:

$$(f^n)' = n(f^{n-1}) \cdot f'$$

CHAIN RULE

Suppose that f and g are both functions of a variable x. The derivative of the composite function (that is, f of g of x), can be found this way:

$$[f(g)]' = f'(g) \cdot g'$$

This is sometimes called by a real tongue-and-brain-twister of a name: the "formula for the derivative of a function of a function." You must first find the derivative of "f of g of x" and then multiply the result by the derivative of "g of x."

QUESTIONS AND PROBLEMS

This is an open-book quiz. You may refer to the text in this chapter (and earlier ones, too, if you want) when figuring out the answers. Take your time! Consider all values given as exact, so you don't have to worry about significant figures. The correct answers are in the back of the book.

1. Refer to Fig. 18-5, which depicts the motion of a heavy weight oscillating with a spring. Assume the maximum positive (upward) and negative (downward) displacements to be precisely six meters (6 m) and −6 m, respectively. Also assume the period of oscillation to be exactly 3 s. At what points in time (values of t) is the velocity of the weight the greatest positively? What is the velocity, in meters per second, at these points? Express your answers to three significant figures.

2. At what points in time is the velocity of the weight in the preceding problem greatest negatively? What is the velocity in meters per second at these points? Express your answer to three significant figures.

3. At what points in time is the acceleration of the weight in the preceding problems greatest positively? What is the acceleration in meters per second squared at these points? Express your answer to three significant figures.

4. At what points in time is the acceleration of the weight in the preceding problems greatest negatively? What is the acceleration in meters per second squared at these points? Express your answer to three significant figures.

5. Suppose the tension of the spring in the previous problems is increased so the weight oscillates at the third harmonic of (three times) the original frequency, but the positive and negative peak amplitudes (excursions) remain exactly the same. At what points in time is the velocity of the weight maximum positively? What is the velocity, in meters per second, at these points? Express your answers to three significant figures.

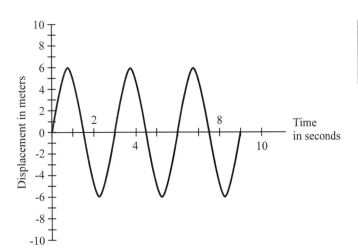

Figure 18-5
Illustration for problems 1 through 8.

Figure 18-6

Illustration for problems 10 and 11.

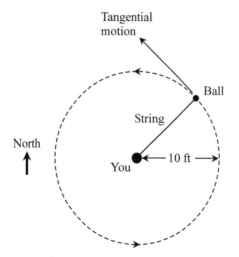

Tangential motion

Ball

String

North

You

←— 10 ft —→

6. In the situation of Prob. 5, at what points in time is the velocity of the weight greatest negatively? What is the velocity in meters per second at these points? Express your answer to three significant figures.

7. In the situation of Prob. 5, at what points in time is the acceleration of the weight greatest positively? What is the acceleration in meters per second squared at these points? Express your answer to three significant figures.

8. In the situation of Prob. 5, at what points in time is the acceleration of the weight greatest negatively? What is the acceleration in meters per second squared at these points? Express your answer to three significant figures.

9. Using the formulas for multiples and powers (Table 18-1), find the following. Assume all angle measures are in radians. You may use a calculator; express your answers to four significant figures.
(a) $\sin^2 (\pi/2)$ (b) $\cos^2 (\pi/3)$
(c) $\sin^3 (\pi/4)$ (d) $\cos^3 (\pi/4)$
(e) $\sin^5 (\pi/3)$ (f) $\cos^6 (2\pi/5)$

10. Suppose you twirl a ball around on a string that is 10 ft long, as shown in Fig. 18-6. The ball makes one complete revolution around your body every 2 s, exactly. What is the tangential speed of the ball in feet per second? Use a calculator and express your answer to four significant figures.

11. Assume the ball whose motion is shown in Fig. 18-6 revolves in a horizontal plane. What is the northward-moving component of the ball's speed at the instant the ball is traveling toward the northwest (exactly 45° west of north)? Express your answer to four significant figures.

19
Introduction to Integrals

Now that you've gained some familiarity with differentiation, it's time to look at the reverse process, called *integration*. A math professor once said, "Differentiation is like navigating down the Mississippi. Integration is like navigating up the Mississippi and its tributaries. It's the same river system, but a different process!" For most people, integration is more difficult than differentiation, and it's easier to "get lost." That professor knew what he was talking about! But with study, integration gets easier.

ADDING UP THE PIECES

Slope, found by differentiation, uses a vanishingly small change in x to produce a vanishingly small change in y. If you add an arbitrarily large number of these tiny changes, you have x and y, respectively.

If $f'(x)$ is equal to y' (two names for the same derivative), then the reverse process gives you the original $f(x)$ or y. This process is expressed with the *integral sign*, which is an old-fashioned long letter S like the ones you see on violins and cellos. They use the long letter S because integration is an exalted form of summation, a sort of super sum. To abbreviate that, you need a super S!

THE ANTIDERIVATIVE

Imagine an object moving in a straight line at variable speed. The function that defines its instantaneous speed is the derivative of the function that defines the distance it has traveled from the starting point (its cumulative displacement). Reversing this, the function that defines the cumulative displacement is the *antiderivative* of the function that defines the instantaneous speed.

For functions denoted by lowercase italic letters such as f, g, or h, the antiderivative is customarily denoted by the uppercase italic counterpart such as F, G, or H. The concept of the antiderivative is the basis for integration. When you take the antiderivative of a particular function f, you find the function F that, when differentiated, will give you the function f again. Integration takes this concept and goes a little bit further, spawning an infinite number of different functions, any one of which will give you the original function back when you differentiate. You'll see why in a minute.

From the differentiating you have done, you can start on integration just by taking the reverse process, or antiderivative. Here is a simple example:

$$\text{When } f(x) = x^n \text{ then } f'(x) = nx^{n-1}$$

so therefore

$$\text{When } f(x) = nx^{n-1} \text{ then } F(x) = x^n$$

Here's another example of a derivative and its corresponding antiderivative:

$$\text{When } f(x) = \sin x \text{ then } f'(x) = \cos x$$

so therefore

$$\text{When } f(x) = \cos x \text{ then } F(x) = \sin x$$

PATTERNS IN CALCULATIONS

When you review the study of mathematics, always begin with a positive process, then reverse it to produce a negative process. After counting, you got into addition, which was reversed to make subtraction. After shortening multiple addition to make multiplication, it was reversed to parallel multiple subtraction, making division. Then indices brought powers, and you reversed that process to find roots.

In each situation, what began as a negative process, searching for a question to produce an answer, later developed into a positive approach to eliminate the search. Integration has a similar relationship to differentiation. They're opposites. One process "undoes" the other (Fig. 19-1).

THE CONSTANT OF INTEGRATION

When you differentiate a function, you find its slope at a point or at a sequence of points. However, saying that a road has a certain slope (1 in 16, for example) doesn't state how high the road is. It could be at sea level or on top of a mountain, or at any one of an infinite number of elevations in between.

POSITIVE PROCESS

Addition	5 and 3 makes	?

Multiplication	5 times 3 is	?

Powers	5 to the 3rd power is	?

Differentiation	$\frac{d}{dx}$ of ax^n is	?

REVERSE

Subtraction	5 and ?	makes 8

Division	5 times ?	is 15

Roots	?	to the 3rd power is 125

Integration	$\frac{d}{dx}$ of ?	is anx^{n-1}

Figure 19-1

In mathematics, every "positive" process has a reverse or "negative" process that undoes it.

Looked at mathematically, the three equations graphed in Fig. 19-2 each begin with $y = x^3 - 12x$. Then there is a constant that is shown as 8, 0 (nothing shown), or -8. All three equations give the same derivative because the derivative of a constant is always equal to 0 (that is, nothing):

$$d/dx\ (x^3 - 12x + 8) = 3x^2 - 12$$

and

$$d/dx\ (x^3 - 12x) = 3x^2 - 12$$

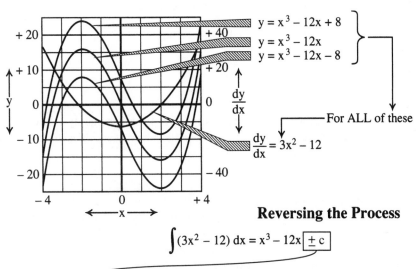

$$y = x^3 - 12x + 8$$
$$y = x^3 - 12x$$
$$y = x^3 - 12x - 8$$

For ALL of these

$$\frac{dy}{dx} = 3x^2 - 12$$

Figure 19-2

The constant of integration determines how high or low a curve is, but does not affect the shape of the curve.

Reversing the Process

$$\int (3x^2 - 12)\ dx = x^3 - 12x \boxed{\pm c}$$

... a constant that fixes the LEVEL of the whole curve called the CONSTANT OF INTEGRATION

and

$$d/dx \, (x^3 - 12x - 8) = 3x^2 - 12$$

By reversing the process, you have no direct means to find the constant. Again, think of the Mississippi and its branches! Going upstream, you don't know which tributary to take. You didn't have that sort of trouble with differentiation. You must leave room for an unknown constant, called the *constant of integration*. If the road began at sea level (constant of integration equal to 0), then integrating over any distance would find the height of the latest point above sea level. But if the road began at 860 ft above sea level, integrating the same way would give you an answer that would be off by 860 ft unless you took that constant into account. In general, you can say this:

$$d/dx \, (x^3 - 12x + c) = 3x^2 - 12$$

where c can be any real number. If you say $f(x) = 3x^2 - 12$, then you can write the antiderivative like this, without the constant:

$$F(x) = x^3 - 12x$$

But you must write the *indefinite integral* with the constant:

$$\int f(x) \, dx = \int (3x^2 - 12) \, dx$$
$$= x^3 - 12x + c$$

The integral is called "indefinite" because you don't know the value of the constant c unless you have more specific, or definite, information about the problem.

Note the infinitesimal dx after the statement of the integral. This little value, called the *differential*, must always be written after integrals. If the variable were something else, such as z, you would write dz.

DEFINITE INTEGRALS

The indefinite integral, shown above and in Fig. 19-2, has mainly a theoretical value. A *definite integral* "adds it up" between specific values. It specifies that the curve starts at a certain point and follows it to a new point.

The definite integral is written the same way as the indefinite integral, but numbers or variables are put against the long S, one to its lower right and the other to its upper right. The figure at the lower right represents where the defined interval begins, and the figure at the upper right represents where the interval ends. Next, the expression for the anti-derivative is put in square brackets with the limits outside, one at the lower right and the other at the upper right. (Alternatively, a vertical bar can be placed to the right of the anti-derivative, and the limits can be written to the right of the bar.) Then you subtract the value at the lower limit from the value at the upper limit. An example is shown in Fig. 19-3.

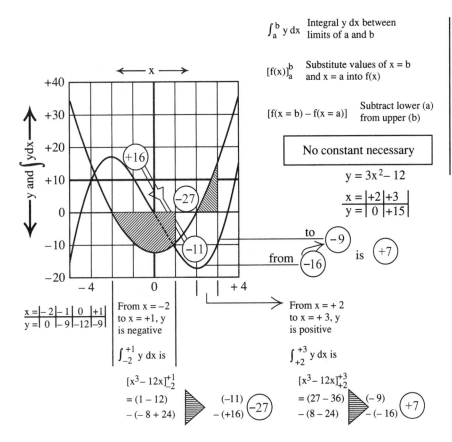

Figure 19-3

The definite integral shows the area between a curve and the independent-variable axis (in this case the x axis) over a defined interval.

In the graph of Fig. 19-3, the indefinite integral is plotted with the constant of integration set at 0. That's the antiderivative function. When substituting in the lower value, you could make the starting point by inserting a constant of integration that would make the point equal to 0. However, it is not necessary, because you subtract this value from the upper value. Whatever you make the constant, it disappears when you subtract one value from the other.

By substituting $x = -2$ and $x = 1$, the second produces -11 and the first produces 16. By subtracting the first from the second, the change is -27. Substituting values $x = 2$ and $x = 3$, the same process produces the change of y in this range as 7.

FINDING AREA BY INTEGRATION

A most useful application for integration is for finding areas. If y is a succession of vanishingly small elements in an area, the sum of these elements over a certain range of the curve that is represented by this function will be equal to the area under the curve, which consists of a huge number of tiny strips (Fig. 19-4A). You go from the arbitrarily

Figure 19-4

At A, the area under a curve can be found by adding up the areas of a huge number of vanishingly small vertical strips. At B, the area of a simple geometric object, in this case a trapezoid, is found by integration.

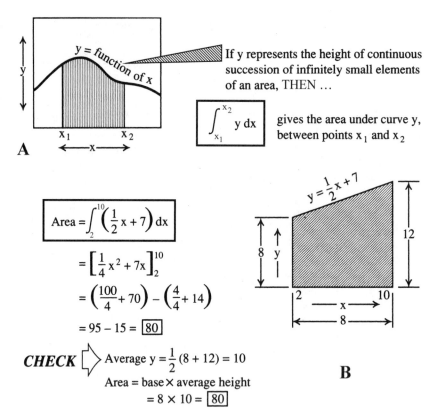

If y represents the height of continuous succession of infinitely small elements of an area, THEN …

$$\int_{x_1}^{x_2} y\, dx$$

gives the area under curve y, between points x_1 and x_2

$$\text{Area} = \int_{2}^{10}\left(\frac{1}{2}x + 7\right) dx$$

$$= \left[\frac{1}{4}x^2 + 7x\right]_{2}^{10}$$

$$= \left(\frac{100}{4} + 70\right) - \left(\frac{4}{4} + 14\right)$$

$$= 95 - 15 = \boxed{80}$$

CHECK ⟹ Average $y = \frac{1}{2}(8 + 12) = 10$

Area = base × average height

$$= 8 \times 10 = \boxed{80}$$

small to the arbitrarily large. In this context, the word "arbitrary" means as extreme as your imagination can make things—and then even a little more extreme than that! It's a way of "sneaking up on infinity."

To demonstrate how this method works, take the shaded area in Fig. 19-4B. The equation of the upper side is

$$y = f(x) = (1/2)x + 7$$

The antiderivative (working the derivative formula backward) is

$$F(x) = (1/4)x^2 + 7x$$

The indefinite integral would be

$$\int f(x)\, dx = \int [(1/2)x + 7]\, dx$$
$$= (1/4)x^2 + 7x + c$$

By making substitutions for $x = 2$ and $x = 10$, the limits for the definite integral, and subtracting, the area is 80. This will be true no matter what the value of c happens to be (it subtracts from itself), so you can just get rid of it when making the subtraction. Checking it by the geometric formula for the area of a trapezoid proves that you have the right answer.

AREA OF A CIRCLE

The area you find by means of integration does not have to be "under a curve." In Fig. 19-5, two methods are illustrated for finding the area of a circle by means of integration.

In the method shown by Fig. 19-5A, the element of area is a wedge from the center of the circle to the circumference, taken at angle (in circular measure) x. The element has an area equal to 1/2 the base length times the height, because as the wedge gets arbitrarily narrow, its shape approaches a triangle. The height of this arbitrarily narrow slice is so close to r that you can think of its as actually being equal to r. The base length, according to similar reasoning, is equal to $r\,dx$. Using the familiar formula for the area of a triangle, you find that the area of the element is $(1/2)r^2\,dx$. Integrating produces $(1/2)r^2 x$ because r can be treated as a constant. (You're not integrating with respect to r, but with respect to the variable x!) For one complete revolution around the circle, the lower limit of x is 0 rad, and the upper limit is 2π rad. Substituting and subtracting give you the well-known formula for the area of a circle, πr^2.

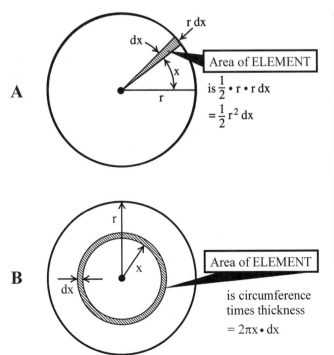

Figure 19-5
At A, the area of a circle can be found by breaking it up into a huge number of tiny wedges and integrating once around. At B, the area can be found by breaking the circle into a huge number of tiny concentric rings and integrating from the center to the circumference.

Figure 19-6

At A, integration is used to find the area of the curved surface of a cylinder. At B, integration is used to find the area of the curved surface of a cone.

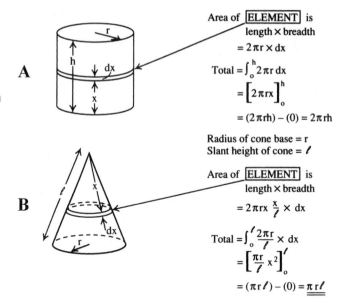

Area of ELEMENT is
length × breadth
$= 2\pi r \times dx$

Total $= \int_0^h 2\pi r \, dx$

$= \left[2\pi rx \right]_0^h$

$= (2\pi rh) - (0) = 2\pi rh$

Radius of cone base $= r$
Slant height of cone $= l$

Area of ELEMENT is
length × breadth
$= 2\pi rx \, \frac{x}{l} \times dx$

Total $= \int_0^l \frac{2\pi r}{l} \times dx$

$= \left[\frac{\pi r}{l} x^2 \right]_0^l$

$= (\pi r l) - (0) = \underline{\underline{\pi r l}}$

The other method, shown in Fig. 19-5B, uses an arbitrarily thin ring at radius x from the center. Here, the area of the element is the length of the ring $2\pi x$ times its thickness dx. Integrating that from 0 to r gives the same well-known result. If you like, you can work through the derivation for yourself and think of it as "extra credit."

CURVED AREAS OF CYLINDERS AND CONES

With cylinders and cones, you can find the area of the curved part of the surface in two ways, as you did with the circle. For the cylinder (Fig. 19-6A), the length of the element is $2\pi r$ and its width is dx. Integrating from 0 to h (the height) produces what, in this case, is fairly obvious: $2\pi rh$. For the cone (Fig. 19-6B), instead of being πrh, it is πrl, where l is the *slant height* of the cone, measured from the apex down to any point on the circumference of the base.

SURFACE AREA OF A SPHERE

To find the surface area of a sphere using integration, imagine a ring that makes its way around the sphere from one pole to the other, as if it were a "moving zone of latitude" going from the earth's north pole to its south pole. Measure the position of the ring according to the angle x with respect to a reference axis from pole to pole as shown in Fig. 19-7. By taking angle x from 0 to π rad, the ring will cover the entire area of the sphere's surface. The circumferential length of the element is $2\pi r \sin x$. Its width is $r \, dx$, so its area, described as a function $a(x)$, is equal to

$$a(x) = 2\pi r^2 \sin x \, dx$$

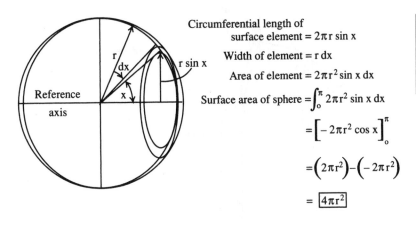

Circumferential length of
surface element = $2\pi r \sin x$

Width of element = $r\,dx$

Area of element = $2\pi r^2 \sin x\,dx$

Surface area of sphere $= \displaystyle\int_0^\pi 2\pi r^2 \sin x\,dx$

$$= \left[-2\pi r^2 \cos x \right]_0^\pi$$

$$= \left(2\pi r^2 \right) - \left(-2\pi r^2 \right)$$

$$= \boxed{4\pi r^2}$$

Figure 19-7

Integration can be used to determine the surface area of a sphere.

Now consider the indefinite integral $\int a(x)\,dx$ as follows:

$$\int 2\pi r^2 \sin x\,dx = -2\pi r^2 \cos x + c$$

You can consider the quantity $2\pi r^2$ as a constant because you're integrating not with respect to r, but with respect to x. Now evaluate the integral from 0 to π. Ignore the constant of integration; it cancels out in the subtraction anyway. When $x = \pi$, $\cos x = -1$, so minus times minus makes a plus. When $x = 0$, $\cos x = 1$, so you end up with $-\cos x = -1$. Again, minus times minus makes a plus. If you follow the calculations in Fig. 19-7, you will see that the final answer is equal to $4\pi r^2$. You should recognize this as the formula for the surface area of a sphere, based on its radius r.

VOLUMES OF WEDGES AND PYRAMIDS

Integration can be used to find the volumes of three-dimensional objects. An example, in which the general formula for the volume of a wedge is determined, is shown in Fig. 19-8. Just as the vanishingly small element of an area is a line or curve of width dx, so the vanishingly small element of a volume is a surface having thickness dx. If we take the volume of this wedge, the area of the element or "slice," based on the distance x from the apex of the wedge, can be defined as a function $a(x)$ such that

$$a(x) = wtx/l$$

where l is the distance from the apex to the "base" (shown as the flat, vertically oriented face at the right in Fig. 19-8), w is the width of the base, and t is the depth of the base. The thickness of the element is dx, so its volume is $a(x)\,dx$, or $(wt/l)x\,dx$. Integrating the function $a(x)$ with respect to x produces this result:

$$\int (wt/l)x\,dx = [wt/(2l)]x^2 + c$$

Figure 19-8

Integration can be used to find the volume of a wedge.

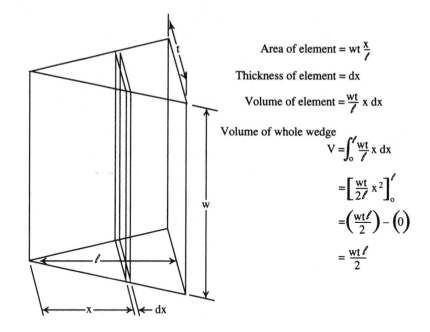

Area of element = $wt\,\dfrac{x}{\ell}$

Thickness of element = dx

Volume of element = $\dfrac{wt}{\ell}\,x\,dx$

Volume of whole wedge

$$V = \int_0^\ell \frac{wt}{\ell}\,x\,dx$$

$$= \left[\frac{wt}{2\ell}\,x^2\right]_0^\ell$$

$$= \left(\frac{wt\ell}{2}\right) - \left(0\right)$$

$$= \frac{wt\,\ell}{2}$$

Evaluating the expression $[wt/(2l)]x^2$ from 0 to 1, you can easily calculate that the volume of the whole wedge is equal to $wtl/2$.

Figure 19-9 illustrates a similar method applied to a pyramid. To make the formula more general, A is used to represent the area of the base, and an element or "slice" is taken at a distance x from the apex. You can follow along with the calculations next to the drawing.

Figure 19-9

Integration can be used to find the volume of a pyramid.

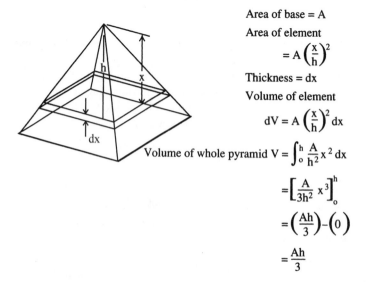

Area of base = A

Area of element

$$= A\left(\frac{x}{h}\right)^2$$

Thickness = dx

Volume of element

$$dV = A\left(\frac{x}{h}\right)^2 dx$$

Volume of whole pyramid $V = \int_0^h \dfrac{A}{h^2}\,x^2\,dx$

$$= \left[\frac{A}{3h^2}\,x^3\right]_0^h$$

$$= \left(\frac{Ah}{3}\right) - \left(0\right)$$

$$= \frac{Ah}{3}$$

VOLUMES OF CONES AND SPHERES

The method for finding the volume of a pyramid can also be applied to find the volume of a cone. Its base is a circle whose area A is πr^2, as shown in Fig. 19-10A. Notice that the height h is the *vertical height*, measured perpendicular to the base, rather than the slant height, which is measured along the curved surface.

The volume of the disk-shaped element dx is $\pi r^2 dx$. The radius of the disk-shaped element is equal to the radius of the base of the whole cone times x/h, or $(r/h)x$. The surface area of one side of this thin disk (either the top or the bottom) can therefore be expressed as a function $a(x)$ like this:

$$a(x) = \pi[(r/h)x]^2$$
$$= (\pi r^2/h^2)x^2$$
$$= (A/h^2)x^2$$

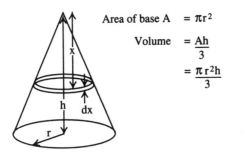

Area of base A $= \pi r^2$

Volume $= \dfrac{Ah}{3}$

$= \dfrac{\pi r^2 h}{3}$

A

Figure 19-10

At A, integration can be used to find the volume of a cone by slicing the cone into thin horizontal disks. At B, the cone is sliced into thin-walled, hollow cylinders.

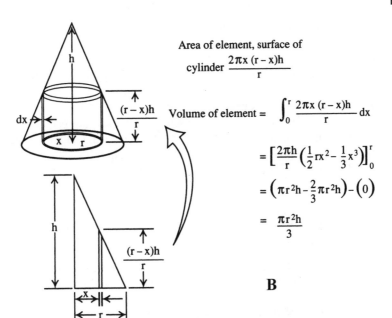

Area of element, surface of

cylinder $\dfrac{2\pi x\,(r-x)h}{r}$

Volume of element $= \displaystyle\int_0^r \dfrac{2\pi x\,(r-x)h}{r}\,dx$

$= \left[\dfrac{2\pi h}{r}\left(\dfrac{1}{2}rx^2 - \dfrac{1}{3}x^3\right)\right]_0^r$

$= \left(\pi r^2 h - \dfrac{2}{3}\pi r^2 h\right) - \left(0\right)$

$= \dfrac{\pi r^2 h}{3}$

B

Figure 19-11

Integration can be used to find the volume of a sphere by slicing the sphere into thin disks perpendicular to the reference axis.

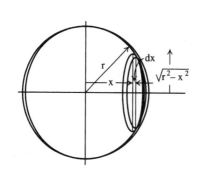

Area of element

$$= \pi\left(\sqrt{r^2 - x^2}\right)^2$$

$$= \pi\left(r^2 - x^2\right)$$

Thickness = dx

Volume of element $= \pi\left(r^2 - x^2\right) dx$

Volume of sphere $= \displaystyle\int_{-r}^{+r} \pi\left(r^2 - x^2\right) dx$

$$= \left[\pi r^2 x - \frac{\pi}{3} x^3\right]_{-r}^{+r}$$

$$= \left(\pi r^3 - \frac{\pi r^3}{3}\right) - \left(-r^3 + \frac{\pi r^3}{3}\right)$$

$$= \frac{4}{3}\pi r^3$$

The volume of the disk-shaped element is $(A/h^2)x^2\, dx$. To find the volume of the whole cone, first note the indefinite integral:

$$\int (A/h^2)x^2\, dx = (1/3)(A/h^2)x^3 + c$$

You know by now that you can get rid of the constant of integration when you evaluate this from $x = 0$ to $x = h$. Therefore the volume V of the entire cone is

$$V = (1/3)(A/h^2)h^3 - (1/3)(A/h^2) \cdot 0^3$$

$$= (1/3)(Ah)$$

$$= Ah/3$$

It checks out! You should recognize this as the familiar formula for the volume of a cone having base radius r and vertical height h.

Another method to find the volume of a cone uses an element that is a cylindrical "shell" of radius x, as shown in Fig. 19-10B. Just follow along with the calculations in the figure.

The simplest method of using integration to find the volume of a sphere (Fig. 19-11) takes thin, disk-shaped "slices" of the sphere, with each slice perpendicular to a reference axis running from pole to pole. The value of x, the variable with respect to which you perform the integration, goes from $-r$ to r. You can follow along with the calculations in the figure to see how it works.

QUESTIONS AND PROBLEMS

This is an open-book quiz. You may refer to the text in this chapter (and earlier ones, too, if you want) when figuring out the answers. Take your time! Consider all values given as exact, so you don't have to worry about significant figures. The correct answers are in the back of the book. This quiz contains questions relevant to both this chapter and the previous one for review.

1. Find the derivatives of the products of functions $u(x)$ and $v(x)$, abbreviated here as u and v, under the following conditions.

    (*a*) $u = 2x + 3$ and $v = x^3$
    (*b*) $u = x^2 - 6x + 4$ and $v = 3 \sin x$
    (*c*) $u = \sin x$ and $v = \cos x$
    (*d*) $u = x^5 - 4$ and $v = -2x^3 + 2$

2. Find the derivatives of the quotients of functions $u(x)$ and $v(x)$, abbreviated here as u and v, under the following conditions.

    (*a*) $u = -3x - 6$ and $v = x^2 + 2$
    (*b*) $u = x^2 + 2x$ and $v = \sin x$
    (*c*) $u = \sin x$ and $v = 3 + \cos x$
    (*d*) $u = 3x^4 - 4$ and $v = -2x^2$

3. In rectangular coordinates, the slope m of a line whose equation is $y = mx + b$, where b is a constant, is equal to dy/dx. Find the slope of a line tangent to a circle centered at the origin and having a radius of 3 units, when the point on the circle through which the tangent line passes corresponds to the following angles counterclockwise from the x axis.

    (*a*) 10°   (*b*) 55°
    (*c*) 105°   (*d*) 190°
    (*e*) 300°   (*f*) 1 rad
    (*g*) 2 rad   (*h*) 4 rad

4. Suppose an oscilloscope shows a *waveform* like the one in Fig. 19-12. Assume each vertical division represents exactly 1 volt (1 V), and each horizontal division represents exactly 1 millisecond (1 ms). The peak signal amplitudes are exactly plus and minus 5 V. The *ramps* (slanted lines) in the waveform are all straight. Assign $t = 0$ for the time at the origin (the center of the display). Now imagine that this signal is passed through a circuit called a *differentiator*, which produces an output waveform that is the derivative of the input waveform. Draw a graph of the output waveform. Include the maximum and minimum amplitudes.

5. Suppose the waveform in Fig. 19-12 is passed through two successive differentiator circuits. Draw a graph of the output waveform from the second differentiator. Include the maximum and minimum amplitudes.

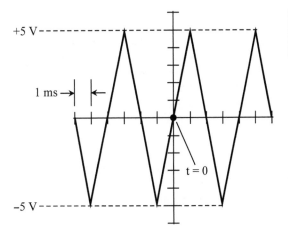

Figure 19-12

Illustration for problems 4 through 7.

Figure 19-13

Illustration for
problems 10
and 11.

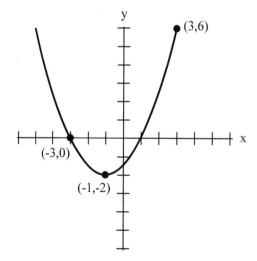

Figure 19-13

Illustration for problems 10 and 11.

6. Suppose the waveform in Fig. 19-12 on page 277 is passed through a circuit called an *integrator,* whose output is the mathematical antiderivative of the input. Draw a graph of the output waveform. Include the maximum and minimum amplitudes.

7. Suppose the waveform in Fig. 19-12 on page 277 is passed through two successive integrators. Draw a graph of the output waveform, including the maximum and minimum amplitudes.

8. Evaluate the following indefinite integrals and check your results by differentiation.
   (*a*) $\int 4x^3 \, dx$ (*b*) $\int 6x^5 \, dx$
   (*c*) $\int 9x^8 \, dx$ (*d*) $\int x^4 \, dx$
   (*e*) $\int 2 \cos x \, dx$ (*f*) $\int -4 \sin x \, dx$

9. Evaluate the integrals from Prob. 8 as definite integrals, with limits of -1 and 1. Assume angle measures in parts (e) and (f) to be in radians.

10. Figure 19-13 shows a quadratic function—that is, a function of the following form:

$$y = ax^2 + bx + c$$

where a, b, and c are constants. You can determine the values of these constants based on the points shown in the graph. Find the area under this curve between the following pairs of vertical lines on the x axis.
   (*a*) $x = -2$ and $x = 0$
   (*b*) $x = 0$ and $x = 1$
   (*c*) $x = -1$ and $x = 2$
   (*d*) $x = -1$ and $x = 4$

11. Find the derivative of the function whose graph is shown in Fig. 19-13. Then find the area under the resulting curve (or straight line, as the case might be) between the same pairs of vertical lines as in Prob. 10.

20
Combining Calculus with Other Tools

In this chapter, you'll learn how calculus can be used along with other techniques to analyze functions, optimize areas and volumes, and evaluate the properties of curves that are sometimes called *conic sections* because they represent cross sections of extended double cones. These curves are the *circle*, the *ellipse*, the *parabola*, and the *hyperbola*.

LOCAL MAXIMUM AND MINIMUM

Derivatives can be used to find "peaks" and "valleys," technically called *local maxima* and *local minima*, of a function that appears as a curve when graphed. When the derivative dy/dx of a curve is equal to 0 at a certain point, the value of the function is momentarily constant. If elsewhere it varies, the value identified by $dy/dx = 0$ is either a local maximum or a local minimum in the function itself, as long as the second derivative, d^2y/dx^2, is different from 0.

If $dy/dx = 0$ at a point but decreases on either side, then $d^2y/dx^2 < 0$ and you have a local maximum at that point, as shown in Fig. 20-1A. If $dy/dx = 0$ at a point but increases on either side, then $d^2y/dx^2 > 0$ and you have a local minimum at that point, as shown in Fig. 20-1B. (In a few moments, you'll see what happens if the first and second derivatives are both equal to 0 at a point.)

As an example of how you can find local maxima and minima by means of derivatives, consider the following function:

$$y = x^4 - x^3 - 18x^2 + 27x$$

You can tabulate some values of x and the corresponding values of y and then mark them out on graph paper. Figure 20-2A lists nine value sets that you can obtain in this way. For some curves, nine points would be enough to draw a good representation of the curve. In this case, you can get the general idea of what the curve will look like by

Figure 20-1

Examples of a local maximum (A) and a local minimum (B) in the graph of a function. Near a local maximum, the derivative decreases. Near a local minimum, the derivative increases.

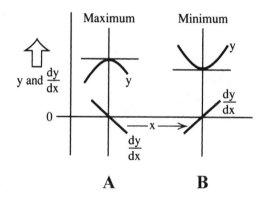

Figure 20-2

Some tabulated points for the function $y = x^4 - x^3 - 18x^2 + 27x$. At B, the points from the table are plotted on graph paper. At C, after finding the two local minima and the single local maximum by differentiation, the curve is filled in.

Values of x	x^4	$-x^3$	$-18x^2$	$+27x$	Values of y
− 4	+ 256	+ 64	− 288	− 108	− 76
− 3	+ 81	+ 27	− 162	− 81	− 135
− 2	+ 16	+ 8	− 72	− 54	− 102
− 1	+ 1	+ 1	− 18	− 27	− 43
0	0	0	0	0	0
+ 1	+ 1	− 1	− 18	+ 27	+ 9
+ 2	+ 16	− 8	− 72	+ 54	− 10
+ 3	+ 81	− 27	− 162	+ 81	− 27
+ 4	+ 256	− 64	− 288	+ 108	+ 12

A

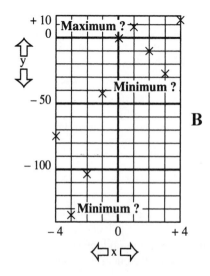

B

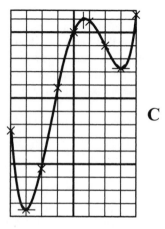

C

plotting nine points, but to get a high degree of precision (not necessary for this exercise) you'd have to use a lot more.

On the graph in Fig. 20-2B, the nine points are marked. Although they give an idea where the curve goes, they are not precise enough to let you be absolutely sure what the curve looks like. A greater number of points might help, but no matter how many points you track in a problem like this, the difficult spots are the local maximum and minimum points. Where, exactly, are they? Apparently, a local minimum exists at or near $x = -3$, a local maximum exists at or near $x = 1$, and another local minimum exists at or near $x = 3$. How can you find out precisely where these points are without spending all day tabulating values and making marks on graph paper, and without using a computer to zero in on these points by "brute force"?

Start out by finding the first derivative of $y = x^4 - x^3 - 18x^2 + 27x$. That's easy. You get this:

$$dy/dx = 4x^3 - 3x^2 - 36x + 27$$

This is a cubic equation in x, so it can have 3 real number roots. Try $x + 3$ as a factor to represent the root at or near $x = -3$. Then the cubic equation factors to

$$dy/dx = (x + 3)(4x^2 - 15x + 9)$$

Factoring the second polynomial, a quadratic, you get $(x - 3)$ and $(4x - 3)$. So the whole cubic factors into

$$dy/dx = (x + 3)(x - 3)(4x - 3)$$

On the basis of these factors, the roots of this cubic equation are easily seen to be $x = -3$, $x = 3$, and $x = 3/4$. Now you know that the two minima are precisely at $x = -3$ and $x = 3$, and the maximum is at $x = 3/4$. From this information, you can trace out the curve on the graph paper quite accurately, as shown in Fig. 20-2C.

POINT OF INFLECTION

Here is a curve-plotting problem in which the second derivative becomes especially significant. Suppose you want to analyze the curve represented by this equation:

$$y = x^4 - 6x^2 + 8x$$

In Fig. 20-3A, six points are tabulated at equal, whole-number increments from $x = -3$ to $x = 2$. You can get an idea of what the graph looks like by plotting these points as shown in Fig. 20-3B. When you first glance at this set of points, you might think there is a mistake in the region between $x = 0$ and $x = 2$. Did you plot enough points? It doesn't look like a curve in that region; it looks more like a straight line. What's going on in this region?

Figure 20-3

Some tabulated points for the function $y = x^4 - 6x^2 + 8x$. At B, the points from the table are plotted on graph paper. At C, after finding the first and second derivatives, the curve is filled in.

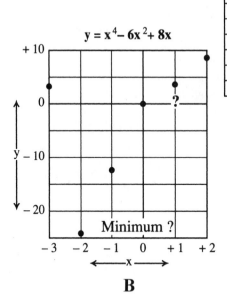

B

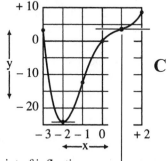

A	Values of x	x^4	$-6x^2$	$+8x$	Values of y
	− 3	81	− 54	− 24	+ 3
	− 2	16	− 24	− 16	− 24
	− 1	1	− 6	− 8	− 13
	0	0	0	0	0
	+ 1	1	− 6	+ 8	+ 3
	+ 2	16	− 24	+ 16	+ 8

C

Point of inflection: concavity of curve reverses

You can tabulate points at smaller intervals and then plot them, but there's another way to find out what's really happening to the graph of the curve in the interval between $x = 0$ and $x = 2$. The derivative of the original function is

$$dy/dx = 4x^3 - 12x + 8$$

This can be factored to

$$dy/dx = 4(x + 2)(x - 1)(x - 1)$$

There are two roots for $x = 1$. What does that mean? Is it possible that there is a local maximum *and* a local minimum at $x = 1$? No, that would be a contradiction! Maybe it's a new characteristic of curves that you haven't seen before. Take the second derivative of the original function:

$$d^2y/dx^2 = 12x^2 - 12$$

When you plug in $x = 1$ to this equation, you get

$$d^2y/dx^2 = (12 \cdot 1 \cdot 1) - 12$$

$$= 0$$

In a situation like this, you have neither a maximum nor a minimum, but something else, called an *inflection point,* or a point of *inflection.* When the second derivative of a function is equal to 0 at a particular point, it means that the *concavity* of the function reverses at that point. That can happen in either of two ways: the curve can change from *concave downward* to *concave upward* as the value of the independent variable increases, or it can go the other way, changing from concave upward to concave downward. If you fill in the graph by connecting the points (Fig. 20-3C), you can see that in this case, the inflection point at $x = 1$ is a place where the curve goes from concave downward to concave upward as you make the value of x more positive.

If you aren't sure what concavity means, look again at Fig. 20-1. The curve in the top part of drawing A is concave downward, and the curve in the top part of drawing B is concave upward. Some mathematicians prefer to use the term "convexity" instead of "concavity." If a curve is concave downward, then it's *convex upward*; if it's concave upward, then it's *convex downward.* Think of concave and convex mirrors!

In the example of Fig. 20-3C, the slope of the function happens to be 0 at the inflection point. That means the curve is momentarily horizontal. This is often the case at inflection points, but not always. It is possible for a curve to have an inflection point where the slope, and therefore the first derivative, is not equal to 0. The graph of the function $y = \tan x$ is a good example. When $x = 0$, the curve is at an inflection point going from concave downward to concave upward; but the slope, represented by dy/dx, is equal to 1. To see how this works, use your calculator to locate points and plot a precision graph of $y = \tan x$ for a couple of dozen values between, say, $x = -\pi/4$ and $x = \pi/4$.

MAXIMUM AND MINIMUM SLOPE

Now, let's return the function considered earlier in this chapter under the heading "Local maximum and minimum":

$$y = x^4 - x^3 - 18x^2 + 27x$$

The second derivative of this is

$$d^2y/dx^2 = 12x^2 - 6x - 36$$

Equating this to 0 and making it into a quadratic equation produce roots of $x = -1.5$ and $x = 2$. The second derivative is equal to 0 where a graph of the first derivative reaches a local maximum or a local minimum. A local maximum in the first derivative is a point of maximum slope (steepest upward) in the original function. Similarly, a local minimum in the first derivative is a point of minimum slope (steepest downward) in the original function. This information can help you plot graphs even more accurately, as shown in Fig. 20-4 for the function $y = x^4 - x^3 - 18x^2 + 27x$.

Figure 20-4

At A, a closer analysis of the graph of $y = x^4 - x^3 - 18x^2 + 27x$, along with its first and second derivatives. At B, some points of signifi-cance in this curve.

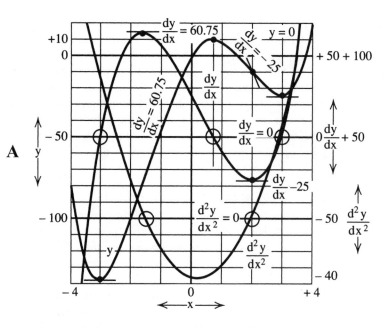

B

X	y	$\dfrac{dy}{dx}$	$\dfrac{d^2y}{dx^2}$	
− 4	− 76	− 133	+ 180	
− 3	− 135 ← − 0		+ 90 ←	Minimum
− 2	− 102	+ 55	+ 24	Maximum
− 1.5	− 72.5625	+60.75 ← 0 ←		Slope +
− 1	− 43	+ 56	− 18	
0	0	+ 27	− 36	Maximum
+ .75	+ 10.01953125 ← 0		−33.75	
+ 1	+ 9	− 8	− 30	Maximum
+ 2	− 10	− 25 ← 0 ←		Slope −
+ 3	− 27 ← − 0		+ 54	
+ 4	+ 12	+ 91	+ 132	Minimum

MAXIMUM AREA WITH CONSTANT PERIMETER

One practical use of differentiation is to find the shape of a rectangle that produces the maximum possible interior area for a given perimeter. You can derive the formula for the area of a rectangle with constant perimeter, as shown in Fig. 20-5. Let p represent one-half of the perimeter of the rectangle. Let x represent the length of one side, and let $p - x$ represent the length of the adjacent side. Then the area A is

$$A = x(p - x)$$
$$= px - x^2$$

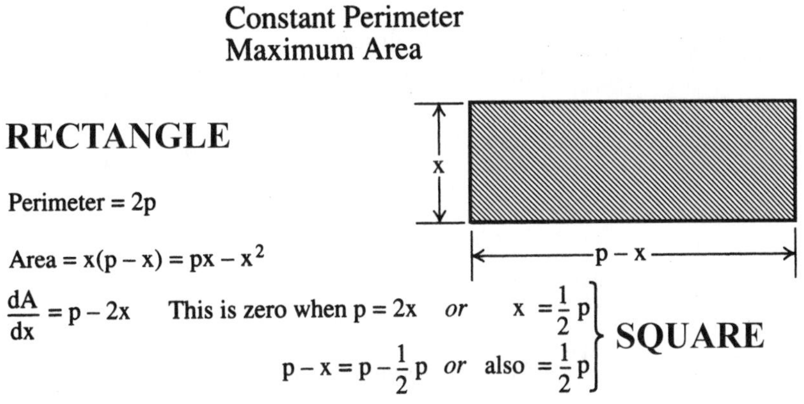

Constant Perimeter
Maximum Area

RECTANGLE

Perimeter = 2p

Area $= x(p - x) = px - x^2$

$\dfrac{dA}{dx} = p - 2x$ This is zero when $p = 2x$ *or* $x = \dfrac{1}{2}p$ $\Big]$

$p - x = p - \dfrac{1}{2}p$ *or* also $= \dfrac{1}{2}p \Big]$ **SQUARE**

Figure 20-5
Differentiation can be used to find the shape of a rectangle with the largest interior area for a constant perimeter. We determine the maximum for the quadratic as shown.

The derivative of the area with respect to x is

$$dA/dx = p - 2x$$

Setting $dA/dx = 0$, you can solve the resulting equation and get $2x = p$ or $x = (1/2)p$. This tells you that the area is maximum when the rectangle is a square with equal dimensions both ways.

If the area of the rectangle is kept constant, the perimeter varies, and the minimum perimeter is solved for, the result is the same. If you like, work it out that way as an "extra credit" exercise!

BOX WITH MINIMUM SURFACE AREA

Now you've seen that a square represents the best shape for a rectangle, assuming you want to get the smallest possible perimeter for a given constant area, or the most area for a given constant perimeter. (Did you already suspect that? Now you've mathematically proved it!) Here's a similar problem. How tall should a square-based box of constant volume be, if you want to have the minimum possible surface area?

Assume that the constant volume of the box is V, and the length of each edge of the base is x, as shown in Fig. 20-6A. The volume is x^3, so the height must be V/x^2. You have a top and bottom that each have an area of x^2, and four sides that each have an area of x times V/x^2, which reduces to V/x. Therefore, the total surface area A is

$$A = 4V/x + 2x^2$$

Differentiating with respect to x while holding V constant gives you this:

$$dA/dx = -4V/x^2 + 4x$$

Figure 20-6

Differentiation can be used to figure out the shape of a box for minimum surface area when the volume is constant. At A, the box has a closed top; at B, the box has an open top.

BOX Constant Volume Minimum Surface Area

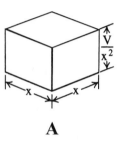

A

Surface area: Top and bottom $2x^2$

Four sides $4x \cdot \dfrac{V}{x^2} = \dfrac{4V}{x}$

Total surface area $= \dfrac{4V}{x} + 2x^2$

$\dfrac{d}{dx}\left(\dfrac{4V}{x} + 2x^2\right) = -\dfrac{4V}{x^2} + 4x$ This is zero when $x^3 = V$

$$\dfrac{V}{x^2} = x$$ **CUBE**

B

Surface area: Bottom x^2

Four sides $4x \cdot \dfrac{V}{x^2} = \dfrac{4V}{x}$

Total surface area $= \dfrac{4V}{x} + x^2$

$\dfrac{d}{dx}\left(\dfrac{4V}{x} + x^2\right) = -\dfrac{4V}{x^2} + 2x$ This is zero when $x^3 = 2V$

$$V = \dfrac{x^3}{2} \qquad \text{Height} = \dfrac{V}{x^2} = \dfrac{x^3}{2x^2} = \dfrac{x}{2}$$

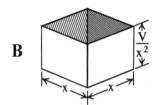

OPEN-TOPPED BOX

If you set this equal to 0 with the idea that you'll get a minimum, you have

$$-4V/x^2 + 4x = 0$$

You want this to be 0 for all possible values of x! That happens if and only if $V = x^3$, making the height equal to the length of the edges of the square base. It follows that the box must be a cube. You're not surprised by this result, are you?

To make this exercise more interesting, suppose that the box has a square base but no top, as shown in Fig. 20-6B. Now, the total surface area of the box is $4V/x + x^2$. Differentiating gives

$$dA/dx = -4V/x^2 + 2x$$

This is equal to 0 if and only if $V = x^3/2$, making the height equal to one-half the length of the edges of the square base.

CYLINDER WITH MINIMUM SURFACE AREA

Here is another optimization problem that differentiation can solve. Suppose you have a *right circular cylinder*, and you want to minimize its surface area for a constant volume. (A right circular cylinder has an axis perpendicular to its base, and a base that is a perfect circle. It is the object that most people imagine when they think of a cylinder.)

The volume V of a right circular cylinder of radius r and height h is

$$V = \pi r^2 h$$

The surface area of the top, which is a disk, is πr^2; the same is true for the bottom. For the curved surface, the area is $2\pi rh$. So the total surface area A of a cylinder such as the one shown in Fig. 20-7A, including the top and the bottom, is

$$A = \pi r^2 + \pi r^2 + 2\pi rh$$
$$= 2\pi r^2 + 2\pi rh$$
$$= 2\pi r(r + h)$$

To obtain only one variable, write h in terms of the constant V and the variable r, so you obtain this:

$$h = V/(\pi r^2)$$

By substituting and rearranging, the equation for total area becomes

$$A = 2\pi r^2 + 2V/r$$

The derivative of the area A with respect to the radius r is

$$dA/dr = 4\pi r - 2V/r^2$$

The value of this function is 0 when $V = 2\pi r^3$. To find h, substitute this in the original formula for V at the beginning of this section, getting

$$2\pi r^3 = \pi r^2 h$$

This solves to $h = 2r$. Twice the radius is the diameter, so you can say that the minimum surface area exists when the height of the cylinder equals its diameter.

Suppose you want a cylindrical container with no top, as shown in Fig. 20-7B. Following the same method as you did for the open box, you will determine that the height needs to be equal to the radius, which is one-half the diameter.

Figure 20-7

Differentiation can be used to figure out the shape of a right circular cylinder for minimum surface area when the volume is constant. At A, the cylinder has a closed top; at B, the cylinder has an open top.

CYLINDRICAL CONTAINER
Minimum Surface Area for Constant Volume

VOLUME $V = \pi r^2 h$

SURFACE AREA $A = 2\pi r^2$ ◁ (top and bottom)
$+ 2\pi rh$ ◁ (curve surface)
$= 2\pi r (r + h)$

A

To get only one variable substitute $h = \dfrac{V}{\pi r^2}$

$$A = 2\pi r \left(r + \frac{V}{\pi r^2}\right) = 2\pi r^2 + \frac{2V}{r}$$

To find minimum $\dfrac{dA}{dr} = 4\pi r - \dfrac{2V}{r^2}$ This is zero when $2\pi r^3 = V$ OR $r^3 = \dfrac{V}{2\pi}$

Substitute to find h: $h = \dfrac{V}{\pi r^2}$

$$= \frac{2\pi r^3}{\pi r^2} = 2r$$

Height equals twice the
radius (or the diameter)

SAME CONTAINER WITHOUT TOP

$$A = \pi r^2 + 2\pi rh$$
$$= \pi r^2 + \frac{2V}{r}$$

B

$\dfrac{dA}{dr} = 2\pi r - \dfrac{2V}{r^2}$ This is zero when $\pi r^3 = V$ OR $r^3 = \dfrac{V}{\pi}$

$$h = \frac{V}{\pi r^2} = \frac{\pi r^3}{\pi r^2} = r$$

Height equals the radius

CONE WITH MINIMUM SURFACE AREA

What is the best shape for an ice cream cone? You want one in which the cone will hold a given volume with minimum surface area.

The volume *V* of a *right circular cone* (one in which the axis is perpendicular to the circular top, as shown in Fig. 20-8) is

$$V = (1/3)\pi r^2 h$$

CONICAL CONTAINER WITHOUT TOP

Constant Volume
Minimum Surface Area (Curved)

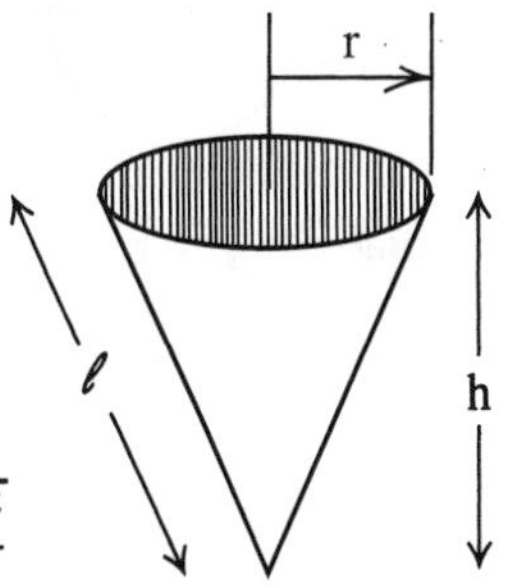

VOLUME $V = \frac{1}{3}\pi r^2 h$

SURFACE AREA $\quad A = \pi r \ell$
$$= \pi r \sqrt{r^2 + h^2}$$

Substitute $h = \frac{3V}{\pi r^2}$: $\quad A = \pi r \sqrt{r^2 + \frac{9V^2}{\pi^2 r^4}} = \sqrt{\pi^2 r^4 + \frac{9V^2}{r^2}}$

$\dfrac{dA}{dr} = \dfrac{1}{2\sqrt{\pi^2 r^4 + \frac{9V^2}{r^2}}} \left(4\pi^2 r^3 - \frac{18V^2}{r^3}\right)$ This is zero when $2\pi^2 r^6 = 9V^2$

or $\sqrt{2}\pi r^3 = 3V$

Substitute to find h: $h = \dfrac{3V}{\pi r^2} = \dfrac{\sqrt{2}\pi r^3}{\pi r^2} = \sqrt{2}r$

Figure 20-8
Differentiation can be used to figure out the shape of a right circular cone, not including the disk-shaped top, for minimum surface area when the volume is constant.

The surface area A, not including the circular top, is

$$A = \pi r l$$

where l is the *slant height* of the cone, measured from a point on the edge of the circular top down to the apex. If you want to write it in terms of h instead, you get

$$A = \pi (r^2 + h^2)^{1/2}$$

Substituting to get h in terms of V and r, then substituting that into the second equation for A, differentiating and equating to zero, you find that h, for minimum surface area, is equal to the square root of 2 multiplied by r. That means the height should be approximately 1.414 times the radius, or 0.707 times the diameter. You can follow along with the calculations in Fig. 20-8 to see how it works.

This problem, looked at in a different way, is one of finding the most volume for a given constant surface area. Do you see many ice cream cones that are wider at the top than they are tall? No? If they were the optimum shape, maybe ice cream vendors would make less profit per cone, but their customers would be happier.

EQUATIONS FOR CIRCLES, ELLIPSES, AND PARABOLAS

Figure 20-9 illustrates three common *second-order curves*: the circle, the ellipse, and the parabola. They are called second-order because they can all be expressed by equations that go no higher than the second power (or square) of the independent variable.

Figure 20-9

Second-order curves. At A, circle centered at the origin. At B, general circle. At C, ellipse centered at the origin. At D, general ellipse. At E, parabola "resting" on the origin. At F, general parabola. In drawings B, D, and F, the origin is indicated by the letter o.

A

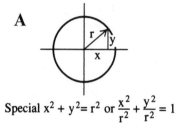

Special $x^2 + y^2 = r^2$ or $\dfrac{x^2}{r^2} + \dfrac{y^2}{r^2} = 1$

B

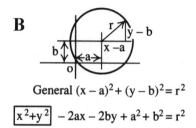

General $(x - a)^2 + (y - b)^2 = r^2$

$\boxed{x^2 + y^2} - 2ax - 2by + a^2 + b^2 = r^2$

C

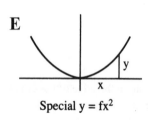

Special $\dfrac{x^2}{q^2} + \dfrac{y^2}{s^2} = 1$

D

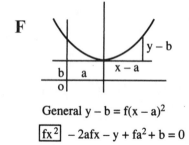

General $\dfrac{(x - a)^2}{q^2} + \dfrac{(y - b)^2}{s^2} = 1$

$\boxed{\dfrac{x}{q^2} + \dfrac{y^2}{s^2}} - \dfrac{2ax}{q^2} - \dfrac{2by}{s^2} + \dfrac{a^2}{q^2} + \dfrac{b^2}{s^2} = 1$

E

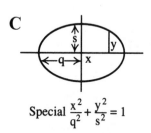

Special $y = fx^2$

F

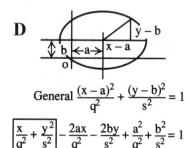

General $y - b = f(x - a)^2$

$\boxed{fx^2} - 2afx - y + fa^2 + b = 0$

At A, C, and E in the figure, you will see the equation, in its simplest form, for each of these curves. In using these equations, both the circle and the ellipse are centered at the *origin*, where $x = 0$ and $y = 0$. The parabola is different. It rests on the origin, but extends upward in the positive y direction for negative or positive values of x. Note that the parabola is symmetric with respect to the y axis. This property is called *bilateral symmetry*.

In drawings B, D, and F in Fig. 20-9, you will see more general equations for the same curves, which are not necessarily centered or resting on the origin. Instead, the point on which the curve is centered or rests is at $x = a$ and $y = b$. Deriving these general forms is a simple matter. You merely substitute $(x - a)$ for x and $(y - b)$ for y in the original simple form, and then multiply out. Notice that the relationship between the second-order terms (those that involve x^2 and y^2) is not affected by this transformation. This fact lets you recognize curves that are circles, ellipses, or parabolas from their respective equations.

In the circle, r represents the radius. In the ellipse, two constants replace the radius, designated q and s. These constants represent one-half the *principal axes* of the ellipse.

These principal axes can be regarded as maximum (major) and minimum (minor) diameters of the ellipse. In the case of the parabola, the curve is symmetrical on either side of a vertical line $x = a$, and rests on the horizontal line defined by $y = b$. The constant f in Fig. 20-9E and F determines the "sharpness" of the curve. If f is positive, the parabola opens upward. If f is negative, the curve opens downward.

DIRECTRIX, FOCUS, AND ECCENTRICITY

When you examine Fig. 20-9, you can see that the parabola is fundamentally different from the circle and the ellipse. The circle and ellipse are closed; the parabola is open. A circle can be drawn with an instrument called a *compass* that has its point at the center. An ellipse can be drawn with a device that uses two centers to elongate it. But how can you generate a parabola?

If you drive around a circular racetrack, you do not have a compass attaching you to the center of the track! You must direct your course with the steering wheel. Paralleling that idea, visualize going along a parabolic course, positioning yourself with respect to two things, a *focus* and a *directrix*. The focus is a point, and the directrix is a straight line, as shown in Fig. 20-10. If you want to follow a parabola, you must stay *equidistant* (equally far away) from the focus and the directrix at all times. That constant distance is designated u. Now imagine a straight line passing through the focus and intersecting the directrix at a right angle. Along this line, the two equal distances are each called the *focal length* and are symbolized f. Drawing a line through the focus parallel to the directrix, you can divide u, measuring your distance from the directrix, into $2f$ and y.

In Fig. 20-10, Fact 1 relates to the u that expresses your distance from the focus. Fact 2 relates to the u that expresses your distance from the directrix. Combine these equations, rearrange the result, and you have an equation of the same form as the equation for the parabola in Fig. 20-9F. Because the two distances denoted by u are equal, this curve has an *eccentricity* of 1, or *unity*.

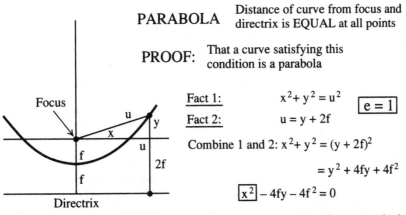

PARABOLA   Distance of curve from focus and directrix is EQUAL at all points

PROOF: That a curve satisfying this condition is a parabola

Fact 1: $x^2 + y^2 = u^2$ $\boxed{e = 1}$

Fact 2: $u = y + 2f$

Combine 1 and 2: $x^2 + y^2 = (y + 2f)^2$

$$= y^2 + 4fy + 4f^2$$

$$\boxed{x^2} - 4fy - 4f^2 = 0$$

Which is an equation for a parabola

Figure 20-10

All the points on a the curve of a parabola are equidistant from the focus and the directrix.

THE ELLIPSE AND THE CIRCLE

Now suppose you have a curve in which the eccentricity e is less than 1 but larger than 0. You can use the same geometric arrangement to determine the eccentricity as you did with the parabola, but now the distance from the focus is eu instead of u, as shown in the two examples of Fig. 20-11. When the value of e is between, but not including, 0 and 1, mathematicians write it this way:

$$0 < e < 1$$

Apply the same two facts and combine them, as you did for the parabola. You will end up with an equation for an ellipse.

As the eccentricity e approaches 0, the center of the ellipse gets farther from the directrix, eu gets closer to f in value, and the ellipse gets less elongated. If you set $e = 0$, the equation simplifies so that it represents a circle, and then $f = r$. Going the other way, as the eccentricity e approaches 1, the center of the ellipse gets closer to the directrix, the value of eu gets farther from f, and the ellipse gets more elongated. If the eccentricity

Figure 20-11

At A, construction of an ellipse based on a defined focus and directrix. At B, an ellipse with smaller eccentricity is more nearly a circle. When the eccentricity becomes equal to 0, the ellipse becomes a circle.

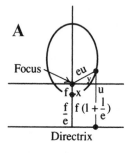

A

Focus

eu

$\dfrac{f}{e}$ $f(1+\dfrac{1}{e})$ u

Directrix

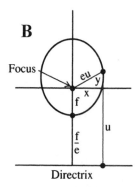

B

Focus

eu y

x

f

$\dfrac{f}{e}$ u

Directrix

ELLIPSE Distance of curve from focus and directrix is constant ratio at all points

Distance from focus on axis $= f$

Distance from directrix on axis $= \dfrac{f}{e}$ $\boxed{e < 1}$

PROOF: That a curve satisfying this condition is an ellipse

Fact 1: $x^2 + y^2 = e^2 u^2$ Fact 2: $y + f(1 + \dfrac{1}{e}) = u$

Combine 1 and 2: $x^2 + y^2 = e^2 \{y + f(1 + \dfrac{1}{e})\}^2$

$$= e^2 y^2 + 2ef(1+e)y + f^2(1+e)^2$$

Rearrange $\boxed{x^2 + (1 - e^2) y^2} - 2ef(1+e)y = f^2(1+e)^2$

Which is an equation for an ellipse

As e becomes smaller, distance from directrix becomes greater; focus is nearer center of ellipse; ellipse becomes more like a circle

A CIRCLE is a second-order curve with an ECCENTRICITY $\boxed{e = 0}$

$x^2 + y^2 = f^2$ (f is then r)

e reaches 1, the ellipse "breaks open" at one end and becomes a parabola. Now you know three things about second-order curves:

- If $e = 0$, you have a circle.
- If $0 < e < 1$, you have an ellipse.
- If $e = 1$, you have a parabola.

What happens if e becomes greater than 1? You'll find out soon!

RELATIONSHIPS BETWEEN FOCUS, DIRECTRIX, AND ECCENTRICITY

Look at these three curves—the circle, the ellipse, and the parabola—in terms of the new parameters. The circle has a single focus, which is the center, and the directrix removes itself to infinity. (You can imagine it as infinitely far away in any direction. Or, if you prefer, you can consider it not to exist at all.) Viewed as an algebraic equation for a circle, the coefficients of the second-order terms, x^2 and y^2, are equal.

The ellipse has two foci at finite distance from each other. The ellipse is bilaterally symmetric with respect to a straight line connecting these foci. As is the case with the circle, the algebraic equation of the ellipse has two second-order terms, x^2 and y^2, but their coefficients are not equal. With two symmetric foci, the ellipse has two directrixes at finite distances.

The parabola can be viewed as having two foci: one finite and one infinite. Its single directrix is at a distance equal to the focal length. Its algebraic equation has only one second-order term.

The properties of these three second-order curves are summarized in Fig. 20-12.

Curve Type	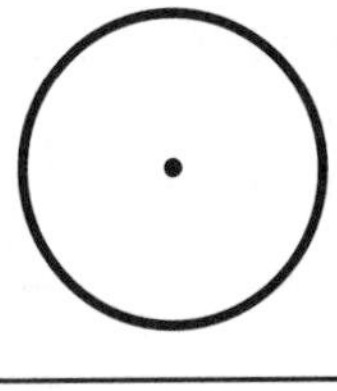 (circle)	 (ellipse)	 (parabola)
Focus	Single	2 Finite	1 Finite 1 Infinite
Directrix	At Infinity	2 at finite distance	Single
Eccentricity	0	< 1	= 1
2nd-Order Terms	Equal	Unequal	Only one

Figure 20-12

Basic characteristics of circles, ellipses, and parabolas.

FOCUS PROPERTY OF PARABOLA

You can rearrange the equation for a parabola to get an equation for x in terms of y and the focal length f. Looking at Fig. 20-13, examine the angles at a specific point (x, y) on the curve: θ is the angle between a line from the focus and the parabola's axis, and ϕ is the slope of the curve expressed as an angle relative to the x axis. Then you can see that the following trigonometric properties hold true:

$$\tan \theta = y/x$$

and

$$\tan \phi = dy/dx$$

Using these facts and by examining the drawing and calculations in Fig. 20-13, you can deduce the relationship $\theta = 2\phi$. This is significant, because it defines the way a *parabolic reflector* can focus or direct beams of radiant energy.

When a ray emerges from the focal point, it always travels in a straight line. Eventually, unless it runs right along the axis of the parabola, the ray strikes the parabola and is reflected from it. Remember from physics that the angle of incidence equals the angle of reflection. If you analyze the drawing in Fig. 20-13 carefully, you will see that any ray emerging from the focus reflects from the parabola parallel to the axis. This is how a lantern, searchlight, or radar transmitting antenna works. In the reverse sense, any ray coming in parallel to the axis will strike the parabola and be reflected to the focal point. This is how a reflecting telescope or dish-type radio receiving antenna works.

Figure 20-13

The focal property of a parabola defines how a reflector of this shape can focus or direct visible light, sound, or radio waves.

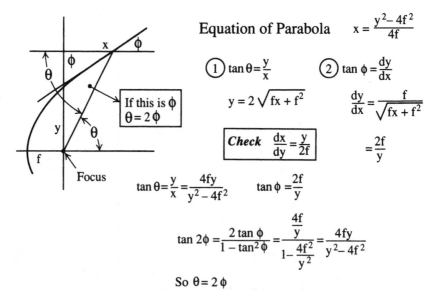

Equation of Parabola $\quad x = \dfrac{y^2 - 4f^2}{4f}$

① $\tan \theta = \dfrac{y}{x}$ $\qquad$ ② $\tan \phi = \dfrac{dy}{dx}$

$$y = 2\sqrt{fx + f^2}$$

$$\dfrac{dy}{dx} = \dfrac{f}{\sqrt{fx + f^2}}$$

$$= \dfrac{2f}{y}$$

Check $\quad \dfrac{dx}{dy} = \dfrac{y}{2f}$

If this is ϕ $\theta = 2\phi$

$$\tan \theta = \dfrac{y}{x} = \dfrac{4fy}{y^2 - 4f^2} \qquad \tan \phi = \dfrac{2f}{y}$$

$$\tan 2\phi = \dfrac{2 \tan \phi}{1 - \tan^2 \phi} = \dfrac{\dfrac{4f}{y}}{1 - \dfrac{4f^2}{y^2}} = \dfrac{4fy}{y^2 - 4f^2}$$

So $\theta = 2\phi$

FOCUS PROPERTY OF ELLIPSE

Applying the same idea to an ellipse, you can write equations based on the axes, as shown in Fig. 20-14. In this example, *a* is one-half the length of the major axis, and *b* is one-half the length of the minor axis. These are called the *major semiaxis* and the *minor semiaxis*, respectively. The eccentricity is *e*. Just follow along with the calculations in the drawing. You will discover that the following two equations hold true:

$$a/b = 1/(1 - e^2)^{1/2}$$

and

$$e = (1 - b^2/a^2)^{1/2}$$

The angular relationship of the parabola explains why a light source at the focus reflects into a parallel outgoing beam, and why a parallel incoming beam converges on the focus. Extending this to the ellipse, instead of focusing into a parallel beam (Fig. 20-15A), the curve focuses rays from one focal point to the other (Fig. 20-15B).

Have you ever been in a room where the walls were curved, intersecting the floor in a huge ellipse? You can stand at one focus and have a friend stand at the other. When your

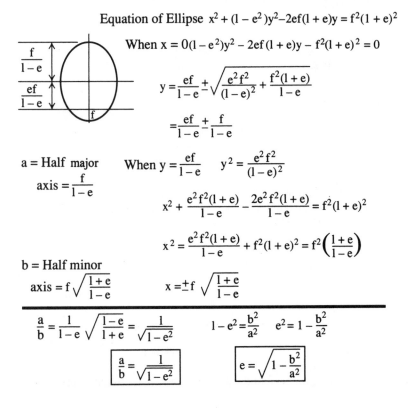

Equation of Ellipse $x^2 + (1 - e^2)y^2 - 2ef(1 + e)y = f^2(1 + e)^2$

When $x = 0$, $(1 - e^2)y^2 - 2ef(1 + e)y - f^2(1 + e)^2 = 0$

$$y = \frac{ef}{1-e} \pm \sqrt{\frac{e^2 f^2}{(1-e)^2} + \frac{f^2(1+e)}{1-e}}$$

$$= \frac{ef}{1-e} \pm \frac{f}{1-e}$$

a = Half major axis $= \dfrac{f}{1-e}$

When $y = \dfrac{ef}{1-e}$ $\quad y^2 = \dfrac{e^2 f^2}{(1-e)^2}$

$$x^2 + \frac{e^2 f^2(1+e)}{1-e} - \frac{2e^2 f^2(1+e)}{1-e} = f^2(1+e)^2$$

$$x^2 = \frac{e^2 f^2(1+e)}{1-e} + f^2(1+e)^2 = f^2\left(\frac{1+e}{1-e}\right)$$

b = Half minor axis $= f\sqrt{\dfrac{1+e}{1-e}}$

$$x = \pm f\sqrt{\frac{1+e}{1-e}}$$

$$\frac{a}{b} = \frac{1}{1-e}\sqrt{\frac{1-e}{1+e}} = \frac{1}{\sqrt{1-e^2}} \qquad 1 - e^2 = \frac{b^2}{a^2} \qquad e^2 = 1 - \frac{b^2}{a^2}$$

$$\boxed{\frac{a}{b} = \frac{1}{\sqrt{1-e^2}}} \qquad \boxed{e = \sqrt{1 - \frac{b^2}{a^2}}}$$

Figure 20-14

Relationships between the eccentricity of an ellipse and the lengths of the major and minor semi-axes.

In the figure:

$\dfrac{f}{1-e}$

$\dfrac{ef}{1-e}$

f

Figure 20-15

Reflection proper-
ties of a parabola
(at A) and an
ellipse (at B).

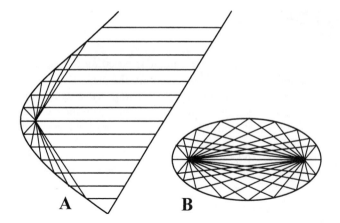

A   B

friend whispers, it sounds as if he is speaking right into both of your ears at the same
time. It works even better in a special room with a three-dimensional, egglike shape
called an *ellipsoid*, which results from the rotation of an ellipse around its major axis.
A suspended walkway runs along the major axis. When two people stand so their heads
are exactly at the foci, they can whisper to each other even if the ellipse is many feet
across. Rooms like this have actually been built to demonstrate the principle!

HYPERBOLA: ECCENTRICITY GREATER THAN 1

The circle has eccentricity equal to 0; the ellipse has eccentricity larger than 0 but less
than 1; the parabola has an eccentricity of exactly 1. What happens if the eccentricity is
greater than 1? You get a curve called a *hyperbola*.

Look at the equations for each type of curve. For the circle, the coefficients of x^2 and y^2
are equal. For the ellipse, they are unequal. For the parabola, one is of the coefficients
is equal to 0. In the case of the hyperbola, one of the coefficients reverses its sign. That
is, one of them is negative and the other is positive.

Let's compare equations for a generalized circle, a generalized ellipse, and a generalized hyperbola, each centered at the origin. For the circle, you have

$$x^2 + y^2 = r^2$$

where r is the radius. For the ellipse, you have

$$x^2/q^2 + y^2/s^2 = 1$$

where q and s are the lengths of the semiaxes. For the hyperbola, you have

$$x^2/q^2 - y^2/s^2 = 1$$

where q and s are again called the semiaxes, but in a different sense from the ellipse.

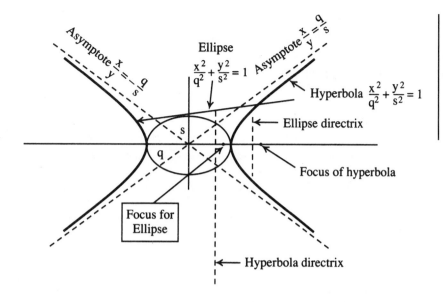

Figure 20-16

Fundamental prop-
erties of a hyper-
bola. The central
ellipse is drawn for
reference only; it is
not part of the
hyperbola itself.

Figure 20-16 shows an example of a hyperbola. (The ellipse between the curves is there for reference only; it defines the semiaxes q and s.) The hyperbola appears, upon casual inspection, like two parabolas back to back. But there's an important difference in the shape of a hyperbola compared with the shape of a parabola. The parameters that help define hyperbolas are straight lines called *asymptotes*. In Fig. 20-16, there are two asymptotes through the origin (the point where $x = 0$ and $y = 0$) whose equations are

$$x/y = q/s$$

and

$$x/y = -q/s$$

The curve approaches the asymptotes as you move away from the origin. However, the curve never quite touches the asymptotes, no matter how far from the origin you go.

Parabolas do not have asymptotes. Hyperbolas always have them. If the asymptotes intersect at right angles to each other, the major and minor semiaxes have equal length. Then $q = s$, and the curve is called a *right hyperbola*.

In the simplest cases, such as the one shown in Fig. 20-16, the asymptotes of a hyperbola intersect at the origin. However, they can intersect at any point $x = a$ and $y = b$. Then the equation of the hyperbola generalizes to this:

$$(x - a)^2/q^2 - (y - b)^2/s^2 = 1$$

This is the counterpart of the generalized equation for the ellipse from Fig. 20-9D:

$$(x - a)^2/q^2 + (y - b)^2/s^2 = 1$$

where $x = a$ and $y = b$ represent the point where the axes (instead of the asymptotes) intersect. If $q = s$, then the ellipse is a circle.

GEOMETRY OF CONIC SECTIONS

At the beginning of this chapter, you were introduced to the notion of a conic section. What, you ask, do cones have to do with circles, ellipses, parabolas, and hyperbolas? The answer can be found in the ways flat planes and double cones intersect.

Imagine a double right circular cone that extends infinitely both upward and downward, as shown in the drawings of Fig. 20-17. Now imagine a flat, infinitely large plane that can be moved around so it slices through the cone in various ways. The intersection

Figure 20-17

The conic sections can be defined by the intersection of a flat plane with a double cone. At A, a circle. At B, an ellipse. At C, a parabola. At D, a hyperbola.

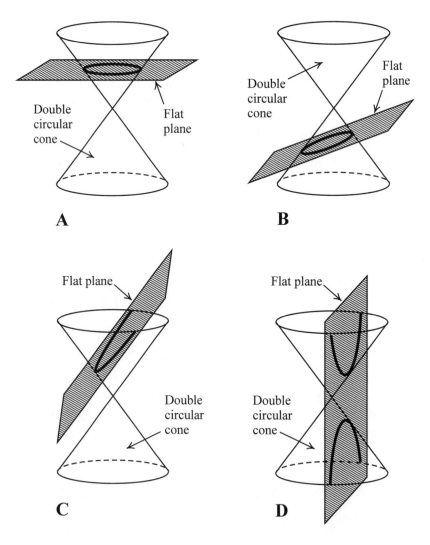

between the plane and the double cone will always be a circle, an ellipse, a parabola, or a hyperbola.

Figure 20-17A shows what happens when the plane is perpendicular to the axis of the double cone. In that case, you get a circle. In Fig. 20-17B, the plane is not perpendicular to the axis of the cone, but it isn't tilted very much. The curve is still closed, but it isn't a circle; it's an ellipse. As the plane tilts farther away from a right angle with respect to the double-cone axis, the ellipse becomes more elongated; its eccentricity increases. Then you reach an angle of tilt where the curve is no longer closed. At this threshold, the intersection between the plane and the cone is a parabola (Fig. 20-17C), and the eccentricity is equal to 1.

So far, the plane has only intersected one-half of the double cone. If you keep tilting the plane beyond the angle at which the intersection is a parabola, the plane will intersect both halves of the cone. Then you get a hyperbola, and the eccentricity is greater than 1. The greatest possible tilt, where the plane is parallel to the axis of the double cone, produces a hyperbola with the maximum possible eccentricity, as shown in the example of Fig. 20-17D.

QUESTIONS AND PROBLEMS

This is an open-book quiz. You may refer to the text in this chapter (and earlier ones, too, if you want) when figuring out the answers. Take your time! Consider all values given as exact, so you don't have to worry about significant figures. The correct answers are in the back of the book.

1. Find the point (x, y) at which the graph of the following function attains its maximum value:

$$y = f(x) = -3x^2 + 2x + 2$$

2. Consider the following function:

$$y = f(x) = 2 \sin x + 2$$

There are infinitely many points (x, y) at which the slope of this function is maximum. Write a general expression for these points.

3. Suppose you are standing on a large, frozen lake at night, holding a flashlight. The beam from the flashlight is cone-shaped. The outer face of the light cone subtends a $20°$ angle with respect to the axis (the ray corresponding to the beam center). How can you aim the flashlight to form a circular region of light on the ice?

4. In the situation described in Prob. 3, how can you point the flashlight so the edge of the region of light forms an ellipse on the ice?

5. In the situation described in Prob. 3, how can you aim the flashlight so the edge of the region of light forms a parabola on the ice?

6. In the situation described in Prob. 3, how can you aim the flashlight so the edge of the region of light forms a half-hyperbola on the ice?

7. Determine x and y intercepts, local maxima, local minima, and inflection points (if any) for these functions. Specify the points as ordered pairs (x, y), where $y = f(x)$.

(a) $f(x) = 2x^3 - 5$ (b) $f(x) = (x^2 + 3)(x - 4)$

(c) $f(x) = 2 \tan x$ (d) $f(x) = -3 \cos x$

8. Using the information you obtained in solving Prob. 7, plot graphs of the functions given. Intercept points, local maxima, local minima, and points of inflection (if any) should be labeled. Otherwise, the curves can be approximately drawn.

21

Alternative Coordinate Systems

The coordinate plane you have seen so far, called the *Cartesian plane* or the *xy plane*, is not the only means by which points can be located and curves graphed on a flat surface. Instead of moving right/left and up/down from an origin or reference point, you can travel outward in a specified direction from that point. You aren't restricted to two dimensions, either. In this chapter, you'll learn about some three-dimensional coordinate systems.

MATHEMATICIAN'S POLAR COORDINATES

The forms of *polar coordinate plane* most often used by mathematicians are shown in Figs. 21-1 and 21-2. The independent variable is plotted as an angle θ counterclockwise from a *reference axis* pointing to the right (or "east"), and the dependent variable is plotted as the distance or radius r from the origin. Coordinate points are denoted as ordered pairs (θ, r). In some texts, it goes the other way; r is the independent variable and coordinates are plotted as ordered pairs (r, θ). This book uses the former method because most people find it easier to think of the angle as the independent variable on which the radius depends, rather than vice versa.

In a polar coordinate plane, specific radius values are shown by concentric circles. The larger the circle, the greater the value of r. In Figs. 21-1 and 21-2, the circles are not labeled in actual units. You can therefore imagine each concentric circle, working outward, as increasing by whatever number of units is convenient. For example, each radial division might represent 1 unit, or 5 units, or 10, or 100. It is important, however, that all the increments be the same. The radius on the plotted graph must be directly proportional to the value of the corresponding variable in the mathematical function. In other words, the radial scale must be linear.

Direction can be expressed in degrees or radians counterclockwise from a reference axis pointing to the right or "east." In Fig, 21-1, the direction angle θ is expressed in degrees. Figure 21-2 shows the same polar plane, using radians to express the direction angle. Regardless of whether degrees or radians are used, the physical angle on the

Figure 21-1

The basic polar
coordinate plane.
The angle θ is in
degrees, and the
radius r is in arbi-
trary units.

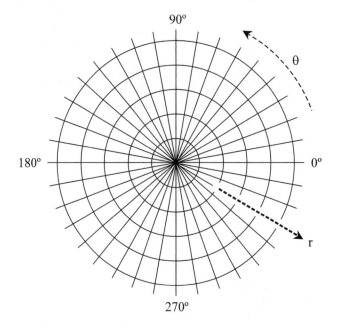

graph must be directly proportional to the value of θ in the mathematical function. The angular scale, like the radial scale, must be linear.

In mathematician's polar coordinates, it is all right to have a negative radius. If some point is specified with $r < 0$, you can multiply r by -1 so it becomes positive, and then add or subtract $180°$ (π rad) to or from the direction. That's like saying, "Proceed 10 km east" instead of "Proceed minus 10 km west." Negative radii make it possible to graph functions whose corresponding variables sometimes attain negative values.

Figure 21-2

Another form of
the polar coordi-
nate plane. The
angle θ is in
radians, and the
radius r is in
arbitrary units.

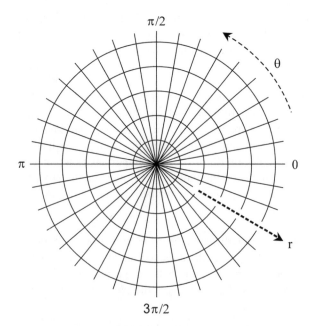

It's also allowed to have nonstandard direction angles. If the value of θ is $360°$ (2π rad) or more, it represents more than one complete counterclockwise revolution from the $0°$ (0 rad) reference axis. If the direction angle is less than $0°$ (0 rad), it represents clockwise revolution instead of counterclockwise revolution. Nonstandard direction angles make it possible to graph functions whose angles go outside the standard angle range, or whose corresponding variables sometimes have negative values.

SOME EXAMPLES

To see how the polar coordinate system works, you can plot the graphs of some familiar objects. Figures whose equations are complicated in the xy plane can often be expressed more simply in polar coordinates. But sometimes, polar graphs are a lot more difficult to express mathematically than their Cartesian-plane counterparts. In this section, you'll see some of the most "polar-coordinate-friendly" functions.

The equation of a circle centered at the origin in the polar plane is given by the following formula:

$$r = a$$

where a is a real number constant greater than 0. An example of this is illustrated in Fig. 21-3. This is a lot simpler than the equation for the same circle in the xy plane!

When the center of the circle moves away from the origin, things get a little messy. Here's one special example that's not too difficult. The general form for the equation of a circle passing through the origin and centered at the point (θ_0, r_0) in the polar plane (Fig. 21-4) is as follows:

$$r = 2r_0 \cos (\theta - \theta_0)$$

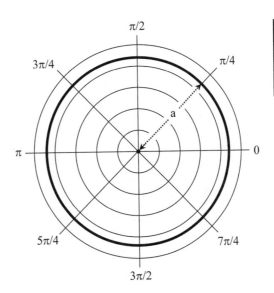

Figure 21-3

Polar graph of a circle centered at the origin, with radius a.

Figure 21-4

Polar graph of a circle passing through the origin, with center at (θ_0, r_0) and radius r_0.

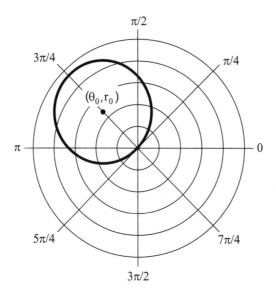

The equation of an ellipse centered at the origin (Fig. 21-5) in the polar plane is given by the following formula:

$$a^2 b^2 / (a^2 \sin^2 \theta + b^2 \cos^2 \theta) = 1$$

where a and b are positive real number constants. In the ellipse, a represents the distance from the origin to the curve as measured along the "horizontal" ray $\theta = 0$, and b represents the distance from the origin to the curve as measured along the "vertical" ray $\theta = \pi/2$. The values a and b therefore represent the lengths of the semiaxes of the ellipse. The greater value is the length of the major semiaxis, and the lesser value is the length of the minor semiaxis.

Figure 21-5

Polar graph of an ellipse centered at the origin, with semi-axes a and b.

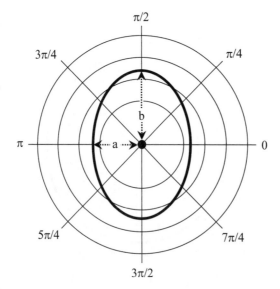

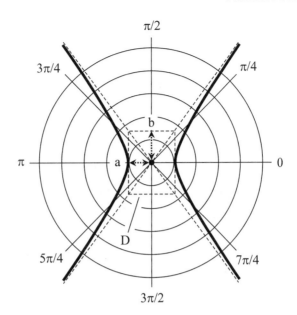

Figure 21-6
Polar graph of a
hyperbola cen-
tered at the origin,
with semi-axes
a and *b*.

The general form of the equation of a hyperbola centered at the origin (Fig. 21-6) in the polar plane is given by the following formula:

$$a^2 b^2 / (a^2 \sin^2 \theta - b^2 \cos^2 \theta) = 1$$

where *a* and *b* are positive real number constants. To define the hyperbola, suppose *D* is a rectangle centered at the origin with vertical edges tangent to the hyperbola, and with vertices (corners) that lie on the asymptotes of the hyperbola as shown. Let *a* represent the distance from the origin to *D* as measured along the "horizontal" ray $\theta = 0$, and let *b* represent the distance from the origin to *D* as measured along the "vertical" ray $\theta = \pi/2$. The values *a* and *b* represent the lengths of the semiaxes of the hyperbola.

The general form of the equation of a *lemniscate* or "sideways figure-8" centered at the origin in the polar plane is given by the following formula:

$$r^2 = a^2 \cos 2\theta$$

where *a* is a real number constant greater than 0, representing the maximum radius. An example is illustrated in Fig. 21-7.

The general form of the equation of a *three-leafed rose* centered at the origin in the polar plane is given by either of the following two formulas:

$$r = a \cos 3\theta$$

$$r = a \sin 3\theta$$

Figure 21-7

Polar graph of a lemniscate centered at the origin, with radius *a*.

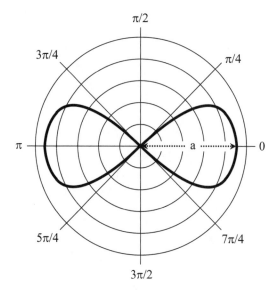

where *a* is a positive real number constant. The cosine curve is illustrated in Fig. 21-8A; the sine curve is illustrated in Fig. 21-8B.

The general form of the equation of a *four-leafed rose* centered at the origin in the polar plane is given by either of the following two formulas:

$$r = a \cos 2\theta$$
$$r = a \sin 2\theta$$

Figure 21-8A

Polar graph of a three-leafed rose with equation $r = a \cos 3\theta$.

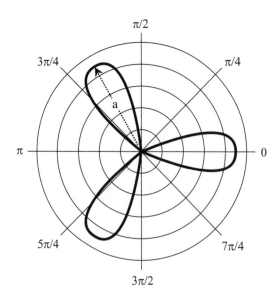

A

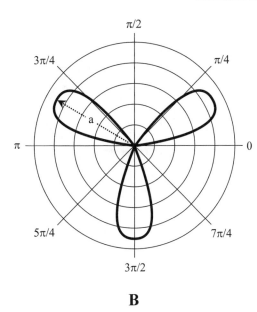

Figure 21-8B

Polar graph of a three-leafed rose with equation $r = a \sin 3\theta$.

B

where a is a positive real number constant. The cosine curve is illustrated in Fig. 21-9A; the sine curve is illustrated in Fig. 21-9B.

Do you wonder why the curves in Figs. 21-8 and 21-9 are not called "clovers"? In mathematics, things are sometimes named in mysterious or confusing ways!

The general form of the equation of a *spiral* centered at the origin in the polar plane is given by this formula:

$$r = a\theta$$

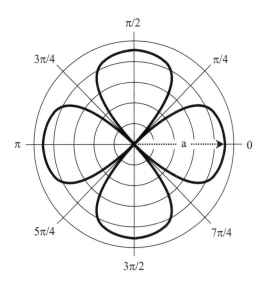

Figure 21-9A

Polar graph of a four-leafed rose with equation $r = a \cos 2\theta$.

A

Figure 21-9B

Polar graph of a four-leafed rose with equation $r = a \sin 2\theta$.

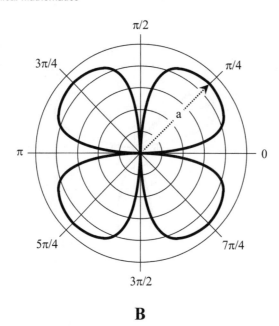

B

where a is a positive real number constant. An example of this type of spiral, called the *spiral of Archimedes* because of the uniform manner in which its radius increases as the angle increases, is shown in Fig. 21-10 for values of θ between 0 and $5\pi/2$.

The general form of the equation of a *cardioid* centered at the origin in the polar plane is given by the following formula:

$$r = 2a\,(1 + \cos\theta)$$

Figure 21-10

Polar graph of a spiral of Archimedes.

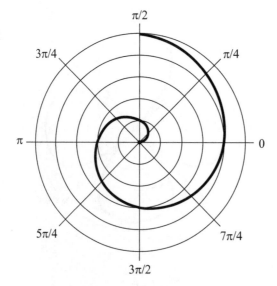

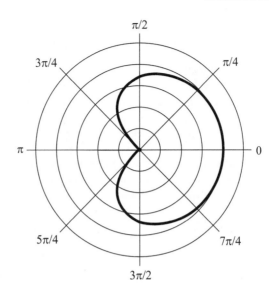

Figure 21-11
Polar graph of
a cardioid.

where *a* is a positive real number constant. An example of this type of curve is illustrated in Fig. 21-11. The term "cardioid" means "heart-shaped curve."

COMPRESSED POLAR COORDINATES

Figure 21-12 shows a variant of the polar coordinate plane with the radial scale graduated geometrically, rather than in a linear fashion. The point corresponding to 1 on the *r* axis is halfway between the origin and the outer periphery, which is labeled ∞ (the "infinity" symbol). Successive whole-number points are placed halfway between previous whole-number points and the outer periphery. In this way, the entire polar

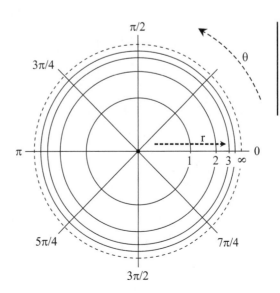

Figure 21-12
A polar coordinate
plane with a
"geometrically
compressed"
radial axis.

coordinate plane can be portrayed inside an open circle having a finite radius, even though it covers a mathematically infinite expanse!

The radial scale of this coordinate system can be magnified or minimized by multiplying all the values on the r axis by a constant. This allows various relations and functions to be plotted, reducing the distortion in particular regions of interest. Distortion relative to the conventional polar coordinate plane is greatest near the periphery and is least near the origin. Plot a few graphs using this system, and see for yourself!

The same sort of axis compression scheme can be used with the xy plane. You won't see this done very often in technical books and papers, but it can be interesting because it provides a "view to infinity" that conventional coordinate systems do not.

MATHEMATICIAN'S POLAR VS. CARTESIAN COORDINATES

Figure 21-13 shows a point graphed on superimposed Cartesian and polar coordinate systems. If you know the Cartesian coordinates (x_0, y_0) of a point P, you can convert to mathematician's polar coordinates (θ_0, r_0) using these formulas:

$$\theta_0 = \arctan(y_0/x_0) \text{ if } x_0 > 0$$

$$\theta_0 = 180° + \arctan(y_0/x_0) \text{ if } x_0 < 0 \text{ (for } \theta_0 \text{ in degrees)}$$

Figure 21-13

Conversion between polar and Cartesian coordinates. Each radial division represents one unit. Each division on the x and y axes also represents one unit.

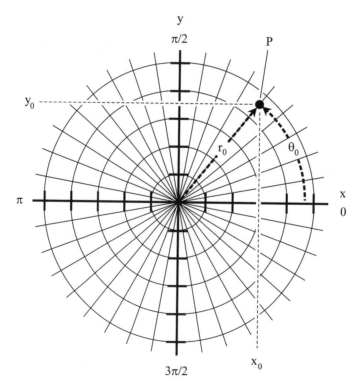

$$\theta_0 = \pi + \arctan (y_0/x_0) \text{ if } x_0 < 0 \text{ (for } \theta_0 \text{ in radians)}$$

$$r_0 = (x_0^2 + y_0^2)^{1/2}$$

The arctangent or arctan function, also symbolized $\tan^{-1}$, is the inverse of the tangent function. For example, arctan $1 = \pi/2$ because $\tan \pi/2 = 1$ for angles in radians. You can't have $x_0 = 0$ because that produces an undefined quotient in the conversion formula to θ_0. If a value of θ_0 thus determined happens to be negative, you can add 360° or 2π rad to get the "legitimate" value.

A point $P = (\theta_0, r_0)$ in mathematician's polar coordinates can be converted to its counterpart (x_0, y_0) in Cartesian coordinates by the following formulas:

$$x_0 = r_0 \cos \theta_0$$

$$y_0 = r_0 \sin \theta_0$$

These same formulas can be used, by means of substitution, to convert Cartesian coordinate relations to polar coordinate relations, and vice versa. The general Cartesian-to-polar conversion formulas can be written by simply getting rid of subscripts so the symbols indicate variables instead of constants. Then you end up with these formulas:

$$\theta = \arctan (y/x) \text{ if } x > 0$$

$$\theta = 180° + \arctan (y/x) \text{ if } x < 0 \text{ (for } \theta \text{ in degrees)}$$

$$\theta = \pi + \arctan (y/x) \text{ if } x < 0 \text{ (for } \theta \text{ in radians)}$$

$$r = (x^2 + y^2)^{1/2}$$

The general polar-to-Cartesian conversion formulas are

$$x = r \cos \theta$$

$$y = r \sin \theta$$

NAVIGATOR'S POLAR COORDINATES

Navigators and military people use a coordinate plane similar to the one preferred by mathematicians. But there's a difference in the way the angles are specified. The radius is called the *range*, and real-world units such as meters (m) or kilometers (km) are commonly used. The angle is called the *azimuth*, *heading*, or *bearing*, and it is measured in degrees clockwise from north, instead of counterclockwise from a reference axis. The basic scheme is shown in Fig. 21-14. The azimuth is symbolized α (the lowercase Greek alpha), and the range is symbolized r. The position of a point is definable by an ordered pair (α_0, r_0).

Figure 21-14

The navigator's polar coordinate plane. The azimuth α is in degrees measured clockwise from north. The range r is in arbitrary units.

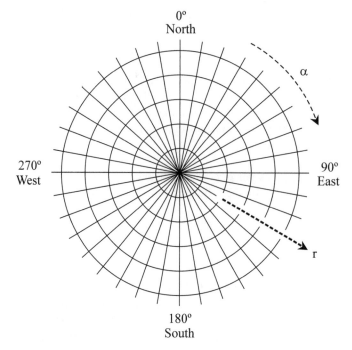

There are two ways of defining north, or 0°. The more accurate standard uses *geographic north*. This is the direction in which you should travel if you want to take the shortest possible route over the earth's surface to the *north geographic pole*. The less accurate standard uses *magnetic north*. This is the direction indicated by the needle in a magnetic compass.

In navigator's polar coordinates, the range can never be negative. No navigator talks about traveling –20 km on a heading of 270°, for example, when she really means to say that her vessel is going 20 km on a heading of 90°. When you are working out certain problems, it's possible that the result might contain a negative range when you're finished calculating. If this happens, the value of r should be multiplied by –1 to make it positive, and then the value of α should be increased or decreased by 180° so the angle is at least 0° but less than 360°.

The azimuth, bearing, or heading in navigator's polar coordinates must conform to certain restrictions. The smallest possible value of α is 0° (representing north). As you turn clockwise as seen from above, the values of α increase through 90° (east), 180° (south), 270° (west), and ultimately approach, but never reach, 360° (north again). You therefore have these limitations on the possible values of an ordered pair (α, r):

$$0° \leq \alpha < 360°$$

$$r \geq 0$$

MATHEMATICIAN'S VS. NAVIGATOR'S POLAR COORDINATES

Sometimes it is necessary to convert from mathematician's polar coordinates (you can call them MPC for short) to navigator's polar coordinates (NPC), or vice versa. When you're making the conversion, the radius or range of a particular point r_0 is the same in both systems, so no change is necessary. But the angles differ.

If you know the direction angle θ_0 of a point in MPC and you want to find the equivalent azimuth angle θ_0 in NPC, first be sure θ_0 is expressed in degrees, not radians. Then you can use whichever of these conversion formulas is applicable:

$$\alpha_0 = 90° - \theta_0 \text{ if } 0° \leq \theta_0 \leq 90°$$

$$\alpha_0 = 450° - \theta_0 \text{ if } 90° < \theta_0 < 360°$$

If you know the azimuth α_0 of a point in NPC and you want to find the equivalent direction angle θ_0 in MPC, then you can use either of the following conversion formulas, whichever applies:

$$\theta_0 = 90° - \alpha_0 \text{ if } 0° \leq \alpha_0 \leq 90°$$

$$\theta_0 = 450° - \alpha_0 \text{ if } 90° < \alpha_0 < 360°$$

Note that the symbol $\leq$ means "less than or equal to," but $<$ means "strictly less than." Similarly, the symbol $\geq$ means "greater than or equal to," but $>$ means "strictly greater than." These distinctions are subtle but important!

NAVIGATOR'S POLAR VS. CARTESIAN COORDINATES

Here are the conversion formulas for translating the coordinates of a point (α_0, r_0) in NPC to its counterpart (x_0, y_0) in the Cartesian xy plane:

$$x_0 = r_0 \sin \alpha_0$$

$$y_0 = r_0 \cos \alpha_0$$

These are similar to the formulas used to convert MPC to Cartesian coordinates, except that the roles of the sine and cosine functions are reversed.

To convert the coordinates of a point (x_0, y_0) in Cartesian coordinates to a point (α_0, r_0) in NPC, use these formulas:

$$\alpha_0 = \arctan (x_0/y_0) \text{ if } y_0 > 0$$

$$\alpha_0 = 180° + \arctan (x_0/y_0) \text{ if } y_0 < 0$$

$$r_0 = (x_0{}^2 + y_0{}^2)^{1/2}$$

You can't have $y_0 = 0$, because that produces an undefined quotient. If a value of α_0 thus determined happens to be negative, add 360° to get the "legitimate" value. These are similar to the formulas used to convert Cartesian coordinates to MPC.

LATITUDE AND LONGITUDE

Latitude and *longitude* are directional angles that uniquely define the positions of points on the surface of a sphere of fixed radius. These angles can also define specific places in the heavens as they appear to an earthbound observer.

The latitude/longitude scheme for defining geographic locations on the surface of the earth is illustrated in Fig. 21-15A. The *polar axis* connects two specified *antipodes*, or points directly opposite each other, on the sphere. These points are assigned latitude $\theta = 90°$ (north pole) and $\theta = -90°$ (south pole). The *equatorial axis* runs outward from the center of the sphere at a right angle to the polar axis, and in a predetermined direction. It is assigned longitude $\phi = 0°$. In Fig. 21-15A, the earth is not shown because it would clutter up the drawing. Imagine the earth's surface as a sphere whose center corresponds to the point where all three of the straight-line axes intersect. The plane of the equator intersects the surface of the earth in a circle that you recognize as the true equator.

Latitude θ is measured positively (north) and negatively (south) relative to the plane of the equator. Longitude ϕ is measured counterclockwise (positively) and clockwise

Figure 21-15A

Latitude and longitude angles for locating points on the earth's surface.

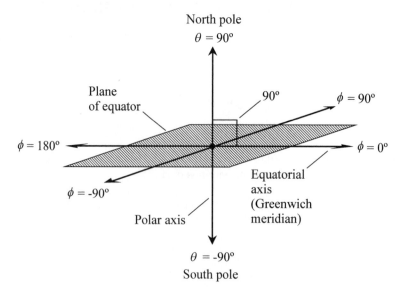

A

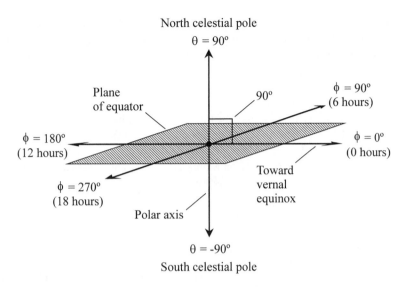

North celestial pole
θ = 90°

Plane
of equator

$\phi = 180°$
(12 hours)

90°

$\phi = 90°$
(6 hours)

$\phi = 0°$
(0 hours)

Toward
vernal
equinox

$\phi = 270°$
(18 hours)

Polar axis

θ = -90°

South celestial pole

B

Figure 21-15B

Declination and
right ascension
angles for locating
points in the sky.

(negatively) in the plane of the equator and relative to the equatorial axis. The angles
are restricted as follows:

$$-90° \leq \theta \leq 90°$$

$$-180° < \phi \leq 180°$$

On the earth's surface, the half-circle connecting the 0° longitude line with the north
and south poles passes through the town of Greenwich, England (near London). This
half-circle is called the *Greenwich meridian,* or the *prime meridian.* Longitude angles
are defined with respect to it.

Pairs of latitude and longitude angles such as $(\theta, \phi) = (-30°, -110°)$ translate into specific points on the earth's surface. But such pairs of angles can also translate into positions in the sky. These positions are not really points, but are rays pointing out from an
earthbound observer's eyes indefinitely into space. A single point defined in this way
revolves through the heavens in a huge imaginary circle as the earth rotates on the earth's
axis, unless the point happens to coincide with either the north pole or the south pole.

CELESTIAL COORDINATES

Celestial latitude and *celestial longitude* are extensions of latitude and longitude angles
into the heavens. The same set of coordinates used for geographic latitude and longitude applies to this system. An object whose celestial latitude and longitude coordinates
are (θ, ϕ) appears at the *zenith* in the sky (directly overhead) to an observer standing at
the point on the earth's surface whose latitude and longitude coordinates are (θ, ϕ).

The big problem with celestial latitude and longitude, as far as astronomers are concerned, is the annoying fact that the whole system turns right along with the earth. You can't define any particular star's coordinates using celestial latitude and longitude unless that star happens to lie on the polar axis. To overcome this problem, astronomers invented coordinates called *declination* and *right ascension* that define the positions of objects in the sky relative to the stars, rather than with respect to the earth. Figure 21-15B illustrates the basic principles of this system. Declination (θ) is identical to celestial latitude. Right ascension (ϕ) is measured eastward from the *vernal equinox*, which is the position of the sun in the heavens at the exact start of the spring season in the northern hemisphere (usually on March 20 or 21). The angles are restricted as follows:

$$-90° \le \theta \le 90°$$

$$0° \le \phi < 360°$$

Astronomers sometimes employ a peculiar scheme to express angles of right ascension. Instead of measuring right ascension in degrees or radians, they use units called *hours*, *minutes*, and *seconds* based on 24 hours in a complete circle, corresponding to the 24 hours in the solar day. That means each hour of right ascension is equivalent to 15°. Does this confuse you? It gets worse! Minutes and seconds of right ascension are different from the angular *minutes of arc* and *seconds of arc* used by mathematicians and scientists. One minute of right ascension is 1/60 of an hour or (1/4)°, and one second of right ascension is 1/60 of a minute or (1/240)°. In Fig. 21-15B, you can divide the degree values of ϕ by 15 to get hours of right ascension.

CARTESIAN THREE-SPACE

An extension of the *xy* plane into three dimensions is called *Cartesian three-space* or *xyz space*. Figure 21-16 shows the basic idea. The independent variables are plotted along the *x* and *y* axes; the dependent variable is plotted along the *z* axis. Each axis is

Figure 21-16

Cartesian three-space, also called *xyz*-space.

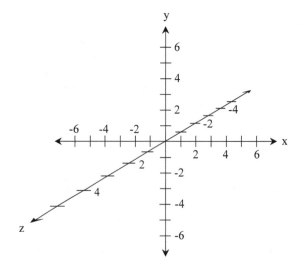

perpendicular to the other two. They all intersect at the origin where $x = 0$, $y = 0$, and $z = 0$. Points are represented by *ordered triples* written in the form (x,y,z). As with ordered pairs, there are no spaces after the commas in ordered triples.

The scales in Cartesian three-space are all linear. But the divisions (that is, the spaces between hash marks) on different axes do not necessarily have to represent the same increments. For example, the x axis might have 1 unit per division, the y axis 10 units per division, and the z axis 5 units per division.

CYLINDRICAL COORDINATES

Figure 21-17 shows two systems of *cylindrical coordinates* for specifying the positions of points in three-space.

In the system shown in Fig. 21-17A, you start with Cartesian three-space. Then an angle θ is defined in the xy plane, measured in degrees or radians (usually radians) counterclockwise from the positive x axis or *reference axis*. Given a point P in space, consider its projection P' onto the xy plane, such that line segment PP' is parallel to the z axis. The position of P is defined by the ordered triple (θ, r, h), where θ represents the angle measured counterclockwise between P' and the positive x axis in the xy plane, r represents the distance or radius from P' to the origin, and h represents the distance (altitude or height) of P above the xy plane. This scheme for cylindrical coordinates is preferred by mathematicians, and it is also used occasionally by engineers and scientists.

In the system shown in Fig. 21-17B, you again start with Cartesian three-space. The xy plane corresponds to the surface of the earth in the vicinity of the origin, and the z axis runs straight up (positive z values) and down (negative z values). The angle θ is defined

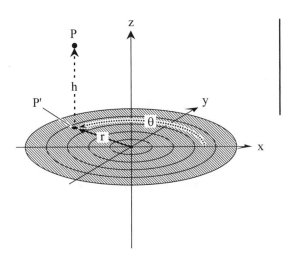

Figure 21-17A

Mathematician's cylindrical coordinates for defining points in three-space.

A

Figure 21-17B

Astronomer's and navigator's cylindrical coordinates for defining points in three-space.

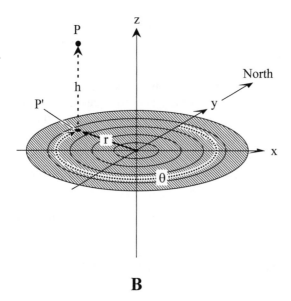

B

in the *xy* plane in degrees *clockwise* from the positive *y* axis, which corresponds to north. Given a point P in space, consider its projection P' onto the *xy* plane, such that line segment PP' is parallel to the *z* axis. The position of P is defined by the ordered triple (θ, r, h), where θ represents the angle measured clockwise between P' and a line running north toward the horizon, r represents the distance or radius from P' to the origin, and h represents the distance (altitude or height) of P above the *xy* plane. This system is often used by navigators, meteorologists, and astronomers.

Negative radii and nonstandard angles are commonly specified in the cylindrical coordinate system of Fig. 21-17A, but rarely or never in the system of Fig. 21-17B.

SPHERICAL COORDINATES

Figure 21-18 shows three systems of *spherical coordinates* for defining points in space. The first two are used by astronomers and aerospace scientists, while the third one is preferred by navigators and surveyors.

In the system shown in Fig. 21-18A, the location of a point P is defined by the ordered triple (θ, ϕ, r), where θ represents the declination of P, ϕ represents the right ascension of P, and r represents the radius or range from P to the origin. In this example, angles are specified in degrees (except in the case of the astronomer's version of right ascension, which is expressed in hours, minutes, and seconds as defined earlier in this chapter). Alternatively, the angles can be expressed in radians. This system is fixed relative to the stars. Negative radii and nonstandard angles are never used.

Instead of declination and right ascension, the variables θ and ϕ can represent celestial latitude and celestial longitude, respectively, as shown in Fig. 21-18B. This system is

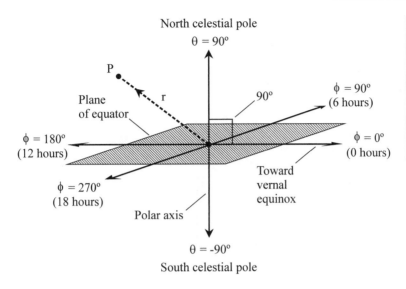

North celestial pole
$\theta = 90°$

South celestial pole
$\theta = -90°$

A

Figure 21-18A
Spherical coordinates in which the angles θ and ϕ represent the declination and right ascension, and r represents the radius or range.

fixed relative to the earth, rather than relative to the stars. This system is sometimes used by aviators, military people, and meteorologists. Again, negative radii and non-standard angles are never used.

There's a third alternative for spherical coordinates. The angle θ can represent the elevation (the angle above or below the horizon), and ϕ can represent the azimuth, bearing, or heading, measured clockwise from north. The radius or range r is, as in the other

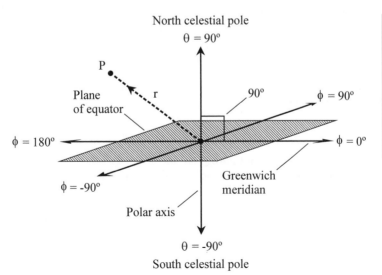

North celestial pole
$\theta = 90°$

South celestial pole
$\theta = -90°$

B

Figure 21-18B
Spherical coordinates in which the angles θ and ϕ represent the latitude and longitude, and r represents the radius or range.

Figure 21-18C

Spherical coordi-
nates in which the
angles θ and ϕ
represent the ele-
vation (angle
above the horizon)
and azimuth, and
r represents the
radius or range.

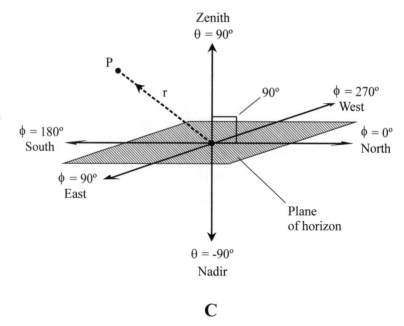

C

two systems, the distance of the specified point from the origin. In this arrangement, the
reference plane corresponds to the horizon as the observer sees it. That means the whole
system is based on where the observer happens to be. The elevation angle can cover the
span of values between and including $-90°$ (the *nadir*, or a distant point directly under-
foot) and $+90°$ (the zenith, or a distant point directly overhead). This system is shown
in Fig. 21-18C. In a variant sometimes used by theoreticians, the angle θ is measured
downward with respect to the zenith rather than upward or downward with respect to
the plane of the horizon. In that system, the angular range is $0° \le \theta \le 180°$. Negative radii
and nonstandard angles are allowed.

QUESTIONS AND PROBLEMS

This is an open-book quiz. You may refer to the text in this chapter (and earlier ones,
too, if you want) when figuring out the answers. Take your time! Consider all values
given as exact, so you don't have to worry about significant figures. The correct answers
are in the back of the book.

1. Figure 21-19 is a graph of a straight line in the Cartesian plane. The x intercept and
y intercept points are shown. Draw a graph of this curve in mathematician's polar coordinates, substituting r for x and θ (in radians) for y. Restrict θ to the range 0 to 2π.

2. Draw a graph of Fig. 21-19 in mathematician's polar coordinates, substituting r for
y and θ (in radians) for x. Restrict θ to the range 0 to 2π.

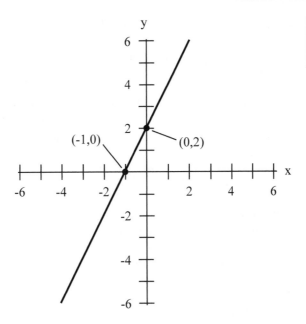

Figure 21-19
Illustration for
problems 1 and 2.

3. Figure 21-20 is a graph of a certain function $r = f(\theta)$ in mathematician's polar coordinates. It happens to be a straight line in this system. A single point (θ_0, r_0) is specified, and the angle of the line relative to the $0°$ reference axis is also given. Create an approximate graph of the corresponding relation $y = f(x)$ in the Cartesian xy plane by directly substituting x for θ and y for r. Consider only positive values for the variables in both sets of coordinates.

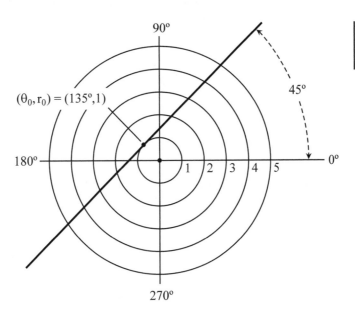

Figure 21-20
Illustration for
problems 3 and 4.

4. Refer again to Fig. 21-20 and the description given in Prob. 3. Create an approximate graph of the corresponding relation $x = f(y)$ in the Cartesian xy plane by directly substituting y for θ and x for r. Consider only positive values for the variables in both sets of coordinates.

5. What are the points (θ_0, r_0) in mathematician's polar coordinates representing the following points (x_0, y_0) in the Cartesian xy plane? In each of the final answers, express the radius as a positive real number, and express the angle as a positive value in radians less than 2π.

(a) (1, 1)   (b) (−1, 1)
(c) (1,−1)   (d) (−1,−1)

6. What are the points (x_0, y_0) in the Cartesian xy plane representing the following points (α_0, r_0) in navigator's polar coordinates?

(a) (90°, 5)   (b) (135°, 2)
(c) (225°, 3)   (d) (270°, 4)

7. What do the following equations represent in Cartesian xyz space, as the system is portrayed in Fig. 21-16?

(a) $x = 5$ (b) $y = 7$
(c) $z = -5$ (d) $x = y = z$

8. What do the following equations represent in cylindrical coordinates, as the system is portrayed in Fig. 21-17A? Allow negative radii and nonstandard angles.

(a) $\theta = \pi/2$ (b) $r = 4$
(c) $h = -3$ (d) $r = h$

9. What do the following equations represent in spherical coordinates, as the system is portrayed in Fig. 21-18C? Allow negative radii and nonstandard angles.

(a) $r = 2$ (b) $\theta = 45°$
(c) $\phi = 45°$

22

Complex Numbers

Long ago, humankind didn't accept fractions, only whole numbers could be counted. Then people could count only positive quantities, because negative numbers were impossible for them to imagine. After negative numbers became familiar, the square roots of negative real numbers were defined, producing imaginary numbers. The resulting quantities were combined with real numbers to get complex numbers. You were introduced to imaginary and complex numbers earlier in this book. Now, you'll learn how to work with these numbers and find out how they can be useful in solving problems that would be much more difficult without them.

IMAGINARY QUANTITIES

Think of all the real numbers as points on a straight line. A certain point is assigned the value 0. Then positive numbers go in one direction from that point, and negative numbers are expressed in a 180° reversal from the positive numbers. For example, if the positive numbers go out on a straight line toward the right, the negative numbers go toward the left. Imaginary numbers can then be placed on another line that runs upward for positive and downward for negative.

Figure 22-1 shows geometric representations of the real (A) and imaginary numbers (B) when portrayed on straight lines. This can help you visualize the relationship between the real numbers and the imaginary numbers (C). Multiplying any number by -1 in the plane of Fig. 22-1C produces rotation through 180°. Multiplying by -1 again produces another rotation through 180°, which brings you back to the original number. Minus times minus equals plus, remember!

You don't need to worry about whether the rotation is clockwise or counterclockwise when you multiply by -1. You reverse the direction, that's all! But when you multiply a number by the unit imaginary operator i, which represents the positive square root of -1, things are a little more interesting. Multiplying by i is the equivalent of rotating counterclockwise through 90° in a geometric representation such as the one shown in Fig. 22-1C.

Figure 22-1

At A, the real number line. At B, the positive imaginary number half-line (i) is perpendicular to the real number line. At C, the positive and negative imaginary numbers ($+i$ and $-i$) lie on a complete geometric line perpendicular to the real number line.

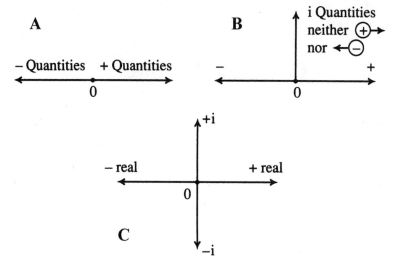

Figure 22-2 shows what happens when you repeatedly multiply a positive real number by i. When you first multiply, you get a positive imaginary number (Fig. 22-2A). When you multiply by i twice, you rotate through 180° as shown in Fig. 22-2B. When you multiply by i three times, the rotation is 270° (Fig. 22-2C). If you multiply by i four times, you go around in a complete circle and return to the original number (Fig. 22-2D).

Figure 22-2

At A, multiplication of a real number by i. At B, multiplication by $i \times i$ or i^2. At C, multiplication by $i \times i \times i$ or i^3. At D, multiplication by $i \times i \times i \times i$ or i^4.

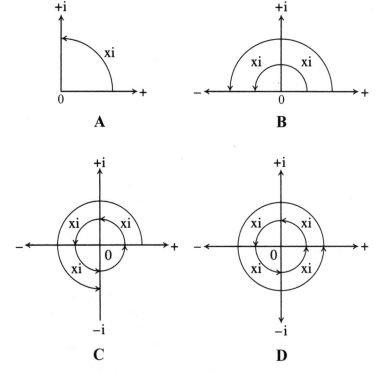

Multiplying repeatedly by i works the same way no matter what number you start out with. You can begin with a negative real, a positive imaginary, a negative imaginary, or even the sum of a real number and an imaginary one. Each multiplication by i always turns the *complex number vector* (a ray connecting the origin with the point in the coordinate plane representing the number) counterclockwise through 90°. The length of the vector does not change unless you multiply a number by some quantity ib, where b is a real number. In that case, the vector length is multiplied by b, and it rotates through 90° also. For example, if you multiply something by $i2$, the vector turns 90° counterclockwise and becomes twice as long; if you multiply by $i/12$ or $i(1/12)$, the vector turns 90° counterclockwise and becomes 1/12 as long.

THE COMPLEX NUMBER PLANE

The set of rectangular coordinates shown in Fig. 22-2A through D is known as the *complex number plane*, in which you can plot quantities that are part real and part imaginary. The real part is expressed toward the right for positive and toward the left for negative. The imaginary part goes upward for positive and downward for negative. Any point in the plane, representing a unique complex number, can be expressed as an ordered pair (a,ib) or written algebraically as $a + ib$, where a and b are real numbers and i is the unit imaginary number.

If $a = 0$, a complex number is called *pure imaginary*. If $b = 0$, a complex number is called *pure real*. If both parts are positive, the quantity is in the first quadrant of the complex number plane. If the real part is negative and the imaginary part is positive, the quantity is in the second quadrant. If both parts are negative, the quantity is in the third quadrant. If the real part is positive but the imaginary part is negative, the quantity is in the fourth quadrant.

MULTIPLYING COMPLEX QUANTITIES

Here is a geometric method by which you can multiply complex numbers. Suppose you have two complex quantities that can be portrayed both in rectangular coordinates as real and imaginary parts and in mathematician's polar coordinates as a vector magnitude and a direction angle. In Fig. 22-3, two complex quantities $a + ib$ and $c + id$ are set out, each beginning at the positive real number axis. In polar form these complex numbers are (θ,r) and (ϕ,s), respectively. Complete the triangles formed by these vectors, the real axis, and vertical lines running from the plotted points down to the real axis. These triangles are shaded in the diagram. Beginning at the magnitude of the first quantity and multiplying each part of the second by the magnitude of the first, you can erect a third, unshaded triangle. This triangle brings you to the product, in magnitude and angle, or it can be read in rectangular coordinates as real and imaginary parts. Study the calculations next to the diagram to see how the quantities that appear in the algebra are reproduced in coordinate geometry.

Figure 22-3

Geometric representation of the product of two complex quantities.

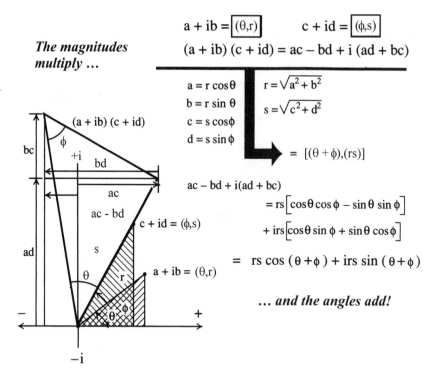

The magnitudes multiply ...

$$a + ib = \boxed{(\theta, r)} \qquad c + id = \boxed{(\phi, s)}$$

$$(a + ib)(c + id) = ac - bd + i(ad + bc)$$

$$a = r\cos\theta \qquad r = \sqrt{a^2 + b^2}$$
$$b = r\sin\theta$$
$$c = s\cos\phi \qquad s = \sqrt{c^2 + d^2}$$
$$d = s\sin\phi$$

$$= [(\theta + \phi), (rs)]$$

$$ac - bd + i(ad + bc)$$

$$= rs\left[\cos\theta\cos\phi - \sin\theta\sin\phi\right]$$
$$+ irs\left[\cos\theta\sin\phi + \sin\theta\cos\phi\right]$$

$$= rs\cos(\theta + \phi) + irs\sin(\theta + \phi)$$

... and the angles add!

A simpler situation exists when you want to square a complex number. In this case, you multiply the magnitude of the original vector by itself, and then you rotate the vector counterclockwise by an angle equal to its original angle with respect to the positive real number axis. That is, you double its counterclockwise direction angle. For example, if you want to square a complex number with a vector length of 4 that points 70° counterclockwise from the positive real axis, you increase the vector length to 4^2, or 16, and you double its angle to 70° · 2, or 140°. If you want to square the complex number 3 + i4, you can draw it on a coordinate plane and figure out that it has a length of exactly 5 (from the Pythagorean theorem) and points at an angle of arctan (4/3), or approximately 53° counterclockwise from the positive real axis. Squaring this gives you a vector with a length of 5^2, or 25, and an angle of approximately 53° · 2, or 106°.

THREE CUBE ROOT EXAMPLES

Now consider the cube roots of 1. You know that within the set of real numbers, the cube root of 1 is equal to 1, because 1 · 1 · 1 = 1. But there are nonreal, complex number roots as well. One of these is a vector with a magnitude of 1 unit, in the second quadrant at an angle of 120° counterclockwise from the positive real number axis. Squaring this puts the product in the third quadrant, at 240°. Cubing it verifies it as a cube root,

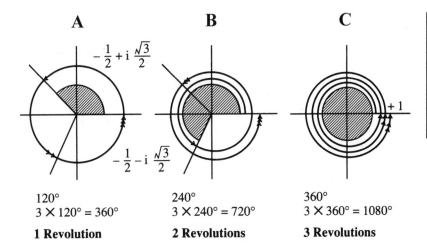

Figure 22-4

The three cube roots of 1. At A and B, the roots are complex and non-real. At C, the root is pure real and equal to 1.

by turning to 360°, so the vector ends up on the positive part of the real number axis with the length still at 1 unit. This is shown in Fig. 22-4A.

Now suppose that you start with a vector of length 1 unit in the third quadrant at 240°. Multiply this by itself to get the square at 480°, and then multiply again by the original vector to get the cube when it is at 720°, or two complete revolutions (Fig. 22-4B). This shows that the quantity you began with is the cube root of 1.

Finally, imagine that you start with a vector of length 1 unit along the positive real axis, but think of it as oriented at 360°. Squaring this takes you around one complete circle to 720°. Then, multiplying again by the original vector, you go another complete circle to 1080°, so you end up with 1 again. This is the "conventional" or real number cube root of 1 that you usually think of (Fig. 22-4C).

How are quantities with magnitudes other than 1 represented? As an example, look at the cube root of 8. Using the same method as before, the first cube root of 8 is $-1 + i(3^{1/2})$, which is at an angle of 120°. Then multiply that quantity by itself and simplify, getting $-2 - i2(3^{1/2})$. Multiplying by the original quantity again returns the answer to 8; the imaginary part disappears to prove that the cube root is correct. This is shown in Fig. 22-5.

Now look at the cube roots of -1. The complex, nonreal root in the first quadrant has a magnitude of 1 and is at 60°. Squaring it puts the product in the second quadrant at 120°. Multiplying again verifies it as a cube root, by turning to 180° so it ends up on the negative real axis with a length of 1 unit. This process is shown in Fig. 22-6. Another cube root of -1 has a length of 1 unit and is oriented at an angle of 180° counterclockwise from the positive real axis. This is the root you usually think of; it is pure real and is equal to -1. Another complex, nonreal cube root of -1 has a length of 1 unit and is oriented at an angle of 300° counterclockwise from the positive real number axis.

Figure 22-5

The three cube roots of 8. Two are complex and non-real; the third is pure real and equal to 2.

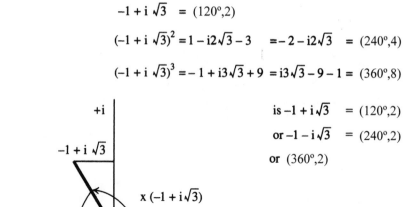

$$-1 + i\sqrt{3} = (120°, 2)$$

$$(-1 + i\sqrt{3})^2 = 1 - i2\sqrt{3} - 3 = -2 - i2\sqrt{3} = (240°, 4)$$

$$(-1 + i\sqrt{3})^3 = -1 + i3\sqrt{3} + 9 = i3\sqrt{3} - 9 - 1 = (360°, 8)$$

$$\text{is } -1 + i\sqrt{3} = (120°, 2)$$

$$\text{or } -1 - i\sqrt{3} = (240°, 2)$$

$$\text{or } (360°, 2)$$

Cube the magnitude ...

... and triple the angle!

Figure 22-6

Geometric construction and algebra describing one of the complex, non-real cube roots of -1.

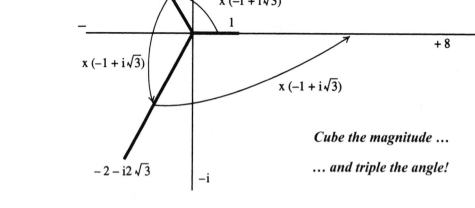

$$\left(\frac{1}{2} + i\,\frac{\sqrt{3}}{2}\right)$$

$$\left(\frac{1}{2} + i\,\frac{\sqrt{3}}{2}\right)^2 = \frac{1}{4} + i\,\frac{\sqrt{3}}{2} - \frac{3}{4}$$

$$= -\frac{1}{2} + i\,\frac{\sqrt{3}}{2}$$

$$\left(\frac{1}{2} + i\,\frac{\sqrt{3}}{2}\right)^3 = \frac{1}{8} + i\,\frac{3}{4}\cdot\frac{\sqrt{3}}{2} - \frac{3}{2}\cdot\frac{3}{4} - i\,\frac{3}{4}\cdot\frac{\sqrt{3}}{2}$$

$$= \frac{1}{8} - \frac{9}{8} = -1$$

When you cube it, you get a vector at an angle of 900°. If you subtract 360° from this twice, you can see that it is in the same direction as 180° which lies exactly along the negative real axis.

RECIPROCAL OF A COMPLEX QUANTITY

If the vector representing a complex number $a + ib$ has magnitude r that is greater than 1, its reciprocal has magnitude $1/r$, which is of course less than 1. An example of this is shown in Fig. 22-7. The triangular portion of the shaded area above the real number axis uses unit magnitude on the positive real axis for its base, and the magnitude r of the quantity $a + ib$ for its top side. Scaling this area down to make the longest side fit unit magnitude on the positive real axis, the side that was 1 in the bigger triangle is now the reciprocal of the original complex quantity, in both magnitude and polar angle. This is represented by the part of the shaded area below the real axis.

The algebra in the figure shows how to calculate these values. When you have the quantity $a + ib$ in the denominator of an expression, it presents a problem that is rather difficult to work out directly. However, if you multiply both the numerator and the denominator of such a fraction by $a - ib$, the denominator becomes the sum of two squares, which is a real number. It's always a lot easier to divide by a real number than it is to divide by a complex number!

In the study and use of complex quantities, the quantity $a - ib$ is called the *conjugate* of $a + ib$, and vice versa. The product of two complex conjugates is always a pure real number.

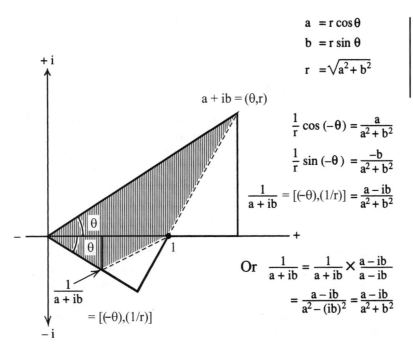

$a = r \cos \theta$

$b = r \sin \theta$

$r = \sqrt{a^2 + b^2}$

$a + ib = (\theta, r)$

$\dfrac{1}{r} \cos(-\theta) = \dfrac{a}{a^2 + b^2}$

$\dfrac{1}{r} \sin(-\theta) = \dfrac{-b}{a^2 + b^2}$

$\dfrac{1}{a + ib} = [(-\theta),(1/r)] = \dfrac{a - ib}{a^2 + b^2}$

Or $\dfrac{1}{a + ib} = \dfrac{1}{a + ib} \times \dfrac{a - ib}{a - ib}$

$= \dfrac{a - ib}{a^2 - (ib)^2} = \dfrac{a - ib}{a^2 + b^2}$

$\dfrac{1}{a + ib}$

$= [(-\theta),(1/r)]$

Figure 22-7

Geometric representation of the reciprocal of a complex quantity.

Figure 22-8

Geometric representation of the quotient of two complex quantities.

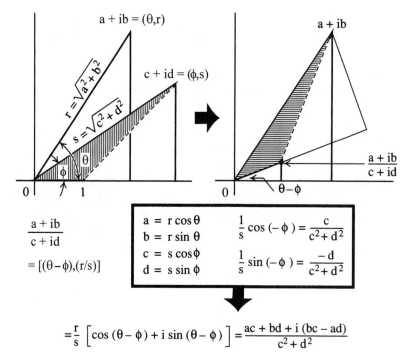

$$\frac{a+ib}{c+id}$$

$$= [(\theta - \phi),(r/s)]$$

$$a = r\cos\theta$$
$$b = r\sin\theta$$
$$c = s\cos\phi$$
$$d = s\sin\phi$$

$$\frac{1}{s}\cos(-\phi) = \frac{c}{c^2+d^2}$$
$$\frac{1}{s}\sin(-\phi) = \frac{-d}{c^2+d^2}$$

$$= \frac{r}{s}\left[\cos(\theta-\phi) + i\sin(\theta-\phi)\right] = \frac{ac+bd+i(bc-ad)}{c^2+d^2}$$

DIVISION OF COMPLEX QUANTITIES

Figure 22-8 is a geometric rendition of how a complex number can be divided by another complex number. From the actual size of the divisor (the shaded area in the upper left part of the figure), change the magnitude of the longest side to fit the longest side of the dividend, maintaining its shape or proportion. The quotient is then the side of the proportionate area of the divisor that was unit positive on the real axis before it was reduced. As the upper right portion of the figure shows, the angle of the quotient is found by subtracting the angle of the divisor from the angle of the dividend $(\theta - \phi)$. The magnitudes are simply divided by each other. This is represented by r/s in the calculations in the lower part of Fig. 22-8.

RATIONALIZATION

In complex number algebra, *rationalization* is the equivalent of the simplification of fractions in ordinary algebra. A complex quantity is the sum of a real number and an imaginary number. Complex quantities can, as numerators, share the same denominator as a matter of convenience. But the denominator in any such expression should always be pure real. This makes complex number fractions much easier to work with.

$$(a + ib)(c + id) = ac - bd + i(bc + ad) \Longrightarrow \boxed{\text{Write } k = ac - bd; \ell = bc + ad}$$
$$= k + i\ell$$
$$(e + if)(g + ih) = eg - fh + i(fg + eh) \Longrightarrow \boxed{\text{Write } m = eg - fh; n = fg + eh}$$
$$= m + in$$

$$\frac{(a + ib)\ (c + id)}{(e + if)\ (g + ih)} = \frac{k + i\ell}{m + in} = \frac{(k + i\ell)(m - in)}{(m + in)(m - in)} = \frac{km + \ell n + i(\ell m - kn)}{m^2 + n^2}$$

Example $\dfrac{(3 + i4)(5 - i6)}{(4 + i3)(1 - i2)}$ $= \dfrac{15 + 24 + i(20 - 18)}{4 + 6 + i(3 - 8)}$

$$= \frac{39 + i2}{10 - i5} = \frac{39 + i2}{5(2 - i1)}$$

$$\times \boxed{\frac{2 + i1}{2 + i1}} \qquad = \frac{(39 + i2)(2 + i1)}{5(2^2 + 1^2)}$$

$$= \frac{78 - 2 + i(4 + 39)}{5 \times 5} = \frac{76 + i43}{25}$$

Consider a simplification that consists of two complex quantities multiplied together in the numerator and two more multiplied together in the denominator, as general-ized in the upper part of Fig. 22-9. When the quantities are multiplied, the numerator and denominator can each be simplified to single real and imaginary parts. To rationalize, the numerator and denominator are each multiplied by the conjugate of the denominator, so only the numerator contains both real and imaginary parts. If desired, the whole quantity can be written separately as a real part and as an imagi-nary part.

A specific example of rationalization, using actual numbers instead of variables, is shown in the lower portion of Fig. 22-9.

CHECKING RESULTS

It is easy to make mistakes when handling complex numbers, even if you use a calcu-lator! Often, the numbers happen to be convenient for making some relatively simple checks. In the example of Fig. 22-10, the first factor in the numerator has the same vector magnitude as the first factor in the denominator, even though the two complex numbers themselves are different. One factor is $3 + i4$ and the other is $4 + i3$. The vector mag-nitude of either one is equal to 5. Because of this, the whole expression will have the same magnitude if these two factors are removed; only the angle is changed. If you like, go ahead and check to see that this is really true.

Figure 22-10

Division of complex numbers can be checked by comparing vector magnitudes.

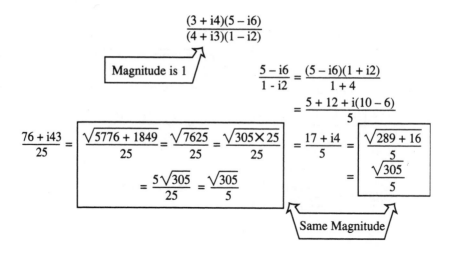

OPERATION SUMMARY

Here is a summary of facts you should remember about combining quantities in complex number algebra using geometric polar coordinate representations.

- When you want to add or subtract two complex numbers, you must always work on the real and imaginary parts separately. There's no polar equivalent that works out conveniently.

- When you want to multiply two complex numbers, multiply the vector magnitudes and add the polar angles.

- When you want to divide two complex numbers, divide the vector magnitude of the numerator by the vector magnitude of the denominator, and subtract the polar angle of the denominator from the polar angle of the numerator.

- When you want to raise a complex number to a real power, take the real power of the vector magnitude, and multiply the polar angle by the value of the real power.

- When you want to find a real root of a complex number, take the root of the vector magnitude, and divide the polar angle by the value (or index) of the real root.

USE OF A COMPLEX PLANE

The invention of complex numbers, and the evolution of the algebra that goes with them, led mathematicians and scientists to use the complex number coordinate plane in various ways. In the conventional graphic representation of quantities—the Cartesian plane or xy plane—the independent variable (usually x) is expressed horizontally, and the dependent variable (usually y) is expressed vertically.

In complex number mathematics, the independent variable requires a plane to be completely represented. The real part is measured left or right, but the imaginary part

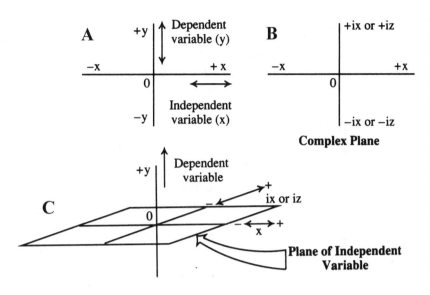

Figure 22-11

At A, the traditional xy-plane. At B, a complex plane as the basis for the independent variable in a real-number function of a complex variable. At C, a three-dimensional set of rectangular coordinates for plotting real-number functions of complex variables.

is measured at right angles to it. The direction for the dependent variable is vertical. This results in a variant of *xyz* space, in which *x* is replaced by the real number part *x* of a complex number, *y* is replaced by the imaginary part *iz*, and *z* is replaced by the dependent variable *y*, which is a function of the complex number $x + iz$. Figure 22-11 shows how this system is put together.

Figure 22-12 illustrates a qualitative example of a real number function of a complex variable, graphed in the system of coordinates from Fig. 22-11C. For each point on the

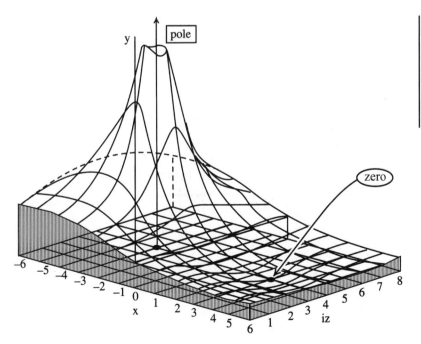

Figure 22-12

An example of a three-dimensional graph representing a real-number function of a complex variable.

complex number plane defined by the axes x and iz, the value of y, which is a real number dependent variable, is plotted vertically. At a point where the denominator is zero, the value of y "goes to infinity" or "blows up." This is called a *pole* of the graph. Where the numerator goes to zero, the value of y is 0. This is called a *zero* of the graph. Not all graphs of this kind have poles or zeros.

ROOTS BY COMPLEX QUANTITIES

You can often simplify the process of finding roots by using complex quantities. For example, you know that 32 has a 5th root that is equal to 2, because $2^5 = 32$. But your advancing knowledge of mathematics ought to help you realize that 32 has four more 5th roots. These roots all have magnitude 2 in the complex polar plane, with angles that divide a full $360°$ revolution into five equal parts (Fig. 22-13). Using the cosines and sines of the angles shown, you can derive the complex roots in the form $a + ib$ from the following expressions:

$$2 (\cos 72° + i \sin 72°)$$

$$2 (\cos 144° + i \sin 144°)$$

$$2 (\cos 216° + i \sin 216°)$$

$$2 (\cos 288° + i \sin 288°)$$

$$2 (\cos 360° + i \sin 360°)$$

The first four of these roots are complex. The last is pure real, as you can determine by working out the cosine and sine values or by looking at Fig. 22-13.

Figure 22-13

The five 5th roots of 32, shown as vectors in the complex polar plane.

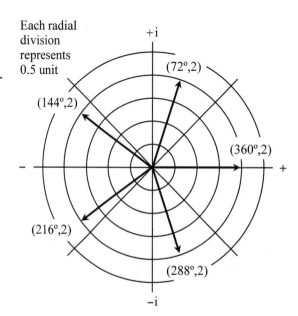

Each radial division represents 0.5 unit

QUESTIONS AND PROBLEMS

This is an open-book quiz. You may refer to the text in this chapter (and earlier ones, too, if you want) when figuring out the answers. Take your time! Consider all values given as exact, so you don't have to worry about significant figures. The correct answers are in the back of the book. *Note:* In these problems, the square root of -1 is denoted by j, not i. You should get used to this notation because it is commonly used by engineers.

1. Find the product $(0.6 + j0.8)(0.8 + j0.6)$. Verify that the result is pure imaginary and that its vector has a magnitude of 1.

2. Square each of the factors in Prob. 1. Then multiply the resulting complex numbers together, and verify that the product is pure real and that its vector has a magnitude of 1.

3. Solve the following quadratic equations. Include all imaginary, complex, and real solutions.
  (a) $x^2 - 2x + 2 = 0$
  (b) $x^2 - 2x + 10 = 0$
  (c) $13x^2 - 4x + 1 = 0$
  (d) $x^2 - j2x - 10 = 0$
  (e) $x^2 - j2x - 8 = 0$

4. Find the six 6th roots of 64. Express the coefficients in decimal form to three significant digits. Plot the corresponding vectors on the complex plane.

5. Find the ten 10th roots of 1,024. Express the coefficients in decimal form to three significant digits. Plot the corresponding vectors on the complex plane.

6. Find the nine 9th roots of 512. Express the coefficients in decimal form to three significant digits. Plot the corresponding vectors on the complex plane.

7. Refer to Fig. 22-14. Find the sum of the two complex numbers shown.

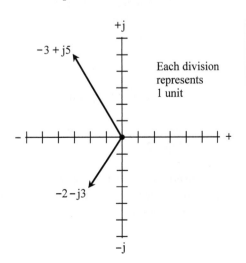

Figure 22-14
Illustration for problems 7 through 14.

Figure 22-15

Illustration for problems 15 through 17.

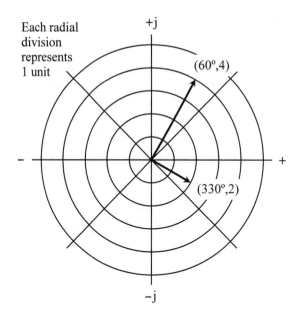

Each radial division represents 1 unit

+j

(60°,4)

−

+

(330°,2)

−j

8. Plot the vector denoting the sum of the complex numbers shown in Fig. 22-14.

9. Find the product of the complex numbers shown in Fig. 22-14.

10. Plot the vector denoting the product of the complex numbers shown in Fig. 22-14.

11. Divide the complex number shown as a second-quadrant vector by the complex number shown as a third-quadrant vector in Fig. 22-14. Express the coefficients in decimal form to three significant digits.

12. Plot the vector denoting the quotient you found in Prob. 11.

13. Divide the complex number shown as a third-quadrant vector by the complex number shown as second-quadrant vector in Fig. 22-14. Express the coefficients in decimal form to three significant digits.

14. Plot the vector denoting the quotient you found in Prob. 13.

15. Refer to Fig. 22-15. Find the product of these two complex numbers.

16. Plot, in polar coordinates, the vector denoting the product of the complex numbers shown in Fig. 22-15.

17. Convert the product vector from Probs. 15 and 16 to the form $a + jb$, where a and b are real numbers and j is the positive square root of -1. Express the coefficients to three significant digits.

Part 4

TOOLS OF APPLIED MATHEMATICS

23
Trigonometry in the Real World

Trigonometry can be used to determine distances by measuring angles. In some cases, the angles are extremely small, requiring precision observational devices. In other cases, angle measurement is less critical, and you can measure or estimate the distances more easily. Trigonometry can also be used to figure out angles, such as headings or bearings, based on known or measured distances between objects.

PARALLAX

When you want to measure distances using trigonometry, you can rely on the fact that rays of light travel in straight lines. This is a "gospel truth" in surveying and basic astronomy. There are exceptions to this rule, but they are of concern only in relativistic astronomy, physics, and cosmology.

Parallax makes it possible to judge distances to objects and to perceive depth. Figure 23-1 shows what parallax looks like. Nearby objects appear displaced, relative to a distant background, when viewed with your left eye as compared to the view seen through your right eye. The extent of the displacement depends on (1) the proportional difference between the distance to the nearby object and the distant reference scale and (2) on the separation distance between your eyes.

Parallax can be used for navigation and guidance. If you are heading toward a point, that point seems stationary while other objects seem to move radially outward from it. You can observe this effect while driving down a flat, straight highway. Signs, trees, and other roadside objects appear to move in straight lines outward from a distant point on the road. Parallax simulation gives video games their realism, and it is used in stereoscopic imaging. It's also useful for creating special effects in science fiction movies and television shows!

Figure 23-1

Parallax allows depth perception. This effect can be used to measure distances.

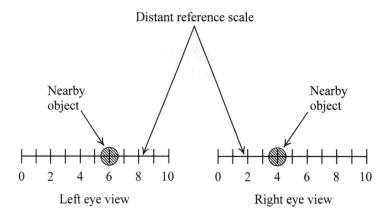

Distant reference scale

Nearby object

Nearby object

Left eye view

Right eye view

THE BASE-LINE

The use of parallax in distance measurement involves establishing a *base-line*. This is a line segment connecting two points of observation. Call the observation points P and Q. If the distant object (to which you want to find the distance) is at point R, then you must choose the base-line such that the triangle defined by the three points P, Q, and R (symbolized ΔPQR and called "triangle PQR") comes as near to being a right triangle as possible. You want the base-line segment PQ to be perpendicular to either line segment PR or line segment QR, as shown in Fig. 23-2. At first thought, it might seem as if getting the base-line oriented perfectly could be a difficult task. But in most cases, a hiker's compass will suffice to set the base-line PQ at a right angle to the line segment connecting your location and the distant object.

ACCURACY

When you want to measure the distance to something within sight, the base-line must be long enough so there is a significant difference in the azimuth of the object (that is, its compass bearing) as seen from opposite ends of the base-line. In this context, "significant" means an angular difference that is well within the ability of the observing apparatus to detect and measure.

Figure 23-2

Choosing a base-line for distance measurement.

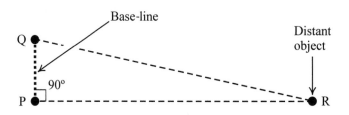

Base-line

Distant object

$90°$

The *absolute accuracy* (in fixed units such as meters) with which the distance to an object can be measured depends on the distance to the object, the length of the base-line, and the accuracy of the equipment you use to measure the angles. As the distance to the object increases, assuming the base-line length stays constant, the absolute accuracy of the distance measurement gets worse. In other words, the *absolute error* increases. As the length of the base-line increases, the absolute accuracy gets better; the absolute error decreases. As the *angular resolution*, or precision of the angle-measuring equipment, gets better, the absolute accuracy improves, if all other factors are held constant.

DISTANCE TO AN OBJECT

Suppose you want to determine the distance to a hovering blimp that you know is a long way off. The base-line for the distance measurement is 500.00 meters (which you can call 0.50000 kilometer) long. The angular difference in azimuth is 0.75000° between opposite ends of the base-line. How far away is the blimp?

It helps to draw a diagram of the situation, even though it cannot be conveniently drawn to scale. (The base-line must be shown out of proportion to its actual relative length.) Figure 23-3 illustrates this situation. The base-line, which is line segment *PQ*, is oriented at right angles to the line segment *PR* connecting one end of the base-line and the distant blimp. The right angle is established approximately, using a hiker's compass, but for purposes of calculation, it can be assumed exact, so Δ*PQR* can be considered a right triangle.

You measure the angle θ between a ray parallel to line segment *PR* and the observation line segment *QR*, and you find this angle to be 0.75000°. The "parallel ray" can be determined either by sighting to an object that is essentially at an infinite distance, or, lacking that, by using an accurate magnetic compass or navigator's sighting apparatus. If it is a clear night, you can use the background of stars.

One of the fundamental principles of plane geometry states that pairs of *alternate interior angles* formed by a *transversal* to parallel lines always have equal measures. In this example, you have line *PR* and an observation ray parallel to it, while line *QR* is a transversal to these parallel lines. Because of this, the two angles labeled θ in Fig. 23-3 have equal measures.

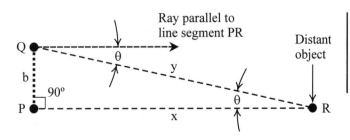

Ray parallel to line segment PR

Distant object

Figure 23-3
An example of a distance-measurement problem.

You can use the right-triangle model for trig functions to calculate the distance to the object. Let b be the length of the base-line (line segment PQ), and let x be the distance to the object (the length of line segment PR). Then

$$\tan \theta = b/x$$

Plugging in known values produces this equation, for which you can solve for x by using your calculator:

$$\tan 0.75000° = 0.50000/x$$

$$0.013090717 = 0.50000/x$$

$$x = 0.50000 / 0.013090717$$

$$= 38.195 \text{ km}$$

The distant object is therefore 38.195 kilometers away from point P.

You can also use the length of line segment QR as the distance to the object, rather than the length of line segment PR in the above example. Observation point Q is just as valid, for determining the distance, as is point P. In this case, the base-line is short compared to the distance being measured. In Fig. 23-3, let y be the length of line segment QR. Then you use the sine function instead of the tangent function:

$$\sin \theta = b/y$$

$$\sin 0.75000° = 0.50000/y$$

$$0.013089596 = 0.50000/y$$

$$y = 0.50000 / 0.013089596$$

$$= 38.198 \text{ km}$$

The percentage difference between this result and the previous result is small. In some situations, the absolute difference between these two determinations (approximately 3 m) could be of concern, and a more precise method of distance measurement, such as *laser ranging*, would be needed. An example of such an application is precise monitoring of the distance between two points at intervals over a period of time, in order to determine minute movements of the earth's crust along a geological fault line.

STADIMETRY

Stadimetry can be used to measure the distance to an object when the object's height or width, oriented at a right angle to the line of sight, is known. The distance is calculated by using trigonometry. This scheme works in the same way as the method described above, except that the base-line is at the opposite end of the triangle from the observer.

Figure 23-4 shows an example of the use of stadimetry to measure the distance d, in meters, to a distant person. Suppose you know the person's height h, in meters, and you

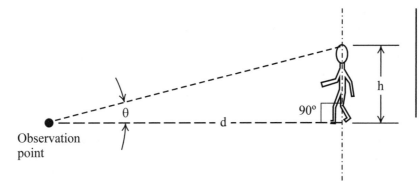

also know that the axis of this dimension is perpendicular to your line of sight to the person. Observational apparatus determines the angle θ that the person subtends in the field of view. From this information, the distance d is calculated according to the following formula:

$$d = h / (\tan \theta)$$

As with base-line surveying, if the distant object's height (or perpendicular dimension such as width or diameter) is small compared to the distance between it and the observer, the sine function can also be used:

$$d = h / (\sin \theta)$$

For stadimetry to be accurate, the linear dimension axis—in this case the axis that depicts the person's height—must be perpendicular to a line between the observation point and one end of the object. If the object is extremely distant, the remote point can be anywhere along the axis of the object perpendicular to the line of sight, and the error will be negligible. If the linear dimension axis is not perpendicular to the line of sight but the actual height or width of the object and its *slant angle* are both known, then the perpendicular component can be determined by using trigonometry. Don't forget that d and h must be expressed in the same units!

INTERSTELLAR DISTANCE MEASUREMENT

The distances to "nearby" stars in the Milky Way (the galaxy containing our solar system) can be measured in a manner similar to the way surveyors measure terrestrial distances. The radius of the earth's orbit around the sun is used as the base-line.

Astronomers often measure and express interplanetary distances in terms of the *astronomical unit* (AU). One AU is equal to the average distance of the earth from the sun, and it is agreed on formally as 1.49597870×10^8 km (this is sometimes rounded off to a figure of 1.50×10^8 km). The distances to other stars and galaxies can be expressed in astronomical units, but the numbers are large.

Astronomers have invented the *light-year*, the distance light travels in one year, to assist in defining interstellar distances so the numbers become reasonable. One light-year is the distance a ray of light travels through space in one earth year. You can figure out how far this is by calculation. Light travels approximately 3.00×10^5 km in 1 s. There are 60 s in a minute (1 min), 60 min in an hour (1 h), 24 h in a solar day, and about 365.25 solar days in a year. Multiplying all this out, you can calculate that a light-year (1 ly) is roughly 9.5×10^{12} km.

Now, think on a cosmic scale. The nearest star to our solar system is a little more than 4 ly away. Our galaxy is about 10^5 ly in diameter. The Andromeda galaxy, a well-known object to anyone who is serious about astronomy, is a little more than 2×10^6 ly away from our solar system. Using powerful telescopes, astronomers can peer out to distances of several billion light-years (where one billion is defined as 10^9 or one thousand million).

The light-year is an interesting unit for expressing the distances to stars and galaxies, but when measurements must be made, it is not the most convenient unit.

The true distances to the stars were unknown until the advent of the telescope, with which it became possible to measure extremely small angles. To determine the distances to the stars, astronomers use *triangulation*, the same way surveyors measure distances on the earth.

Figure 23-5 shows how distances to the stars can be measured. This scheme works only for "nearby" stars, those within a few dozen light-years of our sun. Most stars are too far away to produce measurable parallax against a background of much more distant

Figure 23-5

The distances to "nearby" stars can be determined by measuring the parallax resulting from the revolution of the earth around the sun. It is, in effect, a large-scale survey with a base-line equal to the distance of the earth from the sun.

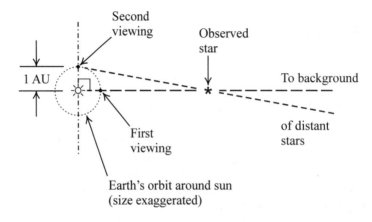

objects, even when they are observed from the earth at different times of the year as it orbits the sun. In Fig. 23-5, the size of the earth's orbit is exaggerated for clarity. The observed star appears in slightly different positions, relative to a background of much more distant objects, at the two observation points shown. The displacement is maximum when the line segment connecting the star and the sun is perpendicular to the line segment connecting the sun with the earth.

Suppose a star thus oriented and, at a certain distance from our solar system, is displaced by one second of arc (1″), which is 1/3600 of an angular degree, when viewed on two occasions 3 months apart in time in the geometric alignment shown in Fig. 23-5. When that is the case, the distance between our solar system and the star is called a *parsec* (a contraction of the term "parallax second"). The word "parsec" is abbreviated pc. Expressed to four significant digits, 1 pc = 3.262 ly = 2.063×10^5 AU.

The nearest visible object outside our solar system is the Alpha Centauri star system, which is 1.4 pc away. There are numerous stars within 20 to 30 pc of our sun. Our galaxy is about 3.0×10^4 pc in diameter.

DISTANCE VS. PARALLAX

Suppose you want to determine the distance to a star. You measure the parallax relative to the background of much more distant stars (or even remote galaxies); that background can be considered infinitely far away. You choose the times for your observations so the earth lies directly between the sun and the star at the first measurement, and a line segment connecting the sun with the star is perpendicular to the line segment connecting the sun with the earth at the time of the second measurement (Fig. 23-6). Suppose the parallax thus determined is precisely 0 degrees, 0 minutes, and 5 seconds of arc (written 0°0′5″). What is the distance to the observed star in astronomical units?

First, you should realize that the star's distance is essentially the same throughout the earth's revolution around the sun, because the star is many astronomical units away from the sun. You want to find the length of the line segment connecting the sun (not the earth)

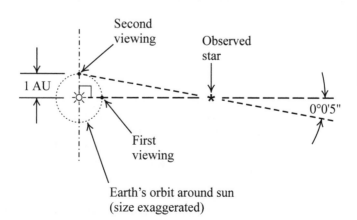

Figure 23-6

Measurement of the distance to a hypothetical star.

with the star. This line segment is perpendicular to the line segment connecting the earth with the sun at the time of the second observation. You therefore have a right triangle and can use trigonometry to find the distance to the star in astronomical units.

The measure of the parallax in Fig. 23-6 is exactly 5″. You can divide this by exactly 3600 to get the number of degrees; call it $(5/3600)°$. Let d be the distance from the sun to the star in astronomical units. Then, using the right-triangle model, you have

$$1/d = \tan (5/3600)°$$

$$1/d = 2.424 \times 10^{-5}$$

$$d = 4.125 \times 10^4 \text{ AU}$$

A POINT OF CONFUSION

The parsec can be a confusing unit. If the distance to a star is doubled, then the parallax observed between two observation points, as shown in Fig. 23-5 or Fig. 23-6, is cut in half. This doesn't mean that the number of parsecs to the star is cut in half; it means the number of parsecs is doubled. If taken literally, the expression "parallax second" is a misleading way of expressing the distances to stars. As the number of parallax seconds decreases, the distance in parsecs increases.

To avoid this confusion, it's best to remember that the parsec is a fixed unit, based on the distance to an object that generates a parallax of one arc second as viewed from two points 1 AU apart. If stadimetry were used in an attempt to measure the distance to a rod 1 AU long and oriented at a right angle to the line of observation (or a person 1 AU tall as shown in Fig. 23-4), then that object would subtend an angle of 1″ as viewed by the observer.

All interstellar objects are so far away, in terms of the astronomical unit, that the number of parsecs can be considered inversely proportional to the parallax in arc seconds. For example, if you multiply the distance in parsecs by 2, the parallax becomes one-half as great; if the parallax doubles, the distance in parsecs is cut in half. These relations are not mathematically perfect, but they're close enough in the real world to be regarded as exact.

RADAR

The term *radar* is an acronym derived from the words "radio detection and ranging." Radio waves having certain frequencies reflect from objects such as aircraft, missiles, and ships. By ascertaining the direction(s) from which radio signals are returned and by measuring the time it takes for a signal pulse to travel from the transmitter location to an object (called a *target*) and back, it is possible to locate that target.

A complete radar set consists of a transmitter, a directional antenna, a receiver, and an indicator or display. The transmitter produces radio-wave pulses that are propagated in

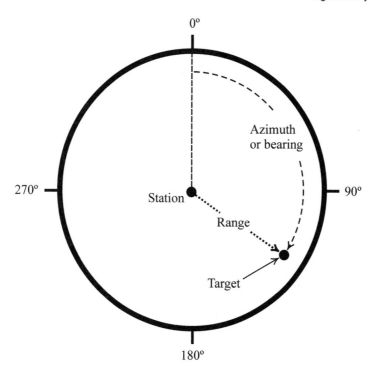

Figure 23-7

A radar display shows azimuth and range in navigator's polar coordinates.

a narrow beam. The waves strike objects at various distances. The greater the distance to the target, the longer the delay before the echo is received. The transmitting antenna is rotated so that all azimuth directions can be observed.

A typical circular radar display is shown in Fig. 23-7. It uses navigator's polar coordinates. The observing station is at the center of the display. Azimuth is indicated in degrees clockwise from north and is marked around the perimeter of the screen. The distance, or range, is indicated by the displacement of the echo from the center of the screen.

LAWS OF SINES AND COSINES

When you want to find the position of an object using trigonometry, or when you want to figure out your own location based on bearings, it helps to know two important rules about triangles.

The first rule is called the *law of sines*. Suppose a triangle is defined by three points P, Q, and R. Let the lengths of the sides opposite the vertices P, Q, and R be called p, q, and r, respectively (Fig. 23-8). Let the angles at vertices P, Q, and R be θ_p, θ_q, and θ_r, respectively. Then the lengths and angles are related in this way:

$$p / (\sin \theta_p) = q / (\sin \theta_q) = r / (\sin \theta_r)$$

Figure 23-8

The law of sines
and the law of
cosines can be
applied to any
triangle with
straight sides.

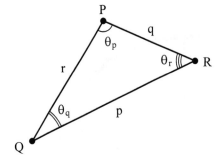

Law of sines ...

$$\frac{p}{\sin \theta_p} = \frac{q}{\sin \theta_q} = \frac{r}{\sin \theta_r}$$

Law of cosines ...

$$r = (p^2 + q^2 - 2pq \cos \theta_r)^{1/2}$$

The lengths of the sides of any triangle are in a constant ratio relative to the sines of the angles opposite those sides.

The second rule is called the *law of cosines*. Suppose a triangle is defined as shown in Fig. 23-8. Suppose you know the lengths of two of the sides, say p and q, and the measure of the angle θ_r between them. Then the length of the third side r can be found by using this formula:

$$r = (p^2 + q^2 - 2pq \cos \theta_r)^{1/2}$$

USE OF RADAR

Now let's get back to radar. Imagine that you use a radar set to observe an aircraft X and a missile Y, as shown in Fig. 23-9A. Suppose aircraft X is at azimuth 240.00° and range 20.00 km, and missile Y is at azimuth 90.00° and range 25.00 km. Both objects are flying directly toward each other. Aircraft X is moving at 1000 km/h while missile Y is moving at 2000 km/h. How long will it be before the missile and the aircraft collide, assuming neither one alters its course or speed?

To begin with, you must determine the distance, in kilometers, between targets X and Y at the time of the initial observation. This is a made-to-order job for the law of cosines.

Consider ΔXSY, formed by the aircraft X, the observing station S, and the missile Y, as shown in Fig. 23-9B. You have been told that $XS = 20.00$ km and $SY = 25.00$ km. You

Each radial division equals 5 km
Each angular division equals 10°

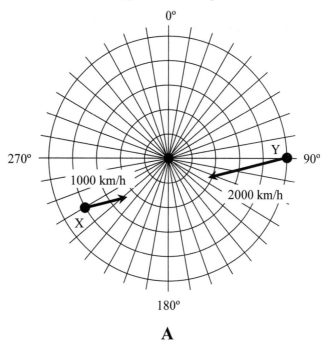

A

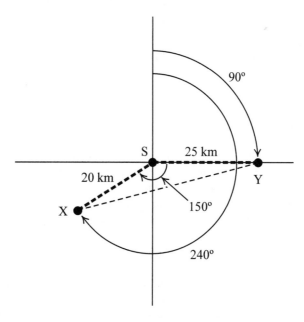

B

Figure 23-9A

An aircraft (shown as *X*) and a missile (*Y*) are on a collision course. How long will it be before they hit each other, if the initial distance between them is known?

Figure 23-9B

Solution to the aircraft-missile problem.

can also deduce that the measure of angle *XSY* (written $\angle XSY$) equals 150.00°, the difference between azimuth 240.00° and azimuth 90.00°. You now have a triangle with a known angle between two sides of known length. When you put these numbers into the formula for the law of cosines, you can figure out that the distance *XY* between the aircraft and the missile is

$$XY = [20.00^2 + 25.00^2 - (2 \cdot 20.00 \cdot 25.00 \cdot \cos 150.00°)]^{1/2}$$

$$= \{400.0 + 625.0 - [1000 \cdot (-0.8660)]\}^{1/2}$$

$$= (1025 + 866.0)^{1/2}$$

$$= 1891^{1/2}$$

$$= 43.49 \text{ km}$$

The two objects are moving directly toward each other, one at 1000 km/h and the other at 2000 km/h. Their mutual speed is therefore 3000 km/h. If neither object changes course or speed, they will collide after a time t_h, in hours, determined as follows:

$$t_h = (43.49 \text{ km}) / (3000 \text{ km/h})$$

$$= 0.01450 \text{ h}$$

You can obtain the time t_s, in seconds, if you multiply the above by 3600, the number of seconds in an hour:

$$t_s = 0.01450 \cdot 3600$$

$$= 52.20 \text{ s}$$

In a real-life scenario of this sort, a computer would take care of these calculations in a split second and send a warning message to the aircraft pilot.

HOW TO HUNT A "FOX"

A radio receiver, equipped with a signal-strength indicator and connected to a special antenna, can be used to determine the direction from which signals are coming. *Radio direction finding* (RDF) equipment can allow people to determine the location of a signal source. Sometimes hidden transmitters are used to train people in the art of finding signal sources. Technical folks call this sort of exercise a *fox hunt*.

An RDF receiver makes use of a loop antenna that is rotated until a minimum occurs in the received signal strength. This minimum, called a *null*, is pronounced and sharp. When the null is found and the antenna is aimed for the weakest possible signal, the axis of the loop lies along a line toward the transmitter. When readings like this are taken from two or more locations separated by a sufficient distance, the transmitter can be pinpointed by finding the intersection point of the azimuth lines drawn on a map.

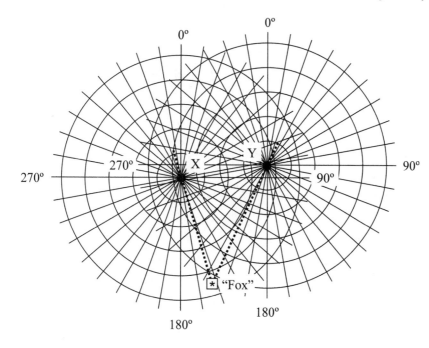

Figure 23-10
The source of a signal can be found using radio direction finding (RDF) equipment and plotting lines on a map or coordinate system.

Figure 23-10 shows an example of an RDF operation. The locations from which readings are taken are indicated by dots labeled X and Y; they form the origins of two navigator's polar coordinate planes. The target, or "fox," is shown as a square with the asterisk inside. The dashed lines show the azimuth orientations of the tracking antennas at points X and Y. These lines intersect at the location of the fox.

THE FOX FINDS ITSELF

The captain of a vessel can find the vessel's position by comparing the signals from two fixed stations whose positions are known, as shown in Fig. 23-11—a "fox hunt in reverse"! The captain of the vessel, shown by the box with the asterisk inside, finds her position by taking directional readings of the signals from sources X and Y. This sort of operation is called *radiolocation*.

Two or more sets of radiolocation readings can be taken, separated by a certain amount of time. On the basis of the information gathered in this way, computers can assist in precisely determining, and displaying, the position and velocity vectors. This process of repeated radiolocation to determine or plot a course is called *radionavigation*.

Here's something you should note: The methods of direction finding and radiolocation described in this chapter work well over small geographic regions, within which the surface of the earth (averaged out so it is a perfect sphere) appears *locally flat*. But things get complicated when the radio signals must travel over distances that represent

Figure 23-11

A vessel can find
its position using
radiolocation.

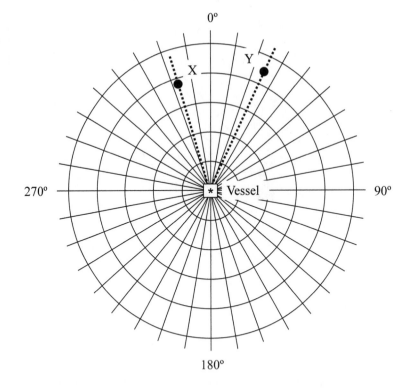

an appreciable fraction of the earth's circumference. Global radionavigation is dealt
with in engineering courses, particularly in the military.

POSITION OF A SHIP

Suppose the navigator of a vessel wishes to find his location in terms of latitude and longitude to the nearest minute of arc. He uses direction-finding equipment to measure the
azimuth bearings of two buoys, called buoy 1 and buoy 2, whose latitude and longitude
coordinates are known. This is shown in Fig. 23-12A. The azimuth of buoy 1 is measured as 350°0′, and the azimuth of buoy 2 is measured as 42°30′, according to the instruments aboard the vessel. What are the latitude and longitude coordinates of the vessel?

This problem ought to make you appreciate computers! It's one thing to plot positions
on maps, as you've seen in the movies, but it's another thing to manually calculate the
values. Computers can do such calculations in a tiny fraction of a second, but it will take
you a while longer if you grind it out "by hand."

In this problem, you are working within a geographic region small enough so the surface of the earth can be considered flat, and the lines of longitude can be considered
parallel. Therefore, you can convert latitude and longitude to a rectangular coordinate

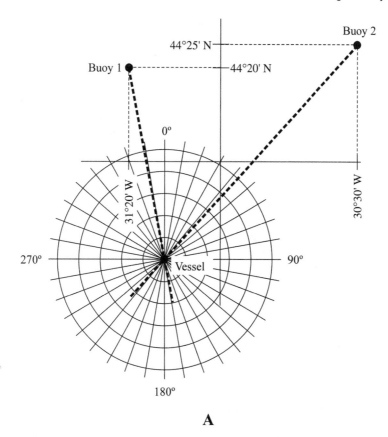

Buoy 2

44°25' N

44°20' N

Buoy 1

31°20' W

30°30' W

0°

270°   Vessel   90°

180°

A

Figure 23-12A

A navigator wishes to pinpoint the location of his vessel. He measures the azimuth bearings of two buoys whose latitudes and longitudes are known.

grid with the origin at buoy 1. Let each division on the axes of this coordinate grid equal 10 minutes of arc of latitude or longitude, as shown in Fig. 23-12B.

Name points P, Q, R, S, T, U, and V, representing intersections among lines and coordinate axes, as shown. Lines TU and SV are perpendicular to the horizontal coordinate axis, and line VU is perpendicular to the vertical coordinate axis. You must find either RV or PS, which will allow you to determine the longitude of the vessel relative to buoy 1, and either PR or SV, which will let you find the latitude of the vessel relative to buoy 1. During the calculation process, consider all values exact. Round off the answer only after you have finished each part of the problem.

Based on the information given, $\angle RPV = 10°$. You know that $PT = 50$ (units) and $TQ = 5$. You know that ΔPTQ is a right triangle, so you can use trigonometry to calculate the measure of $\angle TPQ$ as follows:

$$\tan \angle TPQ = 5/50 = 0.1$$

$$\angle TPQ = \arctan 0.1$$

$$= 5.71059°$$

Figure 23-12B

Solution to the vessel-location problem.

Each horizontal or vertical division equals 10 units

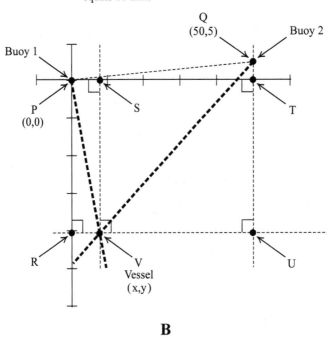

B

Because $\angle RPV = 10°$, you can deduce that $\angle VPT = 90° - 10° = 80°$. Because $\angle TPQ = 5.71059°$, you know that $\angle VPQ = 80° + 5.71059° = 85.71059°$. You now know the measure of one of the interior angles of $\triangle VPQ$, an important triangle in the solution of this problem.

Now find $\angle PVQ$, the angle between the azimuth bearings obtained by the navigator, or $10° + 42°30'$. Remember that $30' = 0.5°$, so $\angle PVQ = 10° + 42.5° = 52.5°$. From this it's easy to figure out $\angle VQP$. It is $180°$ minus the sum of $\angle PVQ$ and $\angle VPQ$:

$$\angle VQP = 180° - (52.5° + 85.71059°)$$

$$= 180° - 138.21059°$$

$$= 41.78941°$$

Next you can find the distance PQ by using the Pythagorean theorem, because $\triangle PTQ$ is a right triangle. You know $PT = 50$ and $TQ = 5$, and that PQ is the hypotenuse of the triangle. Therefore

$$PQ = (50^2 + 5^2)^{1/2}$$

$$= (2500 + 25)^{1/2}$$

$$= 2525^{1/2}$$

$$= 50.2494$$

Now find the distance PV by applying the law of sines to ΔPVQ as follows:

$$PV / (\sin \angle VQP) = PQ / (\sin \angle PVQ)$$

$$PV = PQ (\sin \angle VQP) / (\sin \angle PVQ)$$

$$= 50.2494 (\sin 41.78941°) / (\sin 52.5°)$$

$$= 50.2494 \times 0.6663947 / 0.7933533$$

$$= 42.2081$$

You now know one of the sides, and all the interior angles, of ΔPRV, which is a right triangle. You can use either $\angle RPV$ or $\angle RVP$ as the basis for finding PR and RV. Suppose you use $\angle RPV$, which measures $10°$. Then

$$\cos 10° = PR / PV$$

$$PR = PV \cos 10°$$

$$= 42.2081 \times 0.98481$$

$$= 41.5670$$

$$\sin 10° = RV / PV$$

$$RV = PV \sin 10°$$

$$= 42.2081 \times 0.17365$$

$$= 7.3294$$

The final step involves converting these units back into latitude and longitude. Keep in mind that north latitude increases from the bottom of the page to the top, but west longitude increases from the right to the left. Also remember that there are 60 arc minutes in one angular degree.

Let V_{lat} represent the latitude of the vessel. Subtract the displacement PR from the latitude of buoy 1 and round off to the nearest minute of arc:

$$V_{lat} = 44°20' \text{ N} - 41.5670'$$

$$= 43°38' \text{ N}$$

Let V_{lon} represent the longitude of the vessel. Subtract the displacement RV (which is the same as PS) from the longitude of buoy 1 and round off to the nearest minute of arc:

$$V_{lon} = 31°20' \text{ W} - 7.3294'$$

$$= 31°13' \text{ W}$$

QUESTIONS AND PROBLEMS

This is an open-book quiz. You may refer to the text in this chapter (and earlier ones, too, if you want) when figuring out the answers. Take your time! Consider all values given as exact, so you don't have to worry about significant figures. The correct answers are in the back of the book.

1. On a radar display, a target is 20 mi north of the observing station and 30 mi east of a line running straight north from the station. What is the azimuth of the target to the nearest degree?

2. In the situation of Prob. 1, what is the range of the target to the nearest mile?

3. Suppose several targets can be seen on a radar screen. The azimuth bearings and the ranges of each target are noted at a particular moment in time. To determine the straight-line distances between various pairs of targets, what can the radar operator do?

4. Imagine two stars, one 10 pc away and another 30 pc away. Which of these stars has the smaller parallax with respect to a background of much more distant stars, as the earth travels around the sun carrying the observer with it?

5. Suppose two objects in deep space are the same distance apart as the earth is from the sun (1.00 AU). If these objects are 1.00 pc away, and if a straight-line segment connecting them is oriented at a right angle with respect to a line of sight (Fig. 23-13), what is the approximate angle θ, in degrees, that the objects subtend relative to an arbitrarily (or "infinitely") distant background?

6. Imagine the echo of an aircraft on a radar display. The radar indicates that the aircraft is at azimuth 0° and range 10 km, and is flying on a heading of 90°. After a certain elapsed time period t, the aircraft is at azimuth 45°. What is its range at the end of this time period to the nearest kilometer?

Figure 23-13

Illustration for Prob. 5.

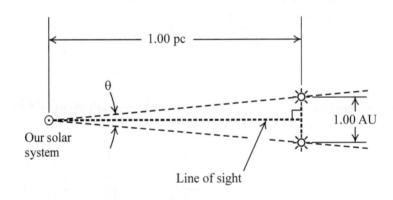

7. What is the range of the aircraft described in Prob. 6 after elapsed time $2t$ to the nearest kilometer, if it continues on the same course at the same speed? What is its azimuth at time $2t$ to the nearest degree?

8. To measure distances by triangulation on the earth's surface over distances small enough so the curvature of the earth is not a significant factor, from how many different locations must a distant object be observed?

9. As the distance to an object increases and all other factors are held constant, what happens to the absolute error (in meters, kilometers, astronomical units, or parsecs) of a distance measurement by triangulation?

10. To use stadimetry to determine the distance to an object, what must be known about that object?

24
Logarithms and Exponentials

A *logarithm* (sometimes called a *log*) of a quantity is a power to which a positive real number constant is raised to obtain that quantity. An *exponential* of a quantity is the result of raising a positive real number constant to a power equal to that quantity. The constant is known as the *base* of the *logarithmic function* or the *exponential function*. The two most common bases for logarithmic and exponential functions are 10 and e, where e is an irrational number approximately equal to 2.71828. The number e is also known as *Euler's constant* and the *exponential constant*.

THE LOGARITHM BASE

Suppose you raise a positive real number b to some real number power y, getting another real number x as the result. A mathematician would say that this relationship exists among the three real numbers b, and x, and y, where $b > 0$:

$$b^y = x$$

The exponent y in this expression is the *base-b logarithm* of x. This expression can also be written as follows:

$$y = \log_b x$$

You can raise nonpositive numbers to real powers, but when it comes to logarithms, this is rarely done. You probably aren't going to come across "base-negative-10" or "base-negative-2" logarithms. The trouble with negative numbers as log bases is that they go outside the set of real numbers. For example,

$$-9^{(1/2)} = i3$$

This is a valid equation, but you won't be likely to hear anyone say that 1/2 is the base-(-9) logarithm of $i3$ unless you're in the company of theoreticians!

COMMON LOGARITHMS

Base-10 logarithms are also known as *common logarithms* or *common logs*. In equations, common logs are denoted by writing "log" followed by a subscript 10, and then the number, called the *argument*, for which you want to find the logarithm. Here are a few examples that you can verify with your calculator:

$$\log_{10} 100 = 2$$

$$\log_{10} 45 \approx 1.653$$

$$\log_{10} 10 = 1$$

$$\log_{10} 6 \approx 0.7782$$

$$\log_{10} 1 = 0$$

$$\log_{10} 0.5 \approx -0.3010$$

$$\log_{10} 0.1 = -1$$

$$\log_{10} 0.07 \approx -1.155$$

$$\log_{10} 0.01 = -2$$

The squiggly equals sign means "is approximately equal to." You will often see it in engineering papers, articles, and books.

The first equation above is simply another way of saying that $10^2 = 100$. In that equation, 100 is the argument, or value on which the log function depends. It's another expression for the independent variable. You would say, "The common log of 100 is equal to 2." In the second equation, you are in effect saying that $10^{1.653} \approx 45$, and you could also say, "The common log of 45 is approximately equal to 1.653." And so it goes. Exponents, when expressed as logarithms, don't have to be whole numbers or fractions. They don't even have to be rational numbers. You can have exponents that are irrational, such as π, e, or the square root of 2.

Figure 24-1 is a partial linear-coordinate graph of the function $y = \log_{10} x$. Figure 24-2 is a partial graph of the same function in *semilog coordinates*, where one graph scale is linear and the other is graduated according to the common log function. Regardless of the base, log functions are defined only for positive arguments in the real numbers.

Here are a couple of function-related terms you should be familiar with. The set of x values for which any function $y = f(x)$ is defined is called the *domain* of the function. The set of all the possible y (or "output") values that you can get when you "input" some x value in the domain is known as the *range* of the function. In any log function with a positive real base, when the domain is the set of positive real numbers, the range is the entire set of real numbers.

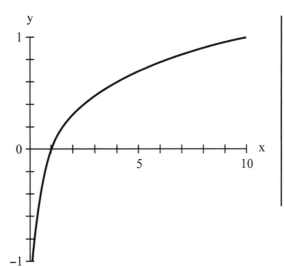

Figure 24-1

Partial linear-coordinate graph of the common logarithm function. As x approaches 0, the value of y becomes arbitrarily large in the negative sense. As x increases without limit, so does y.

NATURAL LOGARITHMS

Base-*e* logarithms are also called *natural logs* or *Napierian logs*. In equations, the natural log function is usually denoted by writing "ln" or "$\log_e$" followed by the argument.

Here's a little trick you can use to get a good display of *e* on your calculator, if it doesn't have a key that gives it directly. Enter the number 1, then hit the inverse function key or put a check in the appropriate box if you're using the calculator in a computer (it might be labeled "Inv" or "Rev"). Then hit the "ln" or "$\log_e$" key. If you have a computer with a scientific calculator, you should get a readout something like this:

2.71828182845904 . . .

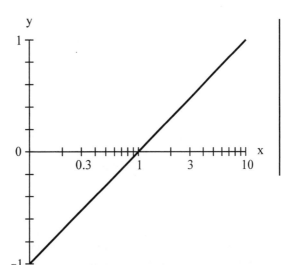

Figure 24-2

Partial semilog-coordinate graph of the common logarithm function. In this illustration, the y axis intersects the x axis at the point where x = 0.1.

And now, a warning! In some texts, the natural log function is denoted by writing "log" (without a subscript) followed by the argument, but in other texts and in most calculators, "log" means the common log function! To avoid confusion, it's a good idea to include the base as a subscript whenever you write "log" to represent a log function. If you aren't sure what the "log" key on a calculator does, conduct some tests to find out. You don't want to get natural logs from your calculator when you want common logs, or vice versa!

Now look at an example of a simple, approximate equation using the natural log function:

$$\ln 54.59815 \approx 4$$

If you want to be theoretically precise with this statement, you can substitute e^4 for 54.59815 and write it as follows:

$$\ln (e^4) = 4$$

These equations reflect the fact that $e^4 \approx 54.59815$. That's an *exponential equation*; you'll learn more about them later in this chapter. Here are some more equations using the natural log function:

$$\ln 100 \approx 4.605$$

$$\ln 45 \approx 3.807$$

$$\ln 10 \approx 2.303$$

$$\ln 6 \approx 1.792$$

$$\ln e = 1$$

$$\ln 1 = 0$$

$$\ln (1/e) = -1$$

$$\ln 0.5 \approx -0.6931$$

$$\ln 0.1 \approx -2.303$$

$$\ln 0.07 \approx -2.659$$

$$\ln 0.01 \approx -4.605$$

Figure 24-3 is a partial linear coordinate graph of $y = \ln x$. Figure 24-4 is a partial graph of the same function in semilog coordinates. As with the base-10 log function, the domain is limited to the set of positive real numbers, and the range extends over the set of all real numbers.

The curves for log functions of different bases look similar. The exact values for specific arguments vary, but the functions always "blow up negatively" (you could also say they "blow down"!) as the argument approaches 0 from the positive direction, and they always increase without limit as the argument approaches positive infinity. If you want

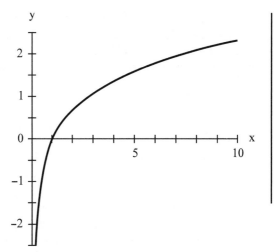

Figure 24-3

Partial linear-coordinate graph of the natural logarithm function. As *x* approaches 0, the value of *y* becomes arbitrarily large in the negative sense. As *x* increases without limit, so does *y*.

visible evidence of this, draw some expanded graphs of common log and natural log functions. Most calculators give you values for both functions, so it's easy to plot the points and connect them to get smooth curves.

BASIC PROPERTIES OF LOGARITHMS

With logarithms, you can convert multiplication problems into addition problems, division into subtraction, powers into products, and roots into quotients.

Suppose that *x* and *y* are positive real numbers. The logarithm of the product is equal to the sum of the logarithms of the individual numbers:

$$\log_b xy = \log_b x + \log_b y$$

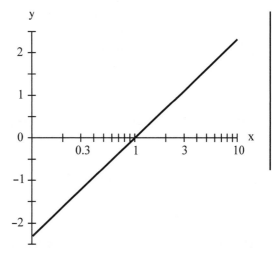

Figure 24-4

Partial semilog-coordinate graph of the natural logarithm function. In this illustration, the *y* axis intersects the *x* axis at the point where *x* = 0.1.

Here, and in some of the formulas that follow, the base b means that it works regardless of the log base. It can be a common log, a natural log, or a logarithm of any other base; it doesn't matter. Now look at a numeric example of the above principle. Consider the arguments exact. Use your calculator to follow along.

$$\log_{10}(3 \cdot 4) = \log_{10} 3 + \log_{10} 4$$

Work out both sides and approximate the results to four significant figures:

$$\log_{10} 12 \approx 0.4771 + 0.6021$$
$$1.079 \approx 1.0792$$

You shouldn't expect to get perfect answers every time you use logarithms, because the results are almost always irrational numbers. That means they are nonterminating, non-repeating decimals. Approximation is the best you can do. Sometimes approximations agree exactly to a certain number of significant figures, but often they do not.

Here's a counterpart of the above property for division. Let x and y be positive real numbers. Then the logarithm of their ratio (or quotient) is equal to the difference of the logarithms of the individual numbers:

$$\log_b(x/y) = \log_b x - \log_b y$$

You can work out an example using the same numerical arguments as before. Again, follow along with your calculator:

$$\log_{10}(3/4) = \log_{10} 3 - \log_{10} 4$$

Working out both sides and approximating the results to four significant figures, we find

$$\log_{10} 0.7500 \approx 0.4771 - 0.6021$$
$$-0.1249 \approx -0.1250$$

Logarithms simplify the raising of a number to a power. This is useful when the argument does not have two whole numbers. Let x be a positive real number, and let y be any real number (positive, negative, or zero). Then the logarithm of x raised to the power y can be reduced to a product:

$$\log_b x^y = y \log_b x$$

Here is a worked-out example using the same arguments as before, carried out to four significant figures:

$$\log_{10}(3^4) = 4 \log_{10} 3$$
$$\log_{10} 81 \approx 4 \cdot 0.4771$$
$$1.908 \approx 1.908$$

This time, the answers agree perfectly to four significant figures!

Now try an example in which neither of the numbers in the input argument is whole. Again, go to four significant figures and follow along with a calculator. But this time, leave the left-hand side just as it is, at least for the moment.

$$\log_{10}(2.635^{1.078}) = 1.078 \log_{10} 2.635$$
$$\approx 1.078 \cdot 0.4208$$
$$\approx 0.4536$$

Set this result aside, but don't put it out of your mind altogether. You'll return to it a little later in this chapter.

The logarithm (to any base b) of the reciprocal of a number is equal to the negative of the logarithm of that number. Stated formally, if x is a positive real number, then

$$\log_b(1/x) = -(\log_b x)$$

This is a special case of division simplifying to subtraction. Here is a numerical example. Suppose $x = 3$ (exactly) and you use natural logs, as follows:

$$\ln(1/3) = -(\ln 3)$$

Using your calculator, you can evaluate both expressions. This time, go to 10 significant figures, just for fun!

$$\ln 0.3333333333 \approx -(\ln 3.000000000)$$
$$-1.098612289 \approx -1.098612289$$

Again, as luck would have it, the results agree exactly.

What happens when you have a reciprocal in an exponent? Suppose x is a positive real number, and y is any real number except zero. Then the logarithm (to any base b) of the yth root of x (also denoted as x to the $1/y$ power) is equal to the log of x, divided by y:

$$\log_b(x^{1/y}) = (\log_b x) / y$$

Try this with $x = 8$ and $y = 1/3$, considering both values exact and using natural logs evaluated to six significant figures. You get

$$\ln(8^{1/3}) = (\ln 8) / 3$$

You know that the cube root of 8 is equal to 2. Therefore,

$$\ln 2 = (\ln 8) / 3$$
$$0.693147 \approx 2.07944 / 3$$
$$0.693147 \approx 0.693147$$

Once again, the error is too small to show up.

Using your calculator, you can invent plenty of examples that show how these equations work with actual number values. Remember that multiplication is a way of adding things over and over, and division is repeated subtraction. Raising to a power is repeated multiplication, and taking a root is repeated division. These facts show up in the behavior of logarithmic functions.

CONVERSION OF LOG FUNCTIONS

Here are a couple of rules you can use to convert from natural logs to common logs and vice versa. You'll sometimes have to do this, especially if you get into physics or engineering.

Let x be a positive real number. The common logarithm of x can be expressed in terms of the natural logarithms of x and 10 as follows:

$$\log_{10} x = (\ln x) / (\ln 10)$$

$$\approx (\ln x) / 2.302585$$

$$\approx 0.4342945 \ln x$$

Try an example. Let $x = 3.537$. Working to four significant figures, use your calculator to find

$$\ln 3.537 \approx 1.263$$

Now multiply by 0.4342945 and round to four significant figures.

$$0.4342945 \cdot 1.263 \approx 0.5485$$

Compare this with the common log of 3.537 as your calculator determines it, rounding off to four significant figures.

$$\log_{10} 3.537 \approx 0.5486$$

There's a little error here—just enough to remind you that you're dealing with irrational numbers, and approximations are the rule.

Now go the other way. Suppose x is a positive real number. The natural logarithm of x can be expressed in terms of the common logarithms of x and e as follows:

$$\ln x = (\log_{10} x) / (\log_{10} e)$$

$$\approx (\log_{10} x) / 0.4342945$$

$$\approx 2.302585 \log_{10} x$$

Let's put a number into this formula and test it. Suppose $x = 238.9$. Working with your calculator to four significant figures, you find

$$\log_{10} 238.9 \approx 2.378$$

Now multiply by 2.302585 and round to four significant figures.

$$2.302585 \cdot 2.378 \approx 5.476$$

Compare this with the natural log of 238.9 as your calculator "sees" it, rounded off to four significant figures. You should get

$$\ln 238.9 \approx 5.476$$

LOGARITHM COMPARISONS

Compare the common logarithms of 0.01, 0.1, 1, 10, and 100. Remember, the common log of a number is the power of 10 that produces that number. Note that $0.01 = 10^{-2}$, $0.1 = 10^{-1}$, $1 = 10^0$, $10 = 10^1$, and $100 = 10^2$. Therefore,

$$\log_{10} 0.01 = -2$$
$$\log_{10} 0.1 = -1$$
$$\log_{10} 1 = 0$$
$$\log_{10} 10 = 1$$
$$\log_{10} 100 = 2$$

These values are all exact! But now suppose you want to compare the base-e logarithms (or natural logarithms) of these same five arguments. The base-e logarithm of a number is the power of e that produces that number. You must use a calculator to find them. The results are as follows, to four significant figures in each case except the natural log of 1, which is an exact value:

$$\ln 0.01 \approx -4.605$$
$$\ln 0.1 \approx -2.303$$
$$\ln 1 = 0$$
$$\ln 10 \approx 2.303$$
$$\ln 100 \approx 4.605$$

The log of 1 is always equal to 0, no matter what the base. This is a reflection of the fact that any positive real number raised to the zeroth power is equal to 1.

What about the logarithm of 0, or of any negative number? These are not defined in the set of real numbers, no matter what the base. To understand why, look at what happens if you try to calculate the base-b logarithm of -2. Remember that b must be a positive real number. Suppose $\log_b (-2) = y$. This can be rewritten in the form $b^y = -2$. What is the value of y? No real number will work here. Regardless of what real number you choose for y, the value of b^y is always positive. If you change -2 to any other negative number, or to 0, you run into the same problem. It's impossible to find any real number y such that b^y is less than or equal to 0.

THE EXPONENTIAL BASE

Logarithms can be "undone" by exponential functions. An exponential function has a base and an exponent, just as a logarithm has. In fact, an exponential function is an "inside-out" way of looking at a logarithmic function! Suppose you have three real numbers b, x, and y, where $b > 0$, and you raise b to the xth power to get y, like this:

$$y = b^x$$

Then y is the *base-b exponential* of x. In the expression $10^2 = 100$, 10 is the base and 100 is the exponential. In the expression $e^3 \approx 20.0855$, e is the base and 20.0855 is the exponential. The two most often used exponential bases are the same as the two most often used log bases: 10 and e.

It's possible to raise imaginary numbers to real number powers and get negative numbers; you've already learned about this. You could talk about base-i exponentials, for example, and make mathematical sense. Consider this:

$$i^2 = -1$$

This is true, as you have already learned. But you won't be likely to hear anybody state this fact as "Minus 1 is the base-i exponential of 2," unless you find yourself in the company of theoreticians! This book does not deal with exponential functions that involve complex numbers.

COMMON EXPONENTIALS

Base-10 exponentials are also known as *common exponentials*. Here are a few examples that you can verify with your calculator:

$$10^2 = 100$$
$$10^{1.478} \approx 30.06$$
$$10^1 = 10$$
$$10^{0.8347} \approx 6.834$$
$$10^0 = 1$$

$$10^{-0.5} \approx 0.3162$$
$$10^{-1} = 0.1$$
$$10^{-1.7} \approx 0.01995$$
$$10^{-2} = 0.01$$

In the first equation, 2 is the argument, or value on which the exponential function depends, and 100 is the resultant. You would say, "The common exponential of 2 is equal to 100." In the second equation, you would say, "The common exponential of 1.478 is approximately 30.06." As with logarithms, the arguments in an exponential function need not be whole numbers. They can even be irrational; you could speak of 10^{π}, for example. It happens to be approximately equal to 1,385.

Figure 24-5 is a partial linear coordinate graph of the function $y = 10^x$. Figure 24-6 is a partial graph of the same function in semilog coordinates. The domain of the common exponential function encompasses the entire set of real numbers. The range is limited to the positive real numbers. This reflects the fact that you can never raise a real number to any real power and get a negative number.

NATURAL EXPONENTIALS

Base-e exponentials are also known as *natural exponentials*. Here are some examples, using the same arguments as in the previous section:

$$e^2 \approx 7.389$$
$$e^{1.478} \approx 4.384$$
$$e^1 \approx 2.718$$

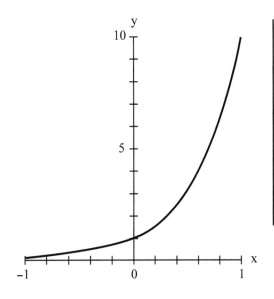

Figure 24-5

Partial linear-coordinate graph of the common exponential function. As x becomes arbitrarily large in the negative sense, the value of y approaches 0. As x increases without limit, so does y.

Figure 24-6

Partial semilog graph of the common exponential function. In this illustration, the x axis intersects the y axis at the point where y = 0.1.

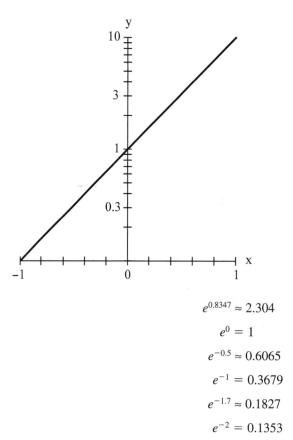

$$e^{0.8347} \approx 2.304$$
$$e^{0} = 1$$
$$e^{-0.5} \approx 0.6065$$
$$e^{-1} = 0.3679$$
$$e^{-1.7} \approx 0.1827$$
$$e^{-2} = 0.1353$$

Figure 24-7 is a partial linear coordinate graph of the function $y = e^x$. Figure 24-8 is a partial graph of the same function in semilog coordinates. As with the base-10

Figure 24-7

Partial linear-coordinate graph of the natural exponential function. As x becomes arbitrarily large in the negative sense, the value of y approaches 0. As x increases without limit, so does y.

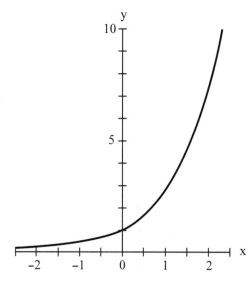

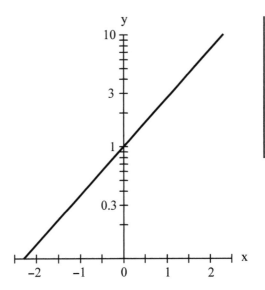

y

Figure 24-8
Partial semilog graph of the natural exponential function. In this illustration, the x axis intersects the y axis at the point where y = 0.1.

exponential function (or an exponential function of any other base), the domain encompasses the entire set of real numbers, and the range is limited to the positive real numbers.

ALTERNATIVE EXPRESSIONS

Sometimes, the common exponential of a quantity is called the *common antilogarithm* (antilog$_{10}$) or the *common inverse logarithm* (log^{-1}) of that number. The natural exponential of a quantity is sometimes called the *natural antilogarithm* (antiln) or the *natural inverse logarithm* (ln^{-1}) of that number.

LOGARITHMS VS. EXPONENTIALS

The exponential function is the *inverse* of the logarithm function, and vice versa. When two functions are inverses, they "undo" each other if both functions are defined for all the arguments of interest. For any real number x, and for any positive real number y, the following equations hold. Let the abbreviation "log" represent a logarithm to any base b for simplicity of notation, avoiding the need for putting a subscript within a superscript in the second equation. Then

$$\log (b^x) = x$$

and

$$b^{(\log y)} = y$$

Now work out an example of how a common antilog can be used to find the value of one nonwhole number raised to the power of another nonwhole number. Recall the example set out earlier in this chapter:

$$\log_{10} (2.635^{1.078}) = 1.078 \log_{10} 2.635$$

$$\approx 1.078 \cdot 0.4208$$

$$\approx 0.4536$$

If you get rid of all the intermediate expressions, you have

$$\log_{10} (2.635^{1.078}) \approx 0.4536$$

Take the common antilog of both sides.

$$\text{antilog}_{10} [\log_{10} (2.635^{1.078})] \approx \text{antilog}_{10} 0.4536$$

Now use your calculator to find the common antilog of 0.4536, and simplify the left-hand side of the equation according to the principle just mentioned in which the antilog "undoes" the log function.

$$2.635^{1.078} \approx 2.842$$

You can use the "x^y" key (or "$x\^{}y$" or whatever it is called, where the first input number is raised to the power of the second) to verify

$$2.635^{1.078} = 2.842$$

Good luck shows itself again here! The calculations agree to four significant figures. This should not come as a surprise when you realize that the "log-antilog" scheme is the way most calculators work to evaluate powers where the numbers are not whole. Before the invention of logs and antilogs, expressions such as $2.635^{1.078}$ were confusing and mysterious!

Try this same exercise using the base-e log and antilog functions instead of the base-10 functions. You'll get the same result. In fact, you can use logs and antilogs of any base to evaluate any number raised to the power of any other number, as long as the log and the antilog are both defined for all the arguments.

BASIC PROPERTIES OF EXPONENTIALS

In all the following rules, b is a positive real number that represents the exponential base. In most real-life applications of exponential functions, b is equal to 10 (in the case of the common exponential function) or e (for the natural exponential function). However, once in a while you'll come across a situation where b is some other number such as 2, which could represent a base for *binary exponential functions*.

Suppose x is some real number. The reciprocal of the exponential of x is equal to the exponential of the negative of x as follows:

$$1 / (b^x) = b^{-x}$$

You should recognize this from your work with powers and roots. Here's a familiar example. You know that 1/8 is equal to $1 / (2^3)$. This is the same as saying that 1/8 is equal to 2^{-3}. You also know that 1/100 equals $1 / (10^2)$, which is the same as saying that 1/100 equals 10^{-2}. Now consider this:

$$1 / (e^3) \approx 1 / (2.718^3)$$

$$\approx 1 / 20.08$$

$$\approx 0.04980$$

Compare the above with the result of entering -3 into a scientific calculator, then hitting "Inv," then hitting "ln," and finally rounding off to four significant figures:

$$e^{-3} \approx 0.04979$$

Exponential functions can express the relationship between sums and products, just as logarithms do. Let x and y be real numbers. The product of the exponentials of x and y is equal to the exponential of the sum of x and y:

$$b^x b^y = b^{(x + y)}$$

To demonstrate, let $b = 10$, $x = 4$, and $y = -6$. Plug in the numbers on the left-hand side of the equation and evaluate.

$$10^4 \cdot 10^{-6} = 10{,}000 \cdot 0.000001$$

$$= 0.01$$

Now evaluate the right-hand side:

$$10^{[4+(-6)]} = 10^{(4-6)}$$

$$= 10^{-2}$$

$$= 0.01$$

The results agree! You'll find that this is always true no matter what base and arguments you use, as long as the base is positive. Of course, if you get nonterminating decimals for any of the values in the calculation, you should expect some approximation error.

Let x and y be real numbers. Then the ratio of the exponential of x to the exponential of y is equal to the exponential of the difference between x and y:

$$b^x / b^y = b^{(x-y)}$$

Using the same numerical values as before, you can demonstrate this. Plug in the numbers on the left-hand side of the equation and evaluate.

$$10^4 / 10^{-6} = 10{,}000 / 0.000001$$

$$= 10{,}000 \cdot 1{,}000{,}000$$

$$= 10{,}000{,}000{,}000$$

$$= 10^{10}$$

Now evaluate the right-hand side:

$$10^{[4-(-6)]} = 10^{(4+6)}$$

$$= 10^{10}$$

Here's a more complicated property of exponentials. At least, it looks complicated until you get used to working it out. Let x and y be real numbers, with the restriction that y cannot be equal to 0. Then the exponential of x/y is equal to the exponential of $1/y$ with base b^x:

$$b^{(x/y)} = (b^x)^{(1/y)}$$

Try an example where the base b is equal to 10, with exponents $x = 4$ and $y = 7$. Evaluating the left-hand side first, letting $4/7 \approx 0.5714$, and using the "x^y" or "$x\wedge y$" function key on your calculator, you should get

$$10^{(4/7)} = 10^{0.5714}$$

$$= 3.727$$

Alternatively, you can enter 0.5714, hit the "Inv" key, and then hit "log" to find 10 to the power of 0.5714. Now plug the numbers into the right-hand side of the general equation and work it out.

$$(10^4)^{(1/7)} = 10{,}000^{(1/7)}$$

To do this on your calculator, first figure 1/7 to four significant figures. You should get 0.1429. Then enter 10,000, hit the "x^y" or "$x\wedge y$" key, and enter 0.1429. The result should be 3.729. There's a significant discrepancy here (2 parts in 10,000) because you've taken a rounding error to the 7th power.

Exponentials can show the relationship between a "power of a power" and a product. Suppose x and y are real numbers. Then the yth power of the exponential of x is equal to the exponential of the product xy.

$$(b^x)^y = b^{(xy)}$$

To demonstrate this, let $b = e$, $x = 2$, and $y = 3$. Evaluate the left-hand side first.

$$(e^2)^3 \approx (2.718^2)^3$$

$$\approx 7.388^3$$

$$\approx 403.3$$

Now the right-hand side:

$$e^{(2 \cdot 3)} = e^6$$

$$\approx 2.718^6$$

$$\approx 403.2$$

If you want a more exact and simple example, use $b = 10$ instead of $b = e$. In that case, the left-hand side works out like this:

$$(10^2)^3 = 100^3$$

$$= 1{,}000{,}000$$

$$= 10^6$$

Now the right-hand side:

$$10^{(2 \cdot 3)} = 10^6$$

Let's mix common and natural exponentials and see what happens! Suppose x is a real number representing the argument of an exponential function in base 10 and also in base e. The product of the common and natural exponentials of x is equal to the exponential of x to the base $10e$, as follows:

$$(10^x)(e^x) = (10e)^x$$

This can be illustrated with a numerical example. Go to six significant figures. Set $x = 4$ (exactly). Then the left-hand side of the above equation works out like this:

$$(10^4)(e^4) \approx 10{,}000.0 \cdot 2.71828^4$$

$$\approx 10{,}000.0 \cdot 54.5980$$

$$\approx 545{,}980$$

Now do the right-hand side, rounding off to six significant figures:

$$(10e)^4 \approx (10.0000 \cdot 2.71828)^4$$

$$= 27.1828^4$$

$$= 545{,}980$$

How about ratios of mixed common and natural exponentials? If x is a real number, then the ratio of the common exponential of x to the natural exponential of x is equal to the exponential of x to the base $10/e$:

$$10^x / e^x = (10/e)^x$$

Work this out using $x = 4$, as in the previous example, and go to six significant figures. First, work on the left-hand side:

$$10^4 / e^4 \approx 10{,}000.0 / 2.71828^4$$

$$\approx 10{,}000.0 / 54.5980$$

$$\approx 183.157$$

Next, do the right-hand side:

$$(10/e)^4 \approx (10.0000 / 2.71828)^4$$

$$\approx 3.67880^4$$

$$\approx 183.158$$

Now invert this ratio. Suppose x is a real number. The ratio of the natural exponential of x to the common exponential of x is equal to the exponential of x to the base $e/10$:

$$e^x / 10^x = (e/10)^x$$

Again, use $x = 4$ and go through this with your calculator, rounding to six significant figures. Here's the left-hand side:

$$e^4 / 10^4 \approx 2.71828^4 / 10{,}000.0$$

$$\approx 54.5980 / 10{,}000.0$$

$$\approx 0.00545980$$

And the right-hand side:

$$(e/10)^4 \approx (2.71828 / 10.0000)^4$$

$$\approx 0.271828^4$$

$$\approx 0.00545980$$

EXPONENTIAL COMPARISONS

Suppose you want to compare the values of e^{-2}, e^{-1}, e^0, e^1, and e^2. Assume the exponents given here are exact. You want to express each answer to five significant figures. To determine the values of natural exponentials, you must use a calculator that has this function. The key is often labeled e^x. With some calculators, it is necessary to hit an

"Inv" key followed by a "ln" or "log$_e$" key. Here are the values of the above exponentials, rounded off to five significant figures:

$$e^{-2} = 0.13534$$

$$e^{-1} = 0.36788$$

$$e^0 = 1.0000$$

$$e^1 = 2.7183$$

$$e^2 = 7.3891$$

Now suppose you want to find the number whose common exponential function value is exactly 1,000,000 and also the number whose common exponential function value is exactly 0.0001. The argument 6 produces the common exponential value 1,000,000. This can be demonstrated by the fact that $10^6 = 1,000,000$. The argument -4 produces the common exponential value 0.0001. This is shown by the fact that $10^{-4} = 0.0001$.

How about the number whose natural exponential function value is exactly 1,000,000 and the number whose natural exponential function value is exactly 0.0001? To solve this problem, you must be sure you know what you're trying to get! Suppose you call the solution x. In the first case, solve the following equation for x.

$$e^x = 1,000,000$$

Taking the natural logarithm of each side, you obtain the following:

$$\ln (e^x) = \ln 1,000,000$$

$$x = \ln 1,000,000$$

This simplifies to a matter of finding a natural logarithm with a calculator. To four significant figures

$$\ln 1,000,000 = 13.82$$

In the second case, you must solve the following equation for x:

$$e^x = 0.0001$$

Taking the natural logarithm of each side gives

$$\ln (e^x) = \ln 0.0001$$

$$x = \ln 0.0001$$

This simplifies, as in the first case, to a matter of finding a natural logarithm with a calculator. Again to four significant figures

$$\ln 0.0001 = -9.210$$

NUMERICAL ORDERS OF MAGNITUDE

You will sometimes hear the expression "order of magnitude" for a positive real number base b. Usually, $b = 10$. Sometimes b can be another number such as 2. An order of magnitude is a difference of 1 in the exponent of the base b.

For example, note that $10^{-2} = 0.01$ and $10^3 = 1000$. This means that in the base 10, the number 1000 is 5 orders of magnitude greater than the number 0.01, because the power of 10 for 1000 is 5 greater than the power of 10 for 0.01.

Here are some power-of-10 numbers, listed in ascending order. Each time you go down one line, the order of magnitude increases by 1:

$$3.7 \times 10^{-5} = 0.000037$$
$$3.7 \times 10^{-4} = 0.00037$$
$$3.7 \times 10^{-3} = 0.0037$$
$$3.7 \times 10^{-2} = 0.037$$
$$3.7 \times 10^{-1} = 0.37$$
$$3.7 \times 10^{0} = 3.7$$
$$3.7 \times 10^{1} = 37$$
$$3.7 \times 10^{2} = 370$$
$$3.7 \times 10^{3} = 3,700$$
$$3.7 \times 10^{4} = 37,000$$
$$3.7 \times 10^{5} = 370,000$$

Now look at powers of 2. An increase of 1 order of magnitude in base 2 is the equivalent of multiplying a given number by 2 (doubling it). Thus, starting with $2^0 = 1$, you can build up like this:

$$2^0 = 1$$
$$2^1 = 2$$
$$2^2 = 4$$
$$2^3 = 8$$
$$2^4 = 16$$
$$2^5 = 32$$
$$2^6 = 64$$
$$2^7 = 128$$

and so on. A decrease of 1 order of magnitude in base 2 is the equivalent of dividing a given number by 2 (halving it). Again starting with $2^0 = 1$, you get

$$2^0 = 1$$

$$2^{-1} = 1/2$$

$$2^{-2} = 1/4$$

$$2^{-3} = 1/8$$

$$2^{-4} = 1/16$$

$$2^{-5} = 1/32$$

$$2^{-6} = 1/64$$

$$2^{-7} = 1/128$$

and so on. Orders of magnitude in base 2 are unique because of their repetitive doubling and halving properties. This makes them useful in digital electronics and computing applications.

QUESTIONS AND PROBLEMS

This is an open-book quiz. You may refer to the text in this chapter (and earlier ones, too, if you want) when figuring out the answers. Take your time! Consider all values given as exact, so you don't have to worry about significant figures. The correct answers are in the back of the book.

1. Consider the two numbers $x = 2.3713018568$ and $y = 0.902780337$. Find the product xy, using common logarithms, to four significant figures. (Ignore, for the moment, the fact that a calculator can easily be used to solve this problem without using logarithms in any form.)

2. Approximate the product of the two numbers xy from Prob. 1, but use natural logarithms instead. Show that the result is the same. Express the answer to four significant figures.

3. The *power gain* of an electronic circuit, in units called decibels (dB), can be calculated according to

$$\text{Gain (dB)} = 10 \log_{10} (P_{out}/P_{in})$$

where P_{out} is the output signal power and P_{in} is the input signal power, both specified in watts. Suppose the audio input to the left channel of a high-fidelity amplifier is 0.535 watt, and the output is 23.7 watts. What is the power gain of this circuit in decibels? Round off the answer to three significant figures.

4. Suppose the audio output signal in the scenario of Prob. 3 is run through a long length of speaker wire, so that instead of the 23.7 watts that appears at the left-channel amplifier output, the speaker only gets 19.3 watts. What is the power gain of the length of speaker wire, in decibels? Round off the answer to three significant figures.

5. If a positive real number increases by a factor of exactly 10, how does its common (base-10) logarithm change?

6. Show that the solution to Prob. 5 is valid for all positive real numbers.

7. If a positive real number decreases by a factor of exactly 100 (becomes 1/100 as great), how does its common logarithm change?

8. Show that the solution to Prob. 7 is valid for all positive real numbers.

9. If a positive real number is divided by a factor of exactly 357, how does its natural (base-e) logarithm change? Express the answer to five significant figures.

10. Show that the solution to Prob. 9 is valid for all positive real numbers.

25

Scientific Notation Tutorial

In science and engineering, huge or tiny quantities can be unwieldy when written out in full as decimal numerals. Scientific notation provides a "shortcut" for writing such quantities. You've already seen scientific notation in this book. Chapter 24 had plenty of it! If any of this notation has confused you in the past, this chapter explains and reviews how scientific notation and significant figures work.

STANDARD FORM

The scientist's and engineer's way of denoting extreme quantities is to write them as real number multiples of integer powers of 10. A numeral in *standard scientific notation*, also called the *American form*, is written as follows:

$$m \, . \, n_1 n_2 n_3 \ldots n_p \times 10^z$$

where the dot (.) is a period written on the base-line, representing the decimal point. The numeral m to the left of the radix point is a single-digit positive or negative integer (it can never be 0). Each of the numerals n_1, n_2, n_3, and so on up to n_p to the right of the radix point is a single-digit nonnegative integer (it can be 0). The entire decimal expression to the left of the multiplication symbol is called the *coefficient*. The value z, which is the power of 10, can be any integer. Here are three examples of numbers written in standard scientific notation:

$$7.63 \times 10^8$$

$$-4.10015 \times 10^{-15}$$

$$4.000 \times 10^0$$

When writing numbers in scientific notation, you must take care to be sure the form is right. For example, take a close look at 245.89×10^8. This represents a perfectly legitimate number, but it is not written in the correct form for scientific notation. The number to the left of the multiplication symbol must always be a *single digit* from the set

$\{-9, -8, -7, -6, -5, -4, -3, -2, -1, 1, 2, 3, 4, 5, 6, 7, 8, 9\}$. To write down this particular number in the proper format for scientific notation, first divide the portion to the left of the multiplication symbol by 100, so it becomes 2.4589. Then multiply the portion to the right of the multiplication symbol by 100, increasing the exponent by 2 so it becomes 10^{10}. This produces the same numerical value but in the correct format for standard scientific notation: 2.4589×10^{10}.

ALTERNATIVE FORM

In some literature, a variation on the above theme is used. You can call it *alternative scientific notation*. Some people call it the *European form*. This system requires that the number m to the left of the radix point *always* be equal to 0. When numbers are expressed this way, the exponent is increased by 1 compared with the same number in standard scientific notation. In alternative scientific notation, the above three quantities would be expressed like this:

$$0.763 \times 10^9$$

$$-0.410015 \times 10^{-14}$$

$$0.4000 \times 10^1$$

Note that when a negative exponent is "increased by 1," it becomes "1 less negatively." For example, when -15 is increased by 1, it becomes -14, not -16.

THE "TIMES" SIGN

The multiplication sign in scientific notation can be denoted in various ways. Most scientists in the United States use the cross symbol ($\times$), as in the examples shown above. But a small "center" dot raised above the base-line ($\cdot$) can also be used to represent multiplication in scientific notation. When written that way, the above numbers look like this in the standard power-of-10 form:

$$7.63 \cdot 10^8$$

$$-4.10015 \cdot 10^{-15}$$

$$4.000 \cdot 10^0$$

This small dot should not be confused with a radix point.

A small dot symbol is preferred when multiplication is required to express the dimensions of a physical unit. An example is the kilogram-meter per second squared, which is symbolized $kg \cdot m/s^2$ or $kg \cdot m \cdot s^{-2}$. As you've already seen, the small dot is also commonly used when two simple numerals are multiplied. For example,

$$35 \cdot 67 = 2,345$$

Another alternative multiplication symbol in scientific notation is the asterisk (*). You will occasionally see numbers written like this in standard scientific notation:

$$7.63 * 10^8$$

$$-4.10015 * 10^{-15}$$

$$4.000 * 10^0$$

PLAIN-TEXT EXPONENTS

Once in awhile, you will have to express numbers in scientific notation using plain, unformatted text. This is the case when you are transmitting information within the body of an e-mail message (rather than as an attachment). Some calculators and computers use this system in their displays. An uppercase letter E indicates that the quantity immediately before it is to be multiplied by a power of 10, and that power is written immediately after the E. In this format, the above quantities are written

$$7.63E+08$$

$$-4.10015E-15$$

$$4.000E+00$$

Still another alternative is the use of an asterisk to indicate multiplication, and the symbol ^ to indicate a superscript, so the expressions look like this:

$$7.63 * 10\text{^}8$$

$$-4.10015 * 10\text{^}-15$$

$$4.000 * 10\text{^}0$$

In all these examples, the numerical values represented are identical. Respectively, if written out in full, they are

$$763,000,000$$

$$-0.00000000000000410015$$

$$4.000$$

ORDERS OF MAGNITUDE

You learned a little about orders of magnitude in the last chapter. Let's take a closer look at this concept. Consider the following two extreme numbers:

$$8.576 \times 10^{89}$$

$$-5.88 \times 10^{-321}$$

Now look at these:

$$8.576 \times 10^{92}$$

$$-5.88 \times 10^{-318}$$

Concerning the first number in each pair, the exponent 92 is 3 larger than 89. Considering the second number in each pair, the exponent -318 is 3 larger than -321. (Again, remember that numbers grow larger in the mathematical sense as they become more positive or less negative!) Both of the numbers in the second pair are therefore 3 orders of magnitude larger than their counterparts in the first pair. That's a factor of 10^3, or 1,000.

The order-of-magnitude concept lets you create number lines and graphs that cover huge spans. Examples are shown in Fig. 25-1. Drawing A shows a log-scale number

Figure 25-1

At A, a number line spanning 3 orders of magnitude. At B, a number line spanning 10 orders of magnitude. At C, a coordinate system whose horizontal scale spans 10 orders of magnitude, and whose vertical scale is linear and extends from 0 to 10.

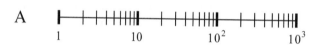

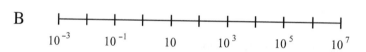

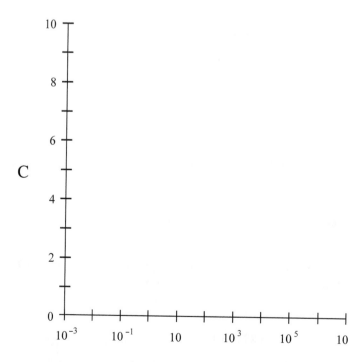

line that covers 3 orders of magnitude from 10^0 (which is equal to 1) to 10^3 (which is 1,000). Illustration B shows a log-scale number line spanning 10 orders of magnitude from 10^{-3} (0.001) to 10^7 (10,000,000). Illustration C shows a coordinate system with a log-scale horizontal axis spanning 10 orders of magnitude from 10^{-3} to 10^7, and a linear-scale vertical axis spanning values from 0 to 10.

WHEN TO USE SCIENTIFIC NOTATION

In most technical documents, scientific notation is used only when the power of 10 is larger or smaller than certain values. If the power of 10 is between -2 and 2 inclusive, numbers are written out in plain decimal form as a rule. If the power of 10 is -3 or 3, numbers may appear in plain decimal form or in scientific notation. If the exponent is -4 or smaller, or if it is 4 or larger, values are expressed in scientific notation as a rule. In number lines and graphs, exceptions are sometimes made for consistency, as is done in Fig. 25-1.

Some calculators, when set for scientific notation, display all numbers that way, even when it is not necessary. This can be confusing, especially when the power of 10 is 0 (or close to it) and the calculator is set to display a lot of digits. Most people understand the expression 8.407 more easily than 8.407000000E+00, for example, even though both represent the same number.

PREFIX MULTIPLIERS

Special verbal prefixes and abbreviations, known as *prefix multipliers*, are used in the physical sciences and in engineering to express certain powers of 10. Table 25-1 shows these multipliers and abbreviations for factors ranging form 10^{-24} to 10^{24}.

Let's take an example. By how many orders of magnitude does a *gigahertz* differ from a *kilohertz*? (The *hertz* is a unit of frequency, equivalent to a cycle per second.) Using Table 25-1, you can figure out that a gigahertz (GHz) represents 10^9 hertz (Hz), and a kilohertz (kHz) represents 10^3 Hz. The exponents differ by 6. That means 1 GHz differs from 1 kHz by 6 orders of magnitude, which is a factor of 10^6, or 1,000,000.

Here's another example. By how many orders of magnitude is a *nanometer* different from a *centimeter*? Look at the table again. The prefix multiplier nano- represents 10^{-9}, and centi- represents 10^{-2}. The exponents -9 and -2 differ by 7. The centimeter is the bigger unit of linear dimension. So a hair measuring a centimeter (1 cm) long is 7 orders of magnitude larger in linear dimension than a submicroscopic wire that is a nanometer (1 nm) long. That's a factor of 10^7, or 10,000,000.

Table 25-1

Power-of-10 prefix multipliers and their symbols.

Designator	Symbol	Multiplier
yocto-	y	10^{-24}
zepto-	z	10^{-21}
atto-	a	10^{-18}
femto-	f	10^{-15}
pico-	p	10^{-12}
nano-	n	10^{-9}
micro-	μ	10^{-6}
milli-	m	10^{-3}
centi-	c	10^{-2}
deci-	d	10^{-1}
(none)	—	10^{0}
deka-	da or D	10^{1}
hecto-	h	10^{2}
kilo-	K or k	10^{3}
mega-	M	10^{6}
giga-	G	10^{9}
tera-	T	10^{12}
peta-	P	10^{15}
exa-	E	10^{18}
zetta-	Z	10^{21}
yotta-	Y	10^{24}

MULTIPLICATION

When you want to multiply two numbers in scientific notation, first multiply the coefficients by each other. Then add the exponents. Finally, reduce the expression to standard form if it doesn't happen to be that way already. Here are three examples:

$$(3.045 \times 10^5)(6.853 \times 10^6) = (3.045 \times 6.853)(10^5 \times 10^6)$$
$$= 20.867385 \times 10^{(5+6)}$$
$$= 20.867385 \times 10^{11}$$
$$= 2.0867385 \times 10^{12}$$

$$(3.045 \times 10^{-4})(-6.853 \times 10^{-7}) = (3.045)(-6.853)(10^{-4} \times 10^{-7})$$
$$= -20.867385 \times 10^{[-4+(-7)]}$$
$$= -20.867385 \times 10^{-11}$$
$$= -2.0867385 \times 10^{-10}$$

$$(-3.045 \times 10^5)(-6.853 \times 10^{-7}) = (-3.045)(-6.853)(10^5 \times 10^{-7})$$

$$= 20.867385 \times 10^{(5-7)}$$

$$= 20.867385 \times 10^{-2}$$

$$= 2.0867385 \times 10^{-1}$$

$$= 0.20867385$$

This last number can be written out in plain decimal form because the exponent is between -2 and 2 inclusive.

Here is the generalized rule for multiplication in scientific notation, using the variables u and v to represent the coefficients and the variables m and n to represent the exponents. Let u and v be real numbers greater than or equal to 1 but less than 10, and let m and n be integers. Then

$$(u \times 10^m)(v \times 10^n) = uv \times 10^{m+n}$$

DIVISION

When you want to divide numbers in scientific notation, first divide the coefficients by each other. Then subtract the exponent in the denominator from the exponent in the numerator. Finally, reduce the expression to standard scientific notation if it doesn't happen to be that way already. Here are three examples of how division is done in scientific notation:

$$(3.045 \times 10^5) / (6.853 \times 10^6) = (3.045 / 6.853)(10^5 / 10^6)$$

$$\approx 0.444331 \times 10^{5-6}$$

$$\approx 0.444331 \times 10^{-1}$$

$$\approx 0.0444331$$

$$(3.045 \times 10^{-4}) / (-6.853 \times 10^{-7}) = [3.045 / (-6.853)](10^{-4} / 10^{-7})$$

$$\approx -0.444331 \times 10^{-4-(-7)}$$

$$\approx -0.444331 \times 10^3$$

$$\approx -4.44331 \times 10^2$$

$$\approx -444.331$$

$$(-3.045 \times 10^5) / (-6.853 \times 10^{-7}) = [-3.045 / (-6.853)](10^5 / 10^{-7})$$

$$\approx 0.444331 \times 10^{5-(-7)}$$

$$\approx 0.444331 \times 10^{12}$$

$$\approx 4.44331 \times 10^{11}$$

The numbers here do not divide out neatly, so the decimal-format portions are approximated.

Here's the generalized rule for division in scientific notation, using the variables u and v to represent the coefficients and the variables m and n to represent the exponents. Let u and v be real numbers greater than or equal to 1 but less than 10, and let m and n be integers. Then

$$(u \times 10^m) / (v \times 10^n) = u/v \times 10^{m-n}$$

RAISING TO A POWER

When a number is raised to a power in scientific notation, both the coefficient and the power of 10 itself must be raised to that power, and the result multiplied. Here is an example:

$$\begin{aligned} (4.33 \times 10^5)^3 &= (4.33)^3 \times (10^5)^3 \\ &= 81.182737 \times 10^{(5 \cdot 3)} \\ &= 81.182737 \times 10^{15} \\ &= 8.1182737 \times 10^{16} \end{aligned}$$

Now look at another example, in which the power of 10 is negative:

$$\begin{aligned} (5.27 \times 10^{-4})^2 &= (5.27)^2 \times (10^{-4})^2 \\ &= 27.7729 \times 10^{(-4 \cdot 2)} \\ &= 27.7729 \times 10^{-8} \\ &= 2.77729 \times 10^{-7} \end{aligned}$$

Here's the generalized rule for taking a number to a power in scientific notation, using the variable u to represent the coefficient and the variables m and n to represent the exponents. Let u be a real number greater than or equal to 1 but less than 10, and let m and n be integers. Then

$$(u \times 10^m)^n = u^n \times 10^{mn}$$

TAKING A ROOT

To find the root of a number in scientific notation, think of it as a fractional exponent. The positive square root is equivalent to the 1/2 power. The cube root is the same thing as the 1/3 power. In general, the positive nth root of a number (where n is a positive integer) is the same thing as the $1/n$ power. When you think of roots in this way, you can multiply things out as you do with whole-number exponents. For example,

$$(5.2700 \times 10^{-4})^{1/2} = (5.2700)^{1/2} \times (10^{-4})^{1/2}$$
$$\approx 2.2956 \times 10^{(-4/2)}$$
$$\approx 2.2956 \times 10^{-2}$$
$$\approx 0.02956$$

Here's the generalized rule for taking the root of a number in scientific notation, using the variable u to represent the coefficient and the variables m and n to represent the exponents. Let u be a real number equal to at least 1 but less than 10. Let m and n be integers. Then

$$(u \times 10^m)^{(1/n)} = u^{(1/n)} \times 10^{(m/n)}$$

ADDITION

Scientific notation can be awkward when you are adding sums, unless all the addends are expressed to the same power of 10. Here are three examples:

$$(3.045 \times 10^5) + (6.853 \times 10^6) = 304{,}500 + 6{,}853{,}000$$
$$= 7{,}157{,}500$$
$$= 7.1575 \times 10^6$$
$$(3.045 \times 10^{-4}) + (6.853 \times 10^{-7}) = 0.0003045 + 0.0000006853$$
$$= 0.0003051853$$
$$= 3.051853 \times 10^{-4}$$
$$(3.045 \times 10^5) + (6.853 \times 10^{-7}) = 304{,}500 + 0.0000006853$$
$$= 304{,}500.0000006853$$
$$= 3.045000000006853 \times 10^5$$

SUBTRACTION

Subtraction follows the same basic rules as addition. It helps to convert the numbers to ordinary decimal format before subtracting.

$$(3.045 \times 10^5) \times (6.853 \times 10^6) = 304{,}500 \times 6{,}853{,}000$$
$$= -6{,}548{,}500$$
$$= -6.548500 \times 10^6$$
$$(3.045 \times 10^{-4}) \times (6.853 \times 10^{-7}) = 0.0003045 \times 0.0000006853$$
$$= 0.0003038147$$
$$= 3.038147 \times 10^{-4}$$

$$(3.045 \times 10^5) - (6.853 \times 10^{-7}) = 304{,}500 - 0.0000006853$$
$$= 304{,}499.9999993147$$
$$= 3.044999999993147 \times 10^5$$

SIGNIFICANT FIGURES

The number of significant figures (or digits) in an expression indicates the degree of accuracy to which you know a numerical value, or to which you have measured, or can measure, a quantity.

When you do multiplication, division, or exponentiation using scientific notation, the number of significant figures in the final calculation result cannot "legally" be greater than the number of significant figures in the least exact expression. Consider the two numbers $x = 2.453 \times 10^4$ and $y = 7.2 \times 10^7$. The following is a perfectly valid statement if the numerical values are exact:

$$xy = (2.453 \times 10^4)(7.2 \times 10^7)$$
$$= (2.453 \cdot 7.2) \times 10^{11}$$
$$= 17.6616 \times 10^{11}$$
$$= 1.76616 \times 10^{12}$$

But if x and y represent measured quantities, as is nearly always the case in experimental science and engineering, the above statement needs qualification. You must pay close attention to how much accuracy you claim.

HOW ACCURATE ARE YOU?

When you see a product or quotient containing quantities expressed in scientific notation, count the number of single digits in the coefficients of each number. Then take the smallest number of digits. This is the number of significant figures you can claim in the final answer or solution.

In the above example, there are four single digits in the coefficient of x and two single digits in the coefficient of y. So you must round off the answer, which appears to contain six significant figures, to two significant figures. It is important to use rounding, and not truncation, as follows:

$$xy = (2.453 \times 10^4)(7.2 \times 10^7)$$
$$= 1.76616 \times 10^{12}$$
$$\approx 1.8 \times 10^{12}$$

In situations of this sort, if you insist on being "mathematically rigorous," you can use approximate-equality symbols (the "squiggly" ones) throughout, because you are always dealing with approximate values. But most folks are content to use ordinary equality symbols. It is universally understood that physical measurements are inherently inexact. So from now on, let's dispense with the "squigglies"!

Suppose you want to find the quotient x/y instead of the product xy. Proceed as follows:

$$x/y = (2.453 \times 10^4) / (7.2 \times 10^7)$$
$$= (2.453 / 7.2) \times 10^{-3}$$
$$= 0.3406944444 \ldots \times 10^{-3}$$
$$= 3.406944444 \ldots \times 10^{-4}$$
$$= 3.4 \times 10^{-4}$$

Again, you must round off to two significant figures at the end. It's best to keep a few extra digits during a calculation to minimize cumulative rounding errors. You saw the results of such errors in some of the problems in Chap. 24.

WHAT ABOUT 0?

Sometimes, when you make a calculation, you'll get an answer that lands on a neat, seemingly whole-number value. Consider $x = 1.41421$ and $y = 1.41422$. Both of these have six significant figures. Taking significant figures into account, we have

$$xy = 1.41421 \cdot 1.41422$$
$$= 2.0000040662$$
$$= 2.00000$$

This appears to be exactly equal to 2. But in the real world (the measurement of a physical quantity, for example), the presence of five zeros after the radix point indicates an uncertainty of up to plus or minus 0.000005 (written ± 0.000005). When you claim a certain number of significant figures, 0 is as important as any other digit.

Now find the product of 1.001×10^5 and 9.9×10^{-6}, taking significant figures into account. First, multiply the coefficients and the powers of 10 separately, and then proceed from there, like this:

$$(1.001 \cdot 9.9) \times (10^5 \times 10^{-6}) = 9.9099 \times 10^{-1}$$
$$= 0.99099$$

You must round this off to two significant figures, because that is the most you can legitimately claim. This particular expression does not have to be written out in power-of-10 form, because the exponent is within the range ± 2 inclusive. Therefore,

$$(1.001 \times 10^5)\,(9.9 \times 10^{-6}) = 0.99$$

WHAT ABOUT EXACT VALUES?

Once in awhile, some of or all the values in physical formulas are intended to be exact. An example is the equation for the area of a triangle

$$A = bh/2$$

where A is the area, b is the base length, and h is the height. In this formula, 2 is a mathematical constant, and its value is exact. It can therefore have as many significant figures as you want, depending on the number of significant figures you are given in the initial statement of the problem.

Sometimes you'll come across constants whose values are exact in a theoretical sense, but which you must round off when you want to assign them a certain number of significant figures. Two common examples are the irrational numbers π and e. Both of these are nonterminating, nonrepeating decimals, and they can never be exactly written down in that form. You *have* to round them off!

Expressed to 10 significant figures and then progressively rounded off (not truncated!) to fewer and fewer significant figures, π has values as follows:

3.141592654

3.14159265

3.1415927

3.141593

3.14159

3.1416

3.142

3.14

3.1

3

You can use however many significant figures you need when you encounter a constant of this type in a formula.

MORE ABOUT ADDITION AND SUBTRACTION

When you add or subtract physical quantities that you have measured (in a lab, for example), determining the number of significant figures can involve subjective judgment. You can minimize the confusion by expanding all the values out to their plain decimal form, making the calculations, and then, at the end of the process, deciding how many significant figures you can claim.

Sometimes, the outcome of determining significant figures in a sum or difference is similar to what happens with multiplication or division. Take this sum $x + y$:

$$x + y = 3.778800 \times 10^{-6} + 9.22 \times 10^{-7}$$

This calculation proceeds as follows:

$$x = 0.000003778800$$
$$y = 0.000000922$$
$$x + y = 0.0000047008$$
$$= 4.7008 \times 10^{-6}$$
$$= 4.70 \times 10^{-6}$$

Occasionally, one of the values in a sum or difference is insignificant with respect to the other. Suppose you want to find this sum $x + y$:

$$x + y = 3.778800 \times 10^4 + 9.22 \times 10^{-7}$$

You calculate this way:

$$x = 37,788$$
$$y = 0.000000922$$
$$x + y = 37,788.000000922$$
$$= 3.7788000000922 \times 10^4$$

Here, y is so much smaller than x that it does not significantly affect the value of the sum. When you round off, you can only go to five significant figures, so the second value vanishes! You can say that the sum is the same as the larger number

$$x + y = 3.7788 \times 10^4$$

If this confuses you, imagine that you set a U.S. dime on the back of an elephant. Does the elephant weigh significantly more with the dime than without it? No! But if you set a U.S. dime on top of a U.S. nickel, the pair weighs significantly more than the nickel alone.

QUESTIONS AND PROBLEMS

This is an open-book quiz. You may refer to the text in this chapter (and earlier ones, too, if you want) when figuring out the answers. Take your time! The correct answers are in the back of the book.

1. Write down the number 238,200,000,000,000 in scientific notation.

2. Write down the number 0.00000000678 in scientific notation.

3. Draw a number line that spans 4 orders of magnitude, from 1 to 10^4.

4. Draw a coordinate system with a horizontal scale that spans 2 orders of magnitude, from 1 to 100, and a vertical scale that spans 4 orders of magnitude, from 0.01 to 100.

5. What does 3.5562E+99 represent? How does it differ from 3.5562E−99?

6. Find the product of 8.0402×10^{64} and 2.73×10^{-63}. Round the answer off to the largest justifiable number of significant figures, and express it properly.

7. Find the quotient -6.7888×10^{34} divided by 8.45453×10^{36}. Round the answer off to the largest justifiable number of significant figures, and express it properly.

8. Calculate the 5th power of 4.57×10^7. Round the answer off to the largest justifiable number of significant figures, and express it properly.

9. Calculate the 4th root of 8.84×10^8. Round the answer off to the largest justifiable number of significant figures, and express it properly.

10. Find $(5.33 \times 10^{34}) - (1.99 \times 10^{-83})$. Round the answer off to the largest justifiable number of significant figures, and express it properly.

26
Vectors

You've already seen a few examples of vectors in this book. You might reasonably ask, "What, exactly, are vectors, and how do they work?" Technically, a vector is a mathematical expression for a quantity that has two independent properties, usually magnitude (or length) and direction (or orientation). Sometimes rectangular coordinates are used instead of magnitude and direction numbers. A *vector quantity* always needs two or more numbers to represent it. A *scalar quantity* needs only one. In this chapter, you'll learn the basic properties of vectors in two and three dimensions.

MAGNITUDE AND DIRECTION

Vectors are denoted by boldface letters of the alphabet. In the xy plane, vectors **a** and **b** can be illustrated as rays from the origin $(0,0)$ to points (x_a, y_a) and (x_b, y_b) as shown in Fig. 26-1. The magnitude, or length, of the vector **a**, written $|\mathbf{a}|$ or a, can be found in the xy plane with a distance formula based on the Pythagorean theorem:

$$|\mathbf{a}| = (x_a^2 + y_a^2)^{1/2}$$

The direction of the vector **a**, written dir **a**, is the angle θ_a that **a** subtends counterclockwise from the positive x axis. This angle is equal to the arctangent of the ratio of y_a to x_a:

$$\text{dir } \mathbf{a} = \theta_a = \arctan (y_a / x_a)$$

All possible directions can be represented by restricting the angle to less than a full circle in the counterclockwise direction:

$$0° \leq \theta_a < 360° \quad \text{(for } \theta_a \text{ in degrees)}$$
$$0 \leq \theta_a < 2\pi \quad \text{(for } \theta_a \text{ in radians)}$$

Figure 26-1

Two vectors in the Cartesian plane. They can be added using the parallelogram method.

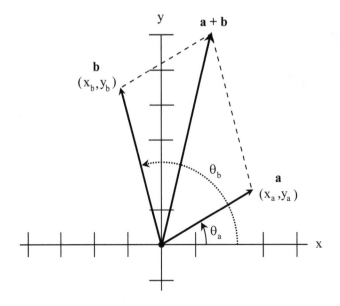

SUM OF TWO VECTORS

The sum of two vectors $\mathbf{a}$ and $\mathbf{b}$, where $\mathbf{a} = (x_a, y_a)$ and $\mathbf{b} = (x_b, y_b)$, can be found by adding their x and y components separately. Use this formula for the vectors $\mathbf{a}$ and $\mathbf{b}$ as shown in Fig. 26-1:

$$\mathbf{a} + \mathbf{b} = [(x_a + x_b),(y_a + y_b)]$$

This sum can be found geometrically by constructing a parallelogram with the vectors $\mathbf{a}$ and $\mathbf{b}$ as adjacent sides. When you do that, the vector $\mathbf{a} + \mathbf{b}$ is the diagonal of the parallelogram. This technique is called the *parallelogram method of vector addition*.

Suppose you want to find the sum of two vectors $\mathbf{a} = (3,-5)$ and $\mathbf{b} = (2,6)$. These are ordered pairs; the first numbers represent the x values, and the second numbers represent the y values. All you must do is to add the x and y components together independently. Writing the vectors as ordered pairs gives

$$\mathbf{a} + \mathbf{b} = (3,-5) + (2,6)$$

$$= [(3 + 2),(-5 + 6)]$$

$$= (5,1)$$

The sum vector $\mathbf{a} + \mathbf{b}$ therefore has the coordinate values $x = 5$ and $y = 1$.

MULTIPLICATION OF A VECTOR BY A SCALAR

To multiply a vector by a scalar quantity such as a real number, multiply both the x and y components of the vector by that scalar. That makes the vector longer or shorter if the real number is positive, but doesn't change its direction. If the real number is negative, the direction of the vector is exactly reversed, and its length may change as well.

Multiplication by a scalar is commutative. This means that it doesn't matter whether the scalar comes before or after the vector in the product. If we have a vector $\mathbf{a} = (x_a, y_a)$ and a scalar k, then

$$k\mathbf{a} = \mathbf{a}k = (kx_a, ky_a)$$

You multiply both the x and y values by the scalar k to get the result.

When you multiply a vector by a scalar, it isn't true vector multiplication because the multiplicand and the multiplier aren't both vectors. You can multiply vectors by other vectors, however—and things get interesting then! There are two ways to multiply a vector by another vector. The first way gives you a scalar as the product. The second (and somewhat more complicated) way gives you a vector as the product. Let's look at the first of these two operations now. Later in this chapter, you'll learn about the second method.

DOT PRODUCT OF TWO VECTORS

Imagine two vectors $\mathbf{a} = (x_a, y_a)$ and $\mathbf{b} = (x_b, y_b)$ such as those shown in Fig. 26-1. If all four of the variables x_a, x_b, y_a, and y_b are real numbers, then the *dot product*, also known as the *scalar product* and written $\mathbf{a} \cdot \mathbf{b}$, of the vectors $\mathbf{a}$ and $\mathbf{b}$ is a real number and can be found by using this formula:

$$\mathbf{a} \cdot \mathbf{b} = x_a x_b + y_a y_b$$

The dot product of $\mathbf{a}$ and $\mathbf{b}$ is called "$\mathbf{a}$ dot $\mathbf{b}$" in informal talk.

As an example, let's find the dot product of the two vectors $\mathbf{a} = (3,-5)$ and $\mathbf{b} = (2,6)$. Use the formula given above.

$$\mathbf{a} \cdot \mathbf{b} = (3 \cdot 2) + (-5 \cdot 6)$$

$$= 6 + (-30)$$

$$= -24$$

Now let's see what happens if the order of the dot product is reversed. Does the value change? If not, the dot product is commutative. Let's try a general proof. Take the formula above for the dot product, and suppose $\mathbf{a} = (x_a, y_a)$ and $\mathbf{b} = (x_b, y_b)$. First consider the dot product with vector $\mathbf{a}$ taken first:

$$\mathbf{a} \cdot \mathbf{b} = x_a x_b + y_a y_b$$

Now evaluate the dot product with $\mathbf{b}$ taken first:

$$\mathbf{b} \cdot \mathbf{a} = x_b x_a + y_b y_a$$

You know that ordinary multiplication is commutative for all real numbers. That means you can reverse the factors in both addends. Because x_a, y_a, x_b, and y_b are all real numbers, the above formula is equivalent to

$$\mathbf{b} \cdot \mathbf{a} = x_a x_b + y_a y_b$$

But $x_a x_b + y_a y_b$ is exactly what you get when you work out the expansion of $\mathbf{a} \cdot \mathbf{b}$. This proves that, for any two vectors $\mathbf{a}$ and $\mathbf{b}$, it is always true that $\mathbf{a} \cdot \mathbf{b} = \mathbf{b} \cdot \mathbf{a}$. The dot product is commutative, just as is plain multiplication of scalars.

MAGNITUDE AND DIRECTION IN THE POLAR PLANE

In the mathematician's polar coordinate plane, vectors $\mathbf{a}$ and $\mathbf{b}$ can be denoted as rays from the origin $(0,0)$ to points (θ_a, r_a) and (θ_b, r_b) as shown in Fig. 26-2. The magnitude and direction are defined directly:

$$|\mathbf{a}| = r_a$$

$$\text{dir } \mathbf{a} = \theta_a$$

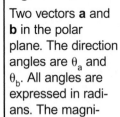

Figure 24-2

Two vectors $\mathbf{a}$ and $\mathbf{b}$ in the polar plane. The direction angles are θ_a and θ_b. All angles are expressed in radians. The magnitudes are r_a and r_b.

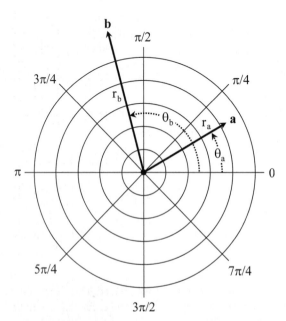

The values of the angle and the length are customarily restricted. The angle is expressed counterclockwise and is less than a full circle. The length is positive or zero. Mathematically, you can write these facts as

$$0° \leq \theta_a < 360° \quad \text{(for } \theta_a \text{ in degrees)}$$
$$0 \leq \theta_a < 2\pi \quad \text{(for } \theta_a \text{ in radians)}$$
$$r_a \geq 0$$

SUM OF TWO VECTORS IN THE POLAR PLANE

The sum of two vectors in polar coordinates can be found by converting them to their equivalents in the xy plane, adding the vectors according to the formula for the xy plane, and then changing the result back to polar coordinates. To convert the vector $\mathbf{a} = (\theta_a, r_a)$ in mathematician's polar coordinates to $\mathbf{a} = (x_a, y_a)$ in rectangular coordinates:

$$x_a = r_a \cos \theta_a$$
$$y_a = r_a \sin \theta_a$$

To convert the vector $\mathbf{a} = (x_a, y_a)$ from rectangular coordinates to polar coordinates so $\mathbf{a} = (\theta_a, r_a)$, use these formulas:

$$\theta_a = \arctan (y_a/x_a) \text{ if } x_a > 0$$
$$\theta_a = 180° + \arctan (y_a/x_a) \text{ if } x_a < 0° \quad \text{(for } \theta_a \text{ in degrees)}$$
$$\theta_a = \pi + \arctan (y_a/x_a) \text{ if } x_a < 0 \quad \text{(for } \theta_a \text{ in radians)}$$
$$r_a = (x_a^2 + y_a^2)^{1/2}$$

Try a practical problem. Consider the vector $\mathbf{a} = (x_a, y_a) = (3,4)$ in the xy plane. Suppose these values are exact. You want to find the equivalent vector $\mathbf{a} = (\theta_a, r_a)$ in mathematician's polar coordinates to the nearest angular degree, and to the nearest hundredth of a linear unit. First find the direction angle θ_a. Because $x_a > 0$, use this formula:

$$\theta_a = \arctan (y_a/x_a)$$
$$= \arctan (4/3)$$
$$= 53°$$

Now solve for r_a as follows:

$$r_a = (x_a^2 + y_a^2)^{1/2}$$
$$= (3^2 + 4^2)^{1/2}$$
$$= 25^{1/2}$$
$$= 5.00$$

Therefore, $\mathbf{a} = (\theta_a, r_a) = (53°, 5.00)$.

Now work out a problem "the other way." Consider the vector $\mathbf{b} = (\theta_b, r_b) = (200°, 4.55)$ in mathematician's polar coordinates. You want to convert this to an equivalent vector $\mathbf{b} = (x_b, y_b)$ in rectangular coordinates to the nearest tenth of a unit. First, solve for x_b:

$$
\begin{aligned}
x_b &= r_b \cos \theta_b \\
&= 4.55 \cos 200° \\
&= 4.55 \cdot (-0.9397) \\
&= -4.3
\end{aligned}
$$

Then solve for y_b:

$$
\begin{aligned}
y_b &= r_b \sin \theta_b \\
&= 4.55 \sin 200° \\
&= 4.55 \cdot (-0.3420) \\
&= -1.6
\end{aligned}
$$

Therefore, $\mathbf{b} = (x_b, y_b) = (-4.3, -1.6)$, with rectangular coordinate values accurate to the nearest tenth of a linear unit.

MULTIPLICATION OF A VECTOR BY A SCALAR IN THE POLAR PLANE

In mathematician's polar coordinates, imagine some vector $\mathbf{a}$ defined as the ordered pair (θ, r) as shown in Fig. 26-3. Suppose $\mathbf{a}$ is multiplied by a positive real scalar k. The result is a longer vector if $k > 1$ and a shorter vector if $0 < k < 1$. The following equation can be used:

$$k\mathbf{a} = (\theta, kr)$$

Figure 26-3

Multiplication of a polar-plane vector **a** by a positive real number k, and by a negative real number $-k$. Angles in this drawing are expressed in radians.

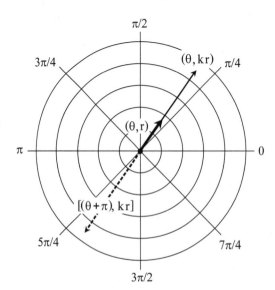

If **a** is multiplied by a negative real scalar –*k*, then the result is a vector of different length in the opposite direction. The vector is longer if –*k* < –1 and shorter if –1 < –*k* < 0. The formula looks like this:

$$-k\mathbf{a} = [(\theta + 180°), kr]$$

for θ in degrees. For θ in radians, the formula is

$$-k\mathbf{a} = [(\theta + \pi), kr]$$

The addition of 180° (π rad or a half-circle) to θ reverses the direction of **a**. The same effect can be produced by subtracting 180° (π rad or a half-circle) from θ. You should add a half-circle if the original angle is less than 180° and subtract a half-circle if the original angle is larger than 180°. That will ensure that the final angle is always defined as something nonnegative, but less than a full circle counterclockwise.

DOT PRODUCT OF TWO VECTORS IN POLAR PLANE

The dot product of two vectors is easy to express in mathematician's polar coordinates if you know their lengths and their directions. Suppose vector **a** is at an angle of θ_a counterclockwise from the reference axis, and vector **b** is at an angle of θ_b. Let |**a**| or r_a represent the magnitude of vector **a**, and let |**b**| or r_b represent the magnitude of vector **b**. Then the dot product of **a** and **b** is given by this formula:

$$\mathbf{a} \cdot \mathbf{b} = |\mathbf{a}|\,|\mathbf{b}|\cos(\theta_b - \theta_a)$$
$$= r_a r_b \cos(\theta_b - \theta_a)$$

Even if you don't have the exact coordinates of the vectors **a** and **b**, you can still find their dot product if you know their lengths r_a and r_b, and if you also know the angle θ between them. Place the vectors so their back endpoints are both at the coordinate origin. Then **a** and **b** define a plane in which the angle can be measured, and the dot product is

$$\mathbf{a} \cdot \mathbf{b} = r_a r_b \cos\theta$$

MAGNITUDE IN *XYZ* SPACE

In rectangular *xyz* space, two vectors **a** and **b** can be denoted as rays from the origin (0,0,0) to points (x_a, y_a, z_a) and (x_b, y_b, z_b) as shown in Fig. 26-4. The magnitude of **a**, written |**a**|, can be found using a three-dimensional extension of the Pythagorean theorem for right triangles. The formula looks like this:

$$|\mathbf{a}| = (x_a^2 + y_a^2 + z_a^2)^{1/2}$$

Figure 26-4

Two vectors **a** and **b** in *xyz*-space. They are added using the parallelo-gram method. This is a perspective drawing, so the parallelogram appears distorted.

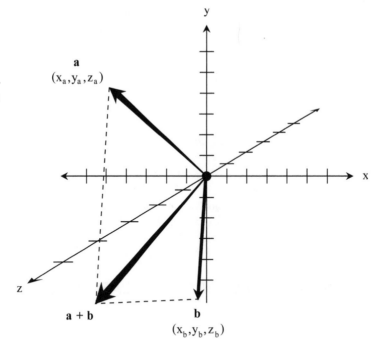

If you want a challenge, prove this fact on the basis of the two-dimensional Pythagorean theorem for the magnitude of a vector **a** in *xy* space:

$$|a| = (x_a^2 + y_a^2)^{1/2}$$

Now imagine that you want to find the magnitude of the vector denoted by $\mathbf{a} = (x_a, y_a, z_a) = (1,2,3)$ in *xyz* space. Suppose you are told that the values 1, 2, and 3 are exact, and you want to get the answer accurate to four digits after the decimal point. Use the distance formula

$$
\begin{aligned}
|a| &= (x_a^2 + y_a^2 + z_a^2)^{1/2} \\
&= (1^2 + 2^2 + 3^2)^{1/2} \\
&= (1 + 4 + 9)^{1/2} \\
&= 14^{1/2} \\
&= 3.7417
\end{aligned}
$$

DIRECTION IN *XYZ* SPACE

In three-dimensional *xyz* space, the direction of a vector **a** is denoted by measuring or expressing the angles θ_x, θ_y, and θ_z that **a** subtends relative to the positive *x*, *y*, and *z* axes, respectively, as shown in Fig. 26-5. These angles, given in radians as an ordered triple $(\theta_x, \theta_y, \theta_z)$, are called the *direction angles* of the vector **a**. So you can write

$$\text{dir } \mathbf{a} = (\theta_x, \theta_y, \theta_z)$$

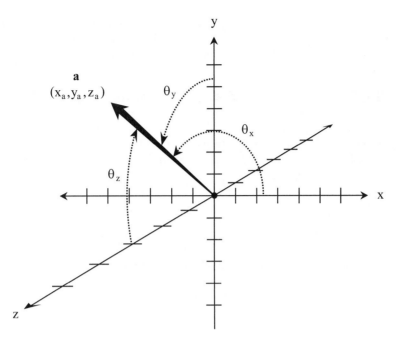

Figure 26-5
Direction angles
of a vector **a** in
xyz-space.

Sometimes the cosines of these angles are specified. These trig ratios are called the *direction cosines* of **a**, and are given as an ordered triple (α, β, γ), like this:

$$\text{dir } \mathbf{a} = (\alpha, \beta, \gamma)$$

where $\alpha = \cos \theta_x$, $\beta = \cos \theta_y$, and $\gamma = \cos \theta_z$.

Direction angles are customarily defined as the smallest possible angle between the vector and the axis in question. Therefore, a direction angle is always positive and cannot exceed π. Formally, the mathematician would say this:

$$0 \le \theta_x \le \pi$$
$$0 \le \theta_y \le \pi$$
$$0 \le \theta_x \le \pi$$

SUM OF TWO VECTORS IN *XYZ* SPACE

In three-dimensional *xyz* space, the sum of two vectors $\mathbf{a} = (x_a, y_a, z_a)$ and $\mathbf{b} = (x_b, y_b, z_b)$ is found by simply adding their *x* values, *y* values, and *z* values together, thereby forming a new ordered triple, like this:

$$\mathbf{a} + \mathbf{b} = [(x_a + x_b), (y_a + y_b), (z_a + z_b)]$$

This sum can, as in the two-dimensional case, be found geometrically by constructing a parallelogram with **a** and **b** as adjacent sides. The sum **a** + **b** is determined by the diagonal of the parallelogram, as shown in Fig. 26-4.

MULTIPLICATION OF A VECTOR BY A SCALAR IN *XYZ* SPACE

Imagine a vector **a** defined by the coordinates (x_a, y_a, z_a) in *xyz* space. Suppose **a** is multiplied by some positive real scalar k. Then the following equation holds, an extension of the two-dimensional case:

$$k\mathbf{a} = k(x_a, y_a, z_a)$$
$$= (kx_a, ky_a, kz_a)$$

If **a** is multiplied by a negative real scalar $-k$, then

$$-k\mathbf{a} = -k(x_a, y_a, z_a)$$
$$= (-kx_a, -ky_a, -kz_a)$$

Now imagine that the direction angles of **a** are represented by $(\theta_x, \theta_y, \theta_z)$. Then the direction angles of $k\mathbf{a}$ are also $(\theta_x, \theta_y, \theta_z)$. When you multiply any vector by a positive real number scalar, its direction does not change, no matter how many dimensions it has. But if the scalar is a negative real number (call it $-k$), the direction angles all change by π, so dir $(-k\mathbf{a})$ is represented by the ordered triple $[(\theta_x \pm \pi), (\theta_y \pm \pi), (\theta_z \pm \pi)]$. Whether π should be added or subtracted from an original angle depends, in each case, on which way will give you a new angle that is positive but not greater than π. The length is multiplied by k (the positive value, with the minus sign removed).

DOT PRODUCT OF TWO VECTORS IN *XYZ* SPACE

The dot product of two vectors $\mathbf{a} = (x_a, y_a, z_a)$ and $\mathbf{b} = (x_b, y_b, z_b)$ in *xyz* space is a real number that can be found by a three-dimensional extension of the dot product formula for vectors in the *xy* plane:

$$\mathbf{a} \cdot \mathbf{b} = x_a x_b + y_a y_b + z_a z_b$$

If you don't have the *xyz* coordinates of the vectors, but instead you know their magnitudes $|\mathbf{a}| = r_a$ and $|\mathbf{b}| = r_b$ as well as the angle θ between them, then you can find their dot product in three dimensions. Position the vectors so their back endpoints are both at the coordinate origin. Then the dot product is found this way:

$$\mathbf{a} \cdot \mathbf{b} = |\mathbf{a}| \, |\mathbf{b}| \cos \theta$$
$$= r_a r_b \cos \theta$$

This formula works in any three-dimensional system of coordinates. It is just as good in spherical or cylindrical coordinates as it is in rectangular *xyz* space. However, finding the vector magnitudes and the angle between them can sometimes be difficult.

CROSS PRODUCT OF TWO VECTORS

The *cross product*, also known as the *vector product* and written $\mathbf{a} \times \mathbf{b}$, of two vectors **a** and **b** is a third vector that is perpendicular to the plane containing both **a** and **b**. Cross products are often used in engineering, physics, and electronics.

Suppose that θ is the angle between vectors **a** and **b** expressed counterclockwise (as viewed from above, or the direction of the positive *z* axis) in the plane containing them both, as shown in Fig. 26-6. Let |a| or r_a represent the magnitude of vector **a**, and let |b| or r_b represent the magnitude of vector **b**. Place the vectors so their back endpoints are both at the coordinate origin. Then the magnitude of $\mathbf{a} \times \mathbf{b}$ is the product of the original vector magnitudes and the sine of the angle between them:

$$|\mathbf{a} \times \mathbf{b}| = |\mathbf{a}|\, |\mathbf{b}| \sin \theta$$

$$= r_a r_b \sin \theta$$

What about the direction of the cross-product vector? You can use a little trick called the *right-hand rule for cross products* to ascertain the direction of $\mathbf{a} \times \mathbf{b}$. Curl the fingers of your right hand in the sense in which θ, the angle between **a** and **b**, is defined. Extend your right thumb. When you hold your hand this way, $\mathbf{a} \times \mathbf{b}$ points in the direction of your thumb. In the example shown by Fig. 26-6, $\mathbf{a} \times \mathbf{b}$ points upward at a right angle to the plane containing both of the original vectors **a** and **b**.

When $180° < \theta < 360°$ ($\pi < \theta < 2\pi$), the cross-product vector reverses direction compared with the situation when $0° < \theta < 180°$ ($0 < \theta < \pi$). This is demonstrated by the

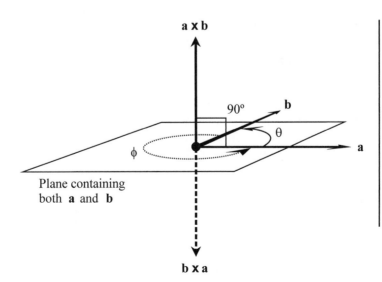

Figure 26-6

Vector cross products. The vector $\mathbf{b} \times \mathbf{a}$ has the same magnitude as vector $\mathbf{a} \times \mathbf{b}$, but points in the opposite direction. Both vectors $\mathbf{b} \times \mathbf{a}$ and $\mathbf{a} \times \mathbf{b}$ are perpendicular to the plane defined by **a** and **b**.

fact that, in the above formula, $\sin \theta$ is positive when $0° < \theta < 180°$ $(0 < \theta < \pi)$, but negative when $180° < \theta < 360°$ $(\pi < \theta < 2\pi)$. When $180° < \theta < 360°$ $(\pi < \theta < 2\pi)$, the right-hand rule doesn't work. Instead, you must use your left hand, and curl your fingers into almost a complete circle! An example is the cross product $\mathbf{b} \times \mathbf{a}$ in Fig. 26-6. The angle ϕ, expressed counterclockwise between these vectors (as viewed from above), is more than 180°. This reverses the direction of the cross-product vector.

Do you see a pattern here? For any two vectors $\mathbf{a}$ and $\mathbf{b}$, the vector $\mathbf{b} \times \mathbf{a}$ is a "mirror image" of $\mathbf{a} \times \mathbf{b}$, where the "mirror" is the plane containing both vectors. One way to imagine the mirror image is to note that $\mathbf{b} \times \mathbf{a}$ has the same magnitude as $\mathbf{a} \times \mathbf{b}$, but points in exactly the opposite direction. A less common (but equally valid, at least in theory) way to look at this situation is to imagine that dir $\mathbf{b} \times \mathbf{a}$ is the same as dir $\mathbf{a} \times \mathbf{b}$, but $|\mathbf{b} \times \mathbf{a}| = -|\mathbf{a} \times \mathbf{b}|$. Either way, it is apparent that the cross-product operation is not commutative. But it is "inversely commutative"! The following relationship always holds for any two vectors $\mathbf{a}$ and $\mathbf{b}$:

$$\mathbf{a} \times \mathbf{b} = -1(\mathbf{b} \times \mathbf{a})$$
$$= -(\mathbf{b} \times \mathbf{a})$$

NEGATIVE VECTOR MAGNITUDES

Are you confused about the concept of vector magnitude and the fact that *absolute-value symbols* (the two vertical lines) are sometimes used to denote vector magnitude? From basic algebra, remember that the absolute value of a real number is always positive. But with vectors, negative magnitudes sometimes appear in the equations. What's going on with that?

Whenever you see a vector whose magnitude is negative, remember that it's the equivalent of a vector pointing in the opposite direction with positive magnitude having the same absolute value. For example, if someone tells you that a force of -20 newtons is exerted upward by a machine, you know it is the equivalent of 20 newtons exerted downward. If you say that a storm is located 50 kilometers to your southwest, you can also say (although you should expect raised eyebrows in response) that the storm is -50 kilometers to your northeast. A statement in mathematics can sound quite weird but be theoretically valid anyhow.

When a vector with negative magnitude occurs in the final answer to a problem, you can *exactly* reverse the direction of the vector, and then assign it a positive magnitude equal to -1 times the negative magnitude. This is always a good idea if you can manage it, because it makes more common sense.

Now imagine two vectors $\mathbf{a}$ and $\mathbf{b}$ in xyz space, represented by the following ordered triples:

$$\mathbf{a} = (3,4,0)$$
$$\mathbf{b} = (0,-5,0)$$

You are told that these values are mathematically exact. You want to find an ordered triple (x,y,z) that represents the vector $\mathbf{a} \times \mathbf{b}$, with values accurate to three significant figures. How do you go about it?

To start out, you should make a rough drawing that shows these two vectors. You'll get something that looks like Fig. 26-7. Note that both of the vectors are in the xy plane, so you don't really have to show the z axis to portray the vectors. In Fig. 26-7, the xy plane coincides with the paper. Imagine the positive z axis coming out of the page directly toward you, and the negative z axis pointing straight away from you on the other side of the page.

Now figure out the direction in which $\mathbf{a} \times \mathbf{b}$ points. Remember that the direction of the cross product of two vectors is always perpendicular to the plane containing the original vectors. Therefore, you know that $\mathbf{a} \times \mathbf{b}$ must point along the z axis. This means that the ordered triple for $\mathbf{a} \times \mathbf{b}$ has to be in the form $(0,0,z)$, where z is some real number. You don't yet know what this number is, and you had better not jump to any conclusions. Is it positive? Negative? Zero? You must proceed further to find out.

Calculate the magnitudes of the two vectors $\mathbf{a}$ and $\mathbf{b}$. To find $|\mathbf{a}|$, use the formula

$$\begin{aligned}
|\mathbf{a}| &= (x_a^2 + y_a^2 + z_a^2)^{1/2} \\
&= (3^2 + 4^2 + 0^2)^{1/2} \\
&= (9 + 16)^{1/2} \\
&= 25^{1/2} \\
&= 5 \text{ (exactly)}
\end{aligned}$$

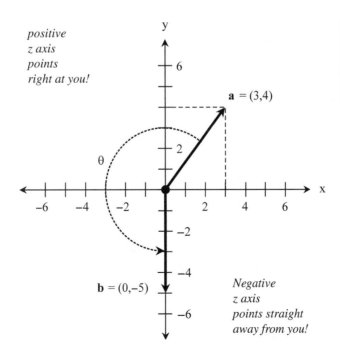

Figure 26-7

What is the cross product $\mathbf{a} \times \mathbf{b}$ of these two vectors in *xyz*-space?

Similarly, for $|\mathbf{b}|$:

$$|\mathbf{b}| = (x_b^2 + y_b^2 + z_b^2)^{1/2}$$
$$= [0^2 + (-5)^2 + 0^2]^{1/2}$$
$$= 25^{1/2}$$
$$= 5 \text{ (exactly)}$$

Therefore, $|\mathbf{a}|\,|\mathbf{b}| = 5 \cdot 5 = 25$—exactly! You want three significant figures in the final answer, so you can call it 25.0.

To determine the magnitude of $\mathbf{a} \times \mathbf{b}$, you must multiply 25.0 by the sine of the angle θ expressed counterclockwise from $\mathbf{a}$ to $\mathbf{b}$. When you examine Fig. 26-7, you can see that θ is equal to exactly 270° (three-quarters of a circle) minus the angle between the x axis and the vector $\mathbf{a}$. The angle between the x axis and vector $\mathbf{a}$ is the arctangent of 4/3, or approximately 53.13° as determined by using a calculator. Therefore,

$$\theta = 270.00° - 53.13°$$
$$= 216.87°$$
$$\sin \theta = \sin 216.87°$$
$$= -0.600$$

This means that the magnitude of $\mathbf{a} \times \mathbf{b}$ is equal to $25.0 \cdot (-0.600)$, or -15.0. The minus sign means that the cross-product vector points negatively along the z axis. The z coordinate of $\mathbf{a} \times \mathbf{b}$ is equal to -15.0. You know that the x and y coordinates of $\mathbf{a} \times \mathbf{b}$ are both exactly equal to 0, because $\mathbf{a} \times \mathbf{b}$ lies exactly along the z axis. From all this information, you can conclude that $\mathbf{a} \times \mathbf{b} = (0.00, 0.00, -15.0)$, with the values each expressed to three significant figures.

QUESTIONS AND PROBLEMS

This is an open-book quiz. You may refer to the text in this chapter (and earlier ones, too, if you want) when figuring out the answers. Take your time! The correct answers are in the back of the book.

1. Find the magnitude of the vector $\mathbf{a} = (-7, -10)$ in the xy plane. Assume the values given are exact. Express the answer to three significant figures.

2. Convert the vector $\mathbf{a} = (-7, -10)$, as expressed in the xy plane, to polar form as a vector $\mathbf{a} = (\theta_a, r_a)$. Assume the values given are exact. Express the answer to three significant figures, with θ_a in degrees.

3. Find the magnitude of the vector $\mathbf{b} = (8, -1, -6)$ in xyz space. Assume the values given are exact. Express the answer to four significant figures.

4. Consider the two vectors $\mathbf{a} = (-7,-10,0)$ and $\mathbf{b} = (8,-1,-6)$ in *xyz* space. What is their dot product?

5. Imagine that two vectors $\mathbf{f}$ and $\mathbf{g}$ point in the same direction along a common line in *xyz* space. Suppose $\mathbf{f}$ has magnitude 4, and $\mathbf{g}$ has magnitude 7. What is $\mathbf{f} \cdot \mathbf{g}$?

6. Imagine that two vectors $\mathbf{f}$ and $\mathbf{g}$ point in opposite directions along a common line in *xyz* space. Suppose $\mathbf{f}$ has magnitude 4, and $\mathbf{g}$ has magnitude 7. What is $\mathbf{f} \cdot \mathbf{g}$?

7. Imagine that two vectors $\mathbf{f}$ and $\mathbf{g}$ are oriented at right angles in *xyz* space. Suppose vector $\mathbf{f}$ has magnitude 4, and vector $\mathbf{g}$ has magnitude 7. What is $\mathbf{f} \cdot \mathbf{g}$? What is $\mathbf{g} \cdot \mathbf{f}$? Does it matter in which sense the rotation from $\mathbf{f}$ to $\mathbf{g}$ is defined?

8. Imagine that two vectors are oriented at right angles with respect to each other in *xyz* space. Suppose one of the vectors has magnitude 4, and the other has magnitude 7. What is the magnitude of the cross-product vector? How is this vector oriented with respect to the original two vectors?

9. Consider two vectors $\mathbf{a}$ and $\mathbf{b}$ that lie in the *xy* plane portion of *xyz* space. Suppose $\mathbf{a} = (2,0,0)$ and is fixed, pointing along the $+x$ axis. Suppose $\mathbf{b}$ has magnitude 2 and rotates counterclockwise in the *xy* plane, starting at $(2,0,0)$, then going around through $(0,2,0)$, $(-2,0,0)$, and $(0,-2,0)$, finally ending up back at $(2,0,0)$. Describe what happens to $\mathbf{a} \times \mathbf{b}$ with one rotation of $\mathbf{b}$, as you watch from some point high above the *xy* plane on the $+z$ axis. What happens if $\mathbf{b}$ keeps going around and around?

27

Logic and Truth Tables

To gain a true mastery of mathematics, you must become familiar with the laws that govern formal reasoning. *Propositional logic* is the simplest system. In this chapter, you'll learn the basics of propositional logic, in which statements are always either "totally true" or "totally false." You'll also learn to use *truth tables* to evaluate complex statements.

SENTENCES

Propositional logic does not involve breaking sentences down into their internal parts. You don't have to worry about how words are interconnected and how they affect one another within a sentence. A *sentence*, also called a *proposition*, is the smallest possible entity in propositional logic.

Sentences are represented by uppercase letters of the alphabet. You might say, "It is raining outside" and represent this by the letter R. Someone else might add, "It's cold outside" and represent this by the letter C. A third person might say, "The weather forecast calls for snow tomorrow" and represent this by the letter S. Still another person might add, "Tomorrow's forecast calls for sunny weather" and represent this by B (for "bright"; you've already used S).

NEGATION (NOT)

When you write down a letter to stand for a sentence, you assert that the sentence is true. So, for example, if Joanna writes down C in the above situation, she means to say, "It is cold outside." You might disagree if you grew up in Alaska and Joanna grew up in Hawaii. You might say, "It's not cold outside." This can be symbolized as the letter C with a *negation* symbol in front of it.

There are several ways in which negation, also called the *logical NOT operation*, can be symbolized. In propositional logic, a common symbol is a drooping minus sign (¬). Let's use it here. Some texts use a tilde (~) to represent negation. Others use a minus sign or em dash (–). Some put a line over the letter representing the sentence; still others use an accent symbol. In our system, the sentence "It's not cold outside" can be denoted as ¬C.

Suppose someone comes along and says, "You are correct to say ¬C. In fact, I'd say it's hot outside!" Suppose this is symbolized H. Does H mean the same thing as ¬C? No! You've seen days that were neither cold nor hot. There can be in-between states such as "cool" (K), "mild" (M), and "warm" (W). But there is no in-between condition when it comes to C and ¬C. In propositional logic, either it is cold, or else it is not cold. Either it's hot, or else it is not hot. A proposition is either true, or else it is false (not true). Of course, temperature opinions like this depend on who you ask, so maybe this is not such a good example. But you should get the general idea!

There are logical systems in which in-between states exist. These go by names such as *trinary logic* or *fuzzy logic*. But discussions of those types of logic belong in a different book. In this chapter, you can assume that any given proposition is either true or false; there is no "neutral" or "don't know" truth state, nor any "continuum" of truth values.

CONJUNCTION (AND)

Propositional logic doesn't bother with how the phrases inside a sentence affect one another, but it is concerned with the ways in which complete sentences interact. Sentences can be combined to make bigger ones, called *compound sentences*. The truth or falsity of a compound sentence depends on the truth or falsity of its components, and on the ways in which those components are connected.

Suppose someone says, "It's cold outside, and it's raining outside." Using the symbols above, you can write this as

C AND R

In logic, it's customary to use a symbol in place of the word AND. There are several symbols in common use, including the ampersand (&), the inverted wedge (∧), the asterisk (*), the period (.), the multiplication sign (×), and the raised dot (·). Let's use the ampersand. Then the above compound sentence becomes

C & R

The formal term for the AND operation is *logical conjunction*. A compound sentence containing one or more conjunctions is true if (but only if) both or all of its components are true. If any one of the components happens to be false, then the whole compound sentence is false.

DISJUNCTION (OR)

Now imagine that a friend comes along and says, "You are correct in your observations about the weather. It's cold and raining. I have been listening to the radio, and I heard the weather forecast for tomorrow. It's supposed to be colder tomorrow than it is today. But it's going to stay wet. So it might snow tomorrow."

You say, "It will rain or it will snow tomorrow, depending on the temperature."

Your friend says, "It might be a mix of rain and snow together, if the temperature is near freezing."

"So we might get rain, we might get snow, and we might get both," you say.

"Sure. But the weather experts say we are certain to get precipitation of some sort," your friend says. "Water is going to fall from the sky tomorrow—maybe liquid, maybe solid, and maybe both."

In this case, suppose we let R represent the sentence "It will rain tomorrow," and we let S represent the sentence "It will snow tomorrow." Then we can say

S OR R

This is an example of *logical disjunction*. There are at least two symbols that can represent disjunction: the plus sign (+) and the wedge (∨). Let's use the wedge. We can now write

S ∨ R

A compound sentence in which both, or all, of the components are joined by disjunctions is true if (but only if) at least one of the components is true. A compound sentence made up of disjunctions is false if (but only if) all the components are false.

Logical disjunction, as defined here, is the *inclusive OR operation*. If all the components are true, then the whole sentence is true. There's another operation called *exclusive OR* in which, if all the components are true, the compound sentence is false. So if you say, "It will rain or snow tomorrow," using the inclusive OR, then your statement is true if you get a mix of rain and snow. But if you use the exclusive OR, you can't have a mix; it must be one or the other, but not both. Sometimes the exclusive OR operation is called the *either-or operation*. It's abbreviated as XOR, and it is important in digital electronic circuit design. From now on, if you see the symbol for the OR operation (or the word "or" in a problem), assume it means the *inclusive OR* operation.

IMPLICATION (IF/THEN)

Imagine that your conversation about the weather continues, getting more strange with each passing minute. You and your friend are trying to figure out if you should get ready for a snow day tomorrow, or whether rain and gloom are all you'll have to contend with.

"Does the weather forecast say anything about snow?" you ask.

"Not exactly," your friend says. "The radio announcer said that there's going to be precipitation through tomorrow night, and that it's going to get colder tomorrow. I looked at my car thermometer as she said that, and the outdoor temperature was only a little bit above freezing."

"If there is precipitation, and if it gets colder, then it will snow," you say.

"Of course."

"Unless we get an ice storm."

"That won't happen."

"Okay," you say. "If there is precipitation tomorrow, and if it is colder tomorrow than it is today, then it will snow tomorrow." (This is a weird way to talk, but you're learning logic here, not the art of conversation. Logically rigorous conversation can sound bizarre, even in the "real world." Have you ever sat in a courtroom during a civil lawsuit between corporations?)

Suppose you use P to represent the sentence "There will be precipitation tomorrow." In addition, let S represent the sentence "It will snow tomorrow," and let C represent the sentence "It will be colder tomorrow." Then in the above conversation, you have made a compound proposition consisting of three sentences, like this:

$$\text{IF (P AND C), THEN S}$$

Another way to write this is

$$\text{(P AND C) IMPLIES S}$$

In this context, "implies" means *always results in*. So in formal logic, "X IMPLIES Y" means "If X, then Y." Symbolically, the above proposition is written this way:

$$(P \ \& \ C) \Rightarrow S$$

The double-shafted arrow pointing to the right represents *logical implication*, also known as the *IF/THEN operation*. In a logical implication, the "implying" sentence (to the left of the double-shafted arrow) is called the *antecedent*. In the above example, the antecedent is (P & C). The "implied" sentence (to the right of the double-shafted arrow) is called the *consequent*. In the above example, the consequent is S.

Some texts use other symbols for logical implication, including the "hook" or "lazy U opening to the left" (⊃), three dots (∴), and a single-shafted arrow pointing to the right (→). Let's keep using the double-shafted arrow pointing to the right.

LOGICAL EQUIVALENCE (IFF)

Suppose your friend changes the subject and says, "If it snows tomorrow, then there will be precipitation and it will be colder."

For a moment you hesitate, because this isn't the way you'd usually think about this kind of situation. But you have to agree. Your friend has made this implication:

$$S \Rightarrow (P \& C)$$

Implication holds in both directions here, but there are plenty of scenarios in which implication holds in one direction but not the other.

You and your friend have agreed that both of the following implications are valid:

$$(P \& C) \Rightarrow S$$

$$S \Rightarrow (P \& C)$$

These two implications can be combined into a conjunction, because you are asserting them both together:

$$[(P \& C) \Rightarrow S] \& [S \Rightarrow (P \& C)]$$

When an implication is valid in both directions, the situation is defined as a case of *logical equivalence*. The above statement can be shortened to

$$(P \& C) \text{ IF AND ONLY IF } S$$

Mathematicians sometimes shorten the phrase "if and only if" to the single "word" "iff." So you can also write

$$(P \& C) \text{ IFF } S$$

The symbol for logical equivalence is a double-shafted, double-headed arrow (⇔). There are other symbols that can be used. Sometimes you'll see an equals sign (=), a three-barred equals sign (≡), or a single-shafted, double-headed arrow (↔). Let's use the double-shafted, double-headed arrow to symbolize logical equivalence. Symbolically, then, you would write this:

$$(P \& C) \Leftrightarrow S$$

Now look at a situation in which logical implication holds in one direction but not in the other. Consider this statement: "If it is overcast, then there are clouds in the sky." This statement is always true. Suppose you let O represent the sentence "It is overcast" and K represent the sentence "There are clouds in the sky." Then you have this, symbolically:

$$O \Rightarrow K$$

If you reverse this, we get a statement that isn't necessarily true:

$$K \Rightarrow O$$

This translates to "If there are clouds in the sky, then it's overcast." You have seen days or nights in which there were clouds in the sky, but there were clear spots too, so it was not overcast.

TRUTH TABLES

The outcome, or *logic value*, of an operation in propositional logic is always either true or false, as you've seen. Truth can be symbolized as T, +, or 1, while falsity can be abbreviated as F, –, or 0. Let's use T and F. They are easy to remember: "T" stands for true and "F" stands for false! When you are performing logic operations, sentences that can attain either T or F logic values (depending on the circumstances) are called *variables*.

A *truth table* is a method of denoting all possible combinations of truth values for the variables in a compound sentence. The values for the individual variables, with all possible arrangements, are shown in vertical columns at the left. The truth values for compound sentences, as they are built up from the single-variable (or *atomic*) propositions, are shown in horizontal rows.

TABLES FOR BASIC LOGIC FUNCTIONS

The simplest truth table is the one for negation, which operates on a single variable. Table 27-1 shows how this works for a single variable called X.

Let X and Y be two logical variables. Conjunction (X & Y) produces results as shown in Table 27-2. This operation produces the truth value T when, but only when, both variables have value T. Otherwise, the operation produces the truth value F.

Table 27-1

Truth table for logical negation.

X	¬X
F	T
T	F

X	Y	X & Y
F	F	F
F	T	F
T	F	F
T	T	T

Table 27-2
Truth table for logical conjunction.

Logical disjunction for two variables ($X \lor Y$) breaks down as in Table 27-3. This operation produces the truth value T when either or both of the variables have the truth value T. If both of the variables have the truth value F, then the operation produces the truth value F. Remember, you're dealing with the inclusive operation (OR) here, not the exclusive operation (XOR)!

A logical implication is valid (that is, it has truth value T) whenever the antecedent is false. It is also valid if the antecedent and the consequent are both true. But implication does not hold (that is, it has truth value F) when the antecedent is true and the consequent is false. Table 27-4 shows the truth values for logical implication.

Let's look at a "word problem" example of a logical implication that is obviously invalid. Let X represent the sentence "The wind is blowing." Let Y represent the sentence "A hurricane is coming." Consider this sentence:

$$X \Rightarrow Y$$

Now imagine that it is a windy day. Therefore, variable X has truth value T. But suppose you are in North Dakota, where there are never any hurricanes. Sentence Y has truth value F. Therefore, the statement "If the wind is blowing, then a hurricane is coming" is not valid.

If X and Y are logical variables, then X IFF Y has truth value T when both variables have value T or when both variables have value F. If the truth values of X and Y are different, then X IFF Y has truth value F. This is broken down fully in Table 27-5.

THE EQUALS SIGN

In logic, you can use an ordinary equals sign to indicate truth value. If you want to say that a particular sentence K is true, you can write K = T. If you want to say that a variable X always has false truth value, you can write X = F. Be careful about this. Don't

X	Y	X ∨ Y
F	F	F
F	T	T
T	F	T
T	T	T

Table 27-3
Truth table for logical disjunction.

Table 27-4

Truth table for
logical implication.

X	Y	X $\Rightarrow$ Y
F	F	T
F	T	T
T	F	F
T	T	T

confuse the meaning of the equals sign with the meaning of the double-shafted, double-headed arrow that stands for logical equivalence!

The truth values shown in Tables 27-1 through 27-4 are defined by convention and are based on common sense. Arguably, the same is true for logical equivalence. It seems reasonable to suppose that two logically equivalent statements have identical truth values. How can you prove this? One method is to derive the truth values for logical equivalence based on the truth tables for conjunction and implication. Let's do it, and show the derivation in the form of a truth table.

Remember that X $\Leftrightarrow$ Y means the same thing as (X $\Rightarrow$ Y) & (Y $\Rightarrow$ X). You can build up X $\Leftrightarrow$ Y in steps, as shown in Table 27-6 as you go from left to right. The four possible combinations of truth values for sentences X and Y are shown in the first (leftmost) and second columns. The truth values for X $\Rightarrow$ Y are shown in the third column, and the truth values for Y $\Rightarrow$ X are shown in the fourth column. To get the truth values for the fifth (rightmost) column, conjunction is applied to the truth values in the third and fourth columns. The *complex logical operation* (also called a *compound logical operation* because it's made up of combinations of the basic ones) in the fifth column is the same thing as X $\Leftrightarrow$ Y.

Q.E.D.

What you have just seen is a mathematical proof of the fact that for any two logical sentences X and Y, the value of X $\Leftrightarrow$ Y is equal to T when X and Y have the same truth value, and the value of X $\Leftrightarrow$ Y is equal to F when X and Y have different truth values. Sometimes, when mathematicians finish proofs, they write "Q.E.D." at the end. This is an abbreviation of the Latin phrase *Quod erat demonstradum*. It translates to "Which was to be demonstrated."

Table 27-5

Truth table for
logical equivalence.

X	Y	X $\Leftrightarrow$ Y
F	F	T
F	T	F
T	F	F
T	T	T

X	Y	X ⇒ Y	Y ⇒ X	(X ⇒ Y) & (Y ⇒ X) which is the same as X ⇔ Y
F	F	T	T	T
F	T	T	F	F
T	F	F	T	F
T	T	T	T	T

Table 27-6

A truth table proof that logically equivalent statements always have identical truth values.

PRECEDENCE

When you are reading or constructing logical statements, the operations within parentheses are always performed first. If there are multilayered combinations of sentences (called *nesting of operations*), then you should use first ordinary parentheses ( ), then brackets [ ], and then braces { }. Alternatively, you can use groups of plain parentheses inside each other, but be sure you end up with the same number of left-hand parentheses and right-hand parentheses in the complete expression.

If there are no parentheses, brackets, or braces in an expression, instances of negation should be performed first. Then conjunctions should be done, then disjunctions, then implications, and finally logical equivalences.

As an example of how precedence works, consider the following compound sentence:

$$A \And \neg B \lor C \Rightarrow D$$

Using parentheses, brackets, and braces to clarify this according to the rules of precedence, you can write it like this:

$$\{[A \And (\neg B)] \lor C\} \Rightarrow D$$

Now consider a more complex compound sentence, which is so messy that you'll run out of grouping symbols if you use the "parentheses, brackets, braces" or PBB scheme:

$$A \And \neg B \lor C \Rightarrow D \And E \Leftrightarrow F \lor G$$

Using plain parentheses only, you can write it this way:

$$(((A \And (\neg B)) \lor C) \Rightarrow (D \And E)) \Leftrightarrow (F \lor G)$$

When you count up the number of left-hand parentheses and the number of right-hand parentheses, you'll see that there are six left-hand ones and six right-hand ones. This should always be true whenever you write, or read, a complicated logical sentence. If the number of opening and closing symbols is not the same, something is wrong.

CONTRADICTION AND DOUBLE NEGATION

A contradiction always results in nonsense, which can be considered to have a truth value of F. This is one of the most interesting and useful laws in all mathematics, and it has been used to prove important facts as well as to construct ridiculous statements and arguments. Symbolically, if X is any logical statement, you can write the rule like this:

$$(X \,\&\, \neg X) \Rightarrow F$$

This rule is sometimes stated as, "From a contradiction, any false statement whatsoever can follow." You can then write it this way:

$$(X \,\&\, \neg X) \Rightarrow Q$$

assuming that Q is false. This sort of direct contradiction is often called an *absurdity*.

Mathematicians sometimes use a process called *reductio ad absurdum* (Latin for "to reduce to absurdity") when faced with difficult proofs. You start by assuming that whatever you want to prove is false. Then, from that assumption, you derive a direct contradiction of the form X & ¬X. This demonstrates that the original assumption (that the thing you set out to prove is false) is false! If something is not false, it must be true. This strange twist of logic gives rise to another rule called the law of *double negation*. The negation of a negation is equivalent to the original expression. That is, if X is any logical variable, then

$$\neg(\neg X) \Leftrightarrow X$$

COMMUTATIVE LAWS

The conjunction of two variables has the same value regardless of the order in which the variables are expressed. If X and Y are logical variables, then X & Y is logically equivalent to Y & X:

$$X \,\&\, Y \Leftrightarrow Y \,\&\, X$$

The same property holds for logical disjunction:

$$X \vee Y \Leftrightarrow Y \vee X$$

These are called the *commutative law for conjunction* and the *commutative law for disjunction*, respectively. The variables can be *commuted* (interchanged or reversed in order), and it doesn't affect the truth value of the resulting sentence.

ASSOCIATIVE LAWS

When there are three variables combined by two conjunctions, it doesn't matter how the variables are grouped. Suppose you have a compound sentence that can be symbolized as

$$X \& Y \& Z$$

where X, Y, and Z represent the truth values of three constituent sentences. Then you can consider X & Y as a single variable and combine it with Z, or you can consider Y & Z as a single variable and combine it with X, and the results are logically equivalent:

$$(X \& Y) \& Z \Leftrightarrow X \& (Y \& Z)$$

The same law holds for logical disjunction:

$$(X \vee Y) \vee Z \Leftrightarrow X \vee (Y \vee Z)$$

These are called the *associative law for conjunction* and the *associative law for disjunction*, respectively.

You must be careful when applying associative laws. All the operations in the compound sentence must be the same. If a compound sentence contains a conjunction and a disjunction, you cannot change the grouping and expect to get the same truth value in all possible cases. For example, the following two compound sentences are not, in general, logically equivalent:

$$(X \& Y) \vee Z$$

$$X \& (Y \vee Z)$$

LAW OF IMPLICATION REVERSAL

When one sentence implies another, you can't reverse the sense of the implication and still expect the result to be valid. When you see $X \Rightarrow Y$, you cannot conclude $Y \Rightarrow X$. Things can work out that way in certain cases, such as when $X \Leftrightarrow Y$. But there are plenty of cases where it doesn't work out.

If you negate both sentences and then reverse the sense of the implication, however, the result is always valid. This can be called the *law of implication reversal*. It is also known as the *law of the contrapositive*. To express it symbolically, suppose you are given two logical variables X and Y. Then the following always holds:

$$(X \Rightarrow Y) \Leftrightarrow (\neg Y \Rightarrow \neg X)$$

Here is an example of the use of words to illustrate an example of the above law "in action." Let V represent the sentence "Jane is a living vertebrate creature." Let B represent the sentence "Jane has a brain." Then V $\Rightarrow$ B reads, "If Jane is a living vertebrate creature, then Jane has a brain." Applying the law of implication reversal, we can also say with certainty that $\neg$B $\Rightarrow$ $\neg$V. That translates to "If Jane does not have a brain, then Jane is not a living vertebrate creature."

DE MORGAN'S LAWS

If the conjunction of two sentences is negated as a whole, the resulting compound sentence can be rewritten as the disjunction of the negations of the original two sentences. Expressed symbolically, if X and Y are two logical variables, then the following holds valid in all cases:

$$\neg(X \& Y) \Leftrightarrow (\neg X \vee \neg Y)$$

This is called *De Morgan's law for conjunction*.

A similar rule holds for disjunction. If a disjunction of two sentences is negated as a whole, the resulting compound sentence can be rewritten as the conjunction of the negations of the original two sentences. Symbolically,

$$\neg(X \vee Y) \Leftrightarrow (\neg X \& \neg Y)$$

This is called *De Morgan's law for disjunction*.

You might now begin to appreciate the use of symbols to express complex statements in logic! The rigorous expression of *De Morgan's laws* in verbal form is quite a mouthful, but it's easy to write these rules down as symbols.

DISTRIBUTIVE LAW

A specific relationship exists between conjunction and disjunction, known as the *distributive law*. It works somewhat like the distributive law that you learned in arithmetic classes—a certain way that multiplication behaves with respect to addition. Do you remember it? It states that if a, b, and c are any three numbers, then

$$a(b + c) = ab + ac$$

Now think of logical conjunction as the analog of multiplication, and logical disjunction as the analog of addition. Then if X, Y, and Z are any three sentences, the following logical equivalence holds:

$$X \& (Y \vee Z) \Leftrightarrow (X \& Y) \vee (X \& Z)$$

This is called the *distributive law of conjunction with respect to disjunction*.

TRUTH TABLE PROOFS

If you claim that two compound sentences are logically equivalent, then you can prove it by showing that their truth tables produce identical results. Also, if you can show that two compound sentences have truth tables that produce identical results, then you can be sure those two sentences are logically equivalent, as long as all possible combinations of truth values are accounted for.

Tables 27-7A and 27-7B show that the following two general sentences are logically equivalent for any two variables X and Y, proving the commutative law of conjunction for two variables:

$$X \& Y$$

$$Y \& X$$

Tables 27-8A and 27-8B show that the following two general sentences are logically equivalent for any two variables X and Y, proving the commutative law of disjunction for two variables:

$$X \vee Y$$

$$Y \vee X$$

If you'd like a challenging "extra credit" exercise, prove that the commutative rules for conjunction and disjunction work out no matter how many variables there are! Here's a hint: Try it by "building things up." Prove that if the rule holds for two variables, then it holds for three. After that, prove that if the rule holds for n variables, then it holds for $n + 1$ variables. If you can do that, you'll have successfully employed a rather sophisticated proof technique called *mathematical induction*.

A		
X	Y	X & Y
F	F	F
F	T	F
T	F	F
T	T	T

B		
X	Y	Y & X
F	F	F
F	T	F
T	F	F
T	T	T

Table 27-7

Truth table proof of the commutative law of conjunction. At A, statement of truth values for X & Y. At B, statement of truth values for Y & X. The outcomes are identical, demonstrating that they are logically equivalent.

Table 27-8

Truth table proof of the commutative law of disjunction. At A, statement of truth values for X ∨ Y. At B, statement of truth values for Y ∨ X. The outcomes are identical, demonstrating that they are logically equivalent.

A

X	Y	X ∨ Y
F	F	F
F	T	T
T	F	T
T	T	T

B

X	Y	Y ∨ X
F	F	F
F	T	T
T	F	T
T	T	T

Tables 27-9A and 27-9B show that the following sentences are logically equivalent for any three variables X, Y, and Z, proving the associative law of conjunction:

$$(X \,\&\, Y) \,\&\, Z$$

$$X \,\&\, (Y \,\&\, Z)$$

Note that in Table 27-9A, the two rightmost columns make use of the commutative law of conjunction, which has already been proved. Once proven, a statement is called a *theorem*, and it can be used in future proofs.

Table 27-9A

Derivation of truth values for (X & Y) & Z. Note that the two rightmost columns of this proof make use of the commutative law for conjunction, which has already been proved.

X	Y	Z	X & Y	Z & (X & Y)	(X & Y) & Z
F	F	F	F	F	F
F	F	T	F	F	F
F	T	F	F	F	F
F	T	T	F	F	F
T	F	F	F	F	F
T	F	T	F	F	F
T	T	F	T	F	F
T	T	T	T	T	T

Table 27-9B

Derivation of truth values for X & (Y & Z). The rightmost column of this table has values that are identical with those in the rightmost column of Table 27-9A, demonstrating that the rightmost expressions in the top rows are logically equivalent.

X	Y	Z	Y & Z	X & (Y & Z)
F	F	F	F	F
F	F	T	F	F
F	T	F	F	F
F	T	T	T	F
T	F	F	F	F
T	F	T	F	F
T	T	F	F	F
T	T	T	T	T

Tables 27-10A and 27-10B show that the following sentences are logically equivalent for any three variables X, Y, and Z, proving the associative law of disjunction:

$$(X \vee Y) \vee Z$$

$$X \vee (Y \vee Z)$$

The two rightmost columns in Table 27-10A take advantage of the commutative law for disjunction, which has already been proved.

Table 27-10A

Derivation of truth values for (X ∨ Y) ∨ Z. Note that the two rightmost columns of this proof make use of the commutative law for disjunction, which has already been proved.

X	Y	Z	X ∨ Y	Z ∨ (X ∨ Y)	(X ∨ Y) ∨ Z
F	F	F	F	F	F
F	F	T	F	T	T
F	T	F	T	T	T
F	T	T	T	T	T
T	F	F	T	T	T
T	F	T	T	T	T
T	T	F	T	T	T
T	T	T	T	T	T

Table 27-10B

Derivation of truth values for $X \vee (Y \vee Z)$. The rightmost column of this table has values that are identical with those in the rightmost column of Table 27-10A, demonstrating that the rightmost expressions in the top rows are logically equivalent.

X	Y	Z	Y ∨ Z	X ∨ (Y ∨ Z)
F	F	F	F	F
F	F	T	T	T
F	T	F	T	T
F	T	T	T	T
T	F	F	F	T
T	F	T	T	T
T	T	F	T	T
T	T	T	T	T

Tables 27-11A and 27-11B show that the following sentences are logically equivalent for any two variables X and Y, proving the law of implication reversal:

$$X \Rightarrow Y$$

$$\neg Y \Rightarrow \neg X$$

Tables 27-12A and 27-12B show that the following sentences are logically equivalent for any two variables X and Y, proving De Morgan's law for conjunction:

$$\neg(X \& Y)$$

$$\neg X \vee \neg Y$$

Table 27-11

Truth table proof of the law of implication reversal. At A, statement of truth values for $X \Rightarrow Y$. At B, derivation of truth values for $\neg Y \Rightarrow \neg X$. The outcomes are identical, demonstrating that they are logically equivalent.

A

X	Y	X ⇒ Y
F	F	T
F	T	T
T	F	F
T	T	T

B

X	Y	¬Y	¬X	¬Y ⇒ ¬X
F	F	T	T	T
F	T	F	T	T
T	F	T	F	F
T	T	F	F	T

A

X	Y	X & Y	¬(X & Y)
F	F	F	T
F	T	F	T
T	F	F	T
T	T	T	F

B

X	Y	¬X	¬Y	¬X ∨ ¬Y
F	F	T	T	T
F	T	T	F	T
T	F	F	T	T
T	T	F	F	F

Table 27-12

Truth table proof of De Morgan's law for conjunction. At A, statement of truth values for ¬(X & Y). At B, derivation of truth values for ¬X ∨ ¬Y. The outcomes are identical, demonstrating that they are logically equivalent.

Tables 27-13A and 27-13B show that the following sentences are logically equivalent for any two variables X and Y, proving De Morgan's law for disjunction:

$$¬(X ∨ Y)$$
$$¬X \; \& \; ¬Y$$

Tables 27-14A and 27-14B show that the following two general sentences are logically equivalent for any three variables X, Y, and Z, proving the distributive law of conjunction with respect to disjunction:

$$X \; \& \; (Y ∨ Z)$$
$$(X \; \& \; Y) ∨ (X \; \& \; Z)$$

A

X	Y	X ∨ Y	¬(X ∨ Y)
F	F	F	T
F	T	T	F
T	F	T	F
T	T	T	F

B

X	Y	¬X	¬Y	¬X & ¬Y
F	F	T	T	T
F	T	T	F	F
T	F	F	T	F
T	T	F	F	F

Table 27-13

Truth table proof of De Morgan's law for disjunction. At A, statement of truth values for ¬(X ∨ Y). At B, derivation of truth values for ¬X & ¬Y. The outcomes are identical, demonstrating that they are logically equivalent.

Table 27-14A

Derivation of truth values for X & (Y ∨ Z).

X	Y	Z	Y ∨ Z	X & (Y ∨ Z)
F	F	F	F	F
F	F	T	T	F
F	T	F	T	F
F	T	T	T	F
T	F	F	F	F
T	F	T	T	T
T	T	F	T	T
T	T	T	T	T

You can use truth tables to prove any logical proposition, as long as it's valid, of course! Here is an example:

$$[(X \& Y) \Rightarrow Z] \Leftrightarrow [\neg Z \Rightarrow (\neg X \vee \neg Y)]$$

Tables 27-15A and 27-15B show that the following sentences are logically equivalent for any three variables X, Y, and Z, proving that the proposition is valid:

$$(X \& Y) \Rightarrow Z$$

$$\neg Z \Rightarrow (\neg X \vee \neg Y)$$

Table 27-14B

Derivation of truth values for (X & Y) ∨ (X & Z). The right-most column of this table has values that are identical with those in the rightmost column of Table 27-14A, demonstrating that the rightmost expressions in the top rows are logically equivalent.

X	Y	Z	X & Y	X & Z	(X & Y) ∨ (X & Z)
F	F	F	F	F	F
F	F	T	F	F	F
F	T	F	F	F	F
F	T	T	F	F	F
T	F	F	F	F	F
T	F	T	F	T	T
T	T	F	T	F	T
T	T	T	T	T	T

X	Y	Z	X & Y	(X & Y) ⟹ Z
F	F	F	F	T
F	F	T	F	T
F	T	F	F	T
F	T	T	F	T
T	F	F	F	T
T	F	T	F	T
T	T	F	T	F
T	T	T	T	T

Table 27-15A
Derivation of truth values for $(X \& Y) \Rightarrow Z$.

PROOFS WITHOUT TABLES

You can use the logical rules presented in this chapter, rather than a direct truth table comparison, to prove the following:

$$[\neg Z \Rightarrow (\neg X \vee \neg Y)] \Rightarrow [(X \& Y) \Rightarrow Z]$$

First, use De Morgan's law for conjunction. This states that the following sentences are logically equivalent for any X and Y:

$$\neg(X \& Y)$$

$$\neg X \vee \neg Y$$

Table 27-15B
Derivation of truth values for $\neg Z \Rightarrow (\neg X \vee \neg Y)$. The rightmost column of this table has values that are identical with those in the rightmost column of Table 27-15A, demonstrating that the rightmost expressions in the top rows are logically equivalent.

X	Y	Z	¬X	¬Y	¬Z	¬X ∨ ¬Y	¬Z ⟹ (¬X ∨ ¬Y)
F	F	F	T	T	T	T	T
F	F	T	T	T	F	T	T
F	T	F	T	F	T	T	T
F	T	T	T	F	F	T	T
T	F	F	F	T	T	T	T
T	F	T	F	T	F	T	T
T	T	F	F	F	T	F	F
T	T	T	F	F	F	F	T

This means that the two expressions are directly interchangeable. Whenever you encounter either of these in any logical sentence, you can "pull it out" and "plug in" the other one. Let's do that here, changing the logical statement $\neg Z \Rightarrow (\neg X \vee \neg Y)$ into the logical statement $\neg Z \Rightarrow \neg(X \& Y)$. According to the law of implication reversal, this is logically equivalent to

$$\neg[\neg(X \& Y)] \Rightarrow \neg(\neg Z)$$

Using the law of double negation on both sides of this expression, you can see that this is logically equivalent to

$$(X \& Y) \Rightarrow Z$$

Q.E.D.!

Now here's another "extra credit" exercise for you. Prove this the other way without using truth tables. That is, show that the following statement is valid:

$$[(X \& Y) \Rightarrow Z] \Rightarrow [\neg Z \Rightarrow (\neg X \vee \neg Y)]$$

QUESTIONS AND PROBLEMS

This is an open-book quiz. You may refer to the text in this chapter (and earlier ones, too, if you want) when figuring out the answers. Take your time! The correct answers are in the back of the book.

1. Under what circumstances is the conjunction of several sentences false? Under what circumstances is the conjunction of several sentences true?

2. Under what circumstances is the disjunction of several sentences false? Under what circumstances is the disjunction of several sentences true?

3. Suppose you see a conjunction, then a double-shafted, double-headed arrow (pointing both left and right), and then a disjunction and a negation. What does the double-shafted, double-headed arrow represent in words?

4. How many possible combinations of truth values are there for a set of four sentences, each of which can attain either the value T or the value F?

5. Suppose you observe, "It is not sunny today, and it's not warm." Your friend says, "The statement that it's sunny or warm today is false." These two sentences are logically equivalent. What rule does this demonstrate?

X	Y	Z	X ∨ Y	(X ∨ Y) & Z
F	F	F	F	T
F	F	T	F	T
F	T	F	T	F
F	T	T	T	T
T	F	F	T	F
T	F	T	T	T
T	T	F	T	F
T	T	T	T	T

Table 27-16

Truth table for Probs. 7 and 8.

6. Suppose someone claims that both of the following statements are valid for all possible truth values of sentences P, Q, and R:

$$P \& (Q \vee R) \text{ implies } (P \& Q) \vee (P \& R)$$

$$(P \& Q) \vee (P \& R) \text{ implies } P \& (Q \vee R)$$

What, if anything, is wrong with this? If something is wrong, show an example demonstrating why.

7. Look at Table 27-16. What is wrong with this truth table?

8. What can be done to make Table 27-16 show a valid derivation?

9. Imagine that someone says to you, "If I am human and I am not human, then the moon is made of Swiss cheese." What rule of logic does this illustrate?

28

Beyond Three Dimensions

As you have seen, the rectangular (also called the *Cartesian*, after the French mathematician René Descartes) coordinate plane is defined by two number lines that intersect at a right angle. Cartesian three-space or *xyz* space is defined by three number lines that intersect at a single point, such that each line is perpendicular to the plane determined by the other two lines. What about *Cartesian four-space*? Or *five-space*? Or *infinity-space*? These can exist too, but they're difficult to visualize.

IMAGINE THAT!

A system of rectangular coordinates in four dimensions—Cartesian four-space—is defined by four number lines that intersect at a single point. That point, the origin, corresponds to the 0 point on each line. Each line is perpendicular to the other three. The lines form axes, representing variables such as w, x, y, and z. Alternatively, the axes can be labeled x_1, x_2, x_3, and x_4. Points are identified by *ordered quadruples* of the form (w,x,y,z) or (x_1,x_2,x_3,x_4). The origin is defined by $(0,0,0,0)$. As with the variables or numbers in ordered pairs and triples, there are no spaces after the commas.

At first you might think that Cartesian four-space isn't difficult to imagine, and you might draw an illustration such as Fig. 28-1 to show it. But when you start trying to plot points in this system, you'll find out that there's a problem. You can't define points in such a rendition of four-space without ambiguity. There are too many possible values of the ordered quadruple (w,x,y,z), and not enough points in our "restrictive" universe of three spatial dimensions to accommodate them all. In *three-space* as we know it, four number lines such as those shown in Fig. 28-1 cannot be oriented so they intersect at a single point with all four lines perpendicular to the other three.

Imagine a point in a room where two walls meet the floor. This point defines three straight lines. One of the lines runs up and down between the two walls, and the other two run horizontally between the two walls and the floor. The lines are all mutually perpendicular at the point where they intersect. They are like the x, y, and z axes in

Figure 28-1

The concept of rectangular four-space. The w, x, y, and z axes are all mutually perpendicular at the origin point (0,0,0,0). This cannot be accurately rendered in three dimensions.

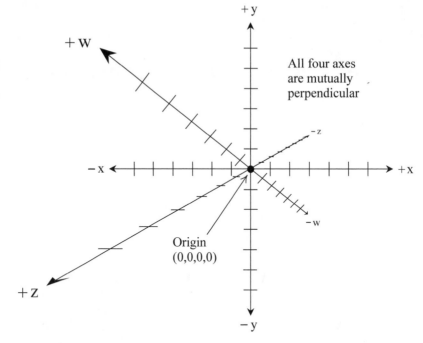

All four axes are mutually perpendicular

Origin (0,0,0,0)

Cartesian three-space. Now think of a fourth line passing through the intersection point of the existing three lines, perpendicular to them all. Such a line segment cannot exist in ordinary space! But in four dimensions, or *hyperspace*, it can. If you were a four-dimensional (4D) creature, you would not be able to understand how three-dimensional (3D) creatures could have trouble envisioning four line segments all coming together at mutual right angles.

Mathematically, you can work with Cartesian four-space, even though you probably can't directly visualize it. (Some people claim they can. I don't know about that.) This makes 4D geometry a powerful mathematical tool. As a matter of fact, the universe you live in requires four or more dimensions to be fully described. Albert Einstein was one of the first scientists to put forth the idea that the fourth dimension really exists.

TIME-SPACE

You've seen *time lines* in history books. You've seen them in graphs of various quantities, such as temperature, barometric pressure, or the Dow Jones industrial average, plotted as functions of time. Isaac Newton, one of the most renowned mathematicians in the history of the Western world, described time as flowing smoothly and unalterably. Time, according to *Newtonian physics*, does not depend on space, nor space on time.

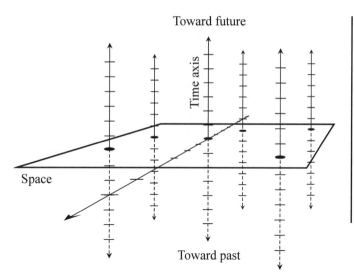

Toward future

Time axis

Space

Toward past

Figure 28-2
Time as a fourth
dimension. Three-
space is shown
dimensionally
reduced, as a
plane. Each
stationary point
in space follows a
time line perpendi-
cular to 3D space
and parallel to the
time axis.

Wherever you are, however fast or slow you travel, and no matter what else you do, the *cosmic clock*, according to Newtonian (or classical) physics, keeps ticking at the same absolute rate. In most practical scenarios, this model works quite well; its imperfections are not evident. It makes the time line a perfect candidate for a "fourth perpendicular axis."

Newton's model of time and space represents an oversimplification; some folks might say it is conceptually flawed. But it is a good approximation of reality under most everyday circumstances.

Envision a time line passing through 3D space, perpendicular to all three spatial axes such as the intersections between two walls and the floor of a room. In 4D *Cartesian time-space* (or simply *time-space*), each point in space follows its own time line. If you assume that none of the points moves relative to the origin, then all the points in 3D space follow time lines parallel to all the other time lines, and they are all constantly "perpendicular" to 3D space! Dimensionally reduced, this situation can be portrayed as shown in Fig. 28-2.

POSITION AND MOTION

Things get more interesting when you think about the paths of moving points in time-space. Suppose, for example, that you choose the center of the sun as the origin point for a Cartesian 3D coordinate system.

Imagine that the x and y axes lie in the plane of the earth's orbit around the sun. Suppose the positive x axis runs from the sun through the earth's position in space on March 21, and then onward into deep space (roughly toward the constellation *Virgo*, for those of you who are interested in astronomy). Then the negative x axis runs from the sun through the earth's position on September 21 (roughly through the constellation *Pisces*), the positive *y* axis runs from the sun through the earth's position on June 21 (roughly toward the constellation *Sagittarius*), and the negative *y* axis runs from the sun through the earth's position on December 21 (roughly toward the constellation *Gemini*). The positive *z* axis runs from the sun toward the north celestial pole (in the direction of *Polaris*, the North Star), and the negative *z* axis runs from the sun toward the south celestial pole. Let each division on the coordinate axes represent one-quarter of an astronomical unit (AU), where 1 AU is the mean distance of the earth from the sun (about 1.50×10^8 km). Figure 28-3A shows this coordinate system, with the earth on the positive *x* axis, at a distance of 1 AU. The coordinates of the earth at this time are (1,0,0) in the *xyz* space thus defined.

Of course, the earth doesn't remain in one place as time passes. It orbits the sun. Take away the *z* axis in Fig. 28-3A and replace it with a time axis called *t* (for "time"). What will the earth's path look like in 3D *xyt* space, if we let each increment on the *t* axis represent exactly one-quarter of a year? The earth's path through this dimensionally reduced time-space is not a straight line, but instead is a helix as shown in Fig. 28-3B. The earth's distance from the *t* axis remains nearly constant, although it varies a little because the earth's orbit around the sun is not a perfect circle. Every one-quarter of a year, the earth advances by one-quarter of a circle (90°) around the helix.

Figure 28-3A

The earth's position at any moment in time can be expressed in *xyz*-space.

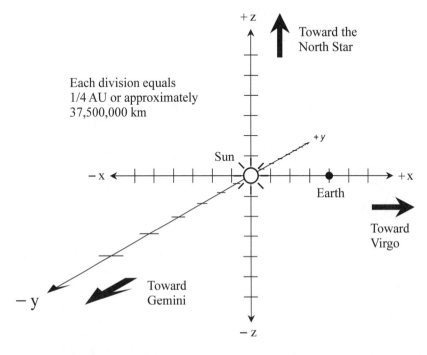

A

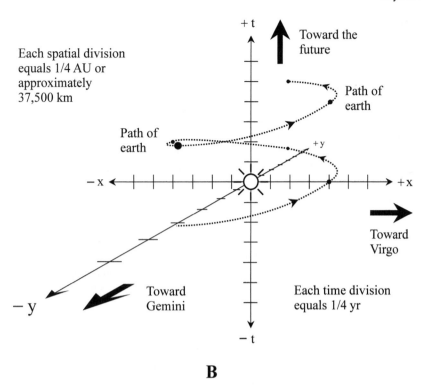

Each spatial division equals 1/4 AU or approximately 37,500 km

+t

Toward the future

Path of earth

Path of earth

+y

−x ←——→ +x

Toward Virgo

Toward Gemini

−y

Each time division equals 1/4 yr

−t

B

Figure 28-3B

A dimensionally reduced rectangular coordinate system for plotting the path of the earth through time-space.

TIME AS DISPLACEMENT

When you consider time as a dimension, it is convenient to have some universal standard that relates time to spatial displacement. How many kilometers are in one second (1 s) of time? At first this seems like a ridiculous question, akin to asking how many gallons are in a mile. But time and displacement can be related to each other by speed in a perfectly sensible way, as long as the speed is known and is constant.

Suppose someone tells you, "Jimsville is an hour (1 h) away from Joesville." You've heard people talk like this, and you understand what they mean. A certain speed is assumed. How fast must you go to get from Jimsville to Joesville in an hour? If Jimsville and Joesville are 50 km from each other, then you must travel 50 km/h in order to say that they are 1 h apart. If they are only 20 km apart, then you need only travel 20 km/h to make the same claim. You learned all about that sort of thing earlier in this book! Now remember the following formula from elementary physics:

$$d = st$$

where d is the distance in kilometers, s is the speed of an object in kilometers per hour, and t is the number of hours elapsed. By using this formula, it is possible to define time in terms of displacement and vice versa.

UNIVERSAL SPEED

Is there any such thing as *universal speed* that can be used as an absolute relating factor between time and displacement? Yes, according to Albert Einstein's special relativity theory. The speed of light in a vacuum, commonly denoted c, is constant, and it is independent of the point of view of the observer (as long as the observer is not accelerating at an extreme rate or in a superintense gravitational field). This constancy of the speed of light is a fundamental principle of the theory of special relativity. The value of c is very close to 2.99792×10^5 km/s; you can round it off to 3.00×10^5 km/s. If d is the distance in kilometers and t is the time in seconds, the following formula is absolute in a certain cosmic sense:

$$d = ct = (3.00 \times 10^5)t$$

According to this model, the moon, which is about 4.00×10^5 km from the earth, is 1.33 *second-equivalents* distant. The sun is about 8.3 *minute-equivalents* away. The Milky Way galaxy is 10^5 *year-equivalents* in diameter. (Astronomers call these units *light-seconds*, *light-minutes*, and *light-years*.) You can also say that any two points in time that are separated by 1 s, but that occupy the same *xyz* coordinates in Cartesian three-space, are separated by 3.00×10^5 *kilometer-equivalents* along the t axis.

At this instant yesterday, if you were in the same location as you are now, your location in 4D time-space was 24 (h/day) times 60 (min/h) times 60 (s/min) times 3.00×10^5 (km/s), or 2.592×10^{10} kilometer-equivalents away. This mode of thinking takes a bit of getting used to. But after awhile, it starts to make sense, even if it's a weird sort of sense. It is just about as difficult to leap 2.592×10^{10} km as it is to change what took place in your room at this time yesterday.

The above formula can be modified for smaller distances. If d is the distance in kilometers and t is the time in milliseconds (units of 0.001 s), then

$$d = 300t$$

This formula also holds for d in meters and t in microseconds (units of 10^{-6} s), or for d in millimeters (units of 0.001 m) and t in nanoseconds (units of 10^{-9} s). So you can speak of *meter-equivalents*, *millimeter-equivalents*, *microsecond-equivalents*, or *nanosecond-equivalents*.

THE FOUR-CUBE

Imagine some simple, regular geometric objects in Cartesian four-space. What are their properties? Think about a *four-cube*, also known as a *tesseract*. This is an object with several identical 3D *hyperfaces*, all of which are cubes.

This is a situation in which time becomes useful as a fourth spatial dimension. You can't make a 4D model of a tesseract out of toothpicks to examine its properties. But you can imagine a cube that pops into existence for a certain length of time and then disappears a little later! It "lives" for a length of time equivalent to the length of any of its spatial edges, and it does not move during its existence. Because you now know an "absolute relation" between time and displacement based on the speed of light, you can graph a tesseract in which each edge is, say, 3.00×10^5 kilometer-equivalents long. It is an ordinary 3D cube that measures 3.00×10^5 km along each edge. It forms at a certain time t_0 and then disappears 1 s later, at $t_0 + 1$. The sides of the cube are each 1 second-equivalent in length, and the cube exists for 3.00×10^5 kilometer-equivalents of time.

Figure 28-4A shows a tesseract in dimensionally reduced form. Each division along the x and y axes represents 10^5 km (the equivalent of 1/3 s), and each division along the t axis represents 1/3 s (the equivalent of 10^5 km). Figure 28-4B is another rendition of

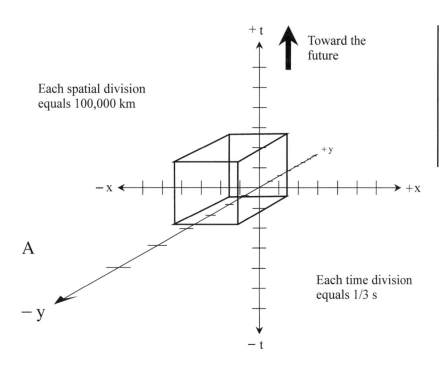

A

Each spatial division equals 100,000 km

Toward the future

Each time division equals 1/3 s

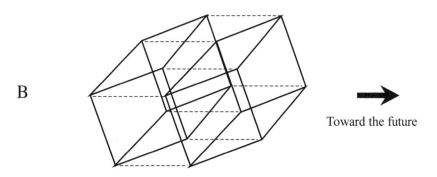

B

Toward the future

Figure 28-4

At A, a dimensionally reduced plot of a time-space tesseract. At B, another rendition of a tesseract, portraying time as lateral motion.

this object, illustrated as two 3D cubes in perspective, connected by dashed lines representing the passage of time, which "flows to the right."

THE RECTANGULAR FOUR-PRISM

A tesseract is a special form of the more general figure, known as a *rectangular four-prism* or *rectangular hyperprism*. Such an object is a 3D rectangular prism that abruptly comes into existence, lasts a certain length of time, disappears all at once, and does not move during its "lifetime." Figure 28-5 shows two examples of rectangular four-prisms in dimensionally reduced time-space.

Suppose the height, width, depth, and lifetime of a rectangular hyperprism, all measured in kilometer-equivalents, are h, w, d, and t, respectively. Then the *4D hypervolume* of this object (call it V_{4D}), in *quartic kilometer-equivalents*, is given by the product of them all:

$$V_{4D} = hwdt$$

The mathematics is the same if you express the height, width, depth, and lifetime of the object in second-equivalents; the 4D hypervolume is then equal to the product $hwdt$ in *quartic second-equivalents*.

Figure 28-5

Dimensionally reduced plots of two rectangular hyperprisms in time-space.

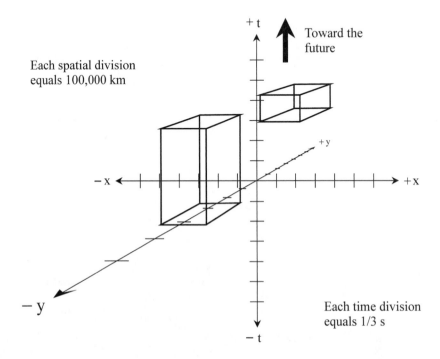

Each spatial division equals 100,000 km

Toward the future

Each time division equals 1/3 s

IMPOSSIBLE PATHS

Certain paths are impossible in Cartesian 4D time-space as defined here. According to the special theory of relativity, nothing can travel faster than the speed of light. This restricts the directions in which line segments, lines, and rays can run when denoting objects in motion.

Consider what happens in 4D Cartesian time-space when a light bulb is switched on. Suppose the bulb is located at the origin and is surrounded by millions of kilometers of empty space. When the switch is closed and the bulb is first illuminated, *photons* (particles of light) emerge. These initial, or leading, photons travel outward from the bulb in expanding spherical paths. If you dimensionally reduce this situation and graph it, you get an expanding circle centered on the time axis, which, as time passes, generates a cone as shown in Fig. 28-6. In true 4D space this is a *hypercone* or *four-cone*. The surface of the four-cone is 3D: two spatial dimensions and one time dimension.

Imagine an object that starts out at the location of the light bulb and then moves away from the bulb as soon as the bulb is switched on. This object must follow a path entirely within the *light cone* defined by the initial photons from the bulb. Figure 28-6 shows one plausible path and one implausible path.

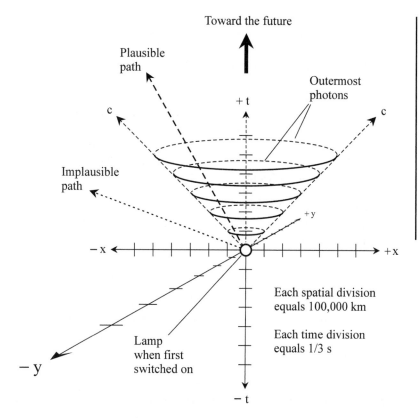

Figure 28-6

Dimensionally reduced plot of the leading photons from a light bulb moments after it has been switched on. Paths outside the light cone represent speeds greater than c (the speed of light) and are therefore implausible.

GENERAL TIME-SPACE HYPERVOLUME

Suppose there is an object—any object—in 3D space. Let its spatial volume in cubic kilometer-equivalents be equal to V_{3D}. Suppose that such an object comes into existence, lasts a certain length of time t in kilometer-equivalents, and then disappears. Also suppose that this object does not move with respect to you, the observer, during its lifetime. Then its 4D time-space hypervolume V_{4D} is given by this formula:

$$V_{4D} = V_{3D}t$$

That is, the 4D time-space hypervolume of any object is equal to its spatial volume multiplied by its lifetime, provided the time and displacement are expressed in equivalent units, and as long as there is no motion involved.

If an object moves, then a correction factor must be included in the above formula. This correction factor does not affect things very much as long as the speed of the object, call it s, is small compared with the speed of light c. But if s is considerable, the above formula becomes

$$V_{4D} = V_{3D}t(1 - s^2/c^2)^{1/2}$$

The correction factor $(1 - s^2/c^2)^{1/2}$ is close to 1 when s is a small fraction of c, and approaches 0 as s approaches c. This correction factor derives from the special theory of relativity. (It can be proved with the help of the Pythagorean theorem. The proof is not complicated, but getting into it here would distract you. Let's just say that objects are "spatially squashed" at extreme speeds and leave it at that.)

In this context, speed is always relative. It depends on the point of view from which it is observed, witnessed, or measured. For speed to have meaning, you should always add the qualifying phrase "relative to a certain observer." In these examples, you can envision motion as taking place relative to the origin of a 3D Cartesian system, which translates into lines, line segments, or rays pitched at various angles with respect to the time axis in a 4D time-space Cartesian system.

If you're still confused about kilometer-equivalents and second-equivalents, you can refer to Table 28-1 for reference. Keep in mind that time and displacement are related according to the speed of light as

$$d = ct$$

where d is the displacement (in linear units), t is the time (in time units), and c is the speed of light in linear units per unit time. Using this conversion formula, you can convert any displacement unit to an equivalent time interval, and any time unit to an equivalent displacement.

Suppose, for example, that you want to know many second-equivalents are in a spatial displacement of 2.50 km. You know that the speed of light is 3.00×10^5 km/s, so it

Displacement	Time	Table 28-1
9.46×10^{12} km	1.00 yr or 365 mean solar days (msd)	
2.59×10^{10} km	1.00 msd or 24.0 h	
1.08×10^9 km	1.00 h or 60 min	
1.50×10^8 km (1 AU)	500 s, or 8 min 20 s	
1.80×10^7 km	1.00 min or 60 s	
3.00×10^5 km	1.00 s	
300 km	0.00100 s (1.00 ms)	
1.00 km	3.33×10^{-6} s (3.33 μs)	
300 m	1.00×10^{-6} s (1.00 μs)	
1.00 m	3.33×10^{-9} s	
300 mm	1.00×10^{-9} s	
1.00 mm	3.33×10^{-12} s	

Table 28-1

Displacement and time equivalents in 4D time-space, accurate to three significant figures.

takes $1/(3.00 \times 10^5)$ s for light to travel 1 km. That is approximately 3.33×10^{-6} s, or 3.33 μs. Multiply by 2.5 to get 8.33 μs, rounded to three significant figures.

CARTESIAN *N*-SPACE

A system of rectangular coordinates in five dimensions defines *rectangular five-space*. There are five number lines, all of which intersect at a point corresponding to the zero point of each line, and such that each of the lines is perpendicular to the other four. The resulting axes can be called v, w, x, y, and z. Alternatively, they can be called x_1, x_2, x_3, x_4, and x_5. Points are identified by *ordered quintuples* such as (v,w,x,y,z) or (x_1,x_2,x_3,x_4,x_5). The origin is defined by $(0,0,0,0,0)$.

A system of rectangular coordinates in *rectangular n-space* (where n is any positive integer) consists of n number lines, all of which intersect at their zero points, such that each of the lines is perpendicular to all the others. The axes can be named x_1, x_2, x_3, . . ., and so on up to x_n. Points in rectangular n-space can be uniquely defined by *ordered n-tuples* of the form $(x_1,x_2,x_3,. . .,x_n)$.

Imagine a tesseract or a rectangular four-prism that pops into existence at a certain time, does not move, and then disappears some time later. This object is a *rectangular five-prism*. If x_1, x_2, x_3, and x_4 represent four spatial dimensions (in kilometer-equivalents or second-equivalents) of a rectangular four-prism in Cartesian four-space, and if t represents its lifetime in the same units, then the *5D hypervolume* (call it V_{5D}) is equal to the product of them all:

$$V_{5D} = x_1 x_2 x_3 x_4 t$$

This holds only as long as there is no motion. If there is motion, then a relativistic correction factor must be included.

DISTANCE FORMULAS

In n-dimensional Cartesian space, the shortest distance between any two points can be found by means of a formula similar to the distance formulas for 2D and 3D space. The distance thus calculated represents the length of a straight-line segment connecting the two points. Suppose there are two points in Cartesian n-space, defined as follows:

$$P = (x_1, x_2, x_3, \ldots, x_n)$$
$$Q = (y_1, y_2, y_3, \ldots, y_n)$$

The length of the shortest possible path between points P and Q, written $|PQ|$, is equal to

$$|PQ| = [(y_1 - x_1)^2 + (y_2 - x_2)^2 + (y_3 - x_3)^2 + \cdots + (y_n - x_n)^2]^{1/2}$$

Now try an example. Find the distance $|PQ|$ between the points $P = (4, -6, -3, 0)$ and $Q = (-3, 5, 0, 8)$ in Cartesian four-space. Assume the coordinate values to be exact, and express the answer to four significant figures.

To solve this problem, you can start by assigning the numbers in the ordered quadruples the following values:

$$x_1 = 4 \quad \text{and} \quad y_1 = -3$$
$$x_2 = -6 \quad \text{and} \quad y_2 = 5$$
$$x_3 = -3 \quad \text{and} \quad y_3 = 0$$
$$x_4 = 0 \quad \text{and} \quad y_4 = 8$$

Then plug these values into the distance formula, like this:

$$|PQ| = \{(-3 - 4)^2 + [5 - (-6)]^2 + [0 - (-3)]^2 + (8 - 0)^2\}^{1/2}$$
$$= [(-7)^2 + 11^2 + 3^2 + 8^2]^{1/2}$$
$$= (49 + 121 + 9 + 64)^{1/2}$$
$$= 243^{1/2}$$
$$= 15.59$$

VERTICES OF A TESSERACT

Imagine a tesseract as a 3D cube that lasts for a length of time equivalent to the linear span of each edge. When you think of a tesseract in this way, and if you think of time as flowing upward from the past toward the future, the tesseract has a "bottom" that represents the instant it is "born" and a "top" that represents the instant it "dies." Both the

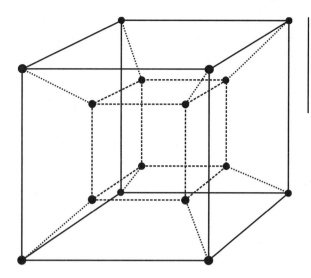

Figure 28-7

The cube-within-a-cube portrayal of a tesseract shows that the figure has 16 vertices.

bottom and the top of the tesseract, thus defined, are cubes. A cube has eight vertices. In the tesseract, there are twice this many vertices. The eight vertices of the bottom cube and the eight vertices of the top cube are connected with line segments that run through time.

You can also envision a tesseract as a cube within a cube (Fig. 28-7). It isn't a true picture because the "inner" and "outer" cubes in a real tesseract are the same size. But this rendition demonstrates that there are 16 vertices in the tesseract. Look at Fig. 28-7 and count them!

HYPERVOLUME OF A FOUR-PRISM

Here is another exercise: Find the 4D hypervolume V_{4D} of a rectangular four-prism consisting of a 3D cube measuring exactly 1 m on each edge, that "lives" for exactly 1 s, and that does not move. Compute the answer to three significant figures.

First, find the 4D hypervolume of a 3D cube measuring $1 \times 1 \times 1$ m that lives for 1 second of time. Remember that light travels 3.00×10^5 km in 1 s, so the four-prism lives for 3.00×10^5 kilometer-equivalents. That can be considered its "length." Its "cross section" is a 3D cube measuring 1 m, or 0.001 km, on each edge; so the 3D volume of this cube is $0.001 \times 0.001 \times 0.001 = 10^{-9}$ cubic kilometer (km^3). Therefore, the 4D hypervolume (V_{4D}) of the rectangular four-prism in quartic kilometer-equivalents is

$$V_{4D} = 3.00 \times 10^5 \times 10^{-9}$$

$$= 3.00 \times 10^{-4} \text{ quartic kilometer-equivalents (km}^4\text{)}$$

Try solving this another way. Note that in 1 microsecond (1 μs), light travels 300 m, so it takes light 1/300 μs to travel 1 m. The 3D volume of the cube in cubic microsecond-equivalents (μs³) is therefore

$$(1/300)^3 = 1 / (2.70 \times 10^7)$$
$$= 3.70 \times 10^{-8} \ \mu s^3$$

The cube lives for exactly 1 s, which is 10^6 μs. Therefore, the 4D hypervolume V_{4D} of the rectangular four-prism in quartic microsecond-equivalents (μs⁴) is

$$V_{4D} = 3.70 \times 10^{-8} \times 10^6$$
$$= 3.70 \times 10^{-2}$$
$$= 0.0370 \ \mu s^4$$

Now suppose that this rectangular four-prism moves, during its brief lifetime, at a speed of 2.70×10^5 km/s relative to an observer. That is 9/10, or 0.900, times the speed of light. If you let s represent its speed, then $s/c = 0.900$, and $s^2/c^2 = 0.810$. You must multiply the answers to the previous problem by $(1 - s^2/c^2)^{1/2}$. This turns out to be

$$(1 - 0.81)^{1/2} = 0.19^{1/2}$$
$$= 0.436$$

In terms of quartic kilometer-equivalents, you get

$$V_{4D} = 3.00 \times 10^4 \times 0.436$$
$$= 1.308 \times 10^4 \ km^4$$

In terms of quartic microsecond-equivalents, you get

$$= 0.0370 \times 0.436$$
$$= 0.0161 \ \mu s^4$$

EUCLID'S AXIOMS

Let's state explicitly the things that Euclid, the ancient Greek mathematician who developed the first rigorous and exhaustive theory of geometry, believed were self-evident truths. Euclid's original wording has been changed slightly, to make the passages sound more contemporary. Examples of each postulate are shown in Fig. 28-8.

- Any two points P and Q can be connected by a straight-line segment (Fig. 28-8A).
- Any straight-line segment can be extended indefinitely and continuously to form a straight line (Fig. 28-8B).

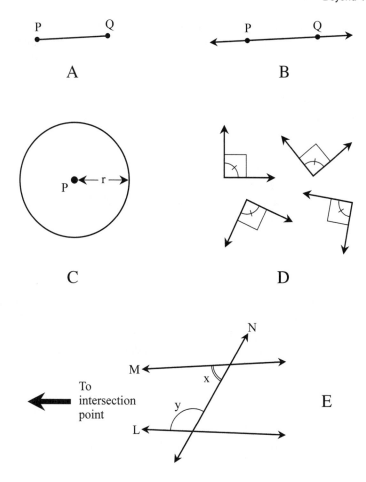

Figure 28-8
Euclid's original
five axioms. See
text for discussion.

- Given any point P, a circle can be defined that has that point as its center and that has a specific radius r (Fig. 28-8C).
- All right angles are congruent; that is, they have equal measure (Fig. 28-8D).
- Suppose two lines L and M lie in the same plane and both lines are crossed by a transversal line N. Suppose the measures of the adjacent interior angles x and y sum up to less than 180° (π rad). Then lines L and M intersect on the same side of line N as angles x and y are defined (Fig. 28-8E).

THE PARALLEL POSTULATE

The last axiom stated above is known as *Euclid's fifth postulate*. It is logically equivalent to the following statement that has become known as the *parallel postulate*:

- Let L be a straight line, and let P be some point not on L. Then there exists one and only one straight line M, in the plane defined by line L and point P, that passes through point P and that is parallel to line L.

This axiom—and in particular its truth or untruth—has received enormous attention. If the parallel postulate is denied, the resulting system of geometry still works. People might find it strange, but it is logically sound! Geometry doesn't need the parallel postulate. There are two ways in which the parallel postulate can be denied:

- There is no line M through point P that is parallel to line L.
- There are two or more lines M_1, M_2, M_3, . . . through point P that are parallel to line L.

When either of these postulates replaces the parallel postulate, we are dealing with a system of *non-Euclidean geometry*. In the 2D case, it is a *non-Euclidean surface*. Visually, such a surface looks warped or curved. There are, as you can imagine, infinitely many ways in which a surface can be non-Euclidean.

GEODESICS

In a non-Euclidean universe, the concept of "straightness" must be modified. Imagine two points P and Q on a non-Euclidean surface. The *geodesic segment* or *geodesic arc* connecting P and Q is the shortest possible path between P and Q that lies on the surface. If the geodesic arc is extended indefinitely in either direction on the surface beyond P and Q, the result is a *geodesic*.

The easiest way to imagine a geodesic arc is to think about the route an airliner takes between two far apart cities as it flies over the surface of the earth. This type of path is called a *great circle*. Now imagine the path that a thin ray of light would travel between two points, if confined to a certain 2D universe. The extended geodesic is the path that the ray would take if allowed to travel over the surface forever without striking any obstructions.

When you restate the parallel postulate as it applies to both Euclidean and non-Euclidean surfaces, you must replace the term "line" with "geodesic." Here is the parallel postulate given above, modified to cover all contingencies.

MODIFIED PARALLEL POSTULATE

Two geodesics G and H on a given surface are parallel if and only if they do not intersect at any point. Let G be a geodesic, let X be a surface, and let P be some point not on geodesic G. Then one of the following three situations holds true:

- There is exactly one geodesic H on the surface X through point P that is parallel to geodesic G.
- There is no geodesic H on the surface X through point P that is parallel to geodesic G.
- There are two or more geodesics H_1, H_2, H_3, . . . on the surface X through point P that are parallel to geodesic G.

NO PARALLEL GEODESICS

Now imagine a universe in which there is no such thing as a pair of parallel geodesics. In this universe, if two geodesics that appear to be parallel on a local scale are extended far enough, they eventually intersect. This type of non-Euclidean geometry is called *elliptic geometry*. It is also known as *Riemannian geometry*, named after Bernhard Riemann, a German mathematician who lived from 1826 until 1866 and who was one of the first mathematicians to recognize that geometry doesn't have to be Euclidean.

A universe in which there are no pairs of parallel geodesics is said to have *positive curvature*. Examples of 2D universes with positive curvature are the surfaces of spheres, oblate (flattened) spheres, and ellipsoids. Figure 28-9 is an illustration of a sphere with a triangle and a quadrilateral on the surface. The sides of polygons in non-Euclidean geometry are always geodesic arcs, just as, in Euclidean geometry, they are always straight-line segments. The interior angles of the triangle and the quadrilateral add up to more than 180° and 360°, respectively. The measures of the interior angles of an *n*-sided polygon on a *Riemannian surface* always add up to more than the sum of the measures of the interior angles of an *n*-sided polygon on a flat plane.

On the surface of the earth, all the lines of longitude, called *meridians*, are geodesics. So is the equator. But latitude circles other than the equator, called *parallels*, are not geodesics. For example, the equator and the parallel representing 10° north latitude do not intersect, but they are not both geodesics.

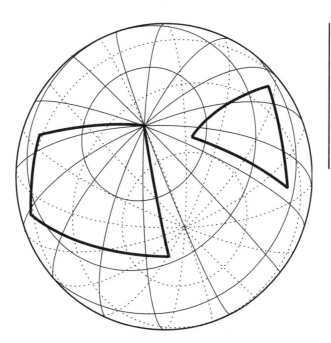

Figure 28-9

A surface with positive curvature, in this case a sphere, showing a triangle and a quadrilateral whose sides are geodesics.

MORE THAN ONE PARALLEL GEODESIC

Now consider a surface in which there can be two or more geodesics through a point, parallel to a given geodesic. This form of non-Euclidean geometry is known as *hyperbolic geometry*. It is also called *Lobachevskian geometry*, named after Nikolai Lobachevsky, a Russian mathematician who lived from 1793 until 1856.

A Lobachevskian universe is said to have *negative curvature*. Examples of 2D universes with negative curvature are extended saddle-shaped and funnel-shaped surfaces. Figure 28-10 is an illustration of a negatively curved surface containing a triangle and a quadrilateral. On this surface, the interior angles of the triangle and the quadrilateral add up to less than 180° and 360°, respectively. The measures of the interior angles of a polygon on a *Lobachevskian surface* always add up to less than the sum of the measures of the interior angles of a similar polygon on a flat plane.

CURVED SPACE

The observable universe seems, upon casual observation, to be Euclidean (that is, "flat"). If you use lasers to "construct" polygons and then measure their interior angles with precision lab equipment, you'll find that the angle measures add up according to the rules of Euclidean geometry. The conventional formulas for the volumes of solids

Figure 28-10

An example of a surface with negative curvature, showing a triangle and a quadrilateral whose sides are geodesics.

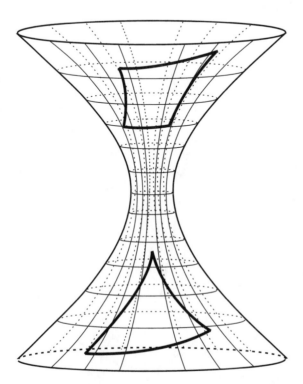

such as the pyramid, cube, and sphere will hold perfectly, as far as you can tell. Now imagine a 3D space in which these rules do not hold! This is called *curved 3D space*, *warped 3D space*, or *non-Euclidean 3D space*. It is the 3D analog of a non-Euclidean 2D surface.

There is evidence that the actual 3D universe is not perfectly Euclidean. Gravitational fields produce effects on light beams that suggest a Lobachevskian sort of warping— negative curvature—of 3D space. Under ordinary circumstances this warping is so subtle that you can't notice it, but it has been detected by astronomers using sensitive equipment.

The behavior of light from distant stars has been carefully observed as the rays pass close to the sun during solar eclipses. The idea is to find out whether the sun's gravitational field, which is strong near the surface, causes light rays to change their paths in a way that would occur on a Lobachevskian surface. Early in the 20th century, Albert Einstein predicted that such bending could be observed and measured, and he calculated the expected angular changes that should be seen in the positions of distant stars as the sun passes almost directly in front of them. Repeated observations have shown Einstein to be correct, not only as to the existence of the spatial curvature, but also as to its extent as a function of distance from the sun. As the distance from the sun increases, the spatial warping decreases. The greatest amount of light-beam bending occurs when the photons graze the sun's surface.

In another experiment, the light from a distant, brilliant object called a *quasar* is observed as it passes close to a compact, dark mass that astronomers think is an intense source of gravitation known as a *black hole*. The light bending is much greater near this type of object than is the case near the sun. The rays are bent enough that multiple images of the quasar appear, with the black hole at the center. One peculiar example, in which four images of the quasar appear, has been called a *gravitational light cross*.

Any source of gravitation, no matter how strong or weak, is attended by curvature of the 3D space in its vicinity, such that light rays follow geodesic paths that are not straight lines. Which causes which? Does spatial curvature cause gravitation, or do gravitational fields cause warping of space? Are both effects the result of some other phenomenon that has yet to be defined and understood? Such questions are of interest to astronomers and cosmologists. For the mathematician, it is enough to know that the curvature exists and can be defined. It's more than a mere product of somebody's wild imagination!

THE "HYPERFUNNEL"

The curvature of space in the presence of a strong gravitational field has been likened to a funnel shape (Fig. 28-11), except that the surface of the funnel is 3D rather than 2D and the entire object is 4D rather than 3D. When the fourth dimension is defined as time, the mathematical result is that time flows more slowly in a gravitational field than it does in interstellar or intergalactic space. This has been experimentally observed.

Figure 28-11

An intense source of gravitation produces negative curvature, or warping, of space in its immediate vicinity.

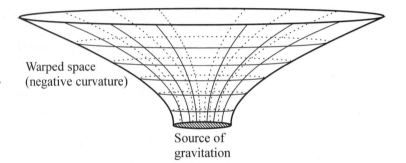

Warped space (negative curvature)

Source of gravitation

The shortest distance in 3D space between any two points near a gravitational source is a geodesic arc, not a straight-line segment. Curvature of space caused by gravitational fields always increases the distances between points in the vicinity of the source of the gravitation. The shortest path between any two points in non-Euclidean space is always greater than the distance would be if the space between the points were Euclidean. As the intensity of the gravitation increases, the extent of the spatial curvature also increases. There is some effect, theoretically, even when gravity is weak.

Any particle that has mass is surrounded by its own gravitational field, so the earth has its own shallow "gravitational hyperfunnel." So does the moon; so do asteroids; so do meteoroids and space dust particles.

Does the entire universe, containing all the stars, galaxies, quasars, and other stuff that exists, have a geometric shape? If so, is it Riemannian, Lobachevskian, or flat? No one knows for sure, but theories abound.

QUESTIONS AND PROBLEMS

This is an open-book quiz. You may refer to the text in this chapter (and earlier ones, too, if you want) when figuring out the answers. Take your time! significant. The correct answers are in the back of the book.

1. Imagine two stationary points P and Q in deep space, separated by exactly 1 second-equivalent (or 1 light-second). At point P there is an astronaut with a pulsed laser. At point Q there is a mirror oriented so the astronaut can shine the laser at it and see the reflected pulse exactly 2 s later. Suppose an extremely dense object passes almost (but not quite) between points P and Q, producing a pronounced negative curvature of space in the region. If the space traveler shines the laser toward point Q while the object is near the line of sight, will the pulse return after more than 2 s, less than 2 s, or exactly 2 s?

2. If there are no pairs of parallel geodesics on a surface, what type of surface is it? Give a real-world example.

3. Einstein developed a theory of the universe that gave it the shape of a 4D object called a *hypersphere* with a 3D surface. This object would consist of all the points equidistant from a given central point in 4D space. (The 3D equivalent is a sphere; the 2D equivalent is a circle.) Einstein said that such a universe would be "finite but unbounded." How could that be possible?

4. Considered with respect to the speed of light, how much distance does 30 minute-equivalents represent?

5. If the universe in the vicinity of our solar system is portrayed in a rectangular 4D system with three spatial dimensions and one time dimension, then what is the 4D path of a stationary point that "lasts forever"? What about the same point moving at constant speed in a straight line through 3D space? What about the same point accelerating in a straight line from a fixed starting position in 3D space?

6. How many line-segment edges does a 4D hypercube (tesseract) have?

7. What is the distance between (0,0,0,0,0,0) and (2,2,2,2,2,2) in Cartesian six-space? Express the answer to four significant digits.

8. Suppose a light bulb is switched on, and the photons travel outward in expanding spherical paths as shown in Fig. 28-6 on page 441. In this graph, what is the flare angle of the cone (the angle between the positive t axis and any ray extending from the origin outward along the cone's surface)?

9. What is the hypervolume in "4D hypercubic kilometers" (km^4) of an object consisting of a 3D cube measuring 25 km on each edge, and whose "life" is represented by a single point in time?

10. Suppose P and Q are points in a positively curved 4D space. How does the length of the straight-line segment PQ in 5D space compare with the length of the geodesic PQ in the warped 4D space?

11. If the curvature of the 4D space in Prob. 10 were negative instead of positive, how would the length of the straight-line segment PQ in 5D space compare with the length of the geodesic PQ in the warped 4D space?

29

A Statistics Primer

Statistics is the analysis of *data*, which is information expressed as measurable or observable quantities. Statistical data is usually obtained by looking at the real world or universe, although it can also be generated by computers in *virtual worlds*.

EXPERIMENT

In statistics, an *experiment* is an act of collecting data with the intent of learning or discovering something. For example, you might conduct an experiment to determine the most popular channels for frequency-modulation (FM) radio broadcast stations whose transmitters are located in U.S. towns having fewer than 5,000 people. Or you might conduct an experiment to determine the lowest barometric pressures inside the eyes of all the Atlantic hurricanes that take place during the next 10 years.

Experiments often, but not always, require specialized instruments to measure quantities. If you conduct an experiment to figure out the average test scores of high school seniors in Wyoming who take a certain standardized test at the end of this school year, the only things you need are the time, energy, and willingness to collect the data. But a measurement of the minimum pressure inside the eye of a hurricane requires sophisticated hardware in addition to time, energy, and courage.

VARIABLE

In mathematics, a variable is a quantity whose value is not necessarily specified, but that can be determined according to certain rules. In statistics, a variable is always associated with one or more experiments. Statistical variables can be either *discrete* or *continuous*.

A *discrete variable* can attain only specific values. The number of possible values is countable. Discrete variables are like the channels of a television set or digital broadcast receiver. It's easy to express the value of a discrete variable, because it can be assumed exact.

When a disc jockey says, "This is radio 97.1," it means that the assigned channel center is at a frequency of 97.1 megahertz (MHz), where 1 MHz represents 10^6 cycles per second. The assigned channels in the FM broadcast band are separated by an *increment* (minimum difference) of 0.2 MHz. The next-lower channel from 97.1 MHz is at 96.9 MHz, and the next-higher one is at 97.3 MHz. There is no "in between." No two channels can be closer together than 0.2 MHz in the set of assigned standard FM broadcast channels in the United States. The lowest channel is at 88.1 MHz and the highest is at 107.9 MHz (Fig. 29-1). Other examples of discrete variables are

- The number of people voting for each of the candidates in a political election
- The scores of students on a standardized test
- The number of car drivers caught speeding every day in your town
- The earned-run averages of pitchers in a baseball league

A *continuous variable* can attain infinitely many values over a certain span or range. Instead of existing as specific values in which there is an increment between any two, a continuous variable can change value to an arbitrarily tiny extent.

Continuous variables are something like the set of radio frequencies to which an old-fashioned analog FM broadcast receiver can be tuned. The radio frequency is adjustable continuously, say from 88 MHz to 108 MHz (Fig. 29-2). If you move the tuning dial only a little bit, you can make the received radio frequency change by something less than 0.2 MHz, the separation between adjacent assigned transmitter channels. There is no limit to how small the increment can get. If you have a light enough touch, you can adjust the received radio frequency by 0.02 MHz, 0.002 MHz, or even 0.0002 MHz. Other examples of continuous variables are

- Temperature in degrees Celsius
- Barometric pressure in millibars
- Brightness of a light source in candelas
- Intensity of the sound from a loudspeaker in decibels

Such quantities can never be determined exactly. There is always some instrument or observation error, even if that error is so small that it does not have a practical effect on the outcome of an experiment.

Figure 29-1

The individual channels in the FM broadcast band constitute values of a discrete variable.

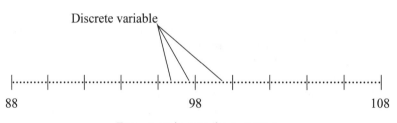

Frequency in megahertz (MHz)

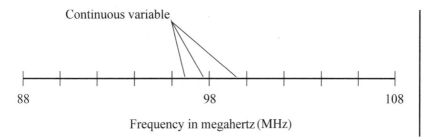

Continuous variable

88   98   108

Frequency in megahertz (MHz)

Figure 29-2
The frequency to which an analog FM broadcast receiver can be set is an example of a continuous variable.

POPULATION

In statistics, the term *population* refers to a particular set of items, objects, phenomena, or people being analyzed. These items, also called *elements*, can be actual subjects such as people or animals, but they can also be numbers or definable quantities expressed in physical units.

Consistent with the above definitions of variables, here are some examples of populations:

- Assigned radio frequencies (in megahertz) of all FM broadcast transmitters in the United States
- Temperature readings (in degrees Celsius) at hourly intervals last Wednesday at various locations around New York City
- Minimum barometric pressure levels (in millibars) at the centers of all the hurricanes in recorded history
- Brightness levels (in candelas) of all the light bulbs in offices in Minneapolis
- Sound-intensity levels (in decibels) of all the electric vacuum cleaners in the world

SAMPLE, EVENT, AND CENSUS

A *sample* of a population is a subset of that population. It can be a set consisting of only one value, reading, or measurement singled out from a population, or it can be a subset that is identified according to certain characteristics. The physical unit (if any) that defines a sample is always the same as the physical unit that defines the main, or parent, population. A single element of a sample is called an *event*. Consistent with the above definitions of variables, here are some examples of samples:

- Assigned radio frequencies of FM broadcast stations whose transmitters are located in the state of Ohio
- Temperature readings at 1:00 P.M. local time last Wednesday at various locations around New York City
- Minimum barometric pressure levels at the centers of Atlantic hurricanes during the decade 1991–2000

- Brightness levels of halogen bulbs in offices in Minneapolis
- Sound-intensity levels of the electric vacuum cleaners used in all the households in Rochester, Minnesota

When a sample consists of the whole population, it is called a *census*. When a sample consists of a subset of a population whose elements are chosen at random, it is called a *random sample*.

RANDOM VARIABLE

A *random variable* is a discrete or continuous variable whose value cannot be predicted in any given instant. Such a variable is usually defined within a certain range of values, such as 1 through 6 in the case of a thrown die, or from 88 MHz to 108 MHz in the case of an FM broadcast channel.

It is often possible to say, in a given scenario, that some values of a random variable are more likely to turn up than others. In the case of a thrown die, assuming the die is not "weighted," all the values 1 through 6 are equally likely to turn up. When one considering the FM broadcast channels of public radio stations, it is tempting to suppose (but would have to be confirmed by observation) that transmissions are made more often at the lower radio-frequency range than at the higher range. Have you noticed that there is a greater concentration of public radio stations in the 4-MHz-wide sample from 88 MHz to 92 MHz than in, say, the equally wide sample from 100 MHz to 104 MHz?

In order for a variable to be random, the only requirement is that it be impossible to predict its value in any single instance. If you contemplate throwing a die one time, you can't predict how it will turn up. If you contemplate throwing a dart one time at a map of the United States while wearing a blindfold, you have no way of knowing, in advance, the lowest radio frequency of all the FM broadcast stations in the town nearest the point where the dart will hit.

FREQUENCY

The *frequency* of a particular outcome (result) of an event is the number of times that outcome occurs within a specific sample of a population. Don't confuse this with radio broadcast or computer processor frequencies! In statistics, frequency refers to how often something happens. There are two species of statistical frequency: *absolute frequency* and *relative frequency*.

Suppose you toss a die 6000 times. If the die is not "weighted," you can expect that the die will turn up showing one dot approximately 1000 times, two dots approximately 1000 times, and so on, up to six dots approximately 1000 times. The absolute frequency in such an experiment is therefore approximately 1000 for each face of the die. The relative frequency for each of the six faces is approximately 1 in 6, which is equivalent to about 16.67%.

PARAMETER

A specific, well-defined characteristic of a population is known as a *parameter* of that population. You might want to know such parameters as the following, concerning the populations mentioned above:

- The most popular assigned FM broadcast frequency in the United States
- The highest temperature reading in New York City as determined at hourly intervals last Wednesday
- The average minimum barometric pressure level or measurement at the centers of all the hurricanes in recorded history
- The lowest brightness level found in all the light bulbs in offices in Minneapolis
- The highest sound-intensity level found in all electric vacuum cleaners in the world

STATISTIC

A specific characteristic of a sample is called a *statistic* of that sample. You might want to know such statistics as these, concerning the samples mentioned above:

- The most popular assigned frequency for FM broadcast stations in Ohio
- The highest temperature reading at 1:00 P.M. local time last Wednesday in New York
- The average minimum barometric pressure level or measurement at the centers of Atlantic hurricanes during the decade 1991–2000
- The lowest brightness level found in all the halogen bulbs in offices in Minneapolis
- The highest sound-intensity level found in electric vacuum cleaners used in households in Rochester, Minnesota

DISTRIBUTION

A *distribution* is a description of the set of possible values that a random variable can take. This can be done by noting the absolute or relative frequency. A distribution can be illustrated in terms of a table or in terms of a graph.

Table 29-1 shows the results of a hypothetical experiment in which a die is tossed 6000 times. Figure 29-3 is a vertical bar graph showing the same data as Table 29-1. Both the table and the graph are distributions that describe the behavior of the die. If the experiment is repeated, the results will differ. If a huge number of experiments are carried out, assuming the die is not weighted, the relative frequency of each face (number) turning up will approach 1 in 6, or approximately 16.67%.

Table 29-2 shows the number of days during the course of a 365-day year in which measurable precipitation occurs within the city limits of five different hypothetical towns.

Table 29-1

Results of a single, hypothetical experiment in which an unweighted die is tossed 6000 times.

Face of Die	Number of Times Face Turns Up
1	968
2	1027
3	1018
4	996
5	1007
6	984

Table 29-2

Number of days on which measurable rain occurs in a specific year, in five hypothetical towns.

Town Name	Number of Days in Year with Measurable Precipitation
Happyville	108
Joytown	86
Wonderdale	198
Sunnywater	259
Rainy Glen	18

Figure 29-3

Results of a single, hypothetical experiment in which an unweighted die is tossed 6,000 times.

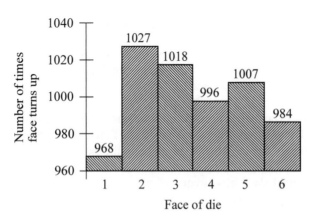

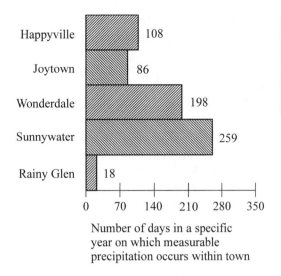

Figure 29-4
Measurable precipitation during a hypothetical year, in five different imaginary towns.

Number of days in a specific year on which measurable precipitation occurs within town

Figure 29-4 is a horizontal bar graph showing the same data as Table 29-2. Again, both the table and the graph are distributions. If the same experiment were carried out for several years in a row, the results would differ from year to year. Over a period of many years, the relative frequencies would converge toward certain values, although long-term climate change might have effects not predictable or knowable in your lifetime.

Both of the preceding examples involve discrete variables. When a distribution is shown for a continuous variable, a graph must be used. Figure 29-5 shows a distribution

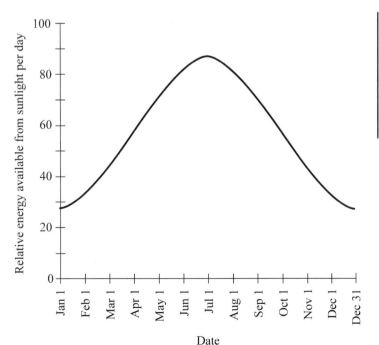

Figure 29-5
Relative energy available from sunlight, per day, during the course of a calendar year at a hypothetical location.

that portrays the relative amount of energy available from sunlight, per day during the course of a calendar year, at a hypothetical city in the northern hemisphere.

FREQUENCY DISTRIBUTION

In both of the above examples (the first showing the results of 6000 die tosses and the second showing the days with precipitation in five hypothetical towns), the scenarios are presented with frequency as the dependent variable. This is true of the tables as well as the graphs. Whenever frequency is portrayed as the dependent variable in a distribution, that distribution is called a *frequency distribution*.

Suppose you complicate the situation involving dice. Instead of one person tossing one die 6000 times, you have five people tossing five different dice, and each person tosses the same die 6000 times. The dice are colored red, orange, yellow, green, and blue, and they are manufactured by five different companies, called Corp. A, Corp. B, Corp. C, Corp. D, and Corp. E, respectively. Four of the dice are weighted, and one is not. There are 30,000 die tosses to tabulate or graph in total. When you conduct this experiment, you can tabulate the data in at least two ways.

The simplest way to tabulate the die toss results as a frequency distribution is to combine all the tosses and show the total frequency for each die face 1 through 6. Table 29-3 shows a hypothetical example of this result, called an *ungrouped frequency distribution*. Here, you have to worry not about the weighting characteristics of each individual die, but only about the potential biasing of the entire set. It appears that, for this particular set

Table 29-3

An ungrouped frequency distribution showing the results of a single, hypothetical experiment in which five different dice, some weighted and others not, are each tossed 6000 times

Face of Die	Toss Results for All Dice
1	4857
2	4999
3	4626
4	5362
5	4947
6	5209

of dice, there is some bias in favor of faces 4 and 6, some bias against faces 1 and 3, and little or no bias either for or against faces 2 and 5.

If you want to be more picayune, you can tabulate the frequency for each die face 1 through 6 separately for each die. A hypothetical product of this effort, called a *grouped frequency distribution*, is shown in Table 29-4. The results are grouped according to manufacturer and die color. From this distribution, it is apparent that some of the dice are heavily weighted! Only the green die, manufactured by Corp. D, seems to lack any bias. If you are astute, you will notice (or at least strongly suspect) that the green die here is the same die, with results gathered from the same experiment, as is portrayed in Table 29-1 and Fig. 29-3.

In Table 29-4, the numbers in each column should add up to 6000. This is the number of times each die (red, orange, yellow, green, or blue) is tossed in the experiment. If the sum of the numbers in any of the columns is not equal to 6000, then you know the experiment was done in a faulty way, or else there has been an error in the compilation of the table. If the experiment is repeated many times, the sums of the numbers in each column should always be 6000.

The sums of the numbers in the rows of Table 29-4 will vary depending on the bias of the set of dice considered as a whole. If, taken all together, the dice show any bias and if the experiment is repeated many times, the sums of the numbers should be consistently lower for some rows than for other rows.

Each entry in a table, corresponding to the intersection of one row with one column, is called a *cell* of the table. In Table 29-4, the individual numbers are absolute frequencies. They represent the actual number of times a particular face of a particular die came up during the course of the experiment.

Table 29-4

A grouped frequency distribution showing the results of a single, hypothetical experiment in which five different dice, some weighted and others not, and manufactured by five different companies, are each tossed 6,000 times

Face of Die	Toss Results by Manufacturer				
	Red Corp. A	Orange Corp. B	Yellow Corp. C	Green Corp. D	Blue Corp. E
1	625	1195	1689	968	380
2	903	1096	1705	1027	268
3	1300	890	1010	1018	408
4	1752	787	540	996	1287
5	577	1076	688	1007	1599
6	843	956	368	984	2058

CUMULATIVE FREQUENCY DATA

When data are tabulated, the absolute frequencies are often shown in one or more columns. Table 29-5 shows the results of the tosses of the blue die in the experiment you looked at a moment ago. The first column shows the number on the die face. The second column shows the absolute frequency for each face, or the number of times each face turned up during the experiment. The third column shows the *cumulative absolute frequency*, which is the sum of all the absolute frequency values in table cells at or above the given position.

The cumulative absolute frequency numbers in a table always ascend (increase) as you go down the column. The highest cumulative absolute frequency value should be equal to the sum of all the individual absolute frequency numbers. In this instance, it is 6000, the number of times the blue die was tossed.

Relative frequency values can be added down the columns of a table in exactly the same way as the absolute frequency values are added up. When this is done, the resulting values, usually expressed as percentages, show the *cumulative relative frequency*.

Now examine Table 29-6. This is a more detailed analysis of what happened with the blue die in the above-mentioned experiment. The first, second, and fourth columns in Table 29-6 are identical with the first, second, and third columns, respectively, in Table 29-5. The third column in Table 29-6 shows the percentage represented by each absolute frequency number. These percentages are obtained by dividing the number in the second column by 6000, the total number of tosses. The fifth column in Table 29-6 shows the cumulative relative frequency, which is the sum of all the relative frequency values in table cells at or above the given position.

The cumulative relative frequency percentages in a table, like the cumulative absolute frequency numbers, always ascend as you go down the column. The total cumulative relative frequency should be equal to 100%. In this sense, the cumulative relative frequency column in a table can serve as a *check sum*, helping to ensure that the entries have been tabulated correctly.

Table 29-5

Results of an experiment in which a weighted die is tossed 6,000 times, showing absolute frequencies and cumulative absolute frequencies.

Face of Die	Absolute Frequency	Cumulative Absolute Frequency
1	380	380
2	268	648
3	408	1056
4	1287	2343
5	1599	3942
6	2058	6000

Figure 29–6

Results of an experiment in which a weighted die is tossed 6,000 times, showing absolute frequencies, relative frequencies, cumulative absolute frequencies, and cumulative relative frequencies.

Face of Die	Absolute Frequency	Relative Frequency	Cumulative Absolute Frequency	Cumulative Relative Frequency
1	380	6.33%	380	6.33%
2	268	4.47%	648	10.80%
3	408	6.80%	1056	17.60%
4	1287	21.45%	2343	39.05%
5	1599	26.65%	3942	65.70%
6	2058	34.30%	6000	100.00%

MEAN

The *mean* for a discrete variable in a distribution is the mathematical average of all the values. If you consider the variable over the entire population, the average is called the *population mean*. If you consider the variable over a particular sample of a population, the average is called the *sample mean*. There can be only one population mean for a population, but there can be many different sample means. The mean is often denoted by the lowercase Greek letter mu, in italics (μ). Sometimes it is also denoted by an italicized lowercase English letter, usually x, with a bar (vinculum) over it.

Table 29-7 shows the results of a 10-question test, given to a class of 100 students. As you can see, every possible score is accounted for. There are some people who answered

Test Score	Absolute Frequency	Letter Grade
10	5	A
9	6	A
8	19	B
7	17	B
6	18	C
5	11	C
4	6	D
3	4	D
2	4	F
1	7	F
0	3	F

Table 29-7

Scores on a 10-question test taken by 100 students.

all 10 questions correctly; there are others who did not get a single answer right. To determine the mean score for the whole class on this test—that is, the population mean, called μ_p—you must add up the scores of each and every student and then divide by 100. First, sum the products of the numbers in the first and second columns. This will give you 100 times the population mean:

$$(10 \cdot 5) + (9 \cdot 6) + (8 \cdot 19) + (7 \cdot 17) + (6 \cdot 18) + (5 \cdot 11) + (4 \cdot 6)$$

$$+ (3 \cdot 4) + (2 \cdot 4) + (1 \cdot 7) + (0 \cdot 3)$$

$$= 50 + 54 + 152 + 119 + 108 + 55 + 24 + 12 + 8 + 7 + 0$$

$$= 589$$

Dividing this by 100, the total number of test scores (one for each student who turns in a paper), you obtain $\mu_p = 589/100 = 5.89$.

The teacher in this class has assigned letter grades to each score. Students who scored 9 or 10 correct received grades of A; students who got scores of 7 or 8 received grades of B; those who got scores of 5 or 6 got grades of C; those who got scores of 3 or 4 got grades of D; those who got less than 3 correct answers received grades of F. The assignment of grades, informally known as the "curve," is a matter of teacher temperament. It would doubtless seem arbitrary to the students who took this test. (Some people might consider the curve in this case to be too lenient, while a few might think it is too severe.)

Now suppose you want to find the sample means for each grade in the test whose results are tabulated in Table 29-7, rounding off the answers to two decimal places. Call the sample means μ_{sa} for the grade of A, μ_{sb} for the grade of B, and so on down to μ_{sf} for the grade of F. To calculate μ_{sa}, note that 5 students received scores of 10, while 6 students got scores of 9, both scores good enough for an A. This is a total of $5 + 6$, or 11, students getting the grade of A. Calculating, you get

$$\mu_{sa} = [(5 \cdot 10) + (6 \cdot 9)] / 11$$

$$= (50 + 54) / 11$$

$$= 104/11$$

$$= 9.45$$

To find μ_{sb}, observe that 19 students scored 8, and 17 students scored 7. Therefore, $19 + 17$, or 36, students received grades of B. Calculating, you get

$$\mu_{sb} = [(19 \cdot 8) + (17 \cdot 7)] / 36$$

$$= (152 + 119) / 36$$

$$= 271/36$$

$$= 7.53$$

To determine μ_{sc}, check the table to see that 18 students scored 6, while 11 students scored 5. Therefore, $18 + 11$, or 29, students did well enough for a C. Grinding out the numbers yields

$$\mu_{sc} = [(18 \cdot 6) + (11 \cdot 5)] / 29$$
$$= (108 + 55) / 29$$
$$= 163/29$$
$$= 5.62$$

To calculate μ_{sd}, note that 6 students scored 4, while 4 students scored 3. This means that $6 + 4$, or 10, students got grades of D:

$$\mu_{sd} = [(6 \cdot 4) + (4 \cdot 3)] / 10$$
$$= (24 + 12) / 10$$
$$= 36/10$$
$$= 3.60$$

Finally, determine μ_{sf}. Observe that 4 students got scores of 2, 7 students got scores of 1, and 3 students got scores of 0. So $4 + 7 + 3$, or 14, students failed the test:

$$\mu_{sf} = [(4 \cdot 2) + (7 \cdot 1) + (3 \cdot 0)] / 14$$
$$= (8 + 7 + 0) / 14$$
$$= 15/14$$
$$= 1.07$$

MEDIAN

If the number of elements in a distribution is even, then the *median* is the value such that one-half the elements have values greater than or equal to it, and one-half the elements have values less than or equal to it. If the number of elements is odd, then the median is the value such that the number of elements having values greater than or equal to it is the same as the number of elements having values less than or equal to it. The word "median" is synonymous with "middle."

Table 29-8 shows the results of the 10-question test described above, but instead of showing letter grades in the third column, the cumulative absolute frequency is shown. The tally is begun with the top-scoring papers and proceeds in order downward. (You could just as well do it the other way, starting with the lowest-scoring papers and proceeding upward.) When the scores of all 100 individual papers are tallied in this way, so that they are in order, the scores of the 50th and 51st papers—the two in the middle —are found to be 6 correct. Thus, the median score is 6, because one-half the students scored 6 or above, and the other half scored 6 or below.

Table 29-8

The median can be determined by tabulating the cumulative absolute frequencies.

Test Score	Absolute Frequency	Cumulative Absolute Frequency
10	5	5
9	6	11
8	19	30
7	17	47
6 (partial)	3	50
6 (partial)	15	65
5	11	76
4	6	82
3	4	86
2	4	90
1	7	97
0	3	100

Now imagine that in another group of 100 students taking this same test, the 50th paper has a score of 6 while the 51st paper has a score of 5. When two values "compete," the median is equal to their average. In this case it would be 5.5.

MODE

The *mode* for a discrete variable is the value that occurs the most often. In the test whose results are shown in Table 29-7, the most "popular" or often-occurring score is 8 correct answers. There are 19 papers with this score. No other score has that many results. Therefore, the mode in this case is 8.

Suppose that another group of students takes this test, and there are two scores that occur equally often. For example, suppose 16 students get 8 answers right, and 16 students also get 6 answers right. In this case there are two modes: 6 and 8. This distribution is an example of a *bimodal distribution*. If still another group of students takes the test, perhaps three different scores will be equally "popular," resulting in a *trimodal distribution*.

Now imagine there are only 99 students in a class, and exactly 9 students get each of the 11 possible scores (from 0 to 10 correct answers). In this distribution, there is no mode. You can also say that the mode is not defined. It is tempting to call this an "11-modal" situation, but that would imply some sort of multiple peaks in the distribution. That's not true here. There is no peak at all. The results are absolutely flat, so the notion of a mode is irrelevant.

Now here's a little problem for you: Draw a vertical bar graph showing all the absolute frequency data from Table 29-5, the results of a weighted die-tossing experiment.

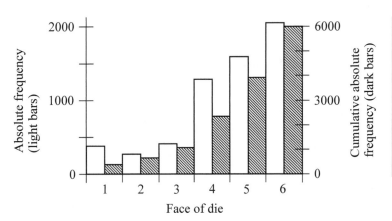

Figure 29-6

Vertical bar graph showing all the absolute-frequency data from Table 29-5, the results of a weighted die-tossing experiment.

Portray each die face on the horizontal axis. Let white vertical bars show the absolute frequency numbers, and let hatched or shaded vertical bars show the cumulative absolute frequency numbers. You should get a graph that looks like Figure 29-6. To minimize clutter, the numerical data is not listed at the tops of the bars.

Now draw a horizontal bar graph showing all the relative-frequency data from Table 29-6, another portrayal of the results of a weighted die-tossing experiment. Show each die face on the vertical axis. Let white horizontal bars show the relative frequency percentages, and hatched or shaded horizontal bars show the cumulative relative frequency percentages. You should get a graph similar to that shown in Figure 29-7. Again, the numerical data is not listed at the ends of the bars, in the interest of neatness.

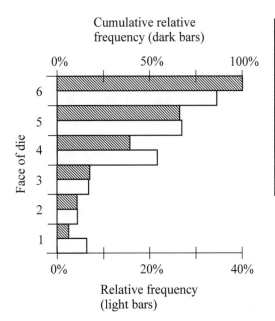

Figure 29-7

Horizontal bar graph showing all the relative-frequency data from Table 29-6, another portrayal of the results of a weighted die-tossing experiment.

Figure 29-8

Point-to-point graph showing the absolute frequencies of the 10-question test described by Table 29-7.

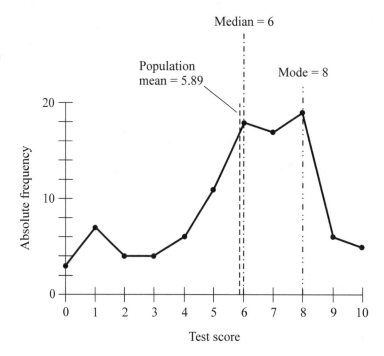

Now try this exercise. Draw a point-to-point graph showing the absolute frequencies of the 10-question test described by Table 29-7. Mark the population mean, the median, and the mode with distinctive vertical lines, and label them. You should get a graph that looks something like Figure 29-8. Numerical data is included for the population mean, median, and mode.

QUESTIONS AND PROBLEMS

This is an open-book quiz. You may refer to the text in this chapter (and earlier ones, too, if you want) when figuring out the answers. Take your time! The correct answers are in the back of the book.

1. Suppose a large number of people take a 200-question test, and every single student gets exactly 100 answers right. In this case, what can you say about the mean, the median, and the mode?

2. Suppose you have two thermometers. Thermometer X is an old-fashioned glass-bulb device with a red column whose height indicates the temperature. You read this thermometer by looking at the top of the column of liquid and comparing it against a graduated scale. Thermometer Y is a digital device that gives you a direct numeric readout to the nearest tenth of a degree Celsius. Temperature is treated as a continuous variable by thermometer X, but as a discrete variable by thermometer Y. Now suppose you could get thermometer Y to read out in smaller and smaller parts of a degree Celsius, thereby making it more and more accurate. How accurate would you have to make thermometer Y so that it would treat temperature as a continuous variable?

3. What is the mean of all the whole numbers between, and including, 10 and 20?

4. What is the median of all the whole numbers between, and including, 10 and 20?

5. What is the mode of all the whole numbers between, and including, 10 and 20?

6. Imagine that 11 people take a 10-question test. Suppose one student gets 10 correct answers, one gets 9 correct, one gets 8 correct, and so on, all the way down to one student getting 0 correct. What is the mean, accurate to three decimal places?

7. In the scenario of Prob. 6, what is the median score?

8. Imagine that 52 people take a 25-question test. Suppose two students gets 25 correct answers, two get 24 correct, two get 23 correct, and so on, all the way down to two students getting 0 correct. What is the mode?

9. What is the largest cumulative absolute frequency in a finite set of individual absolute frequency values?

30
Probability Basics

Probability is the proportion or percentage of the time over which specified things happen. The term "probability" is also used in reference to the art and science of determining the proportion or percentage of the time that specified things happen.

SETS AND SET NOTATION

Before you get into the "meat" of this chapter, you should be familiar with the concept of a set, and you should know some set notation. A *set* is a collection or group of definable *elements* or *members*. If an object or number (call it *a*) is an element of a certain set *A*, this fact is written

$$a \in A$$

The $\in$ symbol means "is an element of."

The *intersection* of two sets *A* and *B*, written $A \cap B$, is the set *C* such that the following statement is true for every element *x*:

$$x \in C \text{ if and only if } x \in A \text{ and } x \in B$$

The $\cap$ symbol is read "intersect." The *union* of two sets *A* and *B*, written $A \cup B$, is the set *C* such that the following statement is true for every element *x*:

$$x \in C \text{ if and only if } x \in A \text{ or } x \in B$$

The $\cup$ symbol is read "union."

A particular set *A* is a *subset* of a set *B*, written $A \subseteq B$, if and only if the following holds true:

$$x \in A \text{ implies that } x \in B$$

The $\subseteq$ symbol is read "is a subset of." In this context, "implies that" is meant in the strongest possible sense. The statement "This implies that" is equivalent to "If this is true, then that is always true." (It's plain old propositional logic!) A certain set A is a *proper subset* of a set B, written $A \subset B$, if and only if both the following hold true:

$$x \in A \text{ implies that } x \in B$$
$$A \neq B$$

The $\subset$ symbol is read "is a proper subset of."

Two sets A and B are *disjoint* if and only if all the following conditions are met:

$$A \neq \varnothing$$
$$B \neq \varnothing$$
$$A \cap B = \varnothing$$

Here $\varnothing$ denotes the *empty set*, also called the *null set*. The empty set doesn't contain any elements, but it's still a set—it's like a basket of apples without the apples. Two nonempty sets A and B are *coincident* if and only if, for all elements x, both of these are true:

$$x \in A \text{ implies that } x \in B$$
$$x \in B \text{ implies that } x \in A$$

EVENT VS. OUTCOME

The terms *event* and *outcome* are easily confused. An event is a single occurrence or trial in the course of an experiment. An outcome is the result of an event.

If you toss a coin 100 times, there are 100 separate events. Each event is a single toss of the coin. If you throw a pair of dice simultaneously 50 times, each act of throwing the pair is an event, so there are 50 events. Now suppose that in the process of tossing coins, you assign "heads" a value of 1 and "tails" a value of 0. Then when you toss a coin and it comes up heads, you can say that the outcome of that event is 1. If you throw a pair of dice and get a sum total of 7, then the outcome of that event is 7.

The outcome of an event depends on the nature of the devices and processes involved in the experiment. The use of a pair of weighted dice produces different outcomes, for an identical set of events, than a pair of unweighted dice. The outcome of an event also depends on how the event is defined. There is a difference between saying that the sum is 7 in a toss of two dice, compared with saying that one die comes up 2 while the other die comes up 5.

Event 1	Event 2	Event 3	Event 4
0	0	0	0
0	0	0	1
0	0	1	0
0	0	1	1
0	1	0	0
0	1	0	1
0	1	1	0
0	1	1	1
1	0	0	0
1	0	0	1
1	0	1	0
1	0	1	1
1	1	0	0
1	1	0	1
1	1	1	0
1	1	1	1

Table 30-1

The sample space for an experiment in which a coin is tossed four times. There are 16 possible outcomes; heads = 1 and tails = 0.

SAMPLE SPACE

A *sample space* is the set of all possible outcomes in the course of an experiment. Even if the number of events is small, a sample space can be large.

If you toss a coin four times, there are 16 possible outcomes. These are listed in Table 30-1, where heads = 1 and tails = 0. (If the coin happens to land on its edge, you disregard that result and toss it again.)

If a pair of dice, one red and one blue, is tossed once, there are 36 possible outcomes in the sample space, as shown in Table 30-2. The outcomes are denoted as ordered pairs, with the face number of the red die listed first and the face number of the blue die listed second.

Table 30-2

The sample space for an experiment consisting of a single event, in which a pair of dice (one red, one blue) is tossed once. There are 36 possible outcomes, shown as ordered pairs (red, blue).

Red →& Blue ↓	1	2	3	4	5	6
1	(1,1)	(2,1)	(3,1)	(4,1)	(5,1)	(6,1)
2	(1,2)	(2,2)	(3,2)	(4,2)	(5,2)	(6,2)
3	(1,3)	(2,3)	(3,3)	(4,3)	(5,3)	(6,3)
4	(1,4)	(2,4)	(3,4)	(4,4)	(5,4)	(6,4)
5	(1,5)	(2,5)	(3,5)	(4,5)	(5,5)	(6,5)
6	(1,6)	(2,6)	(3,6)	(4,6)	(5,6)	(6,6)

MATHEMATICAL PROBABILITY

Imagine that x is a discrete random variable. (If this term confuses you, go back to Chap. 29 and review the definitions of variables.) Suppose x can attain n possible values, all equally likely. Suppose an outcome H results from exactly m different values of x, where $m \leq n$. Then the *mathematical probability* $p_{math}(H)$ that outcome H will result from any given value of x is given by the following formula:

$$p_{math}(H) = m/n$$

Expressed as a percentage, the probability $p_{math\%}(H)$ is equal to $100m/n$.

If you toss an unweighted die once, each of the six faces is just as likely to turn up as each of the others. In this case, there are 6 possible values, so $n = 6$. The mathematical probability of any one of the faces turning up ($m = 1$) is equal to $p_{math}(H) = 1/6$. To calculate the mathematical probability of either of any two different faces turning up (say 3 or 5), you can set $m = 2$ and obtain $p_{math}(H) = 2/6 = 1/3$. If you want to know the mathematical probability that any one of the six faces will turn up, you can set $m = 6$ so the formula gives you $p_{math}(H) = 6/6 = 1$. The respective percentages $p_{math\%}(H)$ in these cases are 16.67% (approximately), 33.33% (approximately), and 100% (exactly).

Mathematical probability can only exist within the range 0 to 1 (or 0% to 100%) inclusive. You cannot have a mathematical probability of 2, or −45%, or −6, or 556%. Things can't happen less often than never, or more often than all the time!

EMPIRICAL PROBABILITY

If you want to determine the likelihood that an event will produce a certain outcome in real life, you must rely on the results of prior experiments. The probability of a particular outcome taking place, based on experience or observation, is called *empirical probability*.

Suppose you are told that a die is unweighted. How does the person, who tells you this, know that it is true? If you want to use this die in some application, such as when you need an object that can help you generate a string of "random numbers" from the set {1, 2, 3, 4, 5, 6}, you can't take the notion that the die is unweighted on faith alone. You have to check it out. How can you do that? You can analyze the die in a lab and figure out where its center of gravity is; you can measure how deep the indentations are where the dots on its faces are inked. You can scan the die electronically, X-ray it, and submerge it in (or float it on) water. But to be absolutely certain that the die is unweighted, you must toss it many times and be sure that each face turns up, on the average, exactly 1/6 of the time. You must conduct an experiment—gather *empirical evidence*—that supports the contention that the die is unweighted.

As with mathematical probability, there are limits to the range an empirical probability figure can attain. If H is an outcome for a particular event, and the empirical probability of H occurring as a result of that event is denoted $p_{emp}(H)$, then

$$0 \leq p_{emp}(H) \leq 1$$

Expressed as a percentage, $0\% \leq p_{emp\%}(H) \leq 100\%$.

Suppose a new cholesterol-lowering drug comes on the market. If the drug is to be approved by the government for public use, it must be shown effective, and it must also be shown not to have too many serious side effects. So it is tested. During the course of testing, 10,000 people, all of whom have been diagnosed with high levels of blood cholesterol, are given this drug. Imagine that 7289 of the people experience a significant drop in cholesterol. Also suppose that 307 of these people experience adverse side effects. (The remaining 2404 people experience no significant effects at all.) If you have high cholesterol and go on this drug, what is the empirical probability $p_{emp}(B)$ that you will derive benefit? What is the empirical probability $p_{emp}(A)$ that you will experience adverse side effects?

You might be tempted to suggest that this question cannot be satisfactorily answered because the experiment is inadequate. Is 10,000 test subjects a large enough number? What physiological factors affect the way the drug works? How about blood type, for example? Ethnicity? Gender? Blood pressure? Diet? What constitutes "high cholesterol"? What constitutes a "significant drop" in cholesterol level? What is an "adverse side effect"? What is the standard drug dose? How long must the drug be taken to figure out if it really works? For convenience, let's ignore all these factors here, even though, in a true scientific experiment, it would be an excellent idea to take them all into consideration.

Based on the above experimental data (shallow as it may be) the relative frequency of effectiveness is $7289/10{,}000 = 0.7289 = 72.89\%$. The relative frequency of ill effects is $307/10{,}000 = 0.0307 = 3.07\%$. You can round these off to 73% and 3%. These are the empirical probabilities that you will derive benefit from, or experience adverse effects, if you take this drug in the hope of lowering your high cholesterol.

LAW OF LARGE NUMBERS

Suppose you toss an unweighted die many times. You get numbers turning up, apparently at random, from the set $\{1, 2, 3, 4, 5, 6\}$. What will the average value be? For example, if you toss the die 100 times, total the numbers on the faces, and then divide by 100, what will you get? Call this number d (for die). It is reasonable to suppose that d will be fairly close to the mean μ.

$$d \approx \mu$$
$$d \approx (1 + 2 + 3 + 4 + 5 + 6)/6$$
$$= 21/6$$
$$= 3.5$$

It's possible, in fact likely, that if you toss a die 100 times, you'll get a value of d that is slightly more or less than 3.5. This is to be expected because of "reality imperfection." But now imagine tossing the die 1000 times, or 100,000 times, or even 100,000,000 times! The reality imperfections will be smoothed out by the fact that the number of tosses is huge. The value of d will converge to 3.5. As the number of tosses increases without limit, the value of d will get closer and closer to 3.5, because the opportunity for repeated coincidences biasing the result will get smaller and smaller.

The foregoing scenario is an example of the *law of large numbers*. In a general, informal way, it can be stated like this:

• As the number of events in an experiment increases, the average value of the outcome approaches the theoretical mean.

INDEPENDENT OUTCOMES

Two outcomes H_1 and H_2 are *independent* if and only if the occurrence of one does not affect the probability that the other will occur. We write it this way:

$$p(H_1 \cap H_2) = p(H_1)\, p(H_2)$$

Figure 30-1 illustrates this situation as a special sort of mathematical drawing called a *Venn diagram*. The intersection is shown by the hatched region.

A good example of independent outcomes is the tossing of a penny and a nickel. The face (heads or tails) that turns up on the penny has no effect on the face (heads or tails) that turns up on the nickel. It does not matter whether the two coins are tossed at the same time or at different times. They never interact with each other.

To illustrate how the above formula works in this situation, let $p(P)$ represent the probability that the penny turns up heads when a penny and a nickel are both tossed once. Clearly, $p(P) = 0.5$ (1 in 2). Let $p(N)$ represent the probability that the nickel turns up heads in the same scenario. It's obvious that $p(N) = 0.5$ (also 1 in 2). The probability

Figure 30-1

Venn diagram showing set intersection.

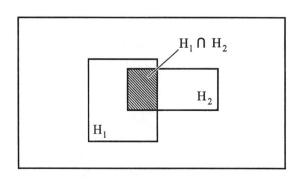

that both coins turn up heads is, as you should be able to guess, 1 in 4, or 0.25. The above formula states it this way, where the intersection symbol ∩ can be translated as "and":

$$p(P \cap N) = p(P)p(N)$$
$$= 0.5 \cdot 0.5$$
$$= 0.25$$

MUTUALLY EXCLUSIVE OUTCOMES

Suppose that H_1 and H_2 are *mutually exclusive outcomes*; that is, they have no elements in common:

$$H_1 \cap H_2 = \varnothing$$

In this type of situation, the probability of either outcome occurring is equal to the sum of the individual probabilities. Here's how we write it, with the union symbol ∪ translated as "either/or":

$$p(H_1 \cup H_2) = p(H_1) + p(H_2)$$

Figure 30-2 shows this as a Venn diagram.

When two outcomes are mutually exclusive, they cannot both occur. A good example is the tossing of a single coin. It's impossible for heads and tails to both turn up on a given toss. But the sum of the two probabilities (0.5 for heads and 0.5 for tails for a typical coin) is equal to the probability that either one outcome or the other will take place —assuming, of course, that the coin lands somewhere and does not get lost!

Another example is the result of a properly run, uncomplicated election for a political office between two candidates. Call the candidates Mrs. Anderson and Mr. Boyd. If Mrs. Anderson wins, call it outcome A, and if Mr. Boyd wins, call it outcome B. Call the respective probabilities of their winning $p(A)$ and $p(B)$. You might argue about the actual values of $p(A)$ and $p(B)$. You might obtain empirical probability figures by

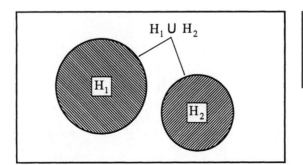

Figure 30-2

Venn diagram showing a pair of mutually exclusive outcomes.

conducting a poll prior to the election and get the idea that, for example, $p_{emp}(A) = 0.29$ and $p_{emp}(B) = 0.71$. In any case, you can be certain of two things: the two candidates won't both win, but one of them will. The probability that either Mrs. Anderson or Mr. Boyd will win is equal to the sum of $p(A)$ and $p(B)$, whatever these values happen to be, and you can be sure it is equal to 1 (assuming neither of the candidates quits during the election and is replaced by a third, unknown person, and assuming there are no write-ins or other election irregularities). Mathematically,

$$p(A \cup B) = p(A) + p(B)$$
$$= p_{emp}(A) + p_{emp}(B)$$
$$= 0.29 + 0.71$$
$$= 1$$

COMPLEMENTARY OUTCOMES

If two outcomes H_1 and H_2 are *complementary*, then the probability of one outcome taking place is equal to 1 (or 100%) minus the probability of the other outcome taking place:

$$p(H_2) = 1 - p(H_1)$$

and

$$p(H_1) = 1 - p(H_2)$$

Expressed as percentages, they become

$$p_\%(H_2) = 100 - p_\%(H_1)$$

and

$$p_\%(H_1) = 100 - p_\%(H_2)$$

Figure 30-3 shows this as a Venn diagram.

Figure 30-3

Venn diagram showing a pair of complementary outcomes.

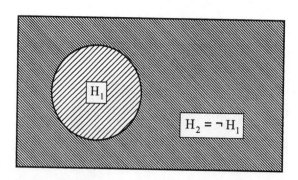

The notion of complementary outcomes is useful when you want to find the probability that an outcome will fail to occur. Consider again the election between Mrs. Anderson and Mr. Boyd. Imagine that you are one of those disillusioned voters who call themselves "contrarians": people who vote entirely *against* candidates in elections. You are interested in the probability that "your candidate" (the one you dislike) will lose. Suppose that, according to the preelection poll, $p_{emp}(A) = 0.29$ and $p_{emp}(B) = 0.71$. You can state this in typical contrarian fashion as

$$p_{emp}(\neg A) = 1 - p_{emp}(A)$$
$$= 1 - 0.29$$
$$= 0.71$$

and

$$p_{emp}(\neg B) = 1 - p_{emp}(B)$$
$$= 1 - 0.71$$
$$= 0.29$$

where the "droopy minus sign" ($\neg$) stands for the NOT operation, also called *logical negation*. If you are fervently wishing for Mr. Boyd to lose, then you can guess from the poll that the likelihood of your being happy (or, at least, not too unhappy) after the election is equal to $p_{emp}(\neg B)$, which is 0.29.

If this confuses you, remember that in this election, there are only two candidates, and one or the other of them must win, so

$$p_{emp}(\neg A) = p_{emp}(B)$$

and

$$p_{emp}(\neg B) = p_{emp}(A)$$

NONDISJOINT OUTCOMES

Outcomes H_1 and H_2 are called *nondisjoint* if and only if they have at least one element in common:

$$H_1 \cap H_2 \neq \emptyset$$

In cases like this, the probability of either outcome is equal to the sum of the probabilities of their occurring separately minus the probability of their occurring simultaneously. The equation looks like this:

$$p(H_1 \cup H_2) = p(H_1) + p(H_2) - p(H_1 \cap H_2)$$

Figure 30-4

Venn diagram
showing a pair
of non-disjoint
outcomes.

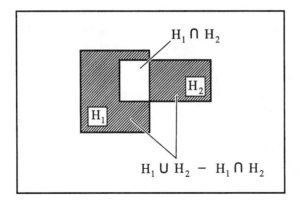

Figure 30-4 shows this as a Venn diagram. The intersection of probabilities is subtracted to ensure that the elements common to both sets (represented by the white region where the two sets overlap) are counted only once.

Imagine that a certain high school has 1000 students. The new swimming and diving coach, during his first day on the job, is looking for team prospects. Suppose that the following are true:

- 200 students can swim well enough to make the swimming team.
- 100 students can dive well enough to make the diving team.
- 30 students can make either team or both teams.

If the coach wanders through the school's hallways blindfolded and picks a student at random, let's determine the probabilities, expressed as ratios, that the coach will pick

- A fast swimmer; call this $p(S)$
- A good diver; call this $p(D)$
- Someone good at both swimming and diving; call this $p(S \cap D)$
- Someone good at either swimming or diving, or both; call this $p(S \cup D)$

This problem is a little tricky. You can assume that the coach has objective criteria for evaluating prospective candidates for his teams! That said, you must note that the outcomes are not mutually exclusive, nor are they independent. There is overlap, and there is interaction. You can find the first three answers immediately, because you are told the numbers:

$$p(S) = 200/1000 = 0.200$$

$$p(D) = 100/1000 = 0.100$$

$$p(S \cap D) = 30/1000 = 0.030$$

To calculate the last answer—the total number of students who can make either team or both teams—you must find $p(S \cup D)$, using this formula:

$$p(S \cup D) = p(S) + p(D) - p(S \cap D)$$
$$= 0.200 + 0.100 - 0.030$$
$$= 0.270$$

This means that 270 of the students in the school are potential candidates for either or both teams. The answer is not 300, as you might at first expect. That would be the case only if there were no students good enough to make both teams. You mustn't count the exceptional students twice. (No matter how well a person can act like a fish, she or he is nevertheless only one person!)

MULTIPLE OUTCOMES

The formulas for determining the probabilities of mutually exclusive and nondisjoint outcomes can be extended to situations in which there are three possible outcomes.

Let H_1, H_2, and H_3 be mutually exclusive outcomes, such that the following facts hold:

$$H_1 \cap H_2 = \varnothing$$
$$H_1 \cap H_3 = \varnothing$$
$$H_2 \cap H_3 = \varnothing$$

The probability of any one of the three outcomes occurring is equal to the sum of their individual probabilities (Fig. 30-5):

$$p(H_1 \cup H_2 \cup H_3) = p(H_1) + p(H_2) + p(H_3)$$

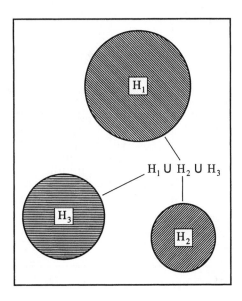

Figure 30-5

Venn diagram showing three mutually exclusive outcomes.

Now suppose that H_1, H_2, and H_3 are nondisjoint outcomes. This means that one or more of the following facts is true:

$$H_1 \cap H_2 \neq \varnothing$$
$$H_1 \cap H_3 \neq \varnothing$$
$$H_2 \cap H_3 \neq \varnothing$$

The probability of any one of the outcomes occurring is equal to the sum of the probabilities of their occurring separately, plus the probabilities of each pair occurring simultaneously, minus the probability of all three occurring simultaneously (see Fig. 30-6 for a Venn diagram):

$$p(H_1 \cup H_2 \cup H_3) = p(H_1) + p(H_2) + p(H_3) - p(H_1 \cap H_2)$$
$$- p(H_1 \cap H_3) - p(H_2 \cap H_3)$$
$$+ p(H_1 \cap H_2 \cap H_3)$$

Consider again the high school with 1000 students. The coach seeks people for the swimming, diving, and water polo teams in the same wandering, blindfolded way as before. Suppose the following is true of the students in the school:

- 200 people can make the swimming team.
- 100 people can make the diving team.
- 150 people can make the water polo team.

Figure 30-6

Venn diagram showing three non-disjoint outcomes.

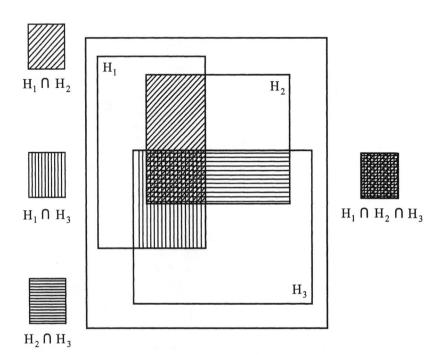

$H_1 \cap H_2$

$H_1 \cap H_3$

$H_2 \cap H_3$

H_1

H_2

H_3

$H_1 \cap H_2 \cap H_3$

- 30 people can make both the swimming and diving teams.
- 110 people can make both the swimming and water polo teams.
- 20 people can make both the diving and water polo teams.
- 10 people can make all three teams.

If the coach tags students at random, what is the probability, expressed as a ratio, that the coach will, on any one tag, select a student who is good enough for at least one of the sports?

Let the following expressions stand for the respective probabilities, all representing the results of random selections by the coach (and all of which you are told):

- Probability that a student can swim fast enough = $p(S) = 200/1000 = 0.200$.
- Probability that a student can dive well enough = $p(D) = 100/1000 = 0.100$.
- Probability that a student can play water polo well enough = $p(W) = 150/1000 = 0.150$.
- Probability that a student can swim fast enough and dive well enough = $p(S \cap D) = 30/1000 = 0.030$.
- Probability that a student can swim fast enough and play water polo well enough = $p(S \cap W) = 110/1000 = 0.110$.
- Probability that a student can dive well enough and play water polo well enough = $p(D \cap W) = 20/1000 = 0.020$.
- Probability that a student can swim fast enough, dive well enough, and play water polo well enough = $p(S \cap D \cap W) = 10/1000 = 0.010$.

To calculate the total number of students who can end up playing at least one sport for this coach, we must find $p(S \cup D \cup W)$ by using this formula:

$$p(S \cup D \cup W) = p(S) + p(D) + p(W) - p(S \cap D) - p(S \cap W) - p(D \cap W)$$
$$+ p(S \cap D \cap W)$$
$$= 0.200 + 0.100 + 0.150 - 0.030 - 0.110 - 0.020 + 0.010$$
$$= 0.300$$

This means that 300 of the students in the school are potential prospects.

THE DENSITY FUNCTION

When you are working with large populations, and especially with continuous random variables, probabilities are defined differently than they are with small populations and discrete random variables. As the number of possible values of a random variable becomes larger and "approaches infinity," it's easier to think of the probability of an outcome within a range of values, rather than the probability of an outcome for a single value.

Imagine that some medical researchers want to find out how people's blood pressure levels compare. At first, a few dozen people are selected at random from the human population, and the numbers of people having each of 10 specific systolic pressure readings are plotted (Fig. 30-7A). The systolic pressure, which is the higher of the two numbers you get when you take your blood pressure, is the random variable. (In this example, exact numbers aren't shown either for blood pressure or for the number of people. This can help you keep in mind that this scenario is make-believe.)

There seems to be a pattern in Fig. 30-7A. This does not come as a surprise to our group of medical research scientists. They expect most people to have "middling" blood pressure, fewer people to have moderately low or high pressure, and only a small number of people to have extremely low or high blood pressure.

In the next phase of the experiment, hundreds of people are tested. Instead of only 10 different pressure levels, 20 discrete readings are specified for the random variable (Fig. 30-7B). A pattern is obvious. Confident that they're onto something significant, the researchers test thousands of people and plot the results at 40 different blood pressure levels.

Figure 30-7

Hypothetical plots of blood pressure. At A, a small population and 10 values of pressure; at B, a large population and 20 values of pressure; at C, a gigantic population and 40 values of pressure; at D, the value of the density function versus blood pressure.

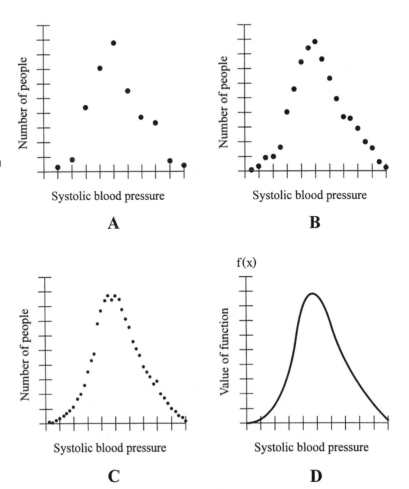

The resulting plot of frequency (number of people) versus the value of the random variable (blood pressure) shows that there is a well-defined pattern. The arrangement of points in Fig. 30-7C is so orderly that the researchers are sure that repeating the experiment with the same number of subjects (but not the same people) will produce the same pattern.

EXPRESSING THE PATTERN

Based on the data in Fig. 30-7C, the researchers can use a technique called *curve fitting* to derive a general rule for the way blood pressure is distributed. Figure 30-7D shows the result. A smooth curve like this is called a *probability density function*, or simply a *density function*. It no longer represents the blood pressure levels of individuals, but is only an expression of how blood pressure varies among the human population. On the vertical axis, instead of the number of people, the function value is portrayed. As the number of possible values of the random variable increases without limit, the point-by-point plot blurs into the density function $f(x)$. The blood pressure of any particular subject is no longer important. Instead, the researchers become concerned with the probability that any randomly chosen subject's blood pressure will fall within a given range of values.

AREA UNDER THE CURVE

Figure 30-8 is an expanded view of the curve shown in Fig. 30-7D. This function, like all density functions, has a special property. If you calculate or measure the total area under the curve, it is always equal to 1. This is true for the same reason that the cumulative

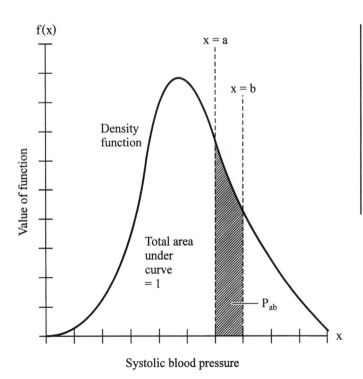

Figure 30-8
The probability of a randomly chosen value k of a random variable x falling between two limit values a and b is equal to the area under the curve between the vertical lines $x = a$ and $x = b$.

relative frequency of the outcomes for a discrete variable always "maxes out" at 1 (or 100%), as you learned in Chap. 29.

Consider two hypothetical systolic blood pressure values, say a and b as shown in Fig. 30-8. The probability P_{ab} that a randomly chosen person has a systolic blood pressure reading k that is between (but not including) a and b is represented by the hatched region under the curve. If you move either or both of the vertical lines $x = a$ and $x = b$ to the left or right, the area of the hatched region changes. This area can never be less than 0 (when the two lines $x = a$ and $x = b$ coincide) or greater than 1 (when the lines are so far apart that they take in the entire area under the curve).

UNIFORM DISTRIBUTION

In a *uniform distribution*, the value of the probability density function is constant. When graphed, it looks "flat" on top. Figure 30-9 shows an example.

Let x be a continuous random variable. Let x_{min} and x_{max} be the minimum and maximum values that x can attain, respectively. In a uniform distribution, x has a density function of the form

$$f(x) = 1 / (x_{max} - x_{min})$$

Figure 30-9

A uniform density function has a constant value when the random variable is between two extremes x_{min} and x_{max}.

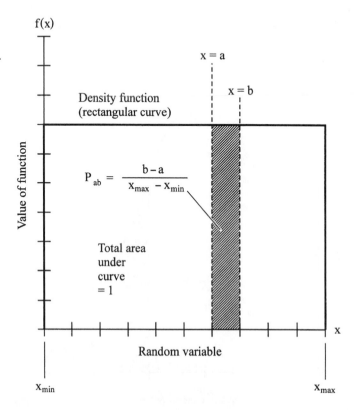

Because the total area under the curve is equal to 1, the probability P_{ab} that any randomly chosen x will be between a and b is

$$P_{ab} = (b - a) / (x_{max} - x_{min})$$

Suppose that the experiment described above reveals that equal numbers of people always have each of the given tested blood pressure numbers between two limiting values, say $x_{min} = 100$ and $x_{max} = 140$. Imagine that this is true no matter how many people are tested, and no matter how many different values of blood pressure are specified within the range of, say, $x_{min} = 100$ and $x_{max} = 140$. This is far-fetched and can't represent the real world, but if you play along with the idea, you can see that such a state of affairs produces a uniform probability distribution.

NORMAL DISTRIBUTION

In a *normal distribution*, the value of the function has a single central peak and tapers off on either side in a symmetric fashion. Because of its shape, a graph of this function is often called a *bell-shaped curve* (Fig. 30-10). The normal distribution isn't just any bell-shaped curve, however. To be a true normal distribution, the curve must conform to specific rules concerning its *standard deviation*, an expression of the extent to which

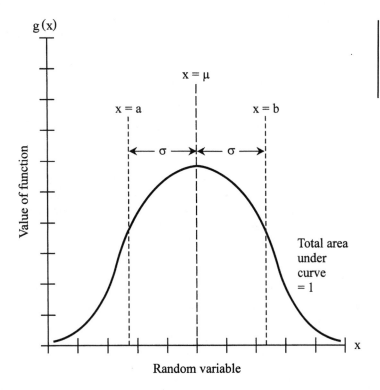

Figure 30-10

The normal distribution is also known as the bell-shaped curve.

the values of the function are concentrated. The symbol for standard deviation is the lowercase Greek letter sigma (σ).

A small value of σ produces a "sharp" curve with a narrow peak and steep sides. A large value of σ produces a "broad" curve with less steep sides. As σ approaches 0, the curve becomes narrower and narrower, closing in on a vertical line. If σ becomes arbitrarily large, the curve becomes almost flat and settles down near the horizontal axis. In any normal distribution, the area under the curve is equal to 1, no matter how steep or shallow the extremes happen to be.

In Fig. 30-10, the symbol μ represents the mean, or average. This is the same mean you learned about in Chap. 29, but generalized for continuous random variables. The value of μ can be found by imagining a movable vertical line that intersects the x axis. When the position of the vertical line is such that the area under the curve to its left is 1/2 (or 50%) and the area under the curve to its right is also 1/2 (50%), then the vertical line intersects the x axis at the point $x = \mu$. In the normal distribution, $x = \mu$ at the same point where the function attains its *peak*, or maximum, value.

THE EMPIRICAL RULE

Imagine two movable vertical lines, one on either side of the vertical line $x = \mu$. Suppose these vertical lines, $x = a$ and $x = b$, are such that the one on the left is the same distance from $x = \mu$ as the one on the right. The proportion of data points in the part of the distribution $a < x < b$ is defined by the proportion of the area under the curve between the two movable lines $x = a$ and $x = b$. Figure 30-10 illustrates this situation. A well-known theorem in statistics, called the *empirical rule*, states that all normal distributions have the following three characteristics:

- Approximately 68% of the data points are within the range $\pm\sigma$ *of* μ.
- Approximately 95% of the data points are within the range $\pm 2\sigma$ of μ.
- Approximately 99.7% of the data points are within the range $\pm 3\sigma$ of μ.

Suppose you want it to rain so your garden will grow. It's a gloomy morning. The weather forecasters, who are a little bit strange in your town, expect a 50% chance that that you'll see up to 1 centimeter (1 cm) of rain in the next 24 hours and a 50% chance that more than 1 cm of rain will fall. They say it is impossible for more than 2 cm to fall (a dubious notion at best), and it is also impossible for less than 0 cm to fall (an absolute certainty!). Suppose the radio disc jockeys (DJs), who are even more weird than the meteorologists, announce the forecast and start talking about a distribution function called $R(x)$ for the rain as predicted by the weather experts. One DJ says that the amount of rain represents a continuous random variable x, and the distribution function $R(x)$ for the precipitation scenario is a normal distribution whose value tails off to 0 at precipitation levels of 0 cm and 2 cm. Figure 30-11 is a crude graph of what they're talking about. The portion of the curve to the left of the

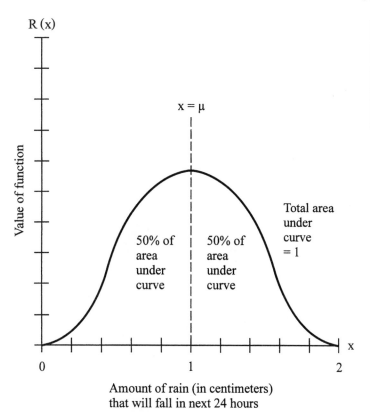

Figure 30-11
A normal distribu-
tion representing
expected rainfall in
the next 24 hours
at a hypothetical
location.

vertical line, which represents the mean, has an area of 0.5. The mean itself is $x = \mu = 1$ cm.

Now suppose that the DJs start discussing the extent to which the distribution function is spread out around the mean value of 1 cm. One of them mentions that there is something called "standard deviation," symbolized by a lowercase Greek letter called "sigma" that looks like a numeral 6 that has fallen over onto its right side. Another DJ says that 68% of the total prospective future "wetness" of the town falls within two values of precipitation defined by sigma on either side of the mean. Figure 30-12 is a graph of what the DJs are talking about now. The hatched region represents the area under the curve between the two vertical lines $x = \mu - \sigma$ and $x = \mu + \sigma$. This is 68% of the total area under the curve, centered at the vertical line representing the mean, $x = \mu$.

Based on the graph of Fig. 30-12, the standard deviation appears to be approximately 0.45, representing the distance of either the vertical line $x = \mu - \sigma$ or the vertical line $x = \mu + \sigma$ from the mean $x = \mu = 1$ cm. This idea is based only on a crude graph. The DJs have not said anything about the actual value of σ. You want to keep listening to the DJs and find out if they'll tell you more, but then lightning hits the broadcast tower and the station goes off the air.

Figure 30-12

In a normal distribution, 68% of the area under the curve lies within the standard deviation on either side of the mean.

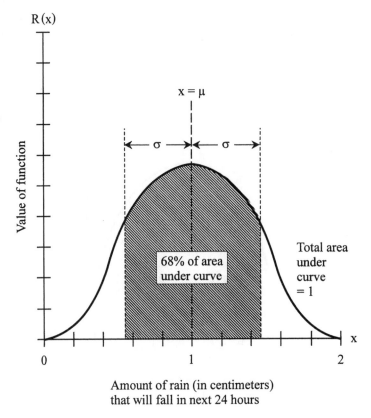

QUESTIONS AND PROBLEMS

This is an open-book quiz. You may refer to the text in this chapter (and earlier ones, too, if you want) when figuring out the answers. Take your time! The correct answers are in the back of the book.

1. What is the difference between empirical probability and mathematical probability?

2. Imagine a perfectly balanced spinning wheel, in which the likelihood of the pointer coming to rest within any particular range of directions (shaped like a thin slice of pie) is the same as the likelihood of its coming to rest within any other range of equal size. If the direction in which the pointer comes to rest, measured in degrees of the compass (clockwise from true north), is the random variable, what type of distribution is the density function for the behavior of this wheel?

3. What is the result of an event called?

4. Suppose some method is devised that quantitatively describes the political views of people on a continuum from "liberal" to "conservative," where the most liberal possible views are represented by a value of –50 (the extreme left end of the scale) and the most conservative possible views are represented by a value of +50 (the extreme right end of the scale). Suppose a density function is graphed and it is discovered, not surprisingly, that most people tend to fall near the "middle" and that the curve is bell-shaped and symmetric to the left and right. What type of distribution does this represent?

5. What is the mathematical probability that a conventional coin, tossed 10 times in a row, will come up tails on all 10 tosses? Express this as a percentage to five significant figures. Assume that the probability of heads is 1 in 2, and the probability of tails is also 1 in 2.

6. Imagine a strange coin that can land any of three ways: heads, tails, or "on edge." Suppose that the chance of the coin landing on the edge for any given toss is 2 in 20, the chance of its landing heads is 9 in 20, and the chance of its landing tails is 9 in 20. If this coin is tossed 10 times in a row, what is the probability of its coming up tails on all 10 tosses? Express this as a percentage to five significant figures.

7. Imagine a normal distribution that produces the classic bell-shaped curve. Suppose some aspect of the experiment is varied so that the curve becomes "sharper"; that is, the peak becomes more pronounced. What happens to the standard deviation? What happens to the total area under the curve?

Final Exam

Do not refer to the text when taking this exam. A good score is at least 75 correct. Answers are in the back of the book. It's best to have a friend check your score the first time, so you won't memorize the answers if you want to take the exam again.

1. The value of 27 to the 2/3 power is
   (a) Undefined
   (b) Equal to 3
   (c) Equal to 6
   (d) Equal to 9
   (e) Equal to 12

2. The value of 5/7 minus 3/5, expressed as a fraction, is exactly
   (a) 4/35
   (b) 2/15
   (c) 1/6
   (d) 2/7
   (e) 5/12

3. What is the area of a right isosceles triangle whose hypotenuse is 10.0 units long?
   (a) 17.7 square units
   (b) 25.0 square units
   (c) 35.4 square units
   (d) 50.0 square units
   (e) More information is needed to answer this.

4. If the length of each side of a square increases by exactly 50%, the interior area of the square increases by
   (a) Exactly 50%
   (b) Approximately 71%
   (c) Exactly 100%
   (d) Exactly 125%
   (e) Approximately 141%

5. Consider the function $f(x) = 4x^3$. The definite integral from $x = -3$ to $x = 3$ is
   (a) -81
   (b) -27
   (c) 9
   (d) 108
   (e) 0

6. A health club's membership dues are $400 per year for full members and $200 per year for student members. There are 200 members in all, with annual total dues receipts of $64,000. This club has

(a) 100 full members and 100 student members
(b) 120 full members and 80 student members
(c) 140 full members and 60 student members
(d) 80 full members and 120 student members
(e) 60 full members and 140 student members

7. Force is directly proportional to

(a) Mass times distance
(b) Mass divided by distance
(c) Mass times acceleration
(d) Mass times speed
(e) Mass divided by speed

8. The duodecimal system has a radix of

(a) 4
(b) 8
(c) 10
(d) 12
(e) 16

9. Suppose that you are observing two people, a woman and a man, through an apparatus designed to measure the distances to objects by means of stadimetry. You know that both people are exactly the same height, and both are standing upright at great distances from you on perfectly level ground. Using the observing apparatus, you see that the man subtends an angle of 0.16° and the woman subtends an angle of 0.04°. From this, you can determine

(a) That the woman is almost exactly four times as far away as the man
(b) That the woman is almost exactly twice as far away as the man
(c) That the woman is almost exactly one-half as far away as the man
(d) That the woman is almost exactly 1/4 as far away as the man
(e) Nothing about their relative distances unless you know the actual distance to either the woman or the man

10. Consider the function $y = ax^n$, where y is the dependent variable, a is a constant, x is the independent variable, and the exponent n is a whole number larger than or equal to 4. How can the second derivative d^2y/dx^2 be expressed?

(a) $d^2y/dx^2 = anx^{n-2}$
(b) $d^2y/dx^2 = an^2x^{n-2}$
(c) $d^2y/dx^2 = a^2n^2x^{n-2}$
(d) $d^2y/dx^2 = a^2nx^{n-2}$
(e) $d^2y/dx^2 = (an^2 - an)x^{n-2}$

11. Concerning positive whole numbers, which of the following statements (a), (b), or (c), if any, is false? (*Hint:* You can use a multiplication table up to 9×9 and the old-fashioned format of "long multiplication" to figure this out.)

(a) An even number times an even number is always even.
(b) An even number times an odd number is always even.
(c) An odd number times an odd number is always odd.
(d) None—that is, all are true.
(e) All are false.

12. As the eccentricity of an ellipse having a major axis of constant length approaches 0, which of the following statements is *inevitably* true?
(a) The foci become closer together.
(b) The foci become farther apart.
(c) The foci remain separated by a constant distance.
(d) The foci converge on the origin.
(e) The foci converge on the curve itself.

13. What is the exponent in the product of 2.0×10^p and 7.0×10^q in scientific notation, reduced to the American standard form where the coefficient is at least equal to 1 but smaller than 10? Assume p and q are both integers.
(a) $p + q$
(b) $p + q + 1$
(c) $p + q - 1$
(d) pq
(e) $pq + 1$

14. Imagine an oscillating-spring system in which a weight moves up and down in a sinusoidal motion according to the formula

$$S = 8 \sin 3t$$

where S is the vertical displacement in meters and t is the time in seconds. Now imagine that the system is changed so the relationship becomes

$$S = 8 \sin 6t$$

How will this affect the frequency of the motion?
(a) It will cause no change in the frequency.
(b) It will cut the frequency in half.
(c) It will cut the frequency to 1/4 its previous value.
(d) It will double the frequency.
(e) It will quadruple the frequency.

15. The distance between (0,0,0,0,0) and (1,1,1,1,1) in rectangular 5D space is equal to
(a) 1
(b) The square root of 2
(c) The fifth root of 2
(d) The square root of 5
(e) The fifth root of 5

16. Which of the following vectors, if any, can be added to any vector (x_0, y_0, z_0) in rectangular xyz space, always producing the vector $(-x_0, -y_0, -z_0)$?

(a) $(0,0,0)$

(b) $(-x_0, -y_0, -z_0)$

(c) $(-1, -1, -1)$

(d) $(-2x_0, -2y_0, -2z_0)$

(e) None of the above vectors can be added to (x_0, y_0, z_0) in rectangular xyz space, always producing the vector $(-x_0, -y_0, -z_0)$.

17. Which of the following equations in (a), (b), or (c), if any, represents a circle in the Cartesian xy plane?

(a) $(x - 3)^2 / 4 + (x + 2)^2 / 7 = 1$

(b) $(x - 3)^2 / 3 - (x + 2)^2 / 3 = 11$

(c) $(x - 3)^2 / 3 + (x + 2)^2 / 3 = 5$

(d) All the above

(e) None of the above

18. Look at Fig. E-1. Suppose points X and Z are fixed, but point Y is free to move along the top line. As Y travels farther and farther toward the right without limit, what happens to the perimeter of triangle XYZ?

(a) It does not change.

(b) It grows larger and larger, but approaches a certain finite limit.

(c) It grows larger and larger without limit.

(d) It grows smaller and smaller.

(e) More information is needed to answer this question.

19. In the situation of Fig. E-1 and Prob. 18, what happens to the interior area of triangle XYZ as point Y travels farther and farther toward the right without limit?

(a) It does not change.

(b) It grows larger and larger, but approaches a certain finite limit.

(c) It grows larger and larger without limit.

(d) It grows smaller and smaller.

(e) More information is needed to answer this question.

Figure E-1

Illustration for Final Exam questions 18 and 19.

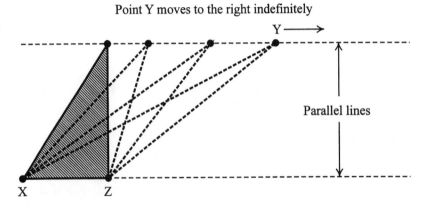

Point Y moves to the right indefinitely

Parallel lines

X   Z

X	Y	Z	?
F	F	F	F
F	F	T	F
F	T	F	F
F	T	T	F
T	F	F	F
T	F	T	F
T	T	F	F
T	T	T	T

Table E-1

Table for Final Exam question 20.

20. Examine Table E -1. Which of the following can be written in the uppermost column header at the extreme right in place of the question mark, making the truth table correct?
  (*a*) X & Y & Z
  (*b*) X ∨ (Y & Z)
  (*c*) X & (Y ∨ Z)
  (*d*) (X ∨ Y) & Z
  (*e*) X ∨ Y ∨ Z

21. Mathematical probabilities can have values
  (*a*) Between −1 and 1 inclusive
  (*b*) Corresponding to any positive real number
  (*c*) Between 0 and 1 inclusive
  (*d*) Corresponding to any integer
  (*e*) Within any defined range

22. In statistics, a discrete variable
  (*a*) Can attain only specific, defined values
  (*b*) Can attain infinitely many values within a defined span
  (*c*) Can attain only positive integer values
  (*d*) Can attain only nonnegative integer values
  (*e*) Can never have a value greater than 1

23. The value of j^{13}, where j is the positive square root of −1, is equal to
  (*a*) 1
  (*b*) −1
  (*c*) j
  (*d*) $-j$
  (*e*) An undefined quantity

24. Suppose you take a day trip by car. You stop once to eat lunch, once to fill up the gas tank, and three times to use the bathroom at rest stops. Which of statements (a), (b), (c), or (d), if any, is *not necessarily true*?
  (*a*) Your instantaneous speed is sometimes less than your average speed.
  (*b*) Your instantaneous speed is sometimes greater than your average speed.
  (*c*) Your instantaneous speed is sometimes zero.
  (*d*) Your instantaneous speed is sometimes three times your average speed.
  (*e*) All the above statements (a), (b), (c), and (d) are true.

Figure E-2

Illustration for Final
Exam questions
26 and 27.

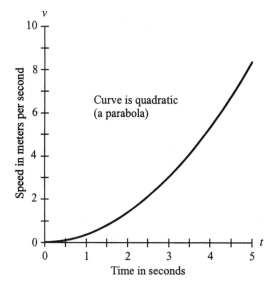

25. In propositional logic, the smallest entities dealt with are
    (*a*) Verbs
    (*b*) Nouns
    (*c*) Adjectives
    (*d*) Predicates
    (*e*) Whole sentences

26. Which of the following equations, when graphed for positive values of *t*, will result
    in the curve shown in Fig. E-2?
    (*a*) $v = 3t^2 + 2t - 1$
    (*b*) $v = t^2 / 3$
    (*c*) $v = 3t^2 - t$
    (*d*) $v = t^{1/2} / 3$
    (*e*) None of the above

27. The graph in Fig. E-2 represents an object that
    (*a*) Accelerates at an increasing rate as time passes
    (*b*) Accelerates at a constant rate
    (*c*) Accelerates at a decreasing rate as time passes
    (*d*) Decelerates at a constant rate
    (*e*) Decelerates at an increasing rate as time passes

28. What is the value of $3 \times 7 + 10 \div 5$? (*Hint:* Use the rules for grouping to put
    parentheses in the correct places before calculating.)
    (*a*) 6-1/5
    (*b*) 42
    (*c*) 23
    (*d*) 27
    (*e*) 10-1/5

29. The sequence 512, 384, 288, 216, 162, . . . is an example of
    (*a*) An arithmetic progression
    (*b*) An exponential progression

(*c*) A harmonic progression

(*d*) A logarithmic progression

(*e*) A geometric progression

30. Imagine a cubic box that measures exactly 1 meter by 1 meter by 1 meter. Now suppose you have an unlimited number of little solid cubes, each of which measures 1 millimeter by 1 millimeter by 1 millimeter, where 1 millimeter is 1/1000 of a meter. If you stack these tiny cubes neatly in the big box so every possible space is accounted for, how many tiny cubes will exactly fill the box?

(*a*) 10^5

(*b*) 10^6

(*c*) 10^9

(*d*) 10^{11}

(*e*) 10^{12}

31. Which of the following equations in the Cartesian *xy* plane represents the same graph as the equation $\theta = \pi/4$ in mathematician's polar coordinates?

(*a*) $x^2 + y^2 = 1$

(*b*) $x + y = 1$

(*c*) $x - y = 0$

(*d*) $x - y = 1$

(*e*) $x^2 - y^2 = 1$

32. The common log of 2 is equal to approximately 0.3. Once you know this fact, what you can say straightaway about the common log of 8?

(*a*) Because $8 = 2 + 6$, the common log of 8 is equal to 6 times the common log of 2, or approximately 1.8.

(*b*) Because $8 = 2 \cdot 4$, the common log of 8 is equal to 4 times the common log of 2, or approximately 1.2.

(*c*) Because $8 = 2 \cdot 4$, the common log of 8 is equal to the 4th power of the common log of 2, or approximately 0.0081.

(*d*) Because $8 = 2^3$, the common log of 8 is equal to 3 times the common log of 2, or approximately 0.9.

(*e*) Because $8 = 2^3$, the common log of 8 is equal to the 3rd power of the common log of 2, or approximately 0.027.

33. What is the magnitude of the cross product of two vectors, both having length 2.000 and pointing in directions separated by $30°0'0''$?

(*a*) 1.000

(*b*) 1.414

(*c*) 1.732

(*d*) 2.000

(*e*) 2.828

34. What is the mathematical probability that an unweighted die, tossed four times, will show the face with 6 dots on all four occasions?

(*a*) 1 in 6

(*b*) 1 in 36

(c) 1 in 64

(d) 1 in 1,296

(e) 1 in 46,656

35. Fill in the blank to make the following sentence true: If the number of elements in a distribution is odd, then the _____ is the value such that the number of elements having values greater than or equal to it is the same as the number of elements having values less than or equal to it.

(a) Mean

(b) Deviation

(c) Frequency

(d) Mode

(e) Median

36. How can the expression $x^3 - 3x^2 + 3x - 1$ be factored?

(a) $(x + 1)^3$

(b) $(x - 1)(x + 1)^2$

(c) $(x + 1)(x - 1)^2$

(d) $(x - 1)^3$

(e) None of the above

37. In Fig. E-3, the value of sin $(A + B)$ is exactly equal to

(a) 0

(b) $2^{1/2} / 2$

(c) 1

(d) $-2^{1/2} / 2$

(e) -1

38. Considered with respect to the speed of light which is about 3.0×10^8 m/s, one minute-equivalent in time-space represents a distance of approximately

(a) 1.8×10^4 km

(b) 1.8×10^5 km

Figure E-3

Illustration for Final Exam question 37.

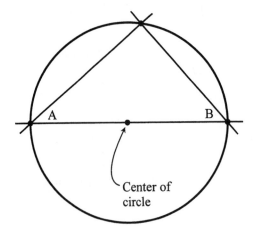

Center of circle

(c) 1.8×10^6 km
(d) 1.8×10^7 km
(e) 1.8×10^8 km

39. If you take the antiderivative of the cosine function, then take the antiderivative of that, and finally take the antiderivative of that, you get
(a) The sine function
(b) Minus the sine function
(c) The cosine function
(d) Minus the cosine function
(e) None of the above

40. The intensity of sunlight in the solar system varies according to the inverse of the square of the distance from the sun. The average distance of the earth from the sun is 93,000,000 mi. The average distance of Mars from the sun is 142,000,000 mi. If the average intensity of sunlight at the top of the earth's atmosphere is represented as 100 percent, what percentage represents the average intensity of sunlight at the top of Mars' atmosphere?
(a) 43 percent
(b) 65 percent
(c) 81 percent
(d) 94 percent
(e) It is impossible to say without more information.

41. In probability theory, if two outcomes are complementary, then the probability (expressed as a ratio) of one outcome is equal to
(a) The probability of the other outcome
(b) One-half the probability of the other outcome
(c) 1 divided by the probability of the other outcome
(d) 1 plus the probability of the other outcome
(e) 1 minus the probability of the other outcome

42. Let q be one of the acute angles in a right triangle. Let a represent the length of the side opposite q, and let b represent the length of the hypotenuse. Which of the following equations is true?
(a) $\sin q = a/b$
(b) $\sin q = b/a$
(c) $\cos q = a/b$
(d) $\cos q = b/a$
(e) $\tan q = a/b$

43. Fill in the blank to make the following sentence true: In a statistical distribution having discrete values for the independent variable, the highest cumulative absolute frequency number should be equal to the _____ of all the individual absolute frequency numbers.
(a) Average
(b) Sum

(c) Factorial
(d) Product
(e) Largest

44. In the graph of Fig. E-4, suppose the wave is a sinusoid. What is the equation of the curve?

(a) $y = 8 \sin x$
(b) $y = -8 \sin x$
(c) $y = 8 \cos x$
(d) $y = -8 \cos x$
(e) None of the above

45. In the graph of Fig. E-4, what is the slope of a line tangent to the curve at point P?

(a) -2
(b) -4
(c) -8
(d) -16
(e) -32

46. Consider the following polynomial:

$$5x^8 + 7x^7 - 12\,x^6 - 12x^5 - x^3 + 7x^2 + 5x - 23$$

Which of statements (a), (b), (c), or (d), if any, is true concerning this polynomial?

(a) It is invalid because there is no term written for x^4.
(b) The coefficient of the term for x^4 is 0.
(c) It is a quadratic polynomial.
(d) If its value is set equal to 0, we get a linear equation.
(e) All the above statements (a), (b), (c), and (d) are true.

Figure E-4

Illustration for Final
Exam questions
44 and 45.

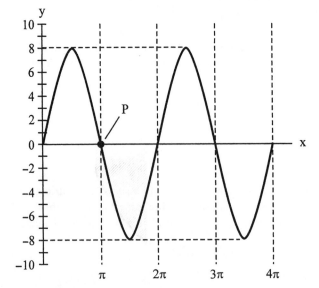

47. Suppose an item at a store costs you $7.58 including sales tax. You pay with a $10.00 bill. Which of the following combinations of currency and coin represents the correct amount of change?

    (*a*) One $1.00 bill, five quarters, two dimes, and two pennies
    (*b*) Two $1.00 bills, four dimes, and two pennies
    (*c*) Two $1.00 bills, one quarter, four nickels, and two pennies
    (*d*) Nine quarters, four nickels, and two pennies
    (*e*) Any of the above

48. In propositional logic, the expression "X implies Y" means that

    (*a*) If X is true, then Y is true
    (*b*) If X is true, then Y is probably true
    (*c*) If X is true, then there is good reason to believe that Y is true
    (*d*) If X is true, it suggests that Y is true, but there are exceptions
    (*e*) If X is true, then Y is true; and if Y is true, then X is true

49. What is the area of the hatched region in Fig. E-5, accurate to two significant figures? (Note that the horizontal and vertical scales differ, so this drawing is distorted and may be visually misleading.)

    (*a*) More information is needed to answer this.
    (*b*) 3.3 square units
    (*c*) 6.3 square units
    (*d*) 6.8 square units
    (*e*) 9.8 square units

50. Suppose a 2.000-kg object is pushed straight upward, directly against the force of gravity, for a distance of 12.00 m. Consider that the acceleration caused by gravity is

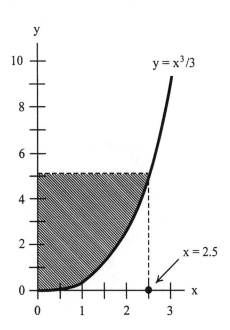

Figure E-5

Illustration for Final Exam question 49.

9.807 m/s^2. How much work, in joules (J), must be done to raise this object for this distance? Express the answer to four significant digits.

(*a*) 235.4 J

(*b*) 2,308 J

(*c*) 2,824 J

(*d*) 4,617 J

(*e*) This cannot be answered without knowing how fast the object is pushed upward.

51. What is the next prime number after 59?

(*a*) 61

(*b*) 63

(*c*) 67

(*d*) 71

(*e*) 73

52. Which of equations (a), (b), or (c), if any, is *false* for all angles A larger than $0°$ but smaller than $90°$ (so the angles $-A$ are, in effect, smaller than $360°$ but larger than $270°$)?

(*a*) $-\sin A = \sin -A$

(*b*) $-\cos A = \cos -A$

(*c*) $-\tan A = \tan -A$

(*d*) Two of equations (a), (b), and (c) are false.

(*e*) All the equations (a), (b), and (c) are false.

53. The quadratic equation $x^2 - x - 12 = 0$ has

(*a*) No real number solutions

(*b*) One real number solution

(*c*) Two different real number solutions

(*d*) Three different real number solutions

(*e*) Four different real number solutions

54. Which, if any, of the infinite series (a), (b), or (c) converges?

(*a*) $8 + 4 + 2 + 1 + 1/2 + 1/4 + 1/8 + \cdots$

(*b*) $1 - 2 + 3 - 4 + 5 - 6 + 7 - \cdots$

(*c*) $1 + 1 - 1 + 1 - 1 + 1 - 1 + \cdots$

(*d*) All the series (a), (b), and (c) converge.

(*e*) None of series (a), (b), or (c) converges.

55. Figure E-6 shows the graph of a straight line in mathematician's cylindrical coordinates. The parameters that define this line are $\theta = 135°$ and $r = 14^{1/2}$ (the positive square root of 14). What parameters define this same line in xyz space, according to the orientation of the x, y, and z axes as shown?

(*a*) $x = -(14^{1/2})$ and $y = 14^{1/2}$

(*b*) $x = y$

(*c*) $z = \pm(14^{1/2})$

(*d*) $x = -(7^{1/2})$ and $y = 7^{1/2}$

(*e*) $x + y = 14^{1/2}$

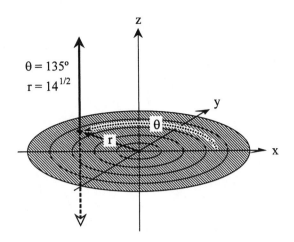

56. To find the distance to the flashing light on top of a television broadcast tower, you use surveying equipment and a baseline b feet long. You observe a parallax angle of θ with respect to the rising moon on the horizon between the two ends of the baseline. You make sure that the baseline is perpendicular to the line of sight to the flashing light. The parallax angle is extremely small, only a fraction of a degree. The distance to the tower, in feet, is equal to approximately

    (*a*) $b / (\tan \theta)$
    (*b*) $b / (\sin \theta)$
    (*c*) $b / (\cos \theta)$
    (*d*) More than one of the above choices
    (*e*) None of choices (a), (b), or (c)

57. In which, if any, of subtraction problems (a), (b), or (c) would you have to borrow when working it out "by hand"?

    (*a*) $8,132 - 7,469$
    (*b*) $244 - 103$
    (*c*) $78,780 - 670$
    (*d*) Borrowing is necessary in all three cases (a), (b), and (c).
    (*e*) Borrowing is not necessary in any of cases (a), (b), or (c).

58. According to one theory in physics, the radius of the smallest electron shell in a hydrogen atom is 5.292×10^{-11} m. This distance is known as the *Bohr radius*. The distance between the northern and southern city limits of the imaginary town of Pingoville, USA, is 5,292 m. How many orders of magnitude larger than the Bohr radius is the north-to-south distance across Pingoville?

    (*a*) 8
    (*b*) 11
    (*c*) 14
    (*d*) 17
    (*e*) More information is necessary to answer this question.

Figure E-7

Illustration for Final Exam questions 59 and 60.

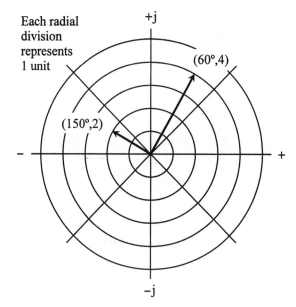

Each radial division represents 1 unit

$+j$

$(60°,4)$

$(150°,2)$

$-$

$+$

$-j$

59. Refer to Fig. E-7. The magnitude of the product of the two complex number vectors shown is

  (*a*) 2 units
  (*b*) The square root of 8 units
  (*c*) The square root of 20 units
  (*d*) 6 units
  (*e*) 8 units

60. In Fig. E-7, the polar angle of the product of the two complex number vectors shown is

  (*a*) Dependent on the order in which the product is performed
  (*b*) 105°
  (*c*) 210°
  (*d*) 90°
  (*e*) 270°

61. Suppose that you need to find the prime factors of 767,376. A good way to get going on this rather tedious process would be to divide a certain number by all the primes less than itself. What is that number?

  (*a*) 383,688
  (*b*) 76,738
  (*c*) 7,674
  (*d*) 876
  (*e*) 768

62. What is the exponent in the quotient 3.29×10^p divided by 1.6787×10^q in scientific notation, reduced to the American standard form where the coefficient is at least equal to 1 but smaller than 10? Assume p and q are both integers.

(a) $p - q$
(b) $p - q + 1$
(c) $p - q + 1$
(d) p/q
(e) $p/q - 1$

63. Suppose the natural log of a certain number x is equal to exactly 10. From this, you know that the natural log of $-x$ is
(a) Equal to exactly 9
(b) Equal to exactly -10
(c) Equal to exactly 1/10
(d) Equal to exactly $-1/10$
(e) Not defined within the set of real numbers

64. Suppose a charter airline makes two round trips between the towns of Thisaway and Thataway every weekday, but only one round trip on Saturday and none on Sunday. If the one-way distance between these two towns is 163 mi, how far does the airline fly in a week on this route?
(a) 1,630 mi
(b) 1,793 mi
(c) 1,956 mi
(d) 3,586 mi
(e) 3,912 mi

65. Which of the following vectors, if any, can be cross-multiplied by any vector (x_0, y_0, z_0) in rectangular xyz space, always producing the vector $(-x_0, -y_0, -z_0)$?
(a) $(0,0,0)$
(b) $(-x_0, -y_0, -z_0)$
(c) $(-1, -1, -1)$
(d) $(-2x_0, -2y_0, -2z_0)$
(e) None of the above

66. What is the kinetic energy, in joules (J), of a 5.00-kg object moving in a straight line at a steady speed of 10.0 m/s?
(a) 0.50 J
(b) 2.00 J
(c) 50.0 J
(d) 250 J
(e) There is no kinetic energy because the object is not accelerating.

67. Which of the following represents the fraction 198/306 reduced to lowest terms?
(a) 11/17
(b) 22/34
(c) 33/51
(d) 66/102
(e) 99/153

68. One of the most familiar Pythagorean triangles has sides of lengths in the precise proportion 3:4:5. For this reason, it is often called a *3-4-5 right triangle*. In this triangle, what is the approximate measure of the angle opposite the shortest side? Use a calculator and determine it to the nearest tenth of a degree.
(*a*) 38.7°
(*b*) 51.3°
(*c*) 31.0°
(*d*) 53.1°
(*e*) 36.9°

69. Which of statements (a), (b), or (c), if any, is true?
(*a*) For all angles A and B, $\sin(A + B) = \sin A + \sin B$.
(*b*) For all angles A and B, $\cos(A + B) = \cos A + \cos B$.
(*c*) For all angles A and B, $\tan(A + B) = \tan A + \tan B$.
(*d*) All statements (a), (b), and (c) are true.
(*e*) None of statements (a), (b), or (c) is true.

70. According to the commutative law,
(*a*) You must do all operations from the outside in.
(*b*) You must do all operations from left to right.
(*c*) You can add two numbers in either order and get the same result.
(*d*) You can subtract two numbers in either order and get the same result.
(*e*) You can divide two numbers in either order and get the same result.

71. The formula for solving quadratics in the form $ax^2 + bx + c = 0$ is

$$x = [-b \pm (b^2 - 4ac)^{1/2}] \,/\, 2a$$

When you use this formula to solve the quadratic $x^2 + 4 = 0$, what solutions do you get for x? Remember that j stands for the positive square root of -1.
(*a*) 2 or -2
(*b*) $j2$ or $-j2$
(*c*) $2, j2, -2$, or $-j2$
(*d*) 2 or $-j2$
(*e*) -2 or $j2$

72. The value of -8 to the 2/3 power is
(*a*) Undefined
(*b*) Defined, but ambiguous
(*c*) Equal to 4
(*d*) Equal to -4
(*e*) Equal to $j4$

73. Suppose you are given the equation $w = xyz$. In this situation, x, y, and z are
(*a*) Factors of w
(*b*) Addends of w
(*c*) Quotients of w

(d) Roots of w

(e) Powers of w

74. The expression $x^2 + y^2$, where x and y are real numbers, can be factored into a product of two imaginary or complex numbers. What are these two factors?

(a) $x + jy$ and $x - jy$

(b) $0 + jx$ and $0 + jy$

(c) $0 + jx$ and $y + j0$

(d) $xy + j$ and $xy - j$

(e) $xy + jx$ and $xy + jy$

75. Figure E-8 is an approximate graph of

(a) An arithmetic function

(b) A geometric function

(c) A logarithmic function

(d) An exponential function

(e) A quadratic function

76. A parallelogram has two sides that measure 20 centimeters (cm) in length and two sides that measure 10 cm in length. What is the area of this figure in square centimeters (cm^2)?

(a) More information is needed to answer this question.

(b) 200 cm^2

(c) 100 cm^2

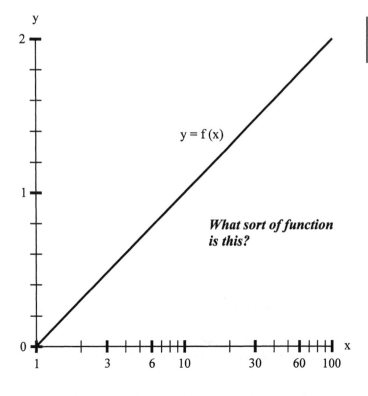

$y = f(x)$

What sort of function is this?

Figure E-8

Illustration for Final Exam question 75.

(*d*) 60 cm²

(*e*) 30 cm²

77. If the order is taken into account, eight different articles be arranged in
- (*a*) 64 different ways
- (*b*) 512 different ways
- (*c*) 4,096 different ways
- (*d*) 40,320 different ways
- (*e*) 16,777,216 different ways

78. Suppose you have a number with a decimal point in it. There are seven digits to the left of the decimal point and four digits to its right. If you multiply this number by 100, you will end up with
- (*a*) The same sequence of digits, but the decimal point one place farther to the right
- (*b*) The same sequence of digits, but the decimal point two places farther to the right
- (*c*) The same sequence of digits, but the decimal point one place farther to the left
- (*d*) The same sequence of digits, but the decimal place two places farther to the left
- (*e*) A different sequence of digits, but with the decimal point in the same place

79. Suppose you see the following expression:

$$\int (6x^2 - 7) \, dx$$

If this expression is worked out, the result is
- (*a*) $2x^3 - 7x + c$, where c is a constant
- (*b*) $12x + c$, where c is a constant
- (*c*) $6x^2 - 7x + c$, where c is a constant
- (*d*) $x^{12} - 7x + c$, where c is a constant
- (*e*) Undefined; when you try to work it out, you must divide by 0

80. When a certain "mystery number" is added to any real number z, the result is the same as if that number were multiplied by $(z + 1)$. What is the mystery number?
- (*a*) i
- (*b*) i^2
- (*c*) i^3
- (*d*) i^4
- (*e*) None of the above

81. In differential calculus, the term "infinitesimal" refers to
- (*a*) A number raised to an extremely large positive power
- (*b*) A number so large that it can be considered as infinity
- (*c*) A number extremely close to zero, but not actually equal to zero
- (*d*) A point on a curve with a line tangent to it
- (*e*) Nothing at all; there is no such term in differential calculus

82. Which of the following fractions is equal to the repeating, nonterminating decimal number 0.123123123 . . . ? You shouldn't need a calculator to solve this.
 (a) 111/123
 (b) 123/321
 (c) 41/333
 (d) 321/999
 (e) 83/666

83. Suppose you are given two functions of a variable x. The functions are called f and g. Suppose you want to find the derivative $(f \cdot g)'$. Which of the following formulas tells you how to do that?
 (a) $(f \cdot g)' = f \cdot g' + g \cdot f'$
 (b) $(f \cdot g)' = f \cdot g' - g \cdot f'$
 (c) $(f \cdot g)' = g \cdot f' - f \cdot g'$
 (d) $(f \cdot g)' = f' \cdot g'$
 (e) $(f \cdot g)' = (f' + g')(f' - g')$

84. The value of 8 to the -3 power is
 (a) Undefined
 (b) Equal to 2 only
 (c) Equal to 2 or -2
 (d) Equal to -2 only
 (e) Equal to 1/512

85. Suppose you have some instructions for finding hidden treasure buried in a flat, level field. You are told, "Find the fencepost with the plastic owl on top. Go exactly 500 ft straight west of this post, and then exactly 200 ft straight south from there. Dig, and you will find the treasure buried 3 ft below the surface." Another way of stating these instructions would be
 (a) "Find the fencepost with the plastic owl on top. Turn to azimuth 21.8°, and then proceed in a straight line for 538 ft 6 in. Dig, and you will find the treasure buried 3 ft below the surface."
 (b) "Find the fencepost with the plastic owl on top. Turn to azimuth 248.2°, and then proceed in a straight line for 538 ft 6 in. Dig, and you will find the treasure buried 3 ft below the surface."
 (c) "Find the fencepost with the plastic owl on top. Turn to azimuth 158.2°, and then proceed in a straight line for 538 ft 6 in. Dig, and you will find the treasure buried 3 ft below the surface."
 (d) "Find the fencepost with the plastic owl on top. Turn to azimuth 201.8°, and then proceed in a straight line for 538 ft 6 in. Dig, and you will find the treasure buried 3 ft below the surface."
 (e) "Find the fencepost with the plastic owl on top. Turn to azimuth 111.8°, and then proceed in a straight line for 538 ft 6 in. Dig, and you will find the treasure buried 3 ft below the surface."

86. Suppose you invest $100.00 in Company X stock on January 10 of year 1. On January 10 of year 2, your investment is worth $150.00, representing an increase of

50% for the year. On January 10 of year 3, the stock has dropped by 50% from its value on January 10 of year 2. What is the stock worth on January 10 of year 3?

(a) $50.00
(b) $75.00
(c) $100.00
(d) $125.00
(e) More information is needed to answer this question.

87. Consider the following function:

$$y = -2x^2 - 12x - 13$$

At what value of x, if any, does the graph of this function reach a local minimum?

(a) $x = 3$
(b) $x = -3$
(c) $x = 0$
(d) $x = 14$
(e) This function does not reach a local minimum for any value of x.

88. Suppose you are confronted with the following equation:

$$u - 13u^{1/2} + 36 = 0$$

Which of the following "longhand" tactics can be used to quickly solve this equation?

(a) Consider v such that $u = v^2$. Substitute v^2 in the equation for u. Once you have solved the resulting quadratic for v, you can solve for u by squaring the values you got for v.
(b) Multiply the whole equation through by u. This will give you an equation that can be solved by using the quadratic formula, letting $a = 1$, $b = -13$, and $c = 36$.
(c) Divide the whole equation through by u. This will give you a linear equation in u that can be solved directly.
(d) Square both sides of the equation. This will give you a quadratic in u that can be solved by using the quadratic formula or by factorizing.
(e) There isn't anything you can do to solve this equation "by hand" in a reasonable time. You will have to use a computer program to do it.

89. The product of two complex conjugates is always

(a) Pure real
(b) Pure imaginary
(c) Equal to 0
(d) Equal to 1
(e) Complex, but not necessarily pure real or pure imaginary

90. When a spring oscillates with a constant mass attached, its back-and-forth excursion gradually decreases because of friction and other imperfections in the real world. What happens to the frequency of the oscillation?

(a) It decreases in direct proportion to the time that has passed after the instant the oscillation began.

(b) It decreases in proportion to the square root of the time that has passed after the instant the oscillation began.

(c) It increases in direct proportion to the time that has passed after the instant the oscillation began.

(d) It increases in direct proportion to the square root of the time that has passed after the instant the oscillation began.

(e) It remains constant.

91. Refer to Fig. E-9. Suppose points P, Q, R and the coordinate origin lie at the vertices of a perfect parallelogram. In that case, how is vector **c** related to vectors **a** and **b**?

(a) $\mathbf{c} = \mathbf{a} + \mathbf{b}$

(b) $\mathbf{c} = \mathbf{a} - \mathbf{b}$

(c) $\mathbf{c} = \mathbf{a} \cdot \mathbf{b}$

(d) $\mathbf{c} = \mathbf{a} \times \mathbf{b}$

(e) $\mathbf{c} = \mathbf{ab}$

92. Fill in the blank to make the following sentence true: Any triangle except a right triangle can be divided into three _____ by constructing perpendicular lines from the midpoints of the sides.

(a) Right triangles

(b) Equilateral triangles

(c) Obtuse triangles

(d) Isosceles triangles

(e) Irregular triangles

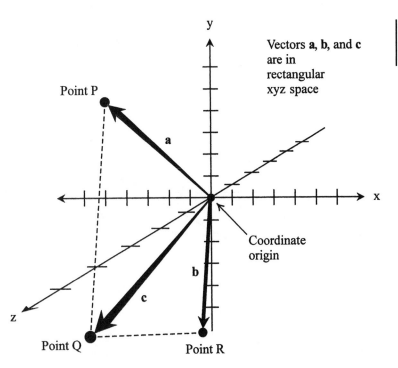

Figure E-9

Illustration for Final Exam question 91.

93. Consider the following function:

$$y = 3 \sin x + 5$$

At what value of x, if any, does the graph of this function have an inflection point?
(a) $x = 0$
(b) $x = \pi/4$
(c) $x = \pi/2$
(d) $x = 3\pi/2$
(e) This function does not have an inflection for any value of x.

94. In the dimensionally reduced illustration Fig. E-10 showing the earth's path through time-space, each vertical division represents 1/4 of a year, or approximately 91.3 days. Suppose the vertical scale is changed so that the pitch of the helix becomes only one-half as great (as if it were a spring compressed by a factor of 2). Further suppose that the horizontal scales remain unchanged. Then each vertical division represents approximately
(a) 22.8 days
(b) 45.7 days
(c) 91.3 days
(d) 183 days
(e) 365 days

Figure E-10

Illustration for Final
Exam questions
94 and 95.

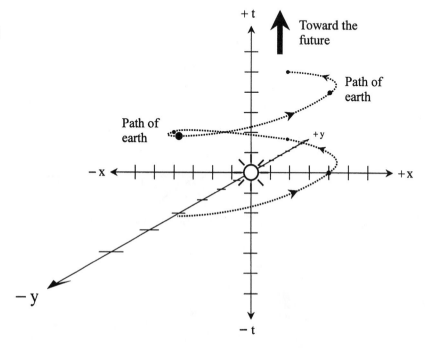

95. In the situation shown by Fig. E-10, suppose the x and y scales are both changed so that the radius of the helix becomes twice as great. Further suppose that the vertical scale remains unchanged. Then each vertical division represents approximately
(*a*) 22.8 days
(*b*) 45.7 days
(*c*) 91.3 days
(*d*) 183 days
(*e*) 365 days

96. Suppose you divide a large whole number M by a smaller whole number m "by hand," using long division. You end up with a whole number Q along with a whole-number remainder of q. That is,

$$M \div m = Q \text{ with a remainder of } q$$

Another way of stating the quotient is
(*a*) $Q + q/m$
(*b*) $Q + q/M$
(*c*) $Q + m/q$
(*d*) $Q + M/q$
(*e*) $Q + M/Q$

97. Which of the following equations in mathematician's polar coordinates represents the same graph as the equation $x^2 + y^2 = 16$ in the Cartesian plane?
(*a*) $r^2 + \theta^2 = 16 + \pi/4$
(*b*) $r = 16$
(*c*) $r = 4$
(*d*) $\theta = \pi/4$
(*e*) $\theta = \pi/16$

98. A computer has a list price of $699.00. A local store offers a mail-in rebate of $59.00 on this item. Assuming you get the rebate, how much will the computer have cost you? Assume there is no tax on any of the transactions.
(*a*) More information is needed to answer this.
(*b*) $758.00
(*c*) $699.00
(*d*) $659.00
(*e*) $640.00

99. Suppose you are given two functions of a variable x. The functions are called f and g. Suppose you want to find the derivative $(f + g)'$. Which of the following formulas tells you how to do that?
(*a*) $(f + g)' = (f + g')(g + f')$
(*b*) $(f + g)' = (f' - g')(g - f')$
(*c*) $(f + g)' = (f' + g')(f' - g')$
(*d*) $(f + g)' = f' + g'$
(*e*) None of the above

100. What is the area in square feet (ft^2) of a rhombus with four sides each 10 feet long and a height of 5 feet, as measured squarely between its base and its top?

    (*a*) 100 ft^2

    (*b*) 50 ft^2

    (*c*) 40 ft^2

    (*d*) 25 ft^2

    (*e*) More information is needed to answer this.

Solution processes are offered for some of the problems answered here. When such a process is explained, it does not necessarily represent the only way the problem can be solved. One or more alternative methods might exist that are easier or more straight-forward for you.

CHAP. 1

1. The final answer is the same regardless of whether counting is done one by one, in groups of ten, or in groups of twelve.

2. Large numbers are counted in groups because it is faster than counting one by one.

3. The numeral zero serves as a placeholder. If the zeros were omitted from a large numeric representation containing zeros, the sequence of digits would no longer represent the same number.

4. The answers are
   (*a*) 10 tens are 100 (one hundred).
   (*b*) 12 twelves are 144 (twelve dozen or one gross).

5. The answers are
   (*a*) 10 hundreds are 1,000 (one thousand).
   (*b*) 10 thousands are 10,000 (ten thousand).
   (*c*) 1000 thousands are 1,000,000 (one million).

6. The answers are
   (*a*) 18   (*b*) 16   (*c*) 12
   (*d*) 18   (*e*) 6   (*f*) 8
   (*g*) 21   (*h*) 24   (*i*) 16

7. The answers are
   (*a*) 130,912   (*b*) 163,123   (*c*) 17,505
   (*d*) 157,549   (*e*) 92,342

8. There is no difference, except that you must always remember to place a decimal point (period) between the second and third digits from the right.

9. The total weight is exactly 10 pounds.

10. To balance 1 pound, 6 ounces, and 14 drams, you would use these weights:
    1 pound
    4 ounces

2 ounces
8 drams
4 drams
2 drams

To balance 2 pounds, 13 ounces, and 11 drams, you would use these weights:
2 pounds
8 ounces
4 ounces
1 ounce
8 drams
2 drams
1 dram

To balance 5 pounds, 11 ounces, and 7 drams, you would use these weights:
4 pounds
1 pound
8 ounces
2 ounces
1 ounce
4 drams
2 drams
1 dram

11. The solutions are as follows.
  (*a*) You would experiment with the ounce weights until the presence or absence of a single ounce weight made the difference between the balance tipping and not tipping. Then you would do the same with the dram weights, until the presence or absence of a single dram weight made the difference between the balance tipping and not tipping. Finally, you would consider the weight of the parcel to be the heavier of the two.
  (*b*) You would experiment with the ounce weights until the presence or absence of a single ounce weight made the difference between the balance tipping and not tipping. You would consider the weight of the parcel to be the heavier of the two.

12. There are 12 inches in a foot, done three times over to get a yard, so $12 + 12 + 12 = 36$ inches. If you do a yard twice over, you get $36 + 36 = 72$ inches. So there are 72 inches in 2 yards.

13. The car owner needs 10 gallons and 3 quarts in all (43 quarts).

14. In general, it is more economical to buy by the gallon. A gallon is 4 quarts. Thus the price is $3.60 per gallon (4 times 90 cents) if the owner buys by the quart, but $3.50 per gallon (given) if bought by the gallon. In the scenario of Prob. 13, the cost of 43 quarts (purchased by the quart) is $38.70, while the cost of 11 gallons (purchased by the gallon) is $38.50. It is still more economical to buy the oil by the gallon in this case.

15. The woman spent $47.51 altogether.

CHAP. 2

1. The answers are

   (*a*) 23   (*b*) 42   (*c*) 506
   (*d*) 391   (*e*) 99   (*f*) 189
   (*g*) 3,087 (*h*) 4,176 (*i*) 2,889

2. You would pay $659.95 − $160.00 = $499.95.

3. The clerk's method yields a total price of $35.00, while the lady's method yields $31.00. She would therefore save $35.00 − $31.00 = $4.00.

4. The cat weighs 93 − 85 = 8 pounds.

5. Place the 4-ounce weight on one pan of the scale, and the 2-pound weight on the other pan. This is a difference in weight of 1 pound and 12 ounces (remember there are 16 ounces in a pound). Then add rice to the scale pan containing the 4-ounce weight, leaving the weight there until the scale balances. The weight of the rice in the pan will then be equal to the difference between the weights, or 1 pound and 12 ounces.

6. The distance between B and C is 293 − 147 = 146 miles, assuming you stay on the road connecting the three towns.

7. The charge is based on the direct distance between towns A and B. Assuming all three towns A, B, and C lie along a straight line, the charge will be based on the difference 1,200 − 250 = 950 miles.

8. The total frontage sold is 1,460 yards. That means there are 1,760 − 1,460 = 300 yards left to sell.

CHAP. 3

1. The answers are

   (*a*) 87,822 (*b*) 216,513 (*c*) 864
   (*d*) 2,738   (*e*) 147,452 (*f*) 49,181

2. The answers are

   (*a*) 48,887 (*b*) 215,404 (*c*) 80,946
   (*d*) 318,364 (*e*) 632,701

3. The answers are

   (*a*) 79,290 (*b*) 238,592 (*c*) 176,064
   (*d*) 574,749 (*e*) 28,672

4. The total number of flights in a week is (4 × 6) + 2 = 24 + 2 = 26.

5. The total number of flights in a year is (26 × 52) − 12 = 1,352 − 12 = 1,340.

6. The answers are

   (*a*) $2.25   (*b*) $4.50   (*c*) $8.25
   (*d*) $27.00   (*e*) $64.50   (*f*) $252.00
   (*g*) $1,252.00 (*h*) $2,502.00

7. If the items cost nothing to manufacture, then packing 500 items in sets of 250 would result in two containers costing $2.00 each to pack, or a total of $2 \times \$2.00 = \$4.00$. Packing the items in sets of 100 would result in five containers costing $2.00 each to pack, or a total of $5 \times \$2.00 = \$10.00$. Therefore, by packing in containers of 250 rather than containers of 100, the savings is $\$10.00 - \$4.00 = \$6.00$.

Interestingly, the savings is the same if the cost of manufacturing each item is taken into account. In that situation, packing the items in two containers of 250 items each would cost

$$[(250 \times \$0.25) + \$2.00] \times 2 = (\$62.50 + \$2.00) \times 2$$
$$= \$64.50 \times 2$$
$$= \$129.00$$

and packing the items in five containers of 100 items each would cost

$$[(100 \times \$0.25) + \$2.00] \times 5 = (\$25.00 + \$2.00) \times 5$$
$$= \$27.00 \times 5$$
$$= \$135.00$$

The savings would therefore be $\$135.00 - \$129.00 = \$6.00$.

8. Ten single tickets cost $\$1.75 \times 10 = \$17.50$. A 10-trip ticket costs $15.75. By purchasing a 10-trip ticket, you save $\$17.50 - \$15.75 = \$1.75$.

9. Taking 22 round trips (which is $22 \times 2 = 44$ one-way trips) with single tickets costs $\$1.75 \times 44 = \$77.00$. A monthly pass costs $55.00. The savings is therefore $\$77.00 - \$55.00 = \$22.00$.

10. Under the employer's plan, an employee would start at $500 a month and get five $50 raises, so at the beginning of the sixth year, the pay would be

$$\$500 + (\$50 \times 5) = \$500 + \$250$$
$$= \$750 \text{ per month}$$

Under the employees' plan, an employee would start at $550 a month and get ten $20 raises, so at the beginning of the sixth year, the pay would be

$$\$550 + (\$20 \times 10) = \$550 + \$200$$
$$= \$750 \text{ per month}$$

which is the same rate as under the employer's plan. Under the employer's plan, an employee would make the following amounts:

$$\$500 \times 12 = \$6,000 \text{ during first year}$$
$$\$550 \times 12 = \$6,600 \text{ during second year}$$
$$\$600 \times 12 = \$7,200 \text{ during third year}$$
$$\$650 \times 12 = \$7,800 \text{ during fourth year}$$
$$\$700 \times 12 = \$8,400 \text{ during fifth year}$$

for a total of $36,000 during the first five years. Under the employees' plan, an employee would make the following amounts:

$550 × 6 = $3,300 during first six months

$570 × 6 = $3,420 during second six months

$590 × 6 = $3,540 during third six months

$610 × 6 = $3,660 during fourth six months

$630 × 6 = $3,780 during fifth six months

$650 × 6 = $3,900 during sixth six months

$670 × 6 = $4,020 during seventh six months

$690 × 6 = $4,140 during eighth six months

$710 × 6 = $4,260 during ninth six months

$730 × 6 = $4,380 during tenth six months

for a total of $38,400 during the first five years. The employees' plan results in more total earnings over five years; the difference is $38,400 − $36,000 = $2,400.

11. If you buy 4,500 items, you must buy 4 lots of 1,000 and one lot of 500. That is a total cost of

$$(4 \times 1,000 \times \$6.25) + (500 \times \$6.75) = \$25,000.00 + \$3,375.00$$

$$= \$28,375.00$$

If you buy 5,000 items, the total cost is

$$5,000 \times \$5.75 = \$28,750.00$$

The difference is therefore $28,750.00 × $28,375.00 = $375.00.

12. You need 30 sets of 100 parts to get 3,000 parts (30 × 100 = 3,000). The total weight is therefore 30 × 2.5 = 75 ounces. This is 4 pounds 11 ounces.

13. The tank's capacity is 4 × 350 = 1,400 gallons.

14. The locomotive must haul 182 × 38 = 6,916 tons. This does not, of course, include the weight of the locomotive itself.

15. The journey was 27 × 260 = 7,020 miles. (The driver would certainly be ready for a good rest at the end of such a trip.)

16. The cost of a trip on the first railroad is 450 × $0.10 = $45.00. The cost of a trip on the second railroad is 320 × $0.15 = $48.00. The difference is therefore $48.00 − $45.00 = $3.00 per one-way trip. The first railroad is cheaper than the second (although the trip will probably take longer because the mileage is greater).

17. A one-way trip on the first airline costs $0.14 × 2,400 = $336.00. A one-way trip on the second airline costs $0.10 × 3,200 = $320.00. The second airline is cheaper by an amount of $336.00 − $320.00 = $16.00 per one-way trip.

18. The answers are determined as follows.

 (*a*) For a family of two on the first airline, a one-way trip will cost

$$(\$0.14 + \$0.09) \times 2,400 = \$0.23 \times 2,400$$
$$= \$552.00$$

On the second airline it will cost

$$(\$0.10 + \$0.09) \times 3,200 = \$0.19 \times 3,200$$
$$= \$608.00$$

Therefore, the first airline is cheaper by $608.00 − $552.00 = $56.00.

 (*b*) For a family of three on the first airline, a one-way trip will cost

$$(\$0.14 + \$0.09 + \$0.09) \times 2,400$$
$$= \$0.32 \times 2,400 = \$768.00$$

On the second airline it will cost

$$(\$0.10 + \$0.09 + \$0.09) \times 3,200 = \$0.28 \times 3,200$$
$$= \$896.00$$

Therefore, the first airline is cheaper by $896.00 − $768.00 = $128.00.

19. The total number of chews is $50 \times 7 \times 3 = 1,050$ chews.

20. There are $9 \times 7 = 63$ flowers around the edge of the plate.

CHAP. 4

1. The answers are

 (*a*) 49 (*b*) 81 (*c*) 616
 (*d*) 653 (*e*) 2,081 (*f*) 3,931
 (*g*) 984 (*h*) 654 (*i*) 237
 (*j*) 1,484 (*k*) 379 (*l*) 439

2. The answers are

 (*a*) 240 (*b*) 753
 (*c*) 2,945 (*d*) 362

3. With the remainders (when they exist) expressed as fractions and reduced to lowest terms, the answers are

 (*a*) 494-1/7 (*b*) 2,928-7/8 (*c*) 1,473-2/3
 (*d*) 4,791   (*e*) 4,889-1/3 (*f*) 484-11/17
 (*g*) 1,226-15/28 (*h*) 3,204-13/29

4. The profit per share is equal to $14,000,000.00 divided by 2,800,000, or $5.00.

5. The fare per passenger should be $8,415.00 divided by 55, or $153.00.

6. The tool costs $5,000.00, and there are 10,000 parts to be made. Each part has a gross manufacturing cost of $5,000.00 divided by 10,000, or $0.50. Parts are made for $0.25 each; therefore the cost of each part is $0.50 − $0.25 = $0.25.

7. The total weight is 58×2000 pounds $= 116,000$ pounds. Each wheel carries $116,000/8$ pounds $= 14,500$ pounds.

8. The total time, 8 hours, is equal to $8 \times 60 \times 60$ seconds $= 28,800$ seconds. There are 1,200 parts. Therefore, the time required to make each part is $28,800/1,200 = 24$ seconds.

9. The full package weighs $1,565 \times 16 = 25,040$ ounces. The empty package weighs $2.5 \times 16 = 40$ ounces. The total weight of the parts is $25,040 - 40 = 25,000$ ounces. The weight of each part is therefore 25,000 ounces divided by 10,000, or 2-1/2 ounces.

10. The full package weighs $2,960 \times 16 = 47,360$ ounces. The empty package weighs $5 \times 16 = 80$ ounces. The total weight of the parts is $47,360 - 80 = 47,280$ ounces. The number of parts in the package is therefore 47,280 ounces divided by 3 ounces, or 15,760.

11. One mile is 5,280 feet. This is to be divided into 33 equal parts. The width of each lot is therefore $5280/33 = 160$ feet.

12. On each gallon of gas, the car travels $462/22 = 21$ miles.

13. The quantities must all be divided by 160 to get the amounts needed for 1 gallon. First, convert all figures to fluid ounces. There are 128 fluid ounces to the gallon. To make 1 gallon of the mixture, therefore, you must have

$$(75 \times 128)/160 = 60 \text{ fluid ounces of ingredient 1}$$
$$(50 \times 128)/160 = 40 \text{ fluid ounces of ingredient 2}$$
$$(25 \times 128)/160 = 20 \text{ fluid ounces of ingredient 3}$$
$$(10 \times 128)/160 = 8 \text{ fluid ounces of ingredient 4}$$

14. The answers are
    (*a*) 1/8   (*b*) 7/10
    (*c*) 3/8   (*d*) 19/20

15. The quickest way to get the answers here is to use the law of the nines.
    (*a*) 416/999   (*b*) 21/99
    (*c*) 189/999   (*d*) 4,892/9,999

CHAP. 5

1. These sets of fractions have the same value:
1/2, 3/6, 4/8, 9/18, and 10/20
1/3, 3/9, 4/12, 5/15, 6/18, and 7/21
2/5, 4/10, 6/15, and 8/20
2/3, 4/6, and 8/12
3/4, 9/12, and 15/20

2. The answers, in order, are as follows. The "complicated" original version is listed to the left of each equals sign, and the simplest form is listed to the right:

$$7/14 = 1/2$$
$$26/91 = 2/7$$

$$21/91 = 3/13$$
$$52/78 = 2/3$$
$$39/65 = 3/5$$
$$22/30 = 11/15$$
$$39/51 = 13/17$$
$$52/64 = 13/16$$
$$34/51 = 2/3$$
$$27/81 = 1/3$$
$$18/45 = 2/5$$
$$57/69 = 19/23$$

3. The answers are
   (a) 10,452 divides by 3 and 4
   (b) 2,088 divides by 3, 4, 8, and 9
   (c) 5,841 divides by 3, 9, and 11
   (d) 41,613 divides by 3 and 11
   (e) 64,572 divides by 3 and 4
   (f) 37,848 divides by 3, 4, and 8

4. The answers are
   (a) Prime factors of 1,829 are 31 and 59.
   (b) Prime factors of 1,517 are 37 and 41.
   (c) Prime factors of 7,387 are 83 and 89.
   (d) Prime factors of 7,031 are 79 and 89.
   (e) Prime factors of 2,059 are 29 and 71.
   (f) Prime factors of 2,491 are 47 and 53.

5. The answers are
   (a) 48/30, reducible to 8/5, properly written as 1-3/5
   (b) 127/72, properly written as 1-55/72
   (c) 48/60, reducible to 4/5
   (d) 262/84, reducible to 131/42, properly written as 3-5/42

6. The answers are
   (a) 7/8   (b) 3/5   (c) 9/16
   (d) 741/1000   (e) 16/125

7. The answers are
   (a) 0.666666 . . .   (b) 0.75
   (c) 0.8   (d) 0.833333 . . .
   (e) 0.857142857142 . . . (f) 0.875
   (g) 0.888888 . . .

8. The answers are
   (a) 0.333333 . . .   (b) 0.25
   (c) 0.2   (d) 0.166666 . . .
   (e) 0.142857142857 . . . (f) 0.125
   (g) 0.111111 . . .

9. The answers are
   (a) 416/999
   (b) 21/99, reducible to 7/33
   (c) 189/999, reducible to 7/37
   (d) 571,428/999,999, reducible to 52/91
   (e) 909/999, reducible to 101/111
   (f) 90/999, reducible to 10/111

10. If a measurement is specified as 158 feet, the actual distance is at least 157.5 feet, but less than 158.5 feet. If a measurement is specified as 857 feet, the actual distance is at least 856.5 feet, but less than 857.5 feet. The margin for error is equal to a span of plus or minus one-half a foot (6 inches) in either case.

11. The quotient 932/173, rounded off to three significant digits, is equal to 5.39. Using a calculator capable of expressing eight digits in the display gives

$$932.5/172.5 = 5.4057971$$
$$931.5/173.5 = 5.3688760$$

The only digit that remains the same over this span is the first one, that is, the numeral 5 to the left of the decimal point. That seems to suggest that this initial 5 is the only digit we can take seriously! However, according to the rules for significant figures, the quotient of 5.39 is accurate that far out, because we are given both the numerator and the denominator to three significant digits.

12. The quotient 93,700/857, rounded off to three significant digits, is equal to 109. Using a calculator capable of expressing eight digits in the display gives

$$93,750/856.5 = 109.45709$$
$$93,650/857.5 = 109.21282$$

The left-hand three digits, 109, are the same over this span. Therefore, in a practical experiment involving this quotient of measured values, one could justify the result to three significant digits:109. The best method of shortening the process and making it less "messy" is to use a calculator (or better yet, a computer programmed to determine significant digits).

13. The answers are determined as follows.
   (a) The prime numbers less than 60 are 2, 3, 7, 11, 13, 17, 19, 23, 29, 31, 37, 41, 43, 47, 53, and 59.
   (b) To test a number to see if it is prime, first take its square root. Then, if the square root is not a whole number, round it up to the next whole number. Call this whole number n. Divide the original (tested) number by every prime starting with 2 and working up the list until you reach n. If none of the quotients is a whole number, then the tested number is prime. If any of the quotients is a whole number, then the tested number is not prime.
   (c) The largest number you can test using this list is $60 \times 60 = 3,600$.

14. The answers are
   (a) 11/16   (b) 64/91   (c) 17/40
   (d) 2-9/20   (e) 1-11/34

CHAP. 6

1. The answers are
    (a) 54 × 78 = 4,212 square inches
    (b) 13 × 17 = 221 square feet
    (c) 250 × 350 = 87,500 square yards
    (d) 3 × 7 = 21 square miles
    (e) 17 × (5 × 12) = 1,020 square inches
    (f) 340 × (1 × 1,760) = 598,400 square yards

2. A right angle measures 1/4 of a circle. It is so named because it represents the best ("right") angle for attaching the legs to a table top.

3. The answers are
    (a) 5 × 1 = 5 square feet
    (b) 5 × 5 = 25 square feet

4. The answers are
    (a) (5 × 6)/2 = 15 square inches
    (b) (12 × 13)/2 = 78 square feet
    (c) (20 × 30)/2 = 300 square yards
    (d) (3 × 4)/2 = 6 square miles
    (e) (20 × 2 × 12)/2 = 240 square inches
    (f) (750 × 1 × 1,760)/2 = 660,000 square yards

5. The area is 220 × 110 = 24,200 square yards, or 24,200/4,840 = 5 acres.

6. The area of a parallelogram is equal to the length of the base times the height. Using a 20-inch side as the base and 12 inches as the height, the area is 20 × 12 = 240 square inches. Using a 15-inch side as the base, the height (straight-across distance between the 15-inch sides) must be such that, when multiplied by 15, the result is 240 (the area). This height is equal to 240/15 = 16 inches.

7. Assuming the figures given are exact, the answers are
    (a) (11 × 16)/2 = 88 square inches
    (b) (31 × 43)/2 = 666.5 square inches
    (c) (27 × 37)/2 = 499.5 square inches

8. The area of a triangle is equal to one-half the length of the base times the height. In this case, it is (39 × 48)/2 = 936 square inches. The product of the base and the height is 1,872 (twice the area). This quantity is a constant, no matter which side of the triangle is considered the base. Therefore, if the 52-inch side is used as the base, the height must be 1,872/52 = 36 inches.

9. The property can be divided into two parts: a rectangle measuring 200 by 106.5 yards and a triangle with base length 200 yards and height 256.5 − 106.5 = 150 yards. The area of the rectangle is 200 × 106.5 = 21,300 square yards. The area of the triangle is (200 × 150)/2 = 15,000 square yards. The total area is 21,300 + 15,000 = 36,300 square yards. An acre is 4,840 square yards, so the area of the property is 36,300/4,840 = 7.5 acres.

10. Assuming the property is rectangular, the total area is $300 \times 440 = 132,000$ square yards. The owner wants to keep a piece that measures $110 \times 44 = 4,840$ square yards. Therefore, the area he wants to sell is $132,000 - 4,840 = 127,160$ square yards.

11. Convert all lengths to inches. Then consider the outside rectangle that measures 207 by 216 inches. Its area is $207 \times 216 = 44,712$ square inches. From this, subtract two rectangles, one measuring 81 by 84 inches and the other measuring 24 by 60 inches. The areas of these rectangles are, respectively, 6,804 and 1,440 inches. The floor area of the room is therefore $44,712 - (6,804 + 1,440) = 44,712 - 8,244 = 36,468$ square inches. You might want to divide by 144 to obtain the answer as 253.25 square feet.

12. Consider the walls as a set of rectangles, each of which is 7 feet (84 inches) high. Then starting at the upper left-hand corner and proceeding clockwise, the areas of the rectangles are

$$84 \times 84 = 7,056 \text{ square inches}$$
$$24 \times 84 = 2,016 \text{ square inches}$$
$$60 \times 84 = 5,040 \text{ square inches}$$
$$24 \times 84 = 2,016 \text{ square inches}$$
$$63 \times 84 = 5,292 \text{ square inches}$$
$$54 \times 84 = 4,536 \text{ square inches}$$
$$\text{Window} = 0 \text{ square inches}$$
$$54 \times 84 = 4,536 \text{ square inches}$$
$$27 \times 84 = 2,268 \text{ square inches}$$
$$\text{Window} = 0 \text{ square inches}$$
$$27 \times 84 = 2,268 \text{ square inches}$$
$$84 \times 84 = 7,056 \text{ square inches}$$
$$81 \times 84 = 6,804 \text{ square inches}$$
$$48 \times 84 = 4,032 \text{ square inches}$$
$$\text{Door} = 0 \text{ square inches}$$
$$48 \times 84 = 4,032 \text{ square inches}$$

The surface area of the walls is the sum of all these areas, or 56,952 square inches. You might wish to divide this figure by 144 to obtain the answer as 395.5 square feet.

CHAP. 7

1. In an hour, a car going 35 mph will go 35 miles, and 36 minutes is 3/5 of an hour. Therefore, in 36 minutes, if the car travels at a constant speed, it will go $35 \times 3/5 = 21$ miles.

2. The answers are
   (*a*) Upstream speed is $10 - 2 = 8$ mph.
   (*b*) Downstream speed is $10 + 2 = 12$ mph.

3. The answers are
   (*a*) To go 96 miles upstream takes 96/8 = 12 hours.
   (*b*) To go 96 miles downstream takes 96/12 = 8 hours.

4. The answers are
   (*a*) The upstream journey takes 12 hours. In this time, at a constant water speed of 10 mph, the riverboat burns 12 × 1/2 = 6 tons of fuel.
   (*b*) The downstream journey takes 8 hours. In this time, at a constant water speed of 10 mph, the riverboat burns 8 × 1/2 = 4 tons of fuel.

5. To make the downstream journey at a land speed of 8 mph, the water speed would have to be cut to 6 mph, or 60 percent (0.6) of its former speed. This would result in the boat burning 1/2 × 0.6 = 0.3 ton of fuel per hour. The journey of 96 miles at 8 mph would take 96/8 = 12 hours, which would burn 12 × 0.3 = 3.6 tons of fuel. That is 0.4 ton less than before. The speed reduction would save fuel. The fuel savings for the downstream journey would be 0.4/4 = 10 percent. Without the speed reduction, the round-trip fuel burned is 10 tons. With the downstream water speed reduction, 0.4 ton is saved going downstream but nothing is saved going upstream (assuming the boat operates in the same way as before during that part of the journey). The total fuel savings for the round trip is therefore 0.4/10 = 4 percent.

6. Suppose the water speed were cut by 2 mph going upstream, from 10 mph to 8 mph. Then the land speed would be reduced to 6 mph, and it would take the boat 96/6 = 16 hours to make the upstream journey. Assuming the fuel consumption decreased in proportion to the water speed reduction, there would be 0.5 × 0.8 = 0.4 ton burned per hour. In 16 hours the ship would burn 16 × 0.4 = 6.4 tons of fuel. Therefore, if the water speed upstream were reduced by 2 mph, the total amount of fuel burned would *increase* from 6 tons to 6.4 tons! This is a change of 100 × 0.4/6 = 6.667 (or 6-2/3) percent.

7. The value increases by $10,000, and the dividend paid on the initial investment is $50,000 × 0.05 = $2,500. Therefore, the total amount made in cash is $10,000 + $2,500 = $12,500. This is $12,500/$50,000 = 0.25, or 25 percent.

8. The percentages of growth break down like this:

First week = (24 − 16)/16 = 8/16 = 0.5 = 50 percent

Second week = (36 − 24)/24 = 12/24 = 0.5 = 50 percent

Third week = (54 − 36)/36 = 18/36 = 0.5 = 50 percent

Fourth week = (81 − 54)/54 = 27/54 = 0.5 = 50 percent

Entire four-week period = (81 − 16)/16 = 65/16 = 4.0625 = 406.25 percent

9. See Fig. A-1. In the middle of the second week, the height is 30 inches as interpolated from the graph.

10. At 90 mph, a car travels 3 miles in 3/90, or 1/30, of an hour, which is 2 minutes. For the whole lap, this car takes 6 + 2 = 8 minutes. That translates to an average speed of 8 × 60/8 = 60 mph. At 120 mph, a car travels 3 miles in 3/120, or 1/40, of an hour; that's 1.5 minutes. For the whole lap, this car takes 6 + 1.5 = 7.5 minutes. That translates to an average speed of 8 × 60/7.5 = 64 mph.

11. At 32 miles per gallon, the car will travel 18 × 32 = 576 miles on 18 gallons.

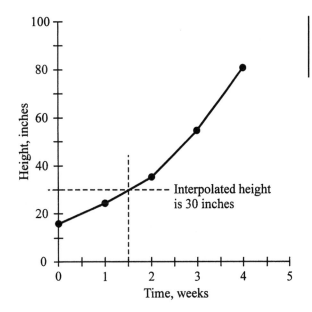

12. Using the first method:

$$100 \text{ boards cost } \$2{,}000 + (\$0.15 \times 100) = \$2{,}015$$
$$500 \text{ boards cost } \$2{,}000 + (\$0.15 \times 500) = \$2{,}075$$
$$1{,}000 \text{ boards cost } \$2{,}000 + (\$0.15 \times 1{,}000) = \$2{,}150$$
$$2{,}000 \text{ boards cost } \$2{,}000 + (\$0.15 \times 2{,}000) = \$2{,}300$$
$$5{,}000 \text{ boards cost } \$2{,}000 + (\$0.15 \times 5{,}000) = \$2{,}750$$
$$10{,}000 \text{ boards cost } \$2{,}000 + (\$0.15 \times 10{,}000) = \$3{,}500$$

Using the second method:

$$100 \text{ boards cost } \$200 + (\$0.65 \times 100) = \$265$$
$$500 \text{ boards cost } \$200 + (\$0.65 \times 500) = \$525$$
$$1{,}000 \text{ boards cost } \$200 + (\$0.65 \times 1{,}000) = \$850$$
$$2{,}000 \text{ boards cost } \$200 + (\$0.65 \times 2{,}000) = \$1{,}500$$
$$5{,}000 \text{ boards cost } \$200 + (\$0.65 \times 5{,}000) = \$3{,}450$$
$$10{,}000 \text{ boards cost } \$200 + (\$0.65 \times 10{,}000) = \$6{,}700$$

13. See Fig. A-2. The figures can only be approximated. Finding the exact answer requires algebra.

14. At 28 mpg, a trip of 594 miles burns $594/28 = 21.21$ gallons of gas (to the nearest hundredth of a gallon). At 24 mpg, the same trip burns $594/24 = 24.75$ gallons. Therefore, by going at the lower speed, the gas savings is $24.75 - 21.21 = 3.54$ gallons. At 40 mph, a trip of 594 miles requires $594/40 = 14.85$ hours. At 60 mph, the same journey will take $594/60 = 9.9$ hours. Therefore, by going at the slower speed, the extra time required is 4.95 hours, or 4 hours and 57 minutes.

Figure A-2

Illustration for solution to Prob. 13 in Chap. 7.

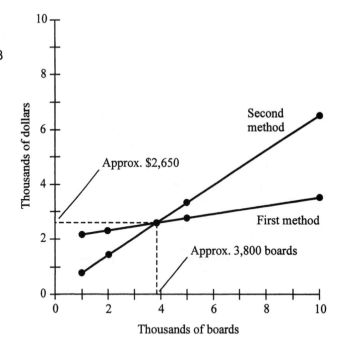

15. At the end of the first year, the property value rises by $200,000 × 25% = $200,000 × 0.25 = $50,000, so it appreciates to a value of $200,000 + $50,000 = $250,000. During the next year it decreases by $250,000 × 10% = $250,000 × 0.1 = $25,000, so it depreciates to a value of $250,000 − $25,000 = $225,000. If the owner sells at that time, he'll make a cash profit of $225,000 − $200,000 = $25,000. Based on the original value of $200,000, that is $25,000/$200,000 = 0.125 = 12.5%. You can't simply subtract the two different percentage changes to calculate the final percentage change, because the percentages are not based on the same amount.

16. The total tax is

$$($30,000 \times 20\%) + [($120,000 \times $30,000) \times 22\%]$$
$$= ($30,000 \times 0.2) + ($90,000 \times 0.22)$$
$$= $6,000 + $19,800$$
$$= $25,800$$

17. The area of the triangle is $(16 \times 12)/2 = 96$ square feet. Therefore, the parallelogram also has an area of 96 square feet. Consider one of the 16-foot sides as the base of the parallelogram. Then the height of the parallelogram (the distance between the 16-foot sides) is equal to $96/16 = 6$ feet.

CHAP. 8

1. The equation is $12 - x = 2x$. To solve, add x to each side, getting $12 = 3x$. Then divide each side by 3, getting $4 = x$.

2. The equation is $19y = y + 36$. Subtract y from each side, getting $18y = 36$. Then divide each side by 18, getting $y = 2$.

3. The problems are solved as follows.

(*a*) The equation is $z/2 + 10 = 20$. Subtract 10 from each side to get $z/2 = 10$. Then multiply each side by 2, so $z = 20$.

(*b*) The equation is $4(z - 7) = 0$. Divide each side through by 4 to get $z - 7 = 0$. Then add 7 to each side, so $z = 7$.

(*c*) The equation is $10/z + 7 = 12$. Subtract 7 from each side, so $10/z = 5$. Multiply each side by z, so $10 = 5z$. Then divide each side by 5, so $2 = z$.

4. The value of xy is 5×7, or 35. It is not 57, because in algebra, xy means x times y. The number 57 is equal to $5 \times 10 + 7 \times 1$, which in this case is $10x + y$. Note that when addition and multiplication operations occur in the same expression without parentheses, you should do all the multiplication operations first, before you do any addition.

5. To simplify, multiply the expressions out, starting with the innermost parts first. This yields the following sums.

(*a*) Proceed by steps in this order:

$$5[3x - 2(5x + 7)] - 9$$
$$5[3x - (10x + 14)] - 9$$
$$5(3x - 10x - 14) - 9$$
$$5(-7x - 14) - 9$$
$$-35x - 70 - 9$$
$$-35x - 79$$

(*b*) Proceed by steps in this order:

$$14 + 2[x + 5(2x + 3)]$$
$$14 + 2(x + 10x + 15)$$
$$14 + 2x + 20x + 30$$
$$2x + 20x + 14 + 30$$
$$22x + 44$$

6. Let the number be x. Then the original expression is

$$3(x + 5) + 4(x + 6) + 50(x + 7)$$

This can be multiplied out to obtain

$$3x + 15 + 4x + 24 + 50x + 350$$

which can be simplified to $57x + 389$.

7. Let the number be x. Then the original expression is

$$5[4(x + 3) + 4x + 3] + 6$$

This can be multiplied out, getting

$$5(4x + 12 + 4x + 3) + 6$$

which can be simplified to

$$5(8x + 15) + 6$$

Multiplying out again gives

$$40x + 75 + 6$$

which can be simplified to $40x + 81$. If the total is 361, then

$$40x + 81 = 361$$

Subtracting 81 from each side of this equation gives

$$40x = 280$$

Then dividing through by 40 yields

$$x = 280/40 = 7$$

8. Let x be the number of full members. Then there are $2,000 - x$ student members. The total annual dues received from full members is $20x$ dollars; the total annual dues received from student members is $8(2,000 - x) = 16,000 - 8x$ dollars. The overall total received is 35,200 dollars. This number is equal to $20x + [8(2,000 - x)]$. We can write the equation

$$20x + (16,000 - 8x) = 35,200$$

Simplifying gives

$$12x + 16,000 = 35,200$$

Subtracting 16,000 from each side, we get

$$12x = 19,200$$

Dividing through by 12 tells us that $x = 1,600$. There are 1,600 full members and $2000 - 1,600 = 400$ student members.

9. Using x for the tens digit, the ones digit is $x + 2$. The problem tells us that $3x = x + 2$. Subtracting x from each side, we get $2x = 2$. Therefore, $x = 1$. The tens digit is 1, and the ones digit is $1 + 2 = 3$, so the number is 13.

10. Letting x represent the tens digit, the ones digit is $x + 1$. The original number is therefore equal to $10x + x + 1$, which is the same as $11x + 1$. Four times this quantity is the new number, which is $4(11x + 1)$, or $44x + 4$. We are further told that the new ones digit is x, and the new tens digit is $3(x + 1)$, or $3x + 3$. The new number is therefore

$$10(3x + 3) + x = 30x + 30 + x$$
$$= 31x + 30$$

Now we know that

$$44x + 4 = 31x + 30$$

Subtracting $31x$ from each side gives

$$13x + 4 = 30$$

Subtracting 4 from each side tells us that

$$13x = 26$$

Dividing through by 13 gives us $x = 2$. Therefore, the original tens digit is 2, and the original ones digit is $2 + 1 = 3$, so the original number is 23.

11. For the original number, let the tens digit be x. Then the ones digit is $x + 1$. The number is therefore equal to $10x + x + 1$, or $11x + 1$. Adding 9 to this yields $11x + 10$. Now suppose that the digits in the original number are reversed. Then the number is equal to $10(x + 1) + x$, or $10x + 10 + x$, which simplifies to $11x + 10$. This is, as we already showed, equal to 9 plus the original number.

12. Let the tens digit of the original number be a. Let the ones digit be $a + x$. Then the original number is equal to $10a + a + x$, or $11a + x$. The difference between the digits is equal to x. Multiplying this by 9 yields $9x$. Adding this to the original number gives us $11a + x + 9x$, or $11a + 10x$. Now consider the original digits being reversed. This number would be $10(a + x) + a$. Multiplying out gives $10a + 10x + a$, which simplifies to $11a + 10x$. This is equal to the original number plus 9 times the difference between the digits, as we showed before.

13. The first expression is meaningless when $x = -1$, because then the denominator is equal to 0, and division by 0 is not defined. The same thing happens in the second expression if $x = -3$. The value of the first expression changes when the value of x changes. But this is not true for the second expression. In this case the numerator is equal to $3(x + 3)$, which is always 3 times the denominator. Therefore, for all values of x (other than -3, for which the quotient is not defined), the second expression is equal to 3.

14. The solution processes are as follows.
   (a) Subtract 5 from each side to get $y - 5 = x$.
   (b) Add 2 to each side, getting $y + 2 = x/3$. Then multiply each side by 3, obtaining $3(y + 2) = x$.
   (c) First, multiply out the right side, getting $y = 6x + 12$. Then subtract 12 from each side, getting $y - 12 = 6x$. Next, divide each side by 6, getting $(y - 12)/6 = x$.
   (d) In this expression, first invert each side, getting $1/y = x - 1$. Then add 1 to each side, getting $1/y + 1 = x$. *Note:* These manipulations require that both x and y be nonzero, and that x never equal 1, so we never end up dividing by 0.
   (e) First, add 7 to each side, getting $y + 7 = 3x$. Then divide each side by 3, getting $(y + 7)/3 = x$.
   (f) First, multiply each side by 3, getting $3y = 5x + 4$. Then subtract 4 from each side to obtain $3y - 4 = 5x$. Next, divide through by 5, getting $(3y - 4)/5 = x$.

15. Each of these three proofs can be done by showing that the "equations" are actually absurd statements because they in effect say that two unequal numbers are equal.
   (a) Multiply out the left-hand side, obtaining $4x - 20 = 4x - 18$. Then add 18 to each side, obtaining $4x + 2 = 4x$. Subtract $4x$ from each side, getting $2 = 0$, which is absurd.

(*b*) Subtract 3 from each side, getting $5x = 5x - 10$. Then subtract $5x$ from each side, getting $0 = -10$, which is absurd.

(*c*) Multiply each side by 4, obtaining $2x + 16 = 2x + 12$. Then subtract $2x$ from each side, getting $16 = 12$, which is absurd.

16. The graphs for (*a*) and (*c*) are straight lines, while the graphs for (*b*) and (*d*) are curves.

17. The sequence of expressions goes like this:

x
$x + 6$
$3(x + 6)$
$3x + 18$
$3x + 6$
$x + 2$
2

You can make up all sorts of problems of this kind by "reverse engineering." Read the above series of expressions from bottom to top, and you'll get the idea!

CHAP. 9

1. The products are as follows. *Warning:* The signs (+ and −) can be tricky!
(*a*) $x^2 - 2x - 3$
(*b*) $x^2 - 8x + 15$
(*c*) $x^2 - 1$
(*d*) $x^3 + 1$

2. The quotients are as follows. *Warning:* The signs (+ and −) can be tricky!
(*a*) $x - 3$
(*b*) $x - 5$
(*c*) $x - y$
(*d*) $x^3 - x + 2$

3. The factors are as follows:
(*a*) $(x + 7)(x - 5)$
(*b*) $(x + 1)(x^2 - 5)$
(*c*) $(x + 3)(x^2 - 2x - 1)$

4. Solutions exist for (*a*), (*b*), and (*c*), but not for (*d*). In the first three cases:
(*a*) $x = 15/2$ and $y = 7/2$
(*b*) $x = 6/25$ and $y = -4/25$
(*c*) $x = -5/2$ and $y = -1$

The last pair of equations, (*d*), implies that x is a number equal to itself plus or minus 1. There exists no such number. Therefore, this pair of simultaneous equations has no solution.

5. Let x represent the length and y represent the width. Then the area of the original rectangle is equal to xy. The following equations hold true according to the information given:

$$(x + 5)(y + 2) = xy + 133$$
$$(x + 8)(y + 3) = xy + 217$$

These equations can be multiplied out to obtain

$$xy + 2x + 5y + 10 = xy + 133$$
$$xy + 3x + 8y + 24 = xy + 217$$

Subtracting $xy + 10$ from each side of the top equation, and $xy + 24$ from each side of the bottom equation, yields

$$2x + 5y = 123$$
$$3x + 8y = 193$$

Multiplying the top equation through by 3 and the bottom equation through by 2 gives

$$6x + 15y = 369$$
$$6x + 16y = 386$$

Subtracting the top equation from the bottom tells us that $y = 17$ feet. "Plugging in" to the top equation previously yields

$$6x + 255 = 369$$

Therefore $6x = 114$, so $x = 19$ feet. The original dimensions are therefore 17 by 19 feet.

6. We are given that $c = 28$, $a = 3$, and $b = 4$. Let the unknown parts be x and y. From the information given, these two simultaneous equations result:

$$x + y = 28$$
$$3x = 4y$$

The first equation can be changed to $y = 28 - x$. Using substitution into the second equation, we get

$$3x = 4(28 - x) = 112 - 4x$$

This can be solved for x, yielding $x = 16$. Therefore, $y = 28 - 16 = 12$.

7. Let the fraction be called x/y. Then, from the information given, we obtain the following two simultaneous equations:

$$(x + 3)/(y + 3) = 4/5$$
$$(x - 4)/(y - 4) = 3/4$$

These can be rearranged to get the following two equations:

$$5x + 15 = 4y + 12$$
$$4x - 16 = 3y - 12$$

Solving these yields $x = 25$ and $y = 32$, so the fraction is 25/32.

8. Let the fraction be called x/y. Then, from the information given, we obtain the following two simultaneous equations:

$$(x + 1)/(y + 1) = 4/7$$
$$(x - 1)/(y - 1) = 5/9$$

These can be rearranged to get the following two equations:

$$7x + 7 = 4y + 4$$
$$9x - 9 = 5y - 5$$

Solving these yields $x = 31$ and $y = 55$, so the fraction is 31/55.

9. Let the fraction be called x/y. Then, from the information given, we obtain the following two simultaneous equations:

$$(x + 1)/(y + 1) = 7/12$$
$$(x - 1)/(y - 1) = 9/16$$

These can be rearranged to get the following two equations:

$$12x + 12 = 7y + 7$$
$$16x - 16 = 9y - 9$$

Solving these yields $x = 23.5$ and $y = 41$, so the fraction is 23.5/41. This is not a fraction in the strict sense; it is merely a quotient or ratio. But it must be in this form to meet the conditions stated in the problem.

10. Let the numbers be x, $x + 1$, $x + 2$, and $x + 3$. Then according to the problem

$$(x + 2)(x + 3) = 90 + x(x + 1)$$

Multiplying out each side of this equation yields

$$x^2 + 5x + 6 = 90 + x^2 + x$$

Subtracting the quantity $(x^2 + x)$ from each side results in this equation:

$$4x + 6 = 90$$

This can be reduced to $4x = 84$, so $x = 21$. The four consecutive numbers are therefore 21, 22, 23, and 24.

11. Let the numbers be x, $x + 1$, $x + 2$, $x + 3$, and $x + 4$. Then according to the problem

$$(x + 1)(x + 2)(x + 3) = 15 + x(x + 2)(x + 4)$$

Multiplying out each side gives

$$x^3 + 6x^2 + 11x + 6 = 15 + x^3 + 6x^2 + 8x$$

Subtracting the quantity $(x^3 + 6x^2)$ from each side gives

$$11x + 6 = 15 + 8x$$

Subtracting $8x - 6$ from each side of this equation gives $3x = 9$. It follows that $x = 3$. Therefore, the numbers are 3, 4, 5, 6, and 7.

12. The man was told the width is x feet, and the length is $x + 50$ feet. This would result in an area of $x(x + 50) = x^2 + 50x$ square feet. However, it turns out that the actual width is $(x - 10)$ feet. The seller offers an extra 10 feet in length, to make the new length $(x + 60)$ feet. This results in a total area of $(x - 10)(x + 60) = x^2 + 50x - 600$ square feet, or 600 square feet less than what he was originally told, no matter what the actual dimensions.

CHAP. 10

1. The solutions are found as follows.
   (*a*) Factors are $(x + 8)(x - 1)$. Solutions are $x = -8$ or $x = 1$.
   (*b*) Factors are $(3x - 13)(x - 1)$. Solutions are $x = 13/3$ or $x = 1$.
   (*c*) Factors are $(7x + 1)(x - 7)$. Solutions are $x = -1/7$ or $x = 7$.
   (*d*) Factors are $(6x - 5)(5x - 8)$. Solutions are $x = 5/6$ or $x = 8/5$.

2. The solutions are as follows, where the 1/2 power denotes the positive square root.
   (*a*) $x = 9$ or $x = -5$
   (*b*) $x = 3$ or $x = -2$
   (*c*) $x = (7 + 21^{1/2})/2$ or $x = (7 - 21^{1/2})/2$
   (*d*) $x = 6 + 40^{1/2}$ or $x = 6 - 40^{1/2}$

3. The solutions are as follows, where the 1/2 power denotes the positive square root. In the second two cases, the equation must be put into standard quadratic form before the formula is used.
   (*a*) $x = 7/5$ or $x = -1$
   (*b*) $x = 1$ or $x = -3/7$
   (*c*) $x = 5$ or $x = -1/5$
   (*d*) $x = 5 + 24^{1/2}$ or $x = 5 - 24^{1/2}$

4. The solutions can be found by using the quadratic formula. The results are as follows. Nothing is particularly unusual about the first two examples; the last one has a single solution that "occurs twice."
   (*a*) $x = 1$ or $x = -2/5$
   (*b*) $x = 0$ or $x = 3/5$
   (*c*) $x = 3/10$

5. Let the unknown quantity be x. Then, according to the problem, $x + 2/x = 4$. This equation can be rearranged to standard quadratic form: $x^2 - 4x + 2 = 0$. The quadratic formula can be used to find the solutions: $x = 2 + 2^{1/2}$ or $x = 2 - 2^{1/2}$.

6. If we let x be the width of the rectangle, then the length is $2x - 10$. The area is 2800 square feet. From this information, we can derive the equation $x(2x - 10) = 2800$, which can be translated to standard quadratic form: $2x^2 - 10x - 2800 = 0$. The quadratic formula yields solutions $x = 40$ feet or $x = -35$ feet. The negative value has no practical meaning. The length of the enclosure is $(2x - 10)$ feet according to the problem. Using the width $x = 40$, this gives us a length of 70 feet. Checking, we find $40 \times 70 = 2800$.

7. Let x be the length of a side of the square. Then the area of this square is equal to x^2 square feet. From the information in the problem, we can derive the equation $(x + 6)^2 = 4x^2$, which translates to standard quadratic form: $-3x^2 + 12x + 36 = 0$. Using the quadratic formula, we obtain the solutions $x = -2$ or $x = 6$. The negative value has no practical meaning. The length of each side of the original square is therefore 6 feet.

8. Let the three numbers be $x - 1$, x, and $x + 1$. Then according to the problem

$$x - 1 + x + x + 1 = (3/8)[x(x - 1)]$$

This can be rearranged into standard quadratic form: $x^2 - 9x = 0$. Using the quadratic formula, we obtain $x = 0$ or $x = 9$. There are two number sequences that fulfill the requirements of the problem: (8,9,10) and (−1,0,1). Both solutions are mathematically valid, although the latter is a bit strange (check it and see!). *Note:* Resist the temptation to divide the equation $x^2 - 9x = 0$ through by x to get $x - 9 = 0$. This finds the solution $x = 9$ all right, but the solution $x = 0$ does not show up. That happens because you have divided by a variable that can attain a value of 0. Be careful whenever you divide each side of an equation by one of its variables in an attempt to solve that equation!

9. Let x represent the width of the strip. The outside rectangle is 60 by 80 feet. The length of the inside rectangle is $(80 - 2x)$ feet, and its width is $(60 - 2x)$ feet. The outside rectangle has an area of $60 \times 80 = 4800$ square feet. We're told that the inside rectangle has an area of one-half that, 2400 square feet. So we know that $(80 - 2x)(60 - 2x) = 2400$. This converts and reduces to standard quadratic form: $x^2 - 70x + 600 = 0$. Using the quadratic formula, we obtain $x = 60$ feet or $x = 10$ feet. The sensible answer is 10 feet for the width of the strip to be mowed. If the mowed strip were 60 feet wide, the area mowed would be greater than the area available to be mowed! (Actually you would end up mowing some of the lawn twice.)

10. Let the number you think of be x. Then the process specified by the man at the party yields this sequence of numbers:

x
$2x$
$2x - 22$
$x(2x - 22)$
$x(2x - 22)/2$
$x(2x - 22)/2 + 70$
$x(2x - 22)/2 + 70 - x$

The man says all this is equal to 35. Therefore, he has in effect given you the following equation to solve:

$$x(2x - 22)/2 + 70 - x = 35$$

This can be simplified to standard quadratic form:

$$x^2 - 12x + 35 = 0$$

Using the quadratic formula, the solutions are found to be $x = 7$ or $x = 5$.

11. Let the width of the box be represented by x inches. Then the height is $x - 1$ inches and the length is $x + 2$ inches. We are told that the total area of all the faces of the box is equal to 108 square inches. Now consider:

- The area of the front plus the area of the back is equal to twice the width times the height, or $2x(x - 1)$.

- The area of the top plus the area of the bottom is equal to twice the width times the length, or $2x(x + 2)$.

- The area of the left side plus the area of the right side is equal to twice the height times the length, or $2(x - 1)(x + 2)$.

The total surface area of the box is equal to the sum of all three of these quantities. Therefore, we know that the following equation holds:

$$2x(x - 1) + 2x(x + 2) + 2(x - 1)(x + 2) = 108$$

This can be converted to standard quadratic form:

$$6x^2 + 4x - 112 = 0$$

Solving with the quadratic formula yields $x = 4$ inches or $x = 14/3 = -4\text{-}2/3$ (minus four and two-thirds) inches. Using the positive answer, the box measures 4 inches wide by 3 inches high by 6 inches long.

12. The "negative solution" to Prob. 11 yields a "width" of $-4\text{-}2/3$ inches. The "height" is 1 inch less than this, or $-5\text{-}2/3$ inches. The "length" is 2 inches more than the "width," or $-2\text{-}2/3$ inches.

CHAP. 11

1. The answers are as follows.
   (*a*) 4,851   (*b*) 8,075   (*c*) 14,391
   (*d*) 39,951   (*e*) 2,499

2. The answers are as follows.
   (*a*) 424   (*b*) 142   (*c*) 676
   (*d*) 99.6   (*e*) 113

3. The answers are as follows when found by extracting the square root, which results in truncation rather than rounding of digits.
   (*a*) 5.477   (*b*) 7.071   (*c*) 7.745
   (*d*) 8.366   (*e*) 8.944

4. The answers can be derived as follows. These solutions do not represent the only approaches, however.
   (*a*) Consider that $y = 20 - x$. Then the equation $xy = 96$ can be rewritten in terms of only x, and rearranged into standard quadratic form as $-x^2 + 20x - 96 = 0$. Using the quadratic formula yields $x = 8$ or $x = 12$. If $x = 8$, then the fact that $xy = 96$ means that $y = 12$. If $x = 12$, then the fact that $xy = 96$ means that $y = 8$. Therefore the solutions, written as ordered pairs, are $(x,y) = (8,12)$ or $(x,y) = (12,8)$.
   (*b*) Consider that $x = 5 + y$. Then the equation $x^2 + y^2 = 53$ can be rewritten in terms of only y, and rearranged into standard quadratic form as $2y^2 + 10y - 28 = 0$. Using the quadratic formula yields $y = -7$ or $y = 2$. If $y = -7$, then the fact that $x = y + 5$ means that $x = -2$. If $y = 2$, then $x = 7$. Written as ordered pairs, the solutions (x,y) are therefore $(-2,-7)$ or $(7,2)$.

(c) Consider that $y = 34 - 3x$. Then the equation $xy = 63$ can be rewritten in terms of only x, and rearranged into standard quadratic form as $-3x^2 + 34x - 63 = 0$. Using the quadratic formula yields $x = 9$ or $x = 7/3$. If $x = 9$, then the fact that $xy = 63$ means that $y = 7$. If $x = 7/3$, then the fact that $xy = 63$ means that $y = 27$. Therefore the solutions, written as ordered pairs, are $(x,y) = (9, 7)$ or $(x,y) = (7/3,27)$.

(d) Consider that $x = 6 + y$. Then the equation $x^2 + y^2 = 26$ can be rewritten in terms of only y, and rearranged into standard quadratic form as $2y^2 + 12y + 10 = 0$. Using the quadratic formula yields $y = -5$ or $y = -1$. If $y = -5$, then the fact that $x = y + 6$ means that $x = 1$. If $y = -1$, then $x = 5$. Written as ordered pairs, the solutions (x,y) are therefore $(1,-5)$ or $(5,-1)$.

5. To obtain relative force figures, let y be the wind speed in miles per hour. Because the force f is proportional to the square of the speed (that's what we're told, anyhow), $f = (y/30)^2 x$. The answers are as follows, to two significant digits.

(a) $y = 10$, therefore $f = (10/30)^2 x = 0.11x$

(b) $y = 20$, therefore $f = (20/30)^2 x = 0.44x$

(c) $y = 40$, therefore $f = (40/30)^2 x = 1.8x$

(d) $y = 60$, therefore $f = (60/30)^2 x = 4.0x$

(e) $y = 100$, therefore $f = (100/30)^2 x = 11x$

6. To obtain the relative brilliance figures, let y be the distance in feet. Because the brightness b is proportional to the inverse of the square of the distance, $b = (10.00/y)^2 x$. The answers are as follows, to four significant digits.

(a) $y = 2.000$, therefore $b = (10.00/2.000)^2 x = 25.00x$

(b) $y = 5.000$, therefore $b = (10.00/5.000)^2 x = 4.000x$

(c) $y = 15.00$, therefore $b = (10.00/15.00)^2 x = 0.4444x$

(d) $y = 25.00$, therefore $b = (10.00/25.00)^2 x = 0.1600x$

(e) $y = 100.0$, therefore $b = (10.00/100.0)^2 x = 0.0100x$

7. Let the dimensions of the box be 6 inches, y inches, and z inches. Then, based on the information given, we have the following two equations:

$$6yz = 480$$

for the volume of the box, and

$$2(6y) + 2(6z) + 2yz = 376$$

for the surface area of the box. These can be simplified to the following two equations:

$$yz = 80$$
$$6y + 6z + yz = 188$$

By substituting from the first equation directly into the second, the second simplifies further (in several steps) to $y + z = 18$. This can be changed to $y = 18 - z$, and then the first equation can be rewritten as $(18 - z)z = 80$. Converting this to standard quadratic form, we obtain

$$-z^2 + 18z - 80 = 0$$

Using the quadratic formula, $z = 8$ or $z = 10$. We know that $yz = 80$. So if $z = 10$, then $y = 8$; if $z = 8$, then $y = 10$. The other two dimensions are therefore 8 inches and 10 inches. It doesn't matter which of these is the width and which is the depth.

8. Let the sides of the rectangular lot be x and y, measured in feet. Based on the information given, we can state the following:

$$xy = 435,600$$

for the area of the lot and

$$2x + (3/2)y = 2,640$$

for the perimeter minus one-half of side y. The second equation can be rewritten as

$$x = 1,320 - 3y/4$$

Substituting into the first equation above, we get

$$(1,320 - 3y/4)y = 435,600$$

Converting to standard quadratic form, we obtain

$$-y^2 + 1,760y - 580,800 = 0$$

Solving by means of the quadratic formula yields the results $y = 1,320$ or $y = 440$. Knowing that $xy = 435,600$, we see that this yields $x = 330$ or $x = 990$. There are thus two possible sets of dimensions for the enclosure: 1,320 by 330 feet, or 440 by 990 feet.

9. As in Prob. 8, let the sides be x and y, measured in feet. Based on the information given, we can state the following

$$xy = 435,600$$

for the area and

$$2x + 2y = 2,750$$

for the available fencing (2,460 feet) plus the extra 110 feet that would be needed to completely surround the enclosure. The second equation can be rewritten as $x = 1,375 - y$. Substituting into the first equation above, we get

$$(1,375 - y)y = 435,600$$

Converting to standard quadratic form, we obtain

$$-y^2 + 1,375y - 435,600 = 0$$

Solving this with the quadratic formula yields the results $y = 880$ or $y = 495$. Knowing that $xy = 435,600$, this yields the results $x = 495$ or $x = 880$. The enclosure therefore measures 495 by 880 feet (x and y are interchangeable). The fact that the dimensions are interchangeable here, but not in the previous situation (Prob. 8), is merely a coincidence.

10. Let the numbers be x and y. Then according to the information given, we get these two equations:

$$xy = 432$$
$$y/x = 3$$

The second equation can be written as $y = 3x$. Substituting into the first equation and rearranging give us $3x^2 = 432$. Dividing each side by 3 tells us that $x^2 = 144$. There are two solutions here: $x = -12$ or $x = 12$. Because $y = 3x$, it follows that $y = -36$ or $y = 36$. Expressed as ordered pairs, the solutions are $(x,y) = (-12,-36)$ or $(x, y) = (12,36)$.

11. The sums are as follows.
  (a) $3 - j2$ (b) $7 - j8$
  (c) 2 (d) 0

12. The differences are as follows.
  (a) $6 + j3$ (b) $1 + j12$
  (c) $j12$ (d) $-j8$

13. The products are as follows.
  (a) $12 - j14$ (b) $-9 + j6$
  (c) $4 + j14$ (d) 45

CHAP. 12

1. The average speed over this 3-min period is 20 mi/h (one-half the final speed) because the acceleration is constant. This is 1/3 mi/min. Therefore, in 3 min, the car travels 1 mi.

2. Because the acceleration rate is uniform, the average speed for the 6-min period is 50 mi/h (midway between the initial and final speed), which is 5/6 mi/min. Therefore, in 6 min, the car will go 5 mi.

3. Because the deceleration is uniform, the average speed for this 30-s period is 30 mi/h (one-half the initial speed), or 1/2 mi/min. Therefore, the car will require 1/4 mi to come to a stop.

4. The speed is 60 mi/h. First, multiply 88 ft/s by 3,600 s/h to get 316,800 ft/h; then divide by 5,280 ft/mi to get 60 mi/h.

5. First, convert 240 mi/h into feet per second. From the solution to Prob. 4, we know that 60 mi/h = 88 ft/s. Therefore, 240 mi/h = (60×4) mi/h = (88×4) ft/s = 352 ft/s. Acceleration (a) is equal to velocity (v) times time (t), provided the units are uniform throughout the calculation. The formula is $a = vt$. Plugging in the values for acceleration and speed gives $16 = 352t$. It follows that $t = 352/16 = 22$ s. That's the length of time from releasing the brakes until the plane lifts into the air. The acceleration rate is constant, so the average speed for this 22-s period is $352/2 = 176$ ft/s. The total runway distance required is therefore 176 ft/s multiplied by 22 s, which is 3,872 ft.

6. The energy is equal to one-half the mass times the speed squared; that is, $e = mv^2/2$. Let the lighter pellet have mass m and the heavier pellet have mass $2m$. Let v be the muzzle speed of the lighter pellet. Then plugging in the numbers from the information given, we obtain

$$2m(150)^2/2 = mv^2/2$$

Mass cancels out from either side. The resulting equation solves for $v = 212$ ft/s (to three significant figures).

7. The time t for a mass m to reach a velocity v, given the application of power p, is given by the formula $t = mv^2/2p$. Both m (the mass of the car) and p (the power) are constants. Therefore, we know that the time t is proportional to v^2. If $v = 60$ mi/h, then $t = 20$ s; we are given this information. If $v = 30$ mi/h (one-half of 60 mi/h), then t is equal to 1/4 of its 60 mi/h value of 20 s, or 5 s. If $v = 45$ mi/h (3/4 of 60 mi/h), then t is equal to 9/16 of its 60 mi/h value of 20 s, or 11.25 s.

8. If m increases from 3,000 lb to 4,000 lb, the right-hand side of the equation $t = mv^2/2p$ increases to 4/3 of its previous value. The left-hand side of the equation (the time t) also increases to 4/3 of its previous value. The new times can therefore be derived by multiplying the solutions to Prob. 7 by 1.33. For 30 mi/h, 45 mi/h, and 60 mi/h, the required times with the additional load are $5 \times 4/3 = 20/3 = 6.67$ s, $11.25 \times 4/3 = 45/3 = 15$ s, and $20 \times 4/3 = 80/3 = 26.7$ s, respectively.

9. Given that the mass of the car is 3,000 lb in the situation of Prob. 8, and remembering that 60 mi/h = 88 ft/s, we can plug numbers into the equation $t = mv^2/2p$ and obtain

$$20 = 3,000 \times 88^2/2p$$

Solving, we obtain $p = 580,800$ foot-poundals per second (ft · pdl/s).

10. Recall the formula $e = mv^2/2$. This formula can be rearranged to get the formula

$$v = (2e/m)^{1/2}$$

which tells us that the velocity is proportional to the inverse of the square root of the mass. If the mass is doubled, the velocity drops to $1/(2^{1/2}) = 0.707$ times its previous value. If the mass is halved, the velocity increases to $2^{1/2} = 1.414$ times its original value.

11. Draw tangent lines to the curve as shown in Fig. A-3. A straightedge, such as a ruler, can be used for this purpose. Then graphically estimate the slopes of the lines at time points 1 s, 2 s, 3 s, and 4 s. These slopes are approximately 1 m/s, 1.5 m/s, 2 m/s, and 3 m/s, respectively. Note that the increments for the horizontal scale (time) are only

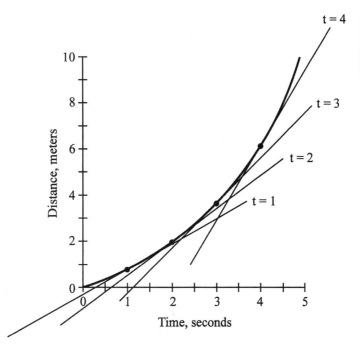

Figure A-3

Illustration for solution to Prob. 11 in Chap. 12.

Figure A-4

Illustration for
solutions to
Probs. 12 and 13
in Chap. 12.

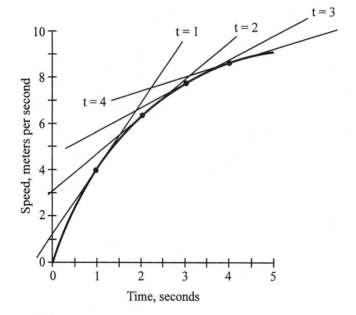

one-half the size of the increments for the vertical scale (distance). For this reason, the "actual" (mathematical) slope is twice the slope as it appears to the casual observer.

12. By using the same method as for the solution to Prob. 11, the slopes of the lines can be estimated as 3 m/s², 2 m/s², 1.5 m/s², and 1 m/s², for time points corresponding to 1 s, 2 s, 3 s, and 4 s respectively. See Fig. A-4. Note that the increments for the horizontal scale (time) are only one-half the size of the increments for the vertical scale (speed). For this reason, the "actual" (mathematical) slope is twice the slope as it appears to the casual observer.

13. The applied force is directly proportional to the acceleration, which is indicated by the slope of the tangent to the curve at any given point in Fig. A-4. Therefore, the applied force decreases with time.

14. The acceleration starts out at a very large value and decreases until the elapsed time is 2 s, whereupon it becomes 0. Beyond 2 s, the acceleration is negative and continues to decrease (that is, become larger negatively) until the elapsed time is about 4 s. After 4 s, the acceleration begins to increase slightly, but remains negative all the way to the end of the horizontal scale indicating an elapsed time of 5 s. The acceleration is greatest at 0 s and is smallest (that is, largest negatively) at about 4 s.

CHAP. 13

1. The wide-screen movie has an aspect ratio of 2:1, which is equivalent to 6:3 (ratio of width to height). The TV picture has an aspect ratio of 4:3. Therefore, the movie image is 6/4, or 3/2, as wide as the TV image when both have the same height. This means that the TV image is only 2/3 as large as the movie image. A total of 1/3 of the movie image is lost (1/6 at either side).

2. In this case, think of the movie aspect ratio (width to height) as 12:6 (2:1 times 6/6), and the TV aspect ratio as 12:9 (4:3 times 3/3). The movie image is therefore 6/9, or 2/3, as high as the TV image when both are the same width. This means that 1/3 of the TV screen's vertical height is not used. If equal-sized blank strips exist at the top and the bottom, therefore, each of them has a vertical height equal to 1/3 times 1/2, or 1/6, of the screen.

3. To simplify this problem, imagine that all the children get money in proportion to their ages, but double the "effective ages" of the boys. The boys' effective ages are therefore $40 \times 2 = 80$, $34 \times 2 = 68$, and $26 \times 2 = 52$. The girls' effective ages are their actual ages of 37 and 23. Add up all the effective ages to get 260 "effective child years." Divide the estate, $22,100, by 260 to get 85 "dollars per effective child year." Each child's share is then determined by multiplying 85 by his or her effective age. The boys receive, in dollars, $85 \times 80 = \$6,800$, $85 \times 68 = \$5,780$, and $85 \times 52 = \$4,420$. The girls get, in dollars, $85 \times 37 = \$3,145$ and $85 \times 23 = \$1,955$.

4. After 10 years, the boys' actual ages will be $40 + 10 = 50$, $34 + 10 = 44$, and $26 + 10 = 36$, so their effective ages will be $2 \times 50 = 100$, $2 \times 44 = 88$, and $2 \times 36 = 72$. The girls' actual (and effective) ages will be $37 + 10 = 47$ and $23 + 10 = 33$. Add up all the effective ages to get 340 effective child years. Divide the estate (still $22,100) by 340 to get 65 dollars per effective child year. Now the boys receive, in dollars, $65 \times 100 = \$6,500$, $65 \times 88 = \$5,720$, and $65 \times 72 = \$4,680$. The girls receive, in dollars, $65 \times 47 = \$3,055$ and $65 \times 33 = \$2,145$.

5. Let the length of the hypotenuse be represented by x. Then according to the Pythagorean theorem, we have

$$x^2 = 12^2 + 5^2$$
$$= 144 + 25$$
$$= 169$$

The value of x is the square root of 169, or 13.

6. Because the gradient is 1:8 in an 8,000-ft span of road, the altitude gained is, by definition, 8,000 ft divided by 8, or 1,000 ft. To find the horizontal distance, construct a right triangle with the base (call its length x) as the horizontal distance, the height equal to 1,000 ft, and the hypotenuse 8,000 ft long. Then $x^2 + 1,000^2 = 8,000^2$. This can be simplified to $x^2 + 1,000,000 = 64,000,000$, and rearranged to get $x^2 = 64,000,000 - 1,000,000 = 63,000,000$. This means that x is equal to the square root of 63,000,000, which is 7,937 ft, expressed to the nearest foot. This falls short of 8,000 ft by $8,000 - 7,937 = 63$ ft, expressed to the nearest foot.

7. Let the height of the mountain, in miles, be represented by x. Then x is the height of a right triangle whose base is 8 mi long. Therefore $x/8 = \tan 9° = 0.1584$, expressed to four significant digits. This easily solves to $x = 0.1584 \times 8 = 1.267$ mi. Now suppose the angle is 5°. Let y be the distance to the mountain; it is the length of the base of a right triangle whose height is 1.267 mi. Therefore, $1.267/y = \tan 5° = 0.08749$. Solving, we get

$$1/y = 0.08749/1.267$$
$$= 0.06905$$

Taking the reciprocal of both sides gives us

$$y = 1/0.06905$$
$$= 14.48 \text{ mi}$$

8. Let's convert all distances to feet and round off to the nearest foot in each calculation. Remember that 1 mi = 5,280 ft. For the first 3 mi (15,840 ft), the altitude gained is 15,840 × 1:42 = 377 ft. For the next 5 mi (26,400 ft), the altitude gained is 26,400 × 1:100 = 264 ft. For the next 2 mi (10,560 ft), there is no change in the altitude. For the next 6 mi (31,680 ft), the altitude gained is 31,680 × (−1:250) = −127 ft, which represents an altitude loss of 127 ft. For the next 4 mi (21,120 ft), there is no change. For the last 5 mi (26,400 ft), the altitude gained is 26,400 × 1:125 = 211 ft. The total altitude gained is the sum of all these figures: 377 + 264 + 0 − 127 + 0 + 211 = 725 ft.

9. Consider that each rafter forms the hypotenuse of a right triangle whose base measures 40/2 + 2 = 22 ft. The angle between the base and the hypotenuse is 30°. Let x be the length of the hypotenuse. Then $22/x = \cos 30° = 0.8660$. Solving, we get $x = 25.4$ ft.

10. Let the rafter length be y when the angle is 20°. Then $22/y = \cos 20° = 0.9397$. Solving gives $y = 23.4$ ft. This is 2 ft shorter than is the case with a 30° angle.

11. Consider a right triangle whose base is the height of the wall (50 ft) and whose hypotenuse is the length of the ladder (60 ft). Then the cosine of the angle between the ladder and the wall is 50/60 = 0.8333. The angle can be determined, using a calculator, to be 33.56°. Let x be the distance between the base of the ladder and the wall; this is the side opposite the 33.56° angle. By trigonometry, $x/50 = \tan 33.56° = 0.6634$. Solving gives $x = 33.17$ ft. Using the Pythagorean theorem, we find $x^2 + 50^2 = 60^2$. Thus $x^2 = 3600 − 2500 = 1100$, so x is the square root of 1100, or 33.17 ft.

12. The Pythagorean theorem tells us that

$$x^2 = 88^2 + 44^2$$
$$= 7,744 + 1,936$$
$$= 9,680$$

Therefore, x is equal to the square root of 9,680, or 98.39 cm.

13. Consider the 45° angle. Its cosine is 0.7071. This is the ratio of the base to side y; that is, $88/y = 0.7071$. Solving, we get $y = 124.5$ cm.

14. The tangent of angle q is the ratio of the 44-cm side to the 88-cm side; "in your head" you can see that this ratio is 1/2. Thus $\tan q = 0.500$. A calculator with inverse trig functions tells us that $q = 26.6°$.

15. The tangent of angle r is the ratio of the 88-cm side to the 44-cm side. Therefore, $\tan r = 2$. A calculator with inverse trig functions tells us that $r = 63.4°$.

16. Remember that the measures of the interior angles of a triangle always add up to 180°. Therefore

$$90 + q + r = 180$$

We know that $q = 26.6$. Therefore

$$r = (180 - 90) - q$$
$$= 90 - q$$
$$= 90 - 26.6$$
$$= 63.4°$$

CHAP. 14

1. Recall that for any angle A, $\sin^2 A + \cos^2 A = 1$. The answers to each part of this problem follow.

    (a), (b) The tangent of any angle is equal to the sine divided by the cosine. Therefore, $\tan A = 0.8/0.6 = 1.333$, and $\tan B = 0.6/0.8 = 0.75$.

    (c) The sine of the sum of the angles is

$$\sin (A + B) = \sin A \cos B + \cos A \sin B$$
$$= 0.8 \times 0.8 + 0.6 \times 0.6$$
$$= 0.64 + 0.36$$
$$= 1$$

    (d) The cosine of the sum of the angles is

$$\cos (A + B) = \cos A \cos B - \sin A \sin B$$
$$= 0.6 \times 0.8 - 0.8 \times 0.6$$
$$= 0.48 - 0.48$$
$$= 0$$

    (e) The sine of the difference of the angles is

$$\sin (A - B) = \sin [A + (-B)]$$
$$= \sin A \cos (-B) + \cos A \sin (-B)$$
$$= 0.8 \times 0.8 + 0.6 \times (-0.6)$$
$$= 0.64 - 0.36$$
$$= 0.28$$

    (f) The cosine of the difference of the angles is

$$\cos (A - B) = \cos [A + (-B)]$$
$$= \cos A \cos (-B) - \sin A \sin (-B)$$
$$= 0.6 \times 0.8 - 0.8 \times (-0.6)$$
$$= 0.48 + 0.48$$
$$= 0.96$$

(g) The tangent of the sum of the angles is

$$\tan (A + B) = (\tan A + \tan B)/(1 - \tan A \tan B)$$
$$= (4/3 + 3/4)/(1 - 4/3 \times 3/4)$$
$$= (25/12)/0$$
$$= \text{undefined!}$$

(h) The tangent of the difference of the angles is

$$\tan (A - B) = \tan [A + (-B)]$$
$$= [(\tan A + \tan (-B)]/[1 - \tan A \tan (-B)]$$
$$= (4/3 - 3/4)/[1 - 4/3 \times (-3/4)]$$
$$= (7/12)/2$$
$$= 7/24$$
$$= 0.2917$$

2. The two places are at the base angles of an isosceles triangle. The apex of the triangle is at the center of the earth. The two equal sides are 4,000 mi long, which is the radius of the earth. The apex angle is the difference in longitude, which is 100.5° − 37.5°, or 63°. Divide this isosceles triangle in half by drawing a perpendicular from the apex to the middle of the base, obtaining a pair of right triangles. Choose one of these two right triangles as shown in Fig. A-5. (It doesn't matter which one; in this case the one with a vertex at Sumatra is used.) The hypotenuse of this triangle measures 4,000 mi. The angle between the hypotenuse and the side representing a "partial earth radius" is 63°/2, or 31.5°. The length of side x, which is one-half the distance asked for in the problem, is such that $x/4,000 = \sin 31.5° = 0.5225$. Therefore, $x = 4,000 \times 0.5225 = 2,090$ mi. The straight-line distance through the earth between Sumatra and Mt. Kenya is $2x$, or 4,180 mi.

Figure A-5

Illustration for solution to Prob. 2 in Chap. 14.

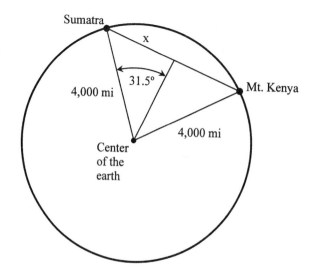

3. The lines of sight would cross at an angle equal to the difference in longitude between them. The two places differ in longitude by 63°. That means the lines would cross at a 63° angle, as shown in Fig. A-6.

4. Refer again to Fig. A-6. The angles between the sighting lines (from the solution to Prob. 3) and the direct line connecting the points (through the earth) are both 31.5°. This is one-half of 63°, the angle between the two sighting lines themselves. Consider a perpendicular line from the satellite to the middle of the direct line connecting Sumatra and Mt. Kenya. This creates two mirror-image right triangles whose bases are each 2,090 mi. The angles adjacent to the bases are $31.5° + 58° = 89.5°$. Let x be the distance to the satellite (either from Sumatra or from Mt. Kenya; they are the same). By trigonometry, $2,090/x = \cos 89.5° = 0.008727$. Solving, we obtain $x = 239,000$ mi, expressed to the nearest 1,000 mi.

5. We are given the fact that $\cos A = 2 \sin A$. Therefore, the sine divided by the cosine is equal to 1/2. That's the tangent of the angle.

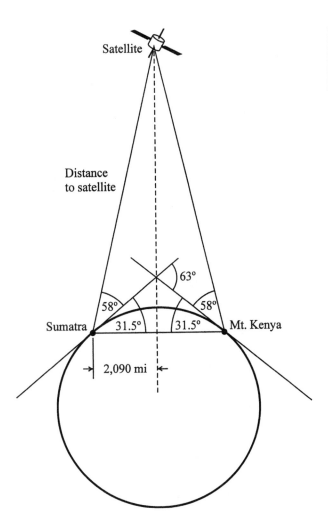

Figure A-6

Illustration for solutions to Probs. 3 and 4 in Chap. 14.

6. To find the cosine, we can use the formula

$$\cos A = (1 - \sin^2 A)^{1/2}$$

We are told that $\sin A = 0.28$. Therefore

$$\cos A = [1 - (0.28)^2]^{1/2}$$
$$= (1 - 0.0784)^{1/2}$$
$$= (0.9216)^{1/2}$$
$$= 0.96$$

The tangent is the sine divided by the cosine, which is 0.28/0.96 (both exact values), giving 0.2917 (accurate to four significant figures).

7. To find the sine of twice the angle, use the formula

$$\sin 2A = 2 \sin A \cos A$$

Substituting for the cosine in terms of the sine (see the solution to Prob. 6 above), we obtain

$$\sin 2A = 2 \sin A (1 - \sin^2 A)^{1/2}$$

We are told that $\sin A = 0.6$. Therefore

$$\sin 2A = 2 \times 0.6 \times [1 - (0.6)^2]^{1/2}$$
$$= 1.2 \times (1 - 0.36)^{1/2}$$
$$= 1.2 \times (0.64)^{1/2}$$
$$= 1.2 \times 0.8$$
$$= 0.96$$

To find the sine of three times the angle, use the formula

$$\sin 3A = 3 \sin A - 4 \sin^3 A$$
$$= 3 \times 0.6 - 4 \times (0.6)^3$$
$$= 1.8 - 4 \times 0.216$$
$$= 1.8 - 0.864$$
$$= 0.936$$

This is a coincidence, but it's no mistake! The sine of three times this angle is almost the same as the sine of twice the angle. If you like, use a calculator to find the measure of the angle A, and draw its multiples as vectors on the coordinate plane. This will provide a graphical demonstration of this situation.

8. Let's call the new angle B. Then $B = 2A$. The sine of angle B was determined to be 0.96 in the solution to Prob. 7. Then we can use the formula

$$\cos B = (1 - \sin^2 B)^{1/2}$$
$$= (1 - 0.96^2)^{1/2}$$
$$= (1 - 0.9216)^{1/2}$$
$$= (0.0784)^{1/2}$$
$$= 0.28$$

9. Use your calculator to determine that sin 15° = 0.259 (to three significant digits). Let $A = 15°$. Then

$$\sin 30° = \sin 2A$$
$$= 2 \sin A\, (1 - \sin^2 A)^{1/2}$$
$$= 2 \times 0.259 \times [1 - (0.259)^2]^{1/2}$$
$$= 0.518 \times (1 - 0.067081)^{1/2}$$
$$= 0.518 \times (0.932919)^{1/2}$$
$$= 0.518 \times 0.965877$$
$$= 0.50$$

rounded off to two significant figures, and

$$\sin 45° = \sin 3A$$
$$= 3 \sin A - 4 \sin^3 A$$
$$= 3 \times 0.259 - 4 \times (0.259)^3$$
$$= 0.777 - 4 \times 0.0173740$$
$$= 0.777 - 0.069496$$
$$= 0.71$$

again rounded off to two significant figures.

10. Use your calculator to determine that sin 30° = 0.500. Let $A = 30°$. Then, calculating and condensing the steps, you should obtain

$$\sin 60° = \sin 2A$$
$$= 2 \sin A\, (1 - \sin^2 A)^{1/2}$$
$$= 1.000 \times (0.75)^{1/2}$$
$$= 0.87$$

rounded off to two significant figures. Calculating again and condensing the steps, you should get

$$\sin 90° = \sin 3A$$
$$= 3 \sin A - 4 \sin^3 A$$
$$= 1.500 - 0.500$$
$$= 1.0$$

again rounded off to two significant figures.

11. Use your calculator to determine that tan 45° = 1.000. Let $A = 45°$. Then, calculating and condensing the steps, you should get

$$\tan 90° = \tan 2A$$
$$= (2 \tan A)/(1 - \tan^2 A)$$
$$= 2.000/0$$
$$= \text{undefined!}$$

and

$$\begin{aligned}
\tan 135° &= \tan 3A \\
&= (3 \tan A - \tan^3 A)/(1 - 3 \tan^2 A) \\
&= 2.000/(-2.000) \\
&= -1.0
\end{aligned}$$

rounded off to two significant figures. Actually, this answer is exact, assuming the measure of angle A is precisely equal to 45°.

12. Use your calculator to determine that $\cos 60° = 0.500$. Let $A = 60°$. Then

$$\begin{aligned}
\cos 120° &= \cos 2A \\
&= \cos^2 A - \sin^2 A
\end{aligned}$$

Recall that $\sin^2 A = 1 - \cos^2 A$. Substitute in the above equation to get

$$\begin{aligned}
\cos 2A &= \cos^2 A - (1 - \cos^2 A) \\
&= \cos^2 A + (\cos^2 A - 1) \\
&= 2 \cos^2 A - 1 \\
&= 2 \times (0.500)^2 - 1 \\
&= 2 \times 0.250 - 1 \\
&= 0.500 - 1 \\
&= -0.50
\end{aligned}$$

rounded to two significant figures, and

$$\begin{aligned}
\cos 180° &= \cos 3A \\
&= 4 \cos^3 A - 3 \cos A \\
&= 4 \times (0.500)^3 - 3 \times 0.500 \\
&= 4 \times 0.125 - 1.500 \\
&= 0.500 - 1.500 \\
&= -1.0
\end{aligned}$$

again rounded to two significant figures. Actually, these answers are exact, assuming the measure of angle A is precisely equal to 60°.

13. Use your calculator to determine that $\cos 90° = 0.0$. Let $A = 90°$. Then, calculating and condensing the steps, you should obtain

$$\begin{aligned}
\cos 180° &= \cos 2A \\
&= 2 \cos^2 A - 1 \\
&= -1.0
\end{aligned}$$

and

$$\begin{aligned}
\cos 270° &= \cos 3A \\
&= 4 \cos^3 A - 3 \cos A \\
&= 0.0
\end{aligned}$$

These answers are exact, assuming that the measure of A is precisely equal to 90°.

14. To find the tangents of the original angles, it is easiest to use a calculator. Let the angles be called A, B, C, and D. Then, to three significant digits,

$$\tan A = \tan 29° = 0.554$$
$$\tan B = \tan 31° = 0.601$$
$$\tan C = \tan 59° = 1.66$$
$$\tan D = \tan 61° = 1.80$$

From the formula for the tangent of three times an angle, you can calculate and condense the steps to determine that

$$\tan 3A = \tan 87°$$
$$= (3 \tan A - \tan^3 A)/(1 - 3 \tan^2 A)$$
$$= (1.662 - 0.170)/(1.000 - 0.921)$$
$$= 1.492/0.079$$
$$= 19$$

and

$$\tan 3B = \tan 93°$$
$$= (3 \tan B - \tan^3 B)/(1 - 3 \tan^2 B)$$
$$= (1.803 - 0.217)/(1.000 - 1.083)$$
$$= 1.586/(-0.083)$$
$$= -19$$

Both answers are rounded off to two significant figures. The change in sign occurs as a result of the angle vector moving from the first to the second quadrant in the coordinate plane. Continuing, we have

$$\tan 3C = \tan 177°$$
$$= (3 \tan C - \tan^3 C)/(1 - 3 \tan^2 C)$$
$$= (4.98 - 4.57)/(1.000 - 8.27)$$
$$= 0.41/(-7.27)$$
$$= -0.056$$

and

$$\tan 3D = \tan 183°$$
$$= (3 \tan D - \tan^3 D)/(1 - 3 \tan^2 D)$$
$$= (5.40 - 5.83)/(1.000 - 9.72)$$
$$= -0.43/(-8.72)$$
$$= 0.049$$

Again, both answers are rounded off to two significant figures. There is a slight discrepancy because of cumulative rounding errors. The change in sign occurs as a result of the angle vector moving from the second to the third quadrant in the coordinate plane.

15. Let the angle be A. Then $\sin A = 0.96$, and

$$\begin{aligned}
\sin 2A &= 2 \sin A \, (1 - \sin^2 A)^{1/2} \\
&= 2 \times 0.96 \times [1 - (0.96)^2]^{1/2} \\
&= 1.92 \times (1 - 0.9216)^{1/2} \\
&= 1.92 \times (0.0784)^{1/2} \\
&= 1.92 \times 0.28 \\
&= 0.538
\end{aligned}$$

You can round this off to 0.54. Now recall that for any angle A, the following holds:

$$\cos^2 A = 1 - \sin^2 A$$

The formula for $\cos 2A$, derived in the solution for Prob. 12, tells us that

$$\cos 2A = 2 \cos^2 A - 1$$

Substituting, you should get

$$\begin{aligned}
\cos 2A &= 2(1 - \sin^2 A) - 1 \\
&= 2 - 2 \sin^2 A - 1 \\
&= 1 - 2 \sin^2 A \\
&= 1 - 2 \times (0.96)^2 \\
&= 1 - 2 \times 0.9216 \\
&= 1 - 1.8432 \\
&= -0.8432
\end{aligned}$$

You can round this off to -0.84.

16. Let $\cos A = x$. Then the equation, reduced to standard quadratic form, becomes

$$8x^2 + x - 3 = 0$$

When you use the quadratic formula, you will obtain expressions containing the square root of 97, which is approximately 9.85. The resulting solutions are $x = 0.553$ or $x = -0.678$. Now recall that $x = \cos A$. You can use your calculator to find the angles, which are, respectively, 56.4° (in the first quadrant) and 132.7° (in the second quadrant).

CHAP. 15

1. Dividing 1 by 37, you obtain 0.0270 (the first three significant digits are 2, 7, 0). Continuing the division yields a repeating decimal: $1/37 = 0.027027027027 \ldots$. The error x is the difference between the three-significant-digit value and the full repeating decimal:

$$\begin{aligned}
x &= 0.027027027027 \ldots - 0.0270 \\
&= 0.0000270270270 \ldots
\end{aligned}$$

2. In the binary system, the decimal number 15 is denoted by 1111, and the decimal number 63 is denoted by 111111. Multiplying these yields 1110110001. When you multiply this out to decimal form, you have

$$(1 \times 1) + (0 \times 2) + (0 \times 4) + (0 \times 8) + (1 \times 16) + (1 \times 32)$$
$$+ (0 \times 64) + (1 \times 128) + (1 \times 256) + (1 \times 512)$$
$$= 1 + 16 + 32 + 128 + 256 + 512 = 945$$

Checking with a calculator, you can verify that $15 \times 63 = 945$.

3. In the binary system, 1,922 is denoted by 11110000010, and 31 is denoted by 11111. The quotient is 111110. When you multiply this out to decimal form, you have

$$(0 \times 1) + (1 \times 2) + (1 \times 4) + (1 \times 8) + (1 \times 16) + (1 \times 32)$$
$$= 2 + 4 + 8 + 16 + 32 = 62$$

Checking with a calculator, you can verify that $1,922/31 = 62$.

4. Fractional exponents involve both powers and roots. Solutions are found as follows.
   (a) Find the 4th root of 16, which is 2; then cube 2 to obtain $16^{3/4} = 8$.
   (b) This expression is equivalent to $243^{4/5}$. First find the 5th root of 243, which is 3; then take the 4th power of 3 to obtain $243^{4/5} = 81$.
   (c) This expression is equivalent to $25^{3/2}$. First find the square root of 25, which is 5; then cube this to obtain $25^{3/2} = 125$. Alternatively, you can cube 25 to get 15,625, and then take the square root of 15,625 to get 125.
   (d) Find the cube root of 64, which is 4; then square 4 to obtain $64^{2/3} = 16$.
   (e) Find the cube root of 343, which is 7; then take the 4th power of 7 to obtain the result $343^{4/3} = 2,401$.

5. The equivalents are as follows.
   (a) Decimal 62 = binary 111110
   (b) Decimal 81 = binary 1010001
   (c) Decimal 111 = binary 1101111
   (d) Decimal 49 = binary 110001
   (e) Decimal 98 = binary 1100010
   (f) Decimal 222 = binary 11011110
   (g) Decimal 650 = binary1010001010
   (h) Decimal 999 = binary 1111100111
   (i) Decimal 2,000 = binary 11111010000

6. The equivalents are as follows.
   (a) Binary 101 = decimal 5
   (b) Binary 1111 = decimal 15
   (c) Binary 10101 = decimal 21
   (d) Binary 111100 = decimal 60
   (e) Binary 110111000110 = decimal 3,526

7. The product $129 \times 31 = 3,999$, and the product $129 \times 41 = 5,289$. This is an error of $5,289 - 3,999 = 1,290$, which is 32 percent of the correct value (3,999). The binary equivalent of 129 is 10000001, and the binary equivalent of 31 is 11111. The product of these numbers is binary 111110011111, which is equivalent to decimal 3,999. If the second-to-last digit in the second binary number is reversed so the number becomes 11101, the binary product becomes binary 111010011101, which is equivalent to decimal 3,741. This is an error of $3{,}999 - 3{,}741 = 258$, which is 6.5 percent of the correct value (3,999).

8. The value of each expression is 1 more than the larger of the two input numbers. For example, in case (f) where $a = 84$ and $b = 13$, you get

$$(a^2 + b^2)^{1/2} = (84^2 + 13^2)^{1/2}$$
$$= (7{,}056 + 169)^{1/2}$$
$$= 7{,}225^{1/2}$$
$$= 85$$

which is 1 more than 84, the larger of the two original numbers. You can verify that the same thing holds true for all the other cases in this problem.

9. The value of each expression is 2 more than the larger of the two input numbers. For example, in case (c) where $a = 26$ and $b = 10$, you get

$$(a^2 + b^2)^{1/2} = (24^2 + 10^2)^{1/2}$$
$$= (576 + 100)^{1/2}$$
$$= 676^{1/2}$$
$$= 26$$

which is 2 more than 24, the larger of the two original numbers. You can verify that the same thing holds true for all the other cases in this problem.

10. The first four expressions are
   (a) $100^2 = 10{,}000$
   (b) $100^{1/2} = 10$
   (c) $100^{-2} = 0.0001$
   (d) $100^{-1/2} = 0.1$

The second four expressions can be derived as whole-number powers of (b) and (d):
   (e) $100^{3/2} = (100^{1/2})^3 = 10^3 = 1{,}000$
   (f) $100^{5/2} = (100^{1/2})^5 = 10^5 = 100{,}000$
   (g) $100^{-3/2} = (100^{-1/2})^3 = (0.1)^3 = 0.001$
   (h) $100^{-5/2} = (100^{-1/2})^5 = (0.1)^5 = 0.00001$

11. The answers can be found by taking successive square roots (1/2 powers) of 100. Note that $100^{1/4} = (100^{1/2})^{1/2}$. That gives you the answer to (a). To find the answer to each of the succeeding parts of the problem, you can repeatedly take the square root of the preceding number. You should get results as follows, displaying each value as a rounded-off quantity to three decimal places.
   (a) $100^{1/4} = (100^{1/2})^{1/2} = 3.162$
   (b) $100^{1/8} = (10^{1/4})^{1/2} = 1.778$
   (c) $100^{1/16} = (10^{1/8})^{1/2} = 1.334$
   (d) $100^{1/32} = (10^{1/16})^{1/2} = 1.155$

If you simply truncate the numbers at three decimal places, you'll get

(a) $100^{1/4} = (100^{1/2})^{1/2} = 3.162$

(b) $100^{1/8} = (10^{1/4})^{1/2} = 1.778$

(c) $100^{1/16} = (10^{1/8})^{1/2} = 1.333$

(d) $100^{1/32} = (10^{1/16})^{1/2} = 1.154$

12. The value will approach 1 because the exponent approaches 0, and any nonzero number to the 0th power equals 1. To see how this works, keep hitting the square root function button on your calculator over and over. The same thing happens when you start with any positive real number and then hit the square root key repeatedly. The value always converges toward 1.

13. Note that $2^5 = 32$, and therefore, $32^{1/5} = 32^{0.2} = 2$. This answers part (b). From this, you can determine the answers to (d), (f), and (h).

(d) $32^{0.4} = (32^{0.2})^2 = 2^2 = 4$

(f) $32^{0.6} = (32^{0.2})^3 = 2^3 = 8$

(h) $32^{0.8} = (32^{0.2})^4 = 2^4 = 16$

The value of $32^{0.1}$ is the same as $(32^{0.2})^{1/2}$, which is $2^{1/2}$. This answers part (a). From this, you can determine the answers to (c), (e), (g), and (i). Here are the answers you should get if you use a calculator for this purpose, rounded off to three decimal places.

(c) $32^{0.3} = (32^{0.1})^3 = (2^{1/2})^3 = 2.828$

(e) $32^{0.5} = (32^{0.1})^5 = (2^{1/2})^5 = 5.657$

(g) $32^{0.7} = (32^{0.1})^7 = (2^{1/2})^7 = 11.314$

(i) $32^{0.9} = (32^{0.1})^9 = (2^{1/2})^9 = 22.627$

14. The expressions can be simplified to the following. Where applicable, the expressions have been rounded off to four decimal places (one more than the minimum the problem asks for).

(a) 6	(b) 28	(c) 7
(d) 2.1147	(e) 2.3012	(f) 81
(g) 0.0123	(h) 9	(i) 0.1111
(j) 3	(k) 0.3333	

CHAP. 16

Note: Starting at this point, multiplication symbols between plain numerals are shown as elevated dots (·) rather than crosses (×), reflecting the alternative notation described in Chap. 16 and used in the text thereafter. For example, $3 \cdot 7 = 21$. In scientific notation, however, the cross is still used. For example, $3 \times 10^7 = 30,000,000$.

1. The progression types and parameters are as follows.

(a) Arithmetic progression with $a = 1, d = 4$

(b) Arithmetic progression with $a = -9, d = 6$

(c) Geometric progression with $a = 16, r = 3/4$

(d) Geometric progression with $a = -81, r = -2/3$

(e) Harmonic progression with $a = 1512, d = 1/5$

(f) Geometric progression with $a = 1, r = 1/2$

(g) Geometric progression with $a = 1, r = 2$

2. The sums are as follows.
    (*a*) 400   (*b*) 1,248   (*c*) 10,235
    (*d*) −1,705 (*e*) 196.875 (*f*) 65.625
    (*g*) 67.1875 (*h*) 1,995

3. The sums of the series, if convergent, are as follows.
    (*a*) 2   (*b*) Does not converge
    (*c*) 1,000   (*d*) 6.25   (*e*) 1,000
    (*f*) 100   (*g*) 100   (*h*) 512

4. Use the formula for the permutations of k articles in an available n articles:

$$_nP_k = n!/(n - k)!$$

The solutions for each part of the problem follow. It's not always necessary to compute the factorials of the numerators. The terms of the factorials in the denominators often cancel out some of the terms of the factorials in the numerators.
    (*a*) $50!/(50 - 3)! = 50!/47! = 50 \cdot 49 \cdot 48 = 117,600$ permutations
    (*b*) $10!/(10 - 5)! = 10!/5! = 10 \cdot 9 \cdot 8 \cdot 7 \cdot 6 = 30,240$ permutations
    (*c*) $12!/(12 - 6)! = 12!/6! = 12 \cdot 11 \cdot 10 \cdot 9 \cdot 8 \cdot 7 = 665,280$ permutations
    (*d*) $10!/(10 - 4)! = 10!/6! = 10 \cdot 9 \cdot 8 \cdot 7 = 5,040$ permutations
    (*e*) $7!(7 - 6)! = 7! = 7 \cdot 6 \cdot 5 \cdot 4 \cdot 3 \cdot 2 = 5,040$ permutations

5. Use the formula for the combinations of k articles in an available n articles:

$$_nC_k = n!/[k!(n - k)!]$$

Notice that the n's and k's in each part of this problem are the same as the n's and k's in each part of the Prob. 4. This is convenient, because the results can be obtained by dividing the answers to Prob. 4 by $k!$ in each case.
    (*a*) $117,600/3! = 19,600$ combinations
    (*b*) $30,240/5! = 252$ combinations
    (*c*) $665,280/6! = 924$ combinations
    (*d*) $5,040/4! = 210$ combinations
    (*e*) $5,040/6! = 7$ combinations

6. Suppose that 0 were, in fact, allowed as the first digit in all telephone numbers. Then there would be $10^7 = 10,000,000$ possible telephone numbers in this area code. There are $10^6 = 1,000,000$ telephone numbers whose first digit is 0 (this in effect creates six-digit numbers). Because the first digit cannot be zero, the total number of available telephone numbers is $10,000,000 - 1,000,000 = 9,000,000$.

7. In this case, there are $10^7 - (3 \times 10^6)$ possible numbers. That's $10,000,000 - 3,000,000 = 7,000,000$.

8. Assuming all 12 horses have equal ability, the following scenarios will prevail.
    (*a*) The probability of your naming the correct winner is 1/12, or about 8.33%.
    (*b*) There are $_{12}C_2 = 12!/[(2!(12 - 2)!] = 66$ different ways in which the horses can come in first and second, in no particular order. Your chance of making a correct guess is therefore 1/66.
    (*c*) There are $_{12}P_2 = 12!/(12 - 2)! = 132$ different ways in which the horses can come in first and second, taking order into account. Your chance of making a correct guess is therefore 1/132.

9. This probability is $1/(12/3) = 1/4$. Another way of thinking of this is to figure the chance that a particular horse will finish in the top quartile (1/4).

10. First, determine the number of sunlike stars in our galaxy. This is

$$0.06 \cdot (2 \times 10^{11}) = 0.12 \times 10^{11}$$
$$= 1.2 \times 10^{10} \text{ stars}$$

Three percent of these stars have at least one earthlike planet; this number is

$$0.03 \cdot (1.2 \times 10^{10}) = 0.036 \times 10^{10}$$
$$= 3.6 \times 10^{8} \text{ stars}$$

One in five, or 20%, of these planets can be expected to have life as we know it; this number is at least

$$0.2 \cdot (3.6 \times 10^{8}) = 0.72 \times 10^{8}$$
$$= 7.2 \times 10^{7} \text{ planets}$$

We say "at least" because some stars might have more than one earthlike planet!

11. This problem can be solved using a bit of trickery. Suppose there are 100 people in the bar, and they are all over 18 years of age. Also suppose there are 50 men and 50 women. Given this information along with the data in the problem, one in 5 men has probably had a ticket in the past 12 months; that is $(1/5) \cdot 50 = 10$ men. One in 10 women has probably had a ticket in the past 112 months; that's $(1/10) \cdot 50 = 5$ women. Therefore, 15 of the 100 people have probably had tickets in the past 12 months. If you choose a person at random, therefore, the probability is 15/100, or 15%, that you will select a person who has had a speeding ticket in the past 12 months.

12. If you choose one man, the probability is 1/5, or 0.2, that he has had a ticket in the past 12 months. If you choose a woman, the probability is 1/10, or 0.1, that she has had a ticket in the past 12 months. Therefore, the probability that they have both had tickets in the past 12 months is $0.2 \cdot 0.1 = 0.02$, or 2%.

13. The probability is 1/2, or 50 percent, that the coin will come up heads, and 50 percent that it will come up tails, on the 15th toss (or on any toss). Every time you toss that coin, it's a brand new game!

CHAP. 17

1. The derivatives are as follows.
   (a) $dy/dx = 10x$
   (b) $dy/dx = 2x + 3$
   (c) $dy/dx = 2x + 3$
   (d) $dy/dx = 2x - \sin x$
   (e) $dy/dx = 12x^2 - 8x - 4$

(f) $dy/dx = \cos x - 2 \sin x$
(g) $dy/dx = 20x^3 + 4x$
(h) $dy/dx = -2x^{-3} - 16x^{-5}$

2. You can calculate that 30 mi/hr = 44 ft/s based on 5,280 ft/mi and 3,600 s/hr. Then note that 5 mi/hr = 30/6 mi/hr = 44/6 ft/s = 7.333 ft/s. The acceleration a is therefore 7.333 ft/s². You know that the initial velocity $v = 0$ ft/s, so you can surmise that v, in terms of the time t in seconds, is a formula whose derivative is

$$dv/dt = a = 7.333$$

Therefore, $v = 7.333t$. Assign the initial displacement $s = 0$ ft (the starting position). Then s, in terms of t, is a formula whose derivative is

$$ds/dt = v = 7.333t$$

If you are willing to think "un-derivative," you should not have much trouble guessing that

$$s = (7.333/2)t^2$$
$$= 3.667t^2$$

So when $t = 10$ s, you have

$$s = 3.667 \cdot 10^2$$
$$= 366.7 \text{ ft}$$

In 10 s the car will move 366.7 ft from its starting position.

3. Plug the value $t = 10$ into the above formula for v in terms of t. This gives you

$$v = 7.333 \cdot 10 = 73.33 \text{ ft/s}$$

4. In the second 10 s (from $t = 10$ to $t = 20$), the car will travel a distance of

$$73.33 \text{ ft/s} \cdot 10 \text{ s} = 733.3 \text{ ft}$$

You already know that in the first 10 s, the car traveled 366.7 ft. Therefore, the total distance traveled in 20 s is 733.3 + 366.7 = 1,100 ft.

5. You know that the acceleration is $a = dv/dt = 32$ ft/s². If you let $t = 0$ at the instant the ball is dropped, you can see (by thinking un-derivative) that $v = 32t$. That means $ds/dt = 32t$, where s is the displacement downward from the point at which the ball is dropped. If you think un-derivative again, you can figure out that $s = 16t^2$. The building is $20 \cdot 10 = 200$ ft high. Substituting 200 for s gives $200 = 16t^2$. Solving, you'll get $t = 3.54$ s. (The other solution to this quadratic, $t = -3.54$ s, is meaningless here.)

6. As you figured out the solution to Prob. 5, you determined that $v = 32t$. When you solved the whole problem, you learned that the ball will hit the ground 3.54 s after the instant it is dropped. Therefore, when the ball strikes the ground, the speed will be

$$v = 32 \cdot 3.54 = 113 \text{ ft/s}$$

7. Let a represent the acceleration in feet per second squared. Based on the slope-intercept form for a straight line, you can figure out from the graph that

$$a = t - 4$$

Because $a = dv/dt$, you can think un-derivative and see that

$$v = t^2/2 - 4t + c$$

where c is a constant. Given that the starting velocity is 44 ft/s, you know that $c = 44$. When you plug in this value, you get

$$v = t^2/2 - 4t + 44$$

8. Because $v = ds/dt$, you can think un-derivative with the results of Prob. 7 to get

$$s = t^3/6 - 2t^2 + 44t + c$$

Given that $s = 0$ when $t = 0$, you know that $c = 0$, so the formula becomes

$$s = t^3/6 - 2t^2 + 44t$$

9. Plug the time values into the equation derived in the solution to Prob. 7. Rounded off to four significant figures, the speeds are as follows.
   (a) 40.50 ft/s   (b) 38.00 ft/s
   (c) 36.50 ft/s   (d) 54.00 ft/s

10. Plug the time values into the equation derived in the solution to Prob. 8. Rounded off to four significant figures, the distances are as follows.
   (a) 42.17 ft   (b) 81.33 ft
   (c) 190.8 ft   (d) 406.7 ft

11. Note that one complete cycle, equivalent to 360°, takes place in 5 s. Then the angle q, in degrees, is equal to $72t$, where t is in seconds. The voltage V peaks positively and negatively at 8.00 times the maximum value of the sine function. Therefore

$$V = 8.00 \sin 72t$$

for angles expressed in degrees.

12. To find these values, you must first find the derivative of the function obtained in the solution to Prob. 11. Because the derivative of the sine is the cosine, you can determine that

$$dV/dt = 8.00 \cos 72t$$

for angles expressed in degrees. Plug in the values for t and use a calculator to get the following results in volts per second (V/s). Assume that all input values are exact. Then you can justify going to four significant figures.
   (a) +8.000 V/s   (b) −6.472 V/s
   (c) −8.000 V/s   (d) +8.000 V/s
   (e) +2.472 V/s   (f) −8.000 V/s

13. When you take advantage of a good scientific calculator with a pi (π) key, you can determine that there is approximately 0.01745329 radian per degree (rad/°), accurate to plenty of decimal places. The angle measures, to four significant digits, are as follows.

(a) 0.1745 rad (b) 0.5236 rad
(c) 1.309 rad (d) 2.531 rad
(e) 3.840 rad (f) 5.236 rad

14. When you employ a calculator with a pi (π) key, you can determine that there are 57.2957795 degrees per radian (°/rad), accurate to plenty of decimal places. The angle measures, to four significant digits, are as follows.

(a) 11.46° (b) 28.65°
(c) 57.30° (d) 97.40°
(e) 126.1° (f) 200.5°

15. To find the answer to (a), divide 360° by 365.25 days, to obtain 0.98563 degree per day (°/d) to five significant digits. To figure out (b), first multiply the answer (a) by exactly 30 to obtain 29.5689° for the month of April, which has 30 days. Then multiply this by the rad/° figure determined in Prob. 13, which yields 29.5689 · 0.01745329 = 0.51607 rad for April, rounded off to five significant figures.

CHAP. 18

1. First, derive a formula for the wave. The period is 3.000 s, corresponding to one complete cycle, or 2π rad. The value of 2π is approximately 6.283. The peak amplitude s of the wave is plus or minus 6.000 m. The wave has a sinusoidal shape and begins at (0,0), increasing positively, so it represents a multiple of a sine function. The formula for s in terms of time t is

$$s = 6.000 \sin (6.283/3.000)t$$
$$= 6.000 \sin 2.094t$$

You can differentiate this as a function of a function using the chain rule, obtaining

$$ds/dt = (6.000 \cos 2.094t) \cdot 2.094$$
$$= 12.56 \cos 2.094t$$

The velocity is greatest positively at $t = 0.000$, $t = 3.000$, $t = 6.000$, etc. This can be ascertained by observing the graph. Take the easiest example where $t = 0.000$. Then

$$\cos 2.094t = \cos 0.000$$
$$= 1.000$$

The speed at this instant is equal to 12.56 · 1.000 = 12.56 m/s. When you round this off to three significant figures, you get 12.6 m/s for the maximum velocity, representing upward movement.

2. You can infer this from the graph, and because you know the sine function is symmetric. The greatest negative (downward) velocity occurs when $t = 1.500$, $t = 4.500$, $t = 7.500$, etc., having the same magnitude as the maximum (upward) velocity but with a negative value. You could say that at these instants in time, the velocity is -12.6 m/s.

3. The acceleration a is greatest positively (upward) at the minimum displacement points on the graph, that is, at $t = 2.250$, $t = 5.250$, $t = 8.250$, etc. Therefore

$$a = d^2s/dt^2$$
$$= (-12.56 \sin 2.094t) \cdot 2.094$$
$$= -26.30 \sin 2.094t$$

The chain rule must be used again here, applying it to the first derivative function you found in the solution to Prob. 1. Plugging in $t = 2.250$ yields

$$a = -26.30 \sin (2.094 \cdot 2.250)$$
$$= -26.30 \cdot (-1.000)$$
$$= 26.30 \text{ m/s}^2$$

Remember that angles are specified in radians here, not in degrees. Rounding off to three significant figures yields $a = 26.3$ m/s^2 as the maximum positive (upward) acceleration.

4. The greatest negative acceleration occurs at the positive displacement peaks, that is, at $t = 0.750$, $t = 3.750$, $t = 6.750$, etc. Because the function is symmetric, you can surmise from the solution to Prob. 3 that at these points, $a = -26.3$ m/s^2.

5. The period is 1/3 as great as before, so it is 1.000 s. The period corresponds to 2π, or approximately 6.283 rad. The formula for s in terms of time t is therefore

$$s = 6.000 \sin (6.283/1.000)t$$
$$= 6.000 \sin 6.283t$$

Differentiating this yields

$$ds/dt = (6.000 \cos 6.283t) \cdot 6.283$$
$$= 37.70 \cos 6.283t$$

Again, note the use of the chain rule to determine the derivative of the composite function. The velocity is greatest positively at $t = 0.000$, $t = 1.000$, $t = 2.000$, etc. Take the easiest example where $t = 0.000$. Then

$$\cos 6.283t = \cos 0.000$$
$$= 1.000$$

The speed at this instant is equal to $37.70 \cdot 1.000 = 37.70$ m/s. Rounded to three significant figures, therefore, the maximum velocity is 37.7 m/s (upward).

6. The greatest negative (downward) velocity occurs at $t = 0.500$, $t = 1.500$, $t = 2.500$, etc. It has the same magnitude as the maximum upward velocity but with a negative value: -37.7 m/sec.

7. The acceleration a is greatest positively (upward) at the minimum displacement points on the graph, that is, at $t = 0.750$, $t = 1.750$, $t = 2.750$, etc. Therefore, once again using the chain rule, you obtain

$$a = d^2s/dt^2 = (-37.70 \sin 6.283t) \cdot 6.283$$
$$= -236.9 \sin 6.283t$$

Plugging in $t = 0.750$ gives you

$$a = -236.9 \sin (6.283 \cdot 0.750)$$
$$= -236.9 \cdot (-1.000)$$
$$= 236.9 \text{ m/s}^2$$

Remember that angles are specified in radians here, not in degrees. Rounding off to three significant figures yields $a = 237$ m/s^2 as the maximum positive (upward) acceleration.

8. The greatest negative (downward) acceleration occurs at the positive displacement peaks, that is, at $t = 0.250$, $t = 1.250$, $t = 3.250$, etc. Because the function is symmetric, you can surmise from the solution to Prob. 7 that at these points, $a = -237$ m/s^2.

9. The answers are derived as follows.

(a) $\sin^2 (\pi/2) = (1/2)(1 - \cos \pi)$
$= (1/2)[1 - (-1)]$
$= (1/2) \cdot 2$
$= 1.000$

(b) $\cos^2 (\pi/3) = (1/2)[1 + \cos (2\pi/3)]$
$= (1/2)(1 - 1/2)$
$= (1/2)(1/2)$
$= 0.2500$

(c) $\sin^3 (\pi/4) = (1/4)[3 \sin (\pi/4) - \sin (3\pi/4)]$
$= (1/4)[(3 \cdot 0.7071) - 0.7071]$
$= (1/4)(1.414)$
$= 0.3535$

(d) $\cos^3 (3\pi/4) = (1/4)[3 \cos (3\pi/4) + \cos (9\pi/4)]$
$= (1/4)[3 \cdot (-0.7071) + 0.7071]$
$= (1/4)(-1.414)$
$= -0.3535$

(e) $\sin^5 (\pi/3) = (1/16)[10 \sin (\pi/3) - 5 \sin \pi + \sin (5\pi/3)]$
$= (1/16)(10 \cdot 0.8660 - 5 \cdot 0 - 0.8660)$
$= (1/16)(7.794)$
$= 0.4871$

(f) $\cos^6 (2\pi/5) = (1/32)[10 + 15 \cos (4\pi/5) + 6 \cos (8\pi/5) + \cos (12\pi/5)]$
$= (1/32)[10 + 15 \cdot (-0.809017) + (6 \cdot 0.309017) + 0.309017]$
$= (1/32)(10 - 12.135255 + 1.854102 + 0.309017)$
$= 0.0008708$

10. You can begin by calculating the circumference of the circle c in terms of the radius r. Because $r = 10$ ft, you can determine that

$$c = 2\pi r = 62.8318 \text{ ft}$$

The ball traverses this distance every 2.000 s. The tangential speed v of the ball is therefore equal to 62.8318 divided by 2.000. Rounded to four significant figures, that's 31.42 ft/s.

11. The northward component of the ball's speed v_n is equal to the tangential speed (31.42 ft/s) multiplied by the cosine of the angle west of north. This angle is 45°, whose cosine is 0.7071. Therefore, v_n at this instant is

$$v_n = 31.42 \cdot 0.7071$$
$$= 22.22 \text{ ft/s}$$

CHAP. 19

1. Use the product formula to derive the answers. This formula can be written in "shorthand" notation as

$$(uv)' = u'v + v'u$$

where the accent symbol (prime) represents the derivative.
  (a) $(uv)' = 2x^3 + (3x^2)(2x + 3)$
       $= 2x^3 + 6x^3 + 9x^2 = 8x^3 + 9x^2$
  (b) $(uv)' = (2x - 6)(3 \sin x) + (3 \cos x)(x^2 - 6x + 4)$
       $= (6x - 18) \sin x + (3x^2 - 18x + 12) \cos x$
  (c) $(uv)' = \cos^2 x - \sin^2 x$
  (d) $(uv)' = (5x^4)(-2x^3 + 2) + (-6x^2)(x^5 - 4)$
       $= -10x^7 + 10x^4 - 6x^7 + 24x^2$
       $= -16x^7 + 10x^4 + 24x^2$

2. Use the quotient formula to derive the answers. In shorthand this formula is

$$(u/v)' = (u'v - v'u)/v^2$$

where the accent symbol (prime) represents the derivative.
  (a) $(u/v)' = [(-3)(x^2 + 2) - (2x)(-3x - 6)]/(x^4 + 4x^2 + 4)$
       $= (3x^2 + 12x - 6)/(x^4 + 4x^2 + 4)$
  (b) $(u/v)' = [(2x + 2)(\sin x) - (\cos x)(x^2 + 2x)]/\sin^2 x$
       $= (2x \sin x + 2 \sin x - x^2 \cos x - 2x \cos x)/\sin^2 x$
  (c) $(u/v)' = [(\cos x)(3 + \cos x) - (-\sin x)(\sin x)]/(\cos^2 x + 6 \cos x + 9)$
       $= (3 \cos x + \cos^2 x + \sin^2 x)/(\cos^2 x + 6 \cos x + 9)$
       $= (1 + 3 \cos x)/(\cos^2 x + 6 \cos x + 9)$
  (d) $(u/v)' = [(12x^3)(-2x^2) - (-4x)(3x^4 - 4)]/4x^4$
       $= (-24x^5 + 12x^4 - 16)/4x^4$
       $= (-6x^5 + 3x^4 - 4)/x^4$

3. Calculus is not necessary to solve this problem, although you can use it if you want. It is only necessary to add 90°, or $\pi/2$ (approximately 1.5708) rad, to the angles given and then find the tangent of the resulting angle. The answers are as follows, accurate to four significant figures, assuming the input values are exact. Angles in (a) through (e) are in degrees; angles in (f) through (h) are in radians.
  (a) slope $= \tan (10 + 90)°$
       $= \tan 100°$
       $= -5.671$

(b) slope $= \tan (55 + 90)°$
$= \tan 145°$
$= -0.7002$

(c) slope $= \tan (105 + 90)°$
$= \tan 195°$
$= 0.2679$

(d) slope $= \tan (190 + 90)°$
$= \tan 280°$
$= -5.671$

(e) slope $= \tan (300 + 90)°$
$= \tan 390°$
$= \tan 30°$
$= 0.5774$

(f) slope $= \tan (1 + 1.5708)$
$= \tan 2.5708$
$= -0.6421$

(g) slope $= \tan (2 + 1.5708)$
$= \tan 3.5708$
$= 0.4577$

(h) slope $= \tan (4 + 1.5708)$
$= \tan 5.5708$
$= -0.8637$

4. The derivative is the slope of a function at a given point. If we let the derivative of this function be the slope of the graph in volts per millisecond (V/ms), then the graph of the output will appear as shown in Fig. A-7. The vertical lines represent instantaneous voltage transitions. Mathematically they are undefined in slope. This waveform is known as a *square wave*.

5. In a practical circuit, the output of the second differentiator will be continuous and will have a value of zero volts (0 V). There will be no output at all! Mathematically, the

Figure A-7

Illustration for solutions to Probs. 4 and 5 in Chap. 19.

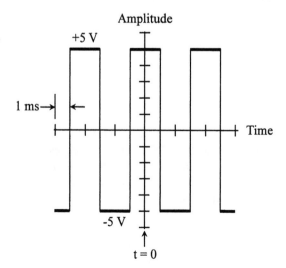

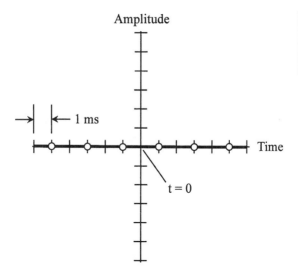

vertical lines in the graph of Fig. A-7 cannot be differentiated because their slope is undefined. The graph of the derivative of the function shown by Fig. A-7 will therefore look like Fig. A-8. The open circles represent points "missing" from the line. In a practical circuit these "missing" points have no effect, because they have zero duration.

6. First, integrate the voltage-vs.-time function represented by

$$f_1(t) = 5t$$

from $t = 0$ to $t = 1$. The indefinite integral of this is $2.5t^2$. The constant of integration can be ignored because we will take the definite integral. The curve has a parabolic (quadratic) shape and attains a value of $+2.5$ V at $t = 1$ ms. The second straight-line part of the function, which slopes downward from $t = 1$ to $t = 3$, has the equation

$$f_2(t) = -5t + 10$$

The indefinite integral of this is

$$-2.5t^2 + 10t + c$$

Evaluated from 1 to 2, the value of this is 2.5, which adds to the existing value of $+2.5$ V to result in $+5$ V at $t = 2$. The curve is parabolic as in the first part, but curved downward rather than upward. Evaluated from 2 to 3, this function is -2.5, so the parabola continues to curve downward, reaching $+2.5$ V at $t = 3$. The next straight-line part of the function slopes up like the first part, so the parabola starts upward again. From this, it is apparent that the waveform is periodic (that is, it repeats itself over and over) and consists of sections of parabolas, having maximum peaks at $+5$ V and minimum peaks at 0 V. The graph is shown in Fig. A-9. (This looks a lot like an elevated sine wave, but it really isn't!)

7. To get the second integral, we must integrate the function graphed in Fig. A-9. Start at the point $t = 0$ and $v = 0$. First integrate the function

$$g_1(t) = 2.5t^2$$

Figure A-9

Illustration for solutions to Probs. 6 and 7 in Chap. 19.

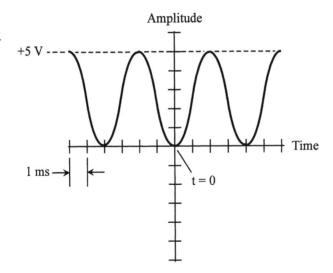

from $t = 0$ to $t = 1$. The indefinite integral is $5t^3/6$. The constant of integration can be ignored because we will take the definite integral. The curve has a cubic function shape, begins at zero, and ends up at $+5/6$ V at $t = 1$ ms. In decimal form this is approximately $+0.8333$ V. Next, we integrate the function

$$g_2(t) = -2.5t^2 + 10t$$

from $t = 1$ to $t = 3$. The indefinite integral is

$$-5t^3/6 + 5t^2 + c$$

Evaluated from $t = 1$ to $t = 3$, the value of this expression is $22.5 - 4.1667 = +18.333$ V, which adds to the initial value of $+0.8333$ V to produce $+19.1667$ V at $t = 3$ ms. The integral (area under the curve) from $t = 3$ to $t = 4$ can be determined, upon visual inspection, to be the same as the first part from $t = 0$ to $t = 1$, that is, $+0.8333$ V. Therefore, at $t = 4$ ms, the function attains a value of $19.1667 + 0.8333 = +20$ V. This completes the evaluation of the first full wave cycle from $t = 0$ to $t = 4$. During the next cycle, the wave increases by 20 V more, to $+40$ V. In theory, the voltage continues to increase without limit as time passes, so the waveform, graphed starting at $t = 0$, will look like the one shown in Fig. A-10. (In a practical circuit, the voltage cannot increase without limit, but must level off at the voltage provided by the power source.)

8. The integral functions, denoted $F(x)$, are as follows. The constant of integration, in each case, is represented by c.

    (a) $F(x) = x^4 + c$ (b) $F(x) = x^6 + c$
    (c) $F(x) = x^9 + c$ (d) $F(x) = x^5/5 + c$
    (e) $F(x) = 2 \sin x + c$ (f) $F(x) = 4 \cos x + c$

9. The above functions are solved for $x = 1$ and for $x = -1$. Then the difference is found, as follows:

$$f(1) - f(-1)$$

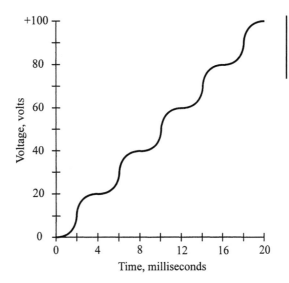

Figure A-10
Illustration for
solution to Prob. 7
in Chap. 19.

In every case, the constants of integration cancel out, so they need not be included in the calculations.

(a) $F(1) = 1$ and $F(-1) = 1$,
so the definite integral is $1 - 1 = 0$.

(b) $F(1) = 1$ and $F(-1) = 1$,
so the definite integral is $1 - 1 = 0$.

(c) $F(1) = 1$ and $F(-1) = -1$,
so the definite integral is $1 - (-1) = 2$.

(d) $F(1) = 1/5$ and $F(-1) = -1/5$,
so the definite integral is $1/5 - (-1/5) = 2/5$.

(e) $F(1) = 1.683$ and $F(-1) = -1.683$,
so the definite integral is $1.683 - (-1.683) = 3.366$.

(f) $F(1) = 2.161$ and $F(-1) = 2.161$,
so the definite integral is $2.161 - 2.161 = 0$.

10. Before it is possible to find the areas under the curve, the function itself must be known. We are told it is quadratic. By checking the points shown on the graph, it can be determined that

$$y = [(x + 1)^2/2] - 2$$
$$= (1/2)x^2 + x - 3/2$$

This function, call it $f(x)$, is integrated to obtain

$$\int [(1/2)x^2 + x - 3/2] \, dx = (1/6)x^3 + (1/2)x^2 - (3/2)x + c$$

where c is the constant of integration, which cancels out when definite integrals are calculated. Call the antiderivative function $F(x)$. Then the solutions are as follows.

(a) $F(0) = 0$ and $F(-2) = 11/3$,
so the definite integral is $0 - 11/3 = -11/3$.

(b) $F(1) = -5/6$ and $F(0) = 0$,
so the definite integral is $-5/6 - 0 = -5/6$.

(c) $F(2) = 1/3$ and $F(-1) = 11/6$,
so the definite integral is $1/3 - 11/6 = -3/2$.
(d) $F(4) = 38/3$ and $F(-1) = 11/6$,
so the definite integral is $38/3 - 11/6 = 65/6$.

11. Again, the function shown in the graph of Fig. 19-13 is

$$y = f(x) = (1/2)x^2 + x - 3/2$$

which is an indefinite integral of its derivative ($-3/2$ being a constant that cancels out in definite integral calculations). To find the areas, we can evaluate the graphed function, modified to get rid of the constant $-3/2$, between the limiting values. Call that function $f_1(x)$, so

$$f_1(x) = (1/2)x^2 + x$$

The results are as follows.
(a) $f_1(0) = 0$ and $f_1(-2) = 0$,
so the definite integral is $0 - 0 = 0$.
(b) $f_1(1) = 3/2$ and $f_1(0) = 0$,
so the definite integral is $3/2 - 0 = 3/2$.
(c) $f_1(2) = 4$ and $f_1(-1) = -1/2$,
so the definite integral is $4 - (-1/2) = 9/2$.
(d) $f_1(4) = 12$ and $f_1(-1) = -1/2$,
so the definite integral is $12 - (-1/2) = 25/2$.

CHAP. 20

1. The maximum is at the point where the derivative is 0. The derivative is

$$f'(x) = -6x + 2$$

Solving the equation $-6x + 2 = 0$ tells us that $x = 1/3$. Then at this point, we have

$$y = -3 \cdot (1/3)^2 + 2 \cdot 1/3 + 2 = 7/3$$

The maximum therefore occurs at $(x,y) = (1/3, 7/3)$. We know it is a maximum rather than a minimum because the coefficient of the x^2 part of the equation is negative, making the parabola open downward.

2. When the slope of the function is maximum, the derivative reaches a maximum. The derivative of the function is

$$f'(x) = 2 \cos x$$

The cosine function attains infinitely many maximum values of 1. Therefore, $2 \cos x$ attains infinitely many maximum values of 2, which occur when x, in radians, is equal to any integer multiplied by 2π. Therefore, the slope of f is maximum at points $(x, y) = (2\pi n, 2)$, where n can be any integer.

3. The edge of the region of light will be a circle if and only if the flashlight is pointed straight down so the axis of the beam is perpendicular to the surface of the ice.

4. The edge of the region of light will be an ellipse if and only if the beam axis subtends an angle of more than $20°$ with respect to the surface of the ice.

5. The edge of the region of light will be a parabola if and only if the beam axis subtends an angle of exactly 20° with respect to the surface of the ice.

6. The edge of the region of light will be a half-hyperbola if and only if one of the following conditions is met:

- The beam axis intersects the lake surface and subtends an angle of less than 20° with respect to the surface of the ice.
- The beam axis is aimed at the horizon.
- The beam axis is aimed into the sky at an angle of less than 20° above the horizon.

7. The characteristics of the functions are as follows.

(a) If $x = 0$, then $f(x) = -5$, so the y intercept is at $(0, -5)$. If $f(x) = 0$, then $2x^3 = 5$, so $x = (5/2)^{1/3} = 1.357$ and the x intercept is at $(1.357, 0)$. When we take the first derivative, we obtain

$$f'(x) = 6x^2$$

The slope of the curve defined by this function is equal to 0 when $6x^2 = 0$. This equation has only one solution, $x = 0$, corresponding to the point $(0, -5)$. It indicates a point of inflection because f is a cubic function. There are no local maxima or minima.

(b) This function is stated in factored form. Multiplied out, it is

$$f(x) = x^3 - 4x^2 + 3x - 12$$

If $x = 0$, then $f(x) = -12$, so the y intercept is at $(0, -12)$. If $f(x) = 0$, then $x = 4$ is the only real number solution (a fact that we can see by looking at the equation in its factored form), so the x intercept is at $(4,0)$. When we take the first derivative, we obtain

$$f'(x) = 3x^2 - 8x + 3$$

Setting $f'(x) = 0$, which will tell us the points at which the slope of the curve is 0, we find there are two real solutions, $x = 2.215$ and $x = 0.4514$ (accurate to four significant figures), which can be found by using the quadratic formula. When we "plug in" numbers and calculate the value of $f(x)$, the points are $(2.215, -14.11)$ and $(0.4514, -11.37)$. The second derivative can reveal whether these points represent local minima (second derivative positive), local maxima (second derivative negative), or inflection points (second derivative equal to 0):

$$f''(x) = 6x - 8$$

When $x = 2.215$, we have $f''(x) = 5.290$, representing a local minimum. When $x = 0.4514$, we have $f''(x) = -5.292$, representing a local maximum. The second derivative is 0 when $x = 4/3 = 1.333$ and $f(x) = -12.74$. The only point of inflection in the curve is $(1.333, -12.74)$. The slope of the curve is not equal to 0 at this inflection point, but the point defines the transition from the range where the curve is concave downward to the range where the curve is concave upward.

(c) This function has no local maxima or minima. Its shape is similar to that of the tangent function, except "stretched" vertically by a factor of 2. Inflection points occur on the x axis at integral multiples of π rad, that is, at points $(\pi n, 0)$, where n can be any integer. The curve does not have a slope of 0 at these inflection points, but the points define transitions from concave downward to concave upward.

Figure A-11

Illustration for solution to Prob. 8 (a) in Chap. 20.

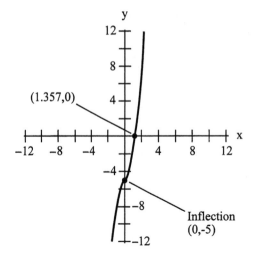

(1.357,0)

Inflection (0,-5)

(*d*) This curve is an inverted cosine wave, "stretched" vertically by a factor of 3. Local minima occur where *x* is an even integral multiple of π and *y* = −3. Local maxima occur where *x* is an odd integral multiple of π and *y* = 3. Inflection points occur where *x* is an odd integral multiple of $\pi/2$ and *y* = 0. The curve does not have a slope of 0 at these inflection points, but the points define transitions from concave downward to concave upward or vice versa.

8. The graphs of the functions given in Prob. 7 are shown here.
   (*a*) Refer to Fig. A-11.
   (*b*) Refer to Fig. A-12.
   (*c*) Refer to Fig. A-13.
   (*d*) Refer to Fig. A-14.

Figure A-12

Illustration for solution to Prob. 8 (b) in Chap. 20.

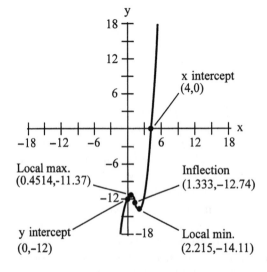

x intercept (4,0)

Local max. (0.4514,−11.37)

Inflection (1.333,−12.74)

y intercept (0,−12)

Local min. (2.215,−14.11)

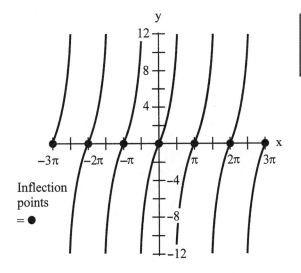

Figure A-13

Illustration for solution to Prob. 8 (c) in Chap. 20.

CHAP. 21

1. Refer to Fig. A-15. Note that each radial division represents 1/2 unit. This polar graph does not show the negative values for y in Fig. 21-19.

2. Refer to Fig. A-16. Note that each radial division represents 3 units. This polar graph does not show the negative values for x in Fig. 21-19, and it accounts for points much farther in the positive x direction than is shown by Fig. 21-19.

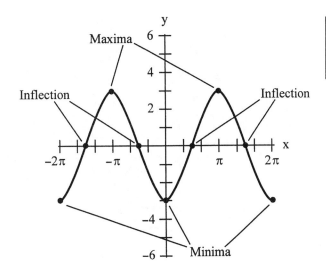

Figure A-14

Illustration for solution to Prob. 8 (d) in Chap. 20.

Figure A-15

Illustration for solution to Prob. 1 in Chap. 21.

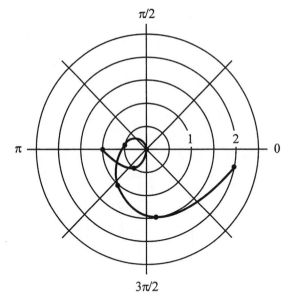

3. Refer to Fig. A-17. Note that x in this graph is the equivalent in angular degrees from Fig. 21-20, so the increments along the x axis each represent 25 units.

4. Refer to Fig. A-18. Note that y in this graph is the equivalent in angular degrees from Fig. 21-20, so the increments along the y axis each represent 25 units.

Figure A-16

Illustration for solution to Prob. 2 in Chap. 21.

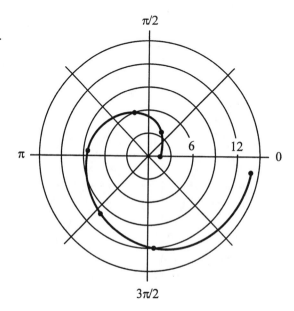

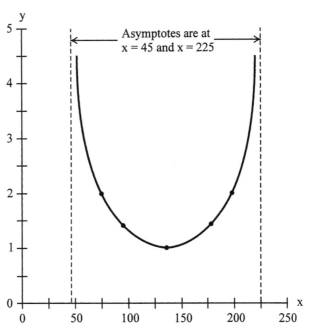

Figure A-17

Illustration for solution to Prob. 3 in Chap. 21.

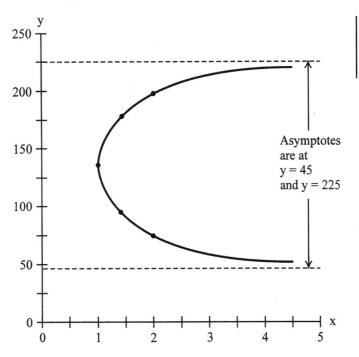

Figure A-18

Illustration for solution to Prob. 4 in Chap. 21.

5. If we know the rectangular coordinates (x_0, y_0), we can convert to mathematician's polar coordinates (θ_0, r_0) by using these formulas:

$$\theta_0 = \arctan (y_0/x_0) \text{ if } x_0 > 0$$
$$\theta_0 = 180° + \arctan (y_0/x_0) \text{ if } x_0 < 0 \text{ (for } \theta_0 \text{ in degrees)}$$
$$\theta_0 = \pi + \arctan (y_0/x_0) \text{ if } x_0 < 0 \text{ (for } \theta_0 \text{ in radians)}$$
$$r_0 = (x_0^2 + y_0^2)^{1/2}$$

The solutions are found by substituting the relevant values x_0 and y_0 into these formulas, as follows. In these examples, it is easy to derive exact solutions, given that the input values are exact.

(a) In this case, $x_0 = 1$ and $y_0 = 1$. We have

$$\theta_0 = \arctan (1/1)$$
$$= \arctan 1$$
$$= \pi/4$$

and

$$r_0 = (1^2 + 1^2)^{1/2}$$
$$= (1 + 1)^{1/2}$$
$$= 2^{1/2}$$

Therefore

$$(\theta_0, r_0) = (\pi/4, 2^{1/2})$$

(b) In this case, $x_0 = -1$ and $y_0 = 1$. We have

$$\theta_0 = \pi + \arctan [1/(-1)]$$
$$= \pi + \arctan (-1)$$
$$= \pi + (-\pi/4)$$
$$= 3\pi/4$$

and

$$r_0 = [(-1)^2 + 1^2]^{1/2}$$
$$= (1 + 1)^{1/2}$$
$$= 2^{1/2}$$

Therefore

$$(\theta_0, r_0) = (3\pi/4, 2^{1/2})$$

(c) In this case, $x_0 = 1$ and $y_0 = -1$. We have

$$\theta_0 = \arctan (-1/1)$$
$$= \arctan (-1)$$
$$= -\pi/4$$
$$= 7\pi/4$$

and

$$r_0 = [1^2 + (-1)^2]^{1/2}$$
$$= (1 + 1)^{1/2}$$
$$= 2^{1/2}$$

Therefore

$$(\theta_0, r_0) = (7\pi/4, 2^{1/2})$$

(d) In this case, $x_0 = -1$ and $y_0 = -1$. We have

$$\theta_0 = \pi + \arctan [(-1)/(-1)]$$
$$= \pi + \arctan 1$$
$$= \pi + \pi/4$$
$$= 5\pi/4$$

and

$$r_0 = [(-1)^2 + (-1)^2]^{1/2}$$
$$= (1 + 1)^{1/2}$$
$$= 2^{1/2}$$

Therefore

$$(\theta_0, r_0) = (5\pi/4, 2^{1/2})$$

6. Recall the conversion formulas for translating the coordinates for a point (α_0, r_0) in navigator's polar coordinates to a point (x_0, y_0) in the rectangular coordinates. Given that the input values are exact, the solutions for (a) and (d) here are exact, while the solutions for (b) and (c) are rounded off to four significant figures.

$$x_0 = r_0 \sin \alpha_0$$

and

$$y_0 = r_0 \cos \alpha_0$$

The solutions are found by substituting the relevant values of α_0 and r_0 into these formulas, as follows.

(a) In this case, $\alpha_0 = 90°$ and $r_0 = 5$. We have

$$x_0 = 5 \sin 90°$$
$$= 5 \cdot 1$$
$$= 5$$

and

$$y_0 = 5 \cos 90°$$
$$= 5 \cdot 0$$
$$= 0$$

Therefore

$$(x_0, y_0) = (5,0)$$

(b) In this case, $\alpha_0 = 135°$ and $r_0 = 2$. We have

$$x_0 = 2 \sin 135°$$
$$= 2.000 \cdot 0.7071$$
$$= 1.414$$

and

$$y_0 = 2 \cos 135°$$
$$= 2.000 \cdot (-0.7071)$$
$$= -1.414$$

Therefore

$$(x_0, y_0) = (1.414, -1.414)$$

(c) In this case, $\alpha_0 = 225°$ and $r_0 = 3$. We have

$$x_0 = 3 \sin 225°$$
$$= 3.000 \cdot (-0.7071)$$
$$= -2.121$$

and

$$y_0 = 3 \cos 225°$$
$$= 3.000 \cdot (-0.7071)$$
$$= -2.121$$

Therefore

$$(x_0, y_0) = (-2.121, -2.121)$$

(d) In this case, $\alpha_0 = 270°$ and $r_0 = 4$. We have

$$x_0 = 4 \sin 270°$$
$$= 4 \cdot (-1)$$
$$= -4$$

and

$$y_0 = 4 \cos 270°$$
$$= 4 \cdot 0$$
$$= 0$$

Therefore

$$(x_0, y_0) = (-4,0)$$

If these answers seem wrong to you at first, remember that we are working in navigator's polar coordinates, not in mathematician's polar coordinates!

7. The equations in (a), (b), and (c) represent Euclidean, or "flat," geometric planes. The equation in (d) represents a straight line.
   (a) The plane defined by $x = 5$ is parallel to the plane formed by the y and z axes and passes through the point $(x,y,z) = (5,0,0)$ on the x axis.
   (b) The plane defined by $y = 7$ is parallel to the plane formed by the x and z axes and passes through the point $(x,y,z) = (0,7,0)$ on the y axis.
   (c) The plane defined by $z = -5$ is parallel to the plane formed by the x and y axes and passes through the point $(x,y,z) = (0,0,-5)$ on the z axis.
   (d) The line defined by $x = y = z$ passes through the origin and subtends equal angles with all six of the semiaxes (positive x, negative x, positive y, negative y, positive z, and negative z). In that sense it can be informally described as running "kitty-corner" through xyz space.

8. Examine Fig. 21-17A carefully. Use your imagination to deduce and envision the following answers. Remember that negative radii and nonstandard direction angles are allowed.
   (a) The equation $\theta = \pi/2$ represents the flat plane defined by the y and z axes as illustrated in Fig. 21-17A. It is a complete plane (not just a half-plane) because negative radius values are allowed.
   (b) The equation $r = 4$ represents a vertically oriented cylinder of radius 4 units, whose axis corresponds with the z axis as shown in Fig. 21-17A. The cylinder extends infinitely upward and downward around the z axis.
   (c) The equation $h = -3$ represents a flat plane parallel to, and 3 units below, the plane formed by the x and y axes as shown in Fig. 21-17A.
   (d) The equation $r = h$ represents a double cone whose apex is at the origin, whose axis corresponds to the z axis as shown in Fig. 21-17A, and whose *flare angle* (the angle between the axis and any line on the cone surface passing through the origin) is 45° or $\pi/4$ rad, assuming the unit increments for r and h are of equal size.

9. Examine Fig. 21-18C carefully. Use your imagination to deduce and envision the following answers. Remember that negative radii and nonstandard direction angles are allowed.
   (a) The equation $r = 2$ represents a sphere of radius 2 units, centered at the origin as shown in Fig. 21-18C.
   (b) The equation $\theta = 45°$ represents a double cone whose apex is at the origin, whose axis corresponds to the zenith-nadir axis as shown in Fig. 21-18C, and whose flare angle is 45°. It is a double cone because negative radii are allowed.
   (c) The equation $\phi = 45°$ represents a flat, vertical plane passing through the zenith-nadir axis as shown in Fig. 21-18C and oriented in a northeast-southwest direction. That is, the plane intersects the plane of the horizon in a straight line running exactly northeast by southwest (azimuth 45° and 225°) and passing through the origin. It is a complete plane, not a half-plane, because negative radius values are allowed.

CHAP. 22

1. The product is multiplied out as follows.

$$(0.6 + j0.8)(0.8 + j0.6) = 0.48 + j0.36 + j0.64 + (j^2 \cdot 0.48)$$
$$= 0.48 + j - 0.48$$
$$= 0 + j1$$
$$= j$$

This number is pure imaginary and has a magnitude of 1, because its point on the complex plane is exactly 1 unit away from the origin, assuming the input values are exact.

2. The squares are

$$(0.6 + j0.8)(0.6 + j0.8) = 0.36 + j0.48 + j0.48 + (j^2 \cdot 0.64)$$
$$= 0.36 + j0.96 - 0.64$$
$$= -0.28 + j0.96$$

and

$$(0.8 + j0.6)(0.8 + j0.6) = 0.64 + j0.48 + j0.48 + (j^2 \cdot 0.36)$$
$$= 0.64 + j0.96 - 0.36$$
$$= 0.28 + j0.96$$

The final product is

$$(-0.28 + j0.96)(0.28 + j0.96) = -0.0784 - j0.2688 + j0.2688 + (j^2 \cdot 0.9216)$$
$$= -0.0784 - 0.9216$$
$$= -(0.0784 + 0.9216)$$
$$= -1$$

This quantity is pure real and has a magnitude (absolute value) equal to precisely 1, assuming the input values are exact.

3. These equations can be solved by using the quadratic formula. The answers are derived as follows, based on the standard quadratic form $ax^2 + bx + c = 0$.
   (a) Here, the coefficients are $a = 1$, $b = -2$, and $c = 2$. When you input these values into the quadratic formula, you get

$$x = [2 + (4 - 8)^{1/2}]/2$$

   or

$$x = [2 - (4 - 8)^{1/2}]/2$$

   which solves to $x = 1 + j$ or $x = 1 - j$ in lowest terms.
   (b) In this case, the coefficients are $a = 1$, $b = -2$, and $c = 10$. When you input these values into the quadratic formula, you get

$$x = [2 + (4 - 40)^{1/2}]/2$$

or

$$x = [2 - (4 - 40)^{1/2}]/2$$

which solves to $x = 1 + j3$ or $x = 1 - j3$ in lowest terms.

(c) Here, the coefficients are $a = 13$, $b = -4$, and $c = 1$. When you input these values into the quadratic formula, you get

$$x = [4 + (16 - 52)^{1/2}]/26$$

or

$$x = [4 - (16 - 52)^{1/2}]/26$$

which solves to $x = (2 + j3)/13$ or $x = (2 - j3)/13$ in lowest terms. You can put these in proper complex form as $x = 2/13 + j3/13$ or $x = 2/13 - j3/13$.

(d) In this equation, $a = 1$, $b = -j2$, and $c = -10$. When you input these values into the quadratic formula, you get

$$x = [j2 + (4 + 40)^{1/2}]/2$$

or

$$x = [j2 - (4 + 40)^{1/2}]/2$$

which solves to $x = 11^{1/2} + j$ or $x = -11^{1/2} + j$ in lowest terms.

(e) Here, the coefficients are $a = 1$, $b = j2$, and $c = -8$. When you input these values into the quadratic formula, you get

$$x = [j2 + (4 + 32)^{1/2}]/2$$

or

$$x = [j2 - (4 + 32)^{1/2}]/2$$

which solves to $x = 3 + j$ or $x = -3 + j$ in lowest terms.

4. Refer to Fig. A-19. The vectors each have radius 2 and are spaced equally around a circle. Therefore, they are separated by $360°/6 = 60°$. Let the roots be represented by r_1 through r_6. (Note that $j2$ in these expressions represents j times 2. These are not typos; they are not meant to indicate j squared!)

$$r_1 = 2 \cos 0° + j2 \sin 0° = 2.00$$

$$r_2 = 2 \cos 60° + j2 \sin 60° = 1.00 + j1.73$$

$$r_3 = 2 \cos 120° + j2 \sin 120° = -1.00 + j1.73$$

$$r_4 = 2 \cos 180° + j2 \sin 180° = -2.00$$

$$r_5 = 2 \cos 240° + j2 \sin 240° = -1.00 - j1.73$$

$$r_6 = 2 \cos 300° + j2 \sin 300° = 1.00 - j1.73$$

Figure A-19

Illustration for solution to Prob. 4 in Chap. 22.

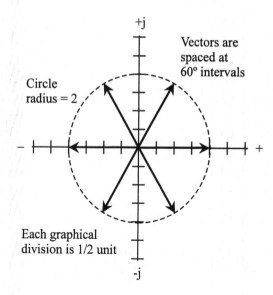

5. Refer to Fig. A-20. The vectors each have radius 2 and are spaced equally around a circle. Therefore, they are separated by 360°/10 = 36°. Let the roots be represented by r_1 through r_{10}. (Note that $j2$ in these expressions represents j times 2. These are not typos; they are not meant to indicate j squared!)

$$r_1 = 2 \cos 0° + j2 \sin 0° = 2.00$$
$$r_2 = 2 \cos 36° + j2 \sin 36° = 1.62 + j1.18$$
$$r_3 = 2 \cos 72° + j2 \sin 72° = 0.618 + j1.90$$
$$r_4 = 2 \cos 108° + j2 \sin 108° = -0.618 + j1.90$$
$$r_5 = 2 \cos 144° + j2 \sin 144° = -1.62 + j1.18$$

Figure A-20

Illustration for solution to Prob. 5 in Chap. 22.

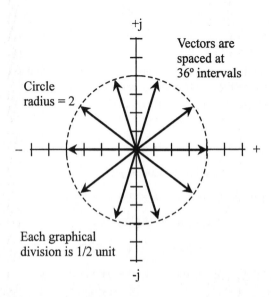

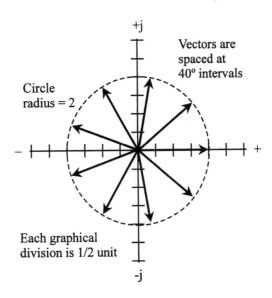

+j

Vectors are
spaced at
40° intervals

Circle
radius = 2

Each graphical
division is 1/2 unit

-j

$r_6 = 2 \cos 180° + j2 \sin 180° = -2.00$

$r_7 = 2 \cos 216° + j2 \sin 216° = -1.62 - j1.18$

$r_8 = 2 \cos 252° + j2 \sin 252° = -0.618 - j1.90$

$r_9 = 2 \cos 288° + j2 \sin 288° = 0.618 - j1.90$

$r_{10} = 2 \cos 324° + j2 \sin 324° = 1.62 - j1.18$

6. Refer to Fig. A-21. The vectors each have radius 2 and are spaced equally around a circle. Therefore, they are separated by 360°/9 = 40°. Let the roots be represented by r_1 through r_9. (Note that $j2$ in these expressions represents j times 2. These are not typos; they are not meant to indicate j squared!)

$r_1 = 2 \cos 0° + j2 \sin 0° = 2.00$

$r_2 = 2 \cos 40° + j2 \sin 40° = 1.53 + j1.29$

$r_3 = 2 \cos 80° + j2 \sin 80° = 0.347 + j1.97$

$r_4 = 2 \cos 120° + j2 \sin 120° = -1.00 + j1.73$

$r_5 = 2 \cos 160° + j2 \sin 160° = -1.88 + j0.684$

$r_6 = 2 \cos 200° + j2 \sin 200° = -1.88 - j0.684$

$r_7 = 2 \cos 240° + j2 \sin 240° = -1.00 - j1.73$

$r_8 = 2 \cos 280° + j2 \sin 280° = 0.347 - j1.97$

$r_9 = 2 \cos 320° + j2 \sin 320° = 1.53 - j1.29$

7. When you want to add two complex numbers of the form $a + jb$, add their real number and imaginary number components individually. In this case the result is

$$(-3 + j5) + (-2 - j3) = [-3 + (-2)] + [j5 + (-j3)]$$
$$= (-3 - 2) + j(5 - 3)$$
$$= -5 + j2$$

Figure A-22

Illustration for
solution to Prob. 8
in Chap. 22.

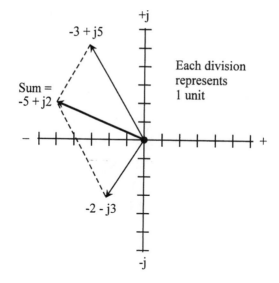

8. Refer to Fig. A-22. This summation obeys the "parallelogram rule" for the addition of vectors in rectangular coordinates. The endpoints of the sum vector and the two addend vectors, along with the origin, lie at the vertices of a parallelogram.

9. To multiply these two complex numbers, treat them as binomials and proceed as follows.

$$(-3 + j5)(-2 - j3) = (-3)(-2) + (-3)(-j3) + (j5)(-2) + (j5)(-j3)$$
$$= 6 + j9 - j10 + 15$$
$$= 21 - j1$$

10. Refer to Fig. A-23. Note that each increment on the coordinate axes represents 4 units.

Figure A-23

Illustration for
solution to Prob.
10 in Chap. 22.

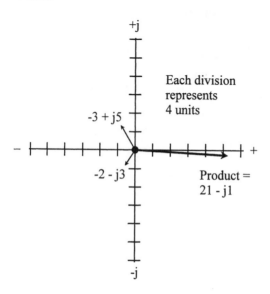

11. First find the reciprocal of the second complex number. Then multiply the result by the first complex number. The reciprocation proceeds as follows.

$$1/(-2 - j3) = (-2 + j3)/[(-2 + j3)(-2 - j3)]$$
$$= (-2 + j3)/(4 + 9)$$
$$= (-2 + j3)/13$$
$$= -2/13 + j(3/13)$$

The multiplication is a little "messy" and the signs are tricky, but the process is straightforward:

$$(-3 + j5)[-2/13 + j(3/13)] = 6/13 + j(-9/13) + j(-10/13) - 15/13$$
$$= -9/13 - j(19/13)$$

12. Refer to Fig. A-24. Note that each increment on the coordinate axes represents 1/4 of a unit.

13. First find the reciprocal of the second complex number. Then multiply the result by the first complex number. The reciprocation proceeds as follows.

$$1/(-3 + j5) = (-3 - j5)/[(-3 - j5)(-3 + j5)]$$
$$= (-3 - j5)/(9 + 25)$$
$$= (-3 - j5)/34$$
$$= -3/34 - j(5/34)$$

The multiplication process goes like this:

$$(-2 - j3)[-3/34 - j(5/34)] = 6/34 + j(10/34) + j(9/34) - 15/34$$
$$= -9/34 + j(19/34)$$

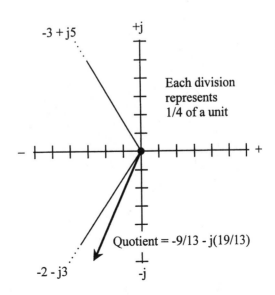

-3 + j5

+j

Each division represents 1/4 of a unit

Quotient = -9/13 - j(19/13)

-2 - j3

-j

Figure A-24

Illustration for solution to Prob. 12 in Chap. 22.

Figure A-25

Illustration for
solution to Prob. 14
in Chap. 22.

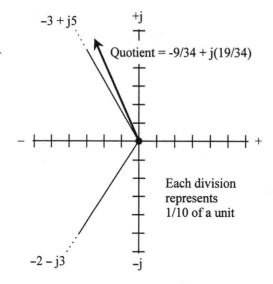

14. Refer to Fig. A-25. Note that each increment on the coordinate axes represents 1/10 of a unit.

15. When you are multiplying complex numbers in polar coordinates, the angles add while the magnitudes multiply. Therefore, the product vector has angle 60° + 330° = 390° (which reduces to 30°) and magnitude 2 · 4 = 8.

16. Refer to Fig. A-26. Note that each radial division in the coordinate system represents 2 units.

Figure A-26

Illustration for
solution to Prob. 16
in Chap. 22.

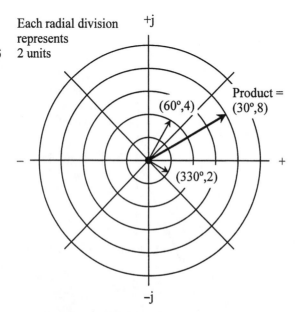

17. Use the formulas from Fig. 22-7. The angle and radius are $\theta = 30°$ and $r = 8$. So the real coefficient a is

$$a = r \cos \theta$$
$$= 8 \cdot 0.866$$
$$= 6.93$$

and the imaginary coefficient b is

$$b = r \sin \theta$$
$$= 8 \cdot 0.500$$
$$= 4.00$$

Therefore, the complex number is $6.93 + j4.00$.

CHAP. 23

1. This situation is illustrated in Fig. A-27. The tangent of the azimuth angle is equal to 30 mi divided by 20 mi, or 1.5. Using a calculator, you can find that the resulting angle is the arctangent of 1.5. This is 56° to the nearest degree.

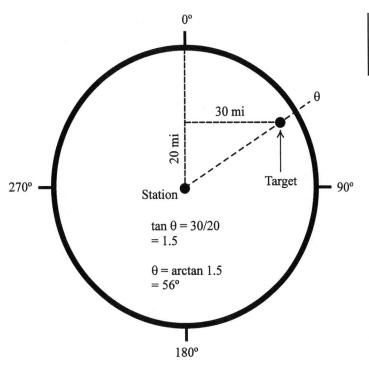

Figure A-27

Illustration for solution to Prob. 1 in Chap. 23.

Figure A-28

Illustration for
solution to Prob. 2
in Chap. 23.

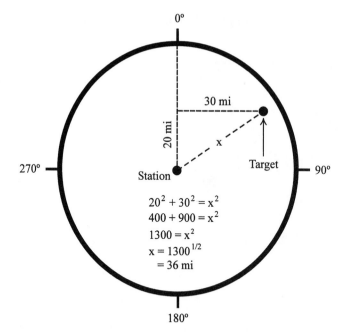

2. To solve this, use the Pythagorean theorem as shown in Fig. A-28. The square of the range is equal to the sum of the squares of 20 mi and 30 mi. When you use a calculator to derive the result (shown as x in the drawing), you should get 36 mi to the nearest mile.

3. This problem can be solved in at least two ways. The more complicated method involves converting all the positions of the targets from navigator's polar coordinates to rectangular coordinates. Then the Pythagorean theorem can be used to derive the distances between pairs of targets. But there's a simpler way. For any particular pair of targets, their ranges can be noted and considered as two sides of a triangle. Then the difference between their azimuth angles can be found. This difference will be the angle between the two sides of the triangle defined by the ranges. The length of the third side of the triangle, which is the distance between the pair of targets, can then be found by using the law of cosines.

4. The star at 30 pc has less parallax, with respect to a background of much more distant objects, than the star at 10 pc. As the distance in parsecs (or any other unit) increases, the actual parallax decreases.

5. Note that 1 pc is approximately equal to 2.06×10^5 AU. In this situation, the two stars are more than 200,000 times as far away from the solar system as they are from each other! Figure 23-13 is obviously distorted. The triangle is actually so elongated that all three line segments from the solar system to the base-line connecting the two distant stars can be considered the same length, and they can each be considered perpendicular to the base-line. (The error introduced by making this assumption is negligible.) On the basis of this information, you can conclude that the measure of the angle θ equals the arctangent of the ratio of 1 AU to 1 pc, as follows.

$$\theta = \arctan (1 \text{ AU}/1 \text{ pc})$$
$$= \arctan [1/(2.06 \times 10^5)]$$

$$= \arctan (4.85 \times 10^{-6})$$
$$= (2.78 \times 10^{-4})°$$

Recall that $1'' = (1/3600)°$. Rounded to three significant figures, this is $(2.78 \times 10^{-4})°$, the same as the answer just derived. The two stars are separated by an angle of $1''$ as seen from our solar system. Perhaps you knew this immediately when the problem was stated, so you did not have to do any calculations at all. (Remember the definition of "parallax second"!)

6. Refer to Fig. A-29. After elapsed time t, the azimuth is $45°$. This means that the aircraft has traveled 10 km, the same distance from its starting point as it originally was from the station. The aircraft's original position, its new position, and the station location lie at the vertices of a right isosceles triangle. Therefore, the range x at time t can be found by using the Pythagorean theorem:

$$10^2 + 10^2 = x^2$$
$$100 + 100 = x^2$$
$$200 = x^2$$
$$x = 200^{1/2}$$
$$= 14 \text{ km}$$

7. Again, see Fig. A-29. After elapsed time $2t$, the aircraft has gone another 10 km because its speed is constant. You now have a right triangle with sides of length 10 km and 20 km adjacent to the right angle. The hypotenuse of this triangle has length y, which is the range when the elapsed time is $2t$. You can use the Pythagorean theorem again:

$$10^2 + 20^2 = y^2$$
$$100 + 400 = y^2$$

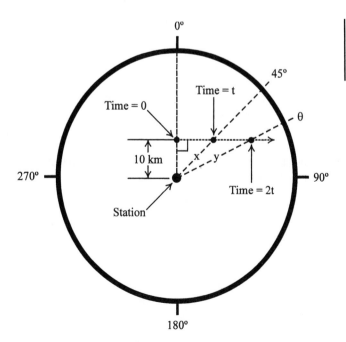

Figure A-29

Illustration for solutions to Probs. 6 and 7 in Chap. 23.

$$500 = y^2$$
$$y = 500^{1/2}$$
$$= 22 \text{ km}$$

The azimuth θ at time $2t$ can be found by means of trigonometry. Because you know the lengths of all three sides of a right triangle, you can use any of the trigonometric functions. Be sure you take the correct ratio for the angle in question! Let's use the tangent function (and its inverse). Then

$$\tan \theta = 20/10$$
$$= 2.0$$

Therefore

$$\theta = \arctan 2.0$$
$$= 63°$$

8. If the earth's curvature is not significant, and provided the observing stations and the object are all on the earth's surface, two observation locations are sufficient to determine the position of, or measure the distance to, that object. If the distance to the object is so great that the earth's curvature is significant, two observing stations are still sufficient; but then you must use a system of *spherical trigonometric functions*.

9. As the distance to an object increases and all other factors are held constant, the absolute error (in meters, kilometers, astronomical units, or parsecs) of a distance measurement by triangulation increases.

10. If you want to measure the distance to an object using stadimetry, you must know, or be able to calculate, the linear dimension of that object along a line perpendicular to the line of sight. For spherical objects such as planets, this is easy; the linear dimension is simply the diameter of the sphere. For irregular objects, it can be difficult or impossible.

CHAP. 24

1. Use the property of common logarithms that converts a product into a sum. That's the following formula:

$$\log_{10} xy = \log_{10} x + \log_{10} y$$

In this case, $x = 2.3713018568$ and $y = 0.902780337$. If you use the calculator found in the Windows computer operating system,

$$\log_{10} (2.3713018568 \cdot 0.902780337) = \log_{10} 2.3713018568 + \log_{10} 0.902780337$$
$$= 0.37498684137 + (-0.0444179086)$$
$$= 0.37498684137 - 0.0444179086$$
$$= 0.3305689328$$

This is the common logarithm of the product we wish to find. If we find the common inverse logarithm of this, we'll get the desired result. Inputting this to a calculator and then rounding to four significant figures

$$\log^{-1} (0.3305689328) = 2.141$$

2. Use the property of natural logarithms that converts a product into a sum. That's the following formula:

$$\ln xy = \ln x + \ln y$$

In this case, $x = 2.3713018568$ and $y = 0.902780337$. Therefore:

$$\ln (2.3713018568 \cdot 0.902780337) = \ln 2.3713018568 + \ln 0.902780337$$
$$= 0.86343911100 + (-0.102276014)$$
$$= 0.86343911100 - 0.102276014$$
$$= 0.761163097$$

This is the natural logarithm of the product we wish to find. If we find the natural inverse logarithm of this, we'll get the desired result. Inputting this to a calculator and then rounding to four significant figures, we find

$$\ln^{-1} (0.761163097) = 2.141$$

3. In this scenario, $P_{out} = 23.7$ and $P_{in} = 0.535$. Plug these numbers into the formula for gain in decibels, and then round off as follows:

$$\text{Gain (dB)} = 10 \log_{10} (P_{out}/P_{in})$$
$$= 10 \log_{10} (23.7/0.535)$$
$$= 10 \log_{10} 44.299$$
$$= 10 \cdot 1.6464$$
$$= 16.5 \text{ dB}$$

4. In this scenario, $P_{out} = 19.3$ and $P_{in} = 23.7$. (We're interested in the power gain of the speaker wire, not the power gain of the amplifier.) Plug these numbers into the formula for gain in decibels, and then round off as follows:

$$\text{Gain (dB)} = 10 \log_{10} (P_{out}/P_{in})$$
$$= 10 \log_{10} (19.3/23.7)$$
$$= 10 \log_{10} (0.81435)$$
$$= 10 \cdot (-0.089189)$$
$$= -0.892 \text{ dB}$$

5. If a positive real number increases by a factor of 10, then its common logarithm increases (it becomes more positive or less negative) by 1.

6. Let x be the original number, and let y be the final number. We are told that $y = 10x$. Taking the common logarithm of each side of this equation gives us this:

$$\log_{10} y = \log_{10} (10x)$$

From the formula for the common logarithm of a product, we can rewrite this as

$$\log_{10} y = \log_{10} 10 + \log_{10} x$$

But $\log_{10} 10 = 1$. Therefore,

$$\log_{10} y = 1 + \log_{10} x$$

7. If a positive real number decreases by a factor of 100 (it becomes 1/100 as great), then its common logarithm changes in value by -2 (it becomes less positive or more negative by 2).

8. Let x be the original number, and let y be the final number. We are told that $y = x/100$. Taking the common logarithm of each side of this equation gives us this:

$$\log_{10} y = \log_{10} (x/100)$$

From the formula for the common logarithm of a product, we can rewrite this as

$$\log_{10} y = \log_{10} x - \log_{10} 100$$

But $\log_{10} 100 = 2$. Therefore

$$\log_{10} y = (\log_{10} x) - 2$$

9. If a positive real number decreases by a factor of exactly 357, then its natural logarithm decreases by ln 357, or approximately 5.8777. In other words, the natural logarithm changes by approximately -5.8777.

10. Let x be the original number, and let y be the final number. We are told that $y = x/357$. Taking the natural logarithm of each side of this equation gives us

$$\ln y = \ln (x/357)$$

From the formula for the natural logarithm of a ratio, we can rewrite this as

$$\ln y = \ln x - \ln 357$$

Using a calculator and rounding to five significant figures, we get $\ln 357 = 5.8777$. Therefore

$$\ln y = (\ln x) - 5.8777$$

CHAP. 25

1. This is the equivalent of 2.382 multiplied by 100,000,000,000,000 (or 10^{14}), so in scientific notation, it is written as 2.382×10^{14}.

2. This is the equivalent of 6.78 multiplied by 0.000000001 (or 10^{-9}), so in scientific notation, it is written as 6.78×10^{-9}.

3. See Fig. A-30. The graph scale is common logarithmic. That is, the actual distance is proportional to the base-10 logarithm of the portrayed numerical value.

4. See Fig. A-31. Both graph scales are common logarithmic. That is, the actual distance is proportional to the base-10 logarithm of the portrayed numerical value. The constant of proportionality is, however, different for the horizontal scale than for the vertical scale.

5. The first expression represents 3.5562×10^{99}, which is a huge positive integer. The letter E means "times 10 to the power of." The digits that follow the E comprise the exponent. The second expression represents 3.5562×10^{-99}. This is an extremely small positive rational number.

Figure A-30

Illustration for solution to Prob. 3 in Chap. 25.

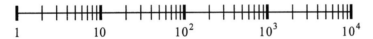

6. The calculation proceeds as follows:

$$(8.0402 \times 10^{64})(2.73 \times 10^{-63}) = (8.0402 \cdot 2.73) \times 10^{[64+(-63)]}$$

$$= 21.949746 \times 10^{64-63}$$

$$= 21.949746 \times 10^{1}$$

$$= 219.49746$$

This rounds off to 219, because the second input number is given to only three significant figures.

7. The calculation proceeds as follows:

$$(-6.7888 \times 10^{34})/(8.45453 \times 10^{36}) = (-6.7888/8.45453) \times 10^{34-36}$$

$$= -0.80297781 \ldots \times 10^{-2}$$

$$= -8.0297781 \ldots \times 10^{-3}$$

This rounds off to -8.0298×10^{-3}, because the first input number is given to only five significant figures. The value can also be expressed as -0.0080298.

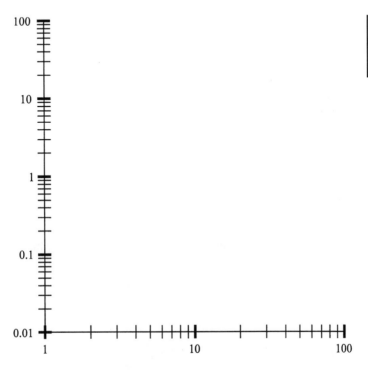

Figure A-31

Illustration for solution to Prob. 4 in Chap. 25.

8. The calculation proceeds as follows.

$$(4.57 \times 10^7)^5 = (4.57)^5 \times 10^{7 \times 5}$$
$$= 1993.338 \ldots \times 10^{35}$$
$$= 1.993338 \ldots \times 10^{38}$$

This rounds off to 1.99×10^{38}, because the input number is given to only three significant figures.

9. The calculation proceeds as follows.

$$(8.84 \times 10^8)^{1/4} = (8.84)^{1/4} \times 10^{(8/4)}$$
$$= 1.724300 \ldots \times 10^2$$
$$= 172.4300 \ldots$$

This rounds off to 172, because the input number is given to only three significant figures.

10. In this case, the second number is insignificant compared with the first, so the difference can be considered as the value of the first number, 5.33×10^{34}.

CHAP. 26

1. Use the formula for vector magnitude in the Cartesian plane:

$$|\mathbf{a}| = (x_a^2 + y_a^2)^{1/2}$$
$$= [(-7)^2 + (-10)^2]^{1/2}$$
$$= (49 + 100)^{1/2}$$
$$= 149^{1/2}$$
$$= 12.2$$

2. First, find the angle θ_a according to the formula for the direction of a vector $\mathbf{a}$ in the Cartesian plane when $x_a < 0$:

$$\text{dir } \mathbf{a} = \theta_a$$
$$= 180° + \arctan (y_a/x_a)$$
$$= 180° + \arctan [(-10)/(-7)]$$
$$= 180° + 55.0°$$
$$= 235°$$

The value r_a is known from the previous problem. It is $|\mathbf{a}|$, which is 12.2. In polar coordinates, therefore, we have

$$\mathbf{a} = (\theta_a, r_a)$$
$$= (235°, 12.2)$$

3. Use the formula for vector magnitude in xyz space:

$$|\mathbf{a}| = (x_a^2 + y_a^2 + z_a^2)^{1/2}$$
$$= [8^2 + (-1)^2 + (-6)^2]^{1/2}$$

$$= (64 + 1 + 36)^{1/2}$$
$$= 101^{1/2}$$
$$= 10.05$$

4. Use the formula for the dot product of two vectors in xyz space:

$$\mathbf{a} \cdot \mathbf{b} = x_a x_b + y_a y_b + z_a z_b$$
$$= (-7 \cdot 8) + [(-10) \cdot (-1)] + [0 \cdot (-6)]$$
$$= -56 + 10 + 0$$
$$= -46$$

5. Use the formula for the dot product of two vectors, based on their magnitudes and the angle between them. In this case, $|\mathbf{f}| = 4$ and $|\mathbf{g}| = 7$, and the angle between them, expressible as θ, measures $0°$ because the vectors point in the same direction along a common line. Therefore

$$\mathbf{f} \cdot \mathbf{g} = |\mathbf{f}| \, |\mathbf{g}| \cos \theta$$
$$= 4 \cdot 7 \cdot \cos 0°$$
$$= 4 \cdot 7 \cdot 1$$
$$= 28$$

6. Use the formula for the dot product of two vectors, based on their magnitudes and the angle between them. In this case, $|\mathbf{f}| = 4$ and $|\mathbf{g}| = 7$, and the angle between them, expressible as θ, measures $180°$ because the vectors point in opposite directions along a common line. Therefore

$$\mathbf{f} \cdot \mathbf{g} = |\mathbf{f}| \, |\mathbf{g}| \cos \theta$$
$$= 4 \cdot 7 \cdot \cos 180°$$
$$= 4 \cdot 7 \cdot (-1)$$
$$= -28$$

7. Use the formula for the dot product of two vectors, based on their magnitudes and the angle between them. In this case, $|\mathbf{f}| = 4$ and $|\mathbf{g}| = 7$. Consider the angle between $\mathbf{f}$ and $\mathbf{g}$, expressed from $\mathbf{f}$ to $\mathbf{g}$ in the counterclockwise sense and denoted as θ_{fg}, to measure $90°$ because the vectors are oriented at right angles with respect to each other. Therefore

$$\mathbf{f} \cdot \mathbf{g} = |\mathbf{f}| \, |\mathbf{g}| \cos \theta_{fg}$$
$$= 4 \cdot 7 \cdot \cos 90°$$
$$= 4 \cdot 7 \cdot 0$$
$$= 0$$

The angle between $\mathbf{g}$ and $\mathbf{f}$, expressed clockwise and denoted as θ_{gf}, measures $-90°$. For the purpose of calculating the cosine function, this is $270°$. Therefore

$$\mathbf{g} \cdot \mathbf{f} = |\mathbf{g}| \, |\mathbf{f}| \cos \theta_{gf}$$
$$= 7 \cdot 4 \cdot \cos 270°$$
$$= 7 \cdot 4 \cdot 0$$
$$= 0$$

The sense in which the angle is defined does not matter in this case, because the cosine of the angle between vectors $\mathbf{f}$ and $\mathbf{g}$ is equal to 0 either way.

8. This problem is solved in two steps. First use the formula for the magnitude of the cross product of two vectors, based on their magnitudes and the angle between them. Let's call the two vectors **a** and **b**. We are told that the angle θ_{ab} between these two vectors measures 90°. Therefore

$$|\mathbf{a} \times \mathbf{b}| = |\mathbf{a}| \, |\mathbf{b}| \sin \theta_{ab}$$
$$= 4 \cdot 7 \cdot \sin 90°$$
$$= 4 \cdot 7 \cdot 1$$
$$= 28$$

The second part of the solution to this problem involves defining the orientation of **a** × **b**. This depends on our reference frame. If the rotational sense of the angle appears counterclockwise, then **a** × **b** points toward us. If the rotational sense of the angle appears clockwise, then **a** × **b** points away from us. Either way, **a** × **b** has magnitude 28 and is perpendicular to the plane containing both **a** and **b**.

9. Vector **b** starts out as (2,0,0), in the same direction as **a**, so **a** × **b** = **0** (the zero vector). As **b** begins to rotate counterclockwise, the cross-product vector **a** × **b** "sprouts and grows" directly toward you along the z axis. When **b** = (0,2,0), having gone through 90° of rotation in the xy plane, |**a** × **b**| = 4 and the vector points toward you. After that, **a** × **b** starts to "shrink," but still points toward you along the z axis. When **b** reaches (−2,0,0), having gone through 180° of rotation, **a** × **b** = **0**. As the rotation continues, **a** × **b** sprouts and grows directly away from you along the z axis. When **b** reaches (0,−2,0), having gone through 270° of rotation, |**a** × **b**| = 4 and the vector points away from you. After that, **a** × **b** shrinks again, all the while continuing to point away from you on the z axis, returning to **0** when **b** = (2,0,0), having passed through a full 360° of rotation in the xy plane. If **b** keeps going around and around, vector **a** × **b** oscillates alternately toward and away from you along the z axis, attaining peak magnitudes of 4 in either direction. If the angular speed of **b** is constant, the fluctuation of |**a** × **b**| takes place according to the sine of the angular displacement, and the oscillation represents a sinusoid.

CHAP. 27

1. The conjunction of several sentences is false if one or more of them is false. The conjunction of several sentences is true if and only if all the sentences are true.

2. The disjunction of several sentences is false if and only if all of them are false. The disjunction of several sentences is true if one or more of them is true. In this context, disjunction refers to the inclusive OR operation.

3. The double-shafted, double-headed arrow can be translated as "if and only if" or "is logically equivalent to." The symbol means that the compound statement on its left always has the same truth value as the compound statement on its right.

4. For a group of four sentences, each of which can attain the value T or F independently of the other three, there are 2^4, or 16, possible combinations of truth values. If you think of F as being equal to 0 and T as being equal to 1, then you can find all the truth values of n independent sentences by counting up to 2^n in the binary numbering system. In this case, with $n = 4$, you would count as follows: FFFF, FFFT, FFTF, FFTT, FTFF, FTFT, FTTF, FTTT, TFFF, TFFT, TFTF, TFTT, TTFF, TTFT, TTTF, TTTT.

5. This logical equivalence is an example of De Morgan's law for disjunction. In general terms, the rule is

$$\neg(X \lor Y) \Leftrightarrow (\neg X \ \& \ \neg Y)$$

for all sentences X and Y.

6. Nothing is wrong with this claim. It is a statement of the distributive law of conjunction with respect to disjunction.

7. In Table 27-16, some of the entries in the rightmost column are incorrect.

8. In the rightmost column header (top of the table), change the ampersand (&) to a double-shafted arrow pointing to the right ($\Rightarrow$).

9. This illustrates the rule of contradiction or absurdity. A direct contradiction implies any false statement that can exist, even if it is ridiculous.

CHAP. 28

1. The pulse will return after more than 2 s because it will have to travel a greater distance between points P and Q in both directions. When the space is "flat" (no curvature), the shortest possible path between P and Q is a straight line. With the introduction of space curvature, the shortest possible path between two points in that space deviates from a straight line, so it gets longer.

2. A surface that has no pairs of parallel geodesics has positive curvature. It is also called a Riemannian surface. An example is the surface of the earth. In this context, the term "parallel" refers to paths that never intersect. Two different lines of latitude on the earth's surface do not intersect each other, but there can never be two different lines of latitude that are both geodesics. (The equator is the only line of latitude that is also a geodesic.)

3. The universe could be "finite but unbounded" if it were shaped like a 4D hypersphere. The surface of the hypersphere would have three spatial dimensions, exactly as we perceive them in the real world. This 3D surface would have no boundaries or edges, just as the 2D surface of the earth has no boundaries or edges. The 3D surface of the hypersphere would have a finite number of cubic kilometers (although a huge number), just as the surface of the earth has a finite number of square kilometers.

4. One second-equivalent represents 3.00×10^5 km (to three significant figures), because the speed of light in free space is 3.00×10^5 km/s. A time interval of 1 min contains 60s. Therefore, in 30 min, light travels 3.00×10^5 multiplied by $60 \cdot 30$, which is 5.40×10^8 km. That is the distance corresponding to 30 minute-equivalents.

5. Refer to Fig. A-32. The 4D path of a stationary point that "lasts forever" is a straight line parallel to the time axis. Path P is an example. The same point moving at constant speed in a straight line through 3D space would follow a straight line in 4D space, but not parallel to the time axis; path Q is an example. The same point starting from a fixed position and then accelerating in a straight line in 3D space would follow a parabolic arc, such as that shown by path R.

Figure A-32

Illustration for solution to Prob. 5 in Chap. 28.

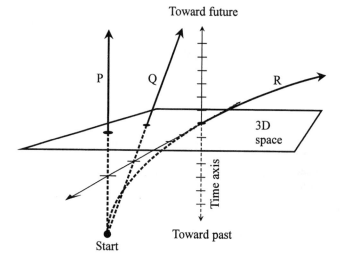

Toward future

P   Q   R

3D space

Time axis

Toward past

Start

6. Draw a tesseract as a cube within a cube (or look at Fig. 28-7), and note the line segments connecting the vertices of the "inner cube" with the corresponding vertices of the "outer cube." There are 8 of these line segments, because each cube has 8 vertices. The tesseract has 12 edges in its inner cube, 12 edges in its outer cube, and 8 edges connecting the vertices of the inner cube with those of the outer cube. That's a total of $12 + 12 + 8 = 32$ edges.

7. Remember the general formula for the distance between two points P and Q in rectangular n-space:

$$|PQ| = [(y_1 - x_1)^2 + (y_2 - x_2)^2 + (y_3 - x_3)^2 + \cdots + (y_n - x_n)^2]^{1/2}$$

where $P = (x_1, x_2, x_3, \ldots, x_n)$ and $Q = (y_1, y_2, y_3, \ldots, y_n)$. We are told that the points are

$$P = (0,0,0,0,0,0)$$
$$Q = (2,2,2,2,2,2)$$

Therefore

$$|PQ| = [(2 - 0)^2 + (2 - 0)^2 + (2 - 0)^2 + (2 - 0)^2 + (2 - 0)^2 + (2 - 0)^2]^{1/2}$$
$$= (4 + 4 + 4 + 4 + 4 + 4)^{1/2}$$
$$= 24^{1/2}$$
$$= 4.899$$

8. In Fig. 28-6, each spatial division is 10^5 km, and each time division is $1/3$ s. Remember that 3.00×10^5 km is equivalent to 1 s of time when the speed of light is used as a reference. This means that the time axis increments shown in Fig. 28-6 are "the same size," in effect, as the space axis equivalents. A photon moving at the speed of light will travel equally far horizontally and vertically from its starting point per unit time. Therefore, the flare angle of the cone is 45°.

9. The hypervolume of this figure in "4D hypercubic kilometers" is equal to 0 km⁴, because the object "lives" for no time at all. The dimensions, in kilometer-equivalents, are 25, 25, 25, and 0. Multiplying all these together gives the 4D hypervolume of 25 · 25 · 25 · 0 = 0 km^4.

10. You can't envision this directly, but you can imagine the 4D space as a positively curved 2D continuum such as the surface of a sphere. For any two distinct points on the 2D surface, the straight line segment connecting them through 3D space is shorter than the geodesic connecting them on the surface itself. This holds true in any number n of dimensions, where n is a positive integer. For any two distinct points in a curved n-space, the straight-line segment connecting them through $(n + 1)$–space is shorter than the geodesic connecting them in n-space.

11. If the 4D space were negatively curved, the straight-line segment through 5D space would still be shorter than the geodesic in the 4D space itself. Any curvature—positive or negative—increases the length of the geodesic between two points compared with the straight-line distance between those same two points in the next-higher dimension.

CHAP. 29

1. This is an unusual situation, but it's theoretically possible. Every single student got exactly one-half of the answers right: 100 correct answers out of 200 questions. The mean, median, and mode are therefore all equal to 100.

2. As long as thermometer Y is a digital device, you can make the readout more and more accurate without limit—say in increments of 10^{-2}, 10^{-3}, 10^{-4}, . . . , 10^{-n} degrees Celsius where n is any whole number—and it will still treat temperature as a discrete variable. Do you see the irony here? A digital thermometer that gives you a reading down to 1/100 of a degree is far more accurate than an analog thermometer with which you have to squint to interpolate a reading to the nearest degree. In the real world, continuous variables are not necessarily more "precise" than discrete ones!

3. Add all the numbers up, and then divide by the number of numbers:

$$\mu = (10 + 11 + 12 + \cdots + 18 + 19 + 20)/11$$
$$= 165/11$$
$$= 15$$

4. In this case, the number of elements is odd. Therefore, the median is the value in the middle, which is 15.

5. As stated, the answer to this question is that there is no mode. Alternatively, we can say that the mode is not defined. Each element occurs exactly as many times as every other element (that is, once).

6. The mean in this situation is the average of all the scores, because one person got each of the 11 possible scores from 0 to 10 inclusive. Calculating gives

$$(10 + 9 + 8 + \cdots + 2 + 1 + 0)/11 = 55/11$$
$$= 5.000$$

accurate to three decimal places.

7. In this case, the number of elements is equal to 11, which is an odd number. Remember that when the number of elements is odd, then the median is the value such that the number of elements having values greater than or equal to it is the same as the number of elements having values less than or equal to it. That element is 5. Checking, note that set of values less than or equal to 5 is {0,1,2,3,4,5}, which has six elements; the set of values greater than or equal to 5 is {5,6,7,8 9,10}, which also has six elements.

8. The mode is undefined because the distribution is perfectly flat.

9. According to the definition of cumulative absolute frequency, this would be the sum of all the individual absolute frequency values.

CHAP. 30

1. Empirical probability is the likelihood of a particular outcome taking place, based on experiments or observations conducted over time. Mathematical probability is based on input data that is not dependent on the real world. Empirical probability data requires hard evidence; mathematical probability does not.

2. This is an example of a uniform distribution, because the chance of the pointer coming to rest within any particular angular range (say, $1°$) is the same as the likelihood of its coming to rest within any other angular range of the same size, all the way round the full circle.

3. In probability theory, the result of an event is an outcome.

4. This is an example of a normal distribution. For any given vertical "slice" of constant size, the area under the curve is largest at the mean, which is located at the center, and diminishes at the same rate on either side of the mean.

5. If a coin is tossed 10 times in a row, the probability P_t of it coming up tails on all 10 tosses is equal to 1/2 to the 10th power. Expressed as a percentage to five significant figures, this is

$$P_t = 100 \cdot (1/2)^{10}$$
$$= 100 \cdot 0.0009766$$
$$= 0.09766 \text{ percent}$$

6. In this case, the probability P_t of the coin coming up tails on all 10 tosses is equal to 9/20 to the 10th power. Expressed as a percentage to five significant figures, this is

$$P_t = 100 \cdot (9/20)^{10}$$
$$= 100 \cdot (0.45000)^{10}$$
$$= 100 \cdot 0.00034051$$
$$= 0.034051 \text{ percent}$$

7. If the bell-shaped curve in a normal distribution becomes "sharper," the standard deviation decreases. The total area under the curve, however, remains constant; it is always equal to 1.

1. d	2. a	3. b	4. d	5. e
6. b	7. c	8. d	9. a	10. e
11. d	12. a	13. b	14. d	15. d
16. d	17. c	18. c	19. a	20. a
21. c	22. a	23. c	24. d	25. e
26. b	27. a	28. c	29. e	30. c
31. c	32. d	33. d	34. d	35. e
36. d	37. c	38. d	39. b	40. a
41. e	42. a	43. b	44. a	45. c
46. b	47. b	48. a	49. d	50. a
51. a	52. b	53. c	54. a	55. d
56. d	57. a	58. c	59. e	60. c
61. d	62. a	63. e	64. d	65. e
66. d	67. a	68. e	69. e	70. c
71. b	72. c	73. a	74. a	75. c
76. a	77. d	78. b	79. a	80. d
81. c	82. c	83. a	84. e	85. b
86. b	87. e	88. a	89. a	90. e
91. a	92. d	93. a	94. d	95. c
96. a	97. c	98. e	99. d	100. b

Appendix 3

Table of Derivatives

Letters a, b, and c denote constants. Letters f, g, and h denote functions; m, n, and p denote integers; w, x, y, and z denote variables. The letter e represents the exponential constant (approximately 2.71828).

FUNCTION	DERIVATIVE
$f(x) = a$	$f'(x) = 0$
$f(x) = ax$	$f'(x) = a$
$f(x) = ax^n$	$f'(x) = nax^{n-1}$
$f(x) = 1/x$	$f'(x) = \ln |x|$
$f(x) = \ln x$	$f'(x) = 1/x$
$f(x) = \ln g(x)$	$f'(x) = g^{-1}(x)\, g'(x)$
$f(x) = 1/x^a$	$f'(x) = -a / (x^{a+1})$
$f(x) = e^x$	$f'(x) = e^x$
$f(x) = a^x$	$f'(x) = a^x \ln a$
$f(x) = a^{g(x)}$	$f'(x) = [a^{g(x)}]\, [g'(x)]\, (\ln a)$
$f(x) = e^{ax}$	$f'(x) = ae^x$
$f(x) = e^{g(x)}$	$f'(x) = [e^{g(x)}]\, [g'(x)]$
$f(x) = \sin x$	$f'(x) = \cos x$
$f(x) = \cos x$	$f'(x) = -\sin x$
$f(x) = \tan x$	$f'(x) = \sec^2 x$
$f(x) = \csc x$	$f'(x) = -\csc x \cot x$
$f(x) = \sec x$	$f'(x) = \sec x \tan x$

$$f(x) = \cot x \qquad\qquad f'(x) = -\csc^2 x$$

$$f(x) = \arcsin x = \sin^{-1} x \qquad\qquad f'(x) = 1 / (1 - x^2)^{1/2}$$

$$f(x) = \arccos x = \cos^{-1} x \qquad\qquad f'(x) = -1 / (1 - x^2)^{1/2}$$

$$f(x) = \arctan x = \tan^{-1} x \qquad\qquad f'(x) = 1 / (1 + x^2)$$

$$f(x) = \text{arccsc } x = \csc^{-1} x \qquad\qquad f'(x) = -1 / [x(x^2 - 1)^{1/2}]$$

$$f(x) = \text{arcsec } x = \sec^{-1} x \qquad\qquad f'(x) = 1 / [x(x^2 - 1)^{1/2}]$$

$$f(x) = \text{arccot } x = \cot^{-1} x \qquad\qquad f'(x) = -1 / (1 + x^2)$$

$$f(x) = \sinh x \qquad\qquad f'(x) = \cosh x$$

$$f(x) = \cosh x \qquad\qquad f'(x) = \sinh x$$

$$f(x) = \tanh x \qquad\qquad f'(x) = \text{sech}^2 x$$

$$f(x) = \text{csch } x \qquad\qquad f'(x) = -\text{csch } x \coth x$$

$$f(x) = \text{sech } x \qquad\qquad f'(x) = \text{sech } x \tanh x$$

$$f(x) = \coth x \qquad\qquad f'(x) = -\text{csch}^2 x$$

$$f(x) = \text{arcsinh } x = \sinh^{-1} x \qquad\qquad f'(x) = 1 / (x^2 + 1)^{1/2}$$

$$f(x) = \text{arccosh } x = \cosh^{-1} x \qquad\qquad f'(x) = 1 / (x^2 - 1)^{1/2}$$

$$f(x) = \text{arctanh } x = \tanh^{-1} x \qquad\qquad f'(x) = 1 / (1 - x^2)$$

$$f(x) = \text{arccsch } x = \text{csch}^{-1} x \qquad\qquad f'(x) = -1 / [x(1 + x^2)^{1/2}] \quad \text{for } x > 0$$

$$f'(x) = 1 / [x(1 + x^2)^{1/2}] \quad \text{for } x < 0$$

$$f(x) = \text{arcsech } x = \text{sech}^{-1} x \qquad\qquad f'(x) = -1 / [x(1 - x^2)^{1/2}] \quad \text{for } x > 0$$

$$f'(x) = 1 / [x(1 - x^2)^{1/2}] \quad \text{for } x < 0$$

$$f(x) = \text{arccoth } x = \coth^{-1} x \qquad\qquad f'(x) = 1 / (1 - x^2)$$

Letters a, b, r, and s denote general constants; c denotes the constant of integration; f, g, and h denote functions; w, x, y, and z denote variables. The letter e represents the exponential constant (approximately 2.71828).

FUNCTION	INDEFINITE INTEGRAL
$f(x) = 0$	$\int f(x)\, dx = c$
$f(x) = 1$	$\int f(x)\, dx = 1 + c$
$f(x) = a$	$\int f(x)\, dx = a + c$
$f(x) = x$	$\int f(x)\, dx = (1/2)x^2 + c$
$f(x) = ax$	$\int f(x)\, dx = (1/2)ax^2 + c$
$f(x) = ax^2$	$\int f(x)\, dx = (1/3)ax^3 + c$
$f(x) = ax^3$	$\int f(x)\, dx = (1/4)ax^4 + c$
$f(x) = ax^4$	$\int f(x)\, dx = (1/5)ax^5 + c$
$f(x) = ax^{-1}$	$\int f(x)\, dx = a \ln |x| + c$
$f(x) = ax^{-2}$	$\int f(x)\, dx = -ax^{-1} + c$
$f(x) = ax^{-3}$	$\int f(x)\, dx = -(1/2)ax^{-2} + c$
$f(x) = ax^{-4}$	$\int f(x)\, dx = (-1/3)ax^{-3} + c$
$f(x) = (ax + b)^{1/2}$	$\int f(x)\, dx = (2/3)(ax + b)^{3/2}a^{-1} + c$
$f(x) = (ax + b)^{-1/2}$	$\int f(x)\, dx = 2(ax + b)^{1/2}a^{-1} + c$
$f(x) = (ax + b)^{-1}$	$\int f(x)\, dx = a^{-1}[\ln (ax + b)] + c$
$f(x) = (ax + b)^{-2}$	$\int f(x)\, dx = -a^{-1}(ax + b)^{-1} + c$
$f(x) = (ax + b)^{-3}$	$\int f(x)\, dx = -(1/2)a^{-1}(ax + b)^{-2} + c$
$f(x) = (ax + b)^n$ where $n \neq -1$	$\int f(x)\, dx = (ax + b)^{n+1}(an + a)^{-1} + c$
$f(x) = x(ax + b)^{1/2}$	$\int f(x)\, dx = (1/15)a^{-2}(6ax - 4b)(ax + b)^{3/2} + c$
$f(x) = x(ax + b)^{-1/2}$	$\int f(x)\, dx = (1/3)a^{-2}(4ax - 4b)(ax + b)^{1/2} + c$

$f(x) = x(ax + b)^{-1}$ $\qquad \int f(x)\,dx = a^{-1}x - a^{-2}b \ln(ax + b) + c$

$f(x) = x(ax + b)^{-2}$ $\qquad \int f(x)\,dx = b(a^3x + a^2b)^{-1} + a^{-2} \ln(ax + b) + c$

$f(x) = x^2(ax + b)^{-1}$ $\qquad \int f(x)\,dx = [(ax + b)^2/2a^3] - a^{-3}(2abx + 2b^2)$
$\qquad\qquad\qquad\qquad\qquad + (a^{-3}b^2) \ln(ax + b) + c$

$f(x) = x^2(ax + b)^{-2}$ $\qquad \int f(x)\,dx = a^{-2}x + a^{-3}b - a^{-3}b^2(ax + b)^{-1}$
$\qquad\qquad\qquad\qquad\qquad - 2a^{-3}b \ln(ax + b) + c$

$f(x) = (ax^2 + bx)^{-1}$ $\qquad \int f(x)\,dx = (b^{-2}a) \ln[(ax + b)\,/\,x)] - (bx)^{-1} + c$

$f(x) = (ax^3 + bx^2)^{-1}$ $\qquad \int f(x)\,dx = \{\ln[x(ax + b)^{-1}]\}/b + c$

$f(x) = (ax + b)(rx + s)$ $\qquad \int f(x)\,dx = (br - as)^{-1} \ln[(ax + b)^{-1}(rx + s)]$
$\qquad\qquad\qquad\qquad\qquad + c$

$f(x) = (ax + b)(rx + s)^{-1}$ $\qquad \int f(x)\,dx = ar^{-1}x + r^{-2}(br - as) \ln(rx + s) + c$

$f(x) = (x^2 + a^2)^{1/2}$ $\qquad \int f(x)\,dx = (x/2)(x^2 + a^2)^{1/2}$
$\qquad\qquad\qquad\qquad\qquad + (1/2)a^2 \ln[x + (x^2 + a^2)^{1/2}] + c$

$f(x) = (x^2 - a^2)^{1/2}$ $\qquad \int f(x)\,dx = (x/2)(x^2 - a^2)^{1/2}$
$\qquad\qquad\qquad\qquad\qquad + (1/2)a^2 \ln[x + (x^2 - a^2)^{1/2}] + c$

$f(x) = (a^2 - x^2)^{1/2}$ $\qquad \int f(x)\,dx = (x/2)(a^2 - x^2)^{1/2}$
$\qquad\qquad\qquad\qquad\qquad + (1/2)a^2 \sin^{-1}(a^{-1}x) + c$

$f(x) = (x^2 + a^2)^{-1/2}$ $\qquad \int f(x)\,dx = \ln[x + (x^2 + a^2)^{1/2}] + c$

$f(x) = (x^2 - a^2)^{-1/2}$ $\qquad \int f(x)\,dx = \ln[x + (x^2 - a^2)^{1/2}] + c$

$f(x) = (a^2 - x^2)^{-1/2}$ $\qquad \int f(x)\,dx = \sin^{-1}(a^{-1}x) + c$

$f(x) = (x^2 + a^2)^{-1}$ $\qquad \int f(x)\,dx = a^{-1} \tan^{-1}(a^{-1}x) + c$

$f(x) = (x^2 - a^2)^{-1}$ $\qquad \int f(x)\,dx = (1/2)a^{-1} \ln[(x + a)^{-1}(x - a)] + c$

$f(x) = (a^2 - x^2)^{-1}$ where $|a| > |x|$ $\qquad \int f(x)\,dx = (1/2)a^{-1} \ln[(a - x)^{-1}(a + x)] + c$

$f(x) = (x^2 + a^2)^{-2}$ $\qquad \int f(x)\,dx = (2a^2x^2 + 2a^4)^{-1}x$
$\qquad\qquad\qquad\qquad\qquad + (1/2)a^{-3} \tan^{-1}(a^{-1}x) + c$

$f(x) = (x^2 - a^2)^{-2}$ $\qquad \int f(x)\,dx = (-x)(2a^2x^2 - 2a^4)^{-1}$
$\qquad\qquad\qquad\qquad\qquad - (1/4)a^{-3} \ln[(x + a)^{-1}(x - a)] + c$

$f(x) = (a^2 - x^2)^{-2}$ where $|a| > |x|$ $\qquad \int f(x)\,dx = -x(2a^4 - 2a^2x^2)^{-1}$
$\qquad\qquad\qquad\qquad\qquad + (1/4)a^{-3} \ln[(a - x)^{-1}(a + x)] + c$

$f(x) = x(x^2 + a^2)^{1/2}$ $\qquad \int f(x)\,dx = (1/3)(x^2 + a^2)^{3/2} + c$

$f(x) = x(x^2 - a^2)^{1/2}$ $\int f(x)\,dx = (1/3)(x^2 - a^2)^{3/2} + c$

$f(x) = x(a^2 - x^2)^{1/2}$ $\int f(x)\,dx = (-1/3)(a^2 - x^2)^{3/2} + c$

$f(x) = x(x^2 + a^2)^{-1/2}$ $\int f(x)\,dx = (x^2 + a^2)^{1/2} + c$

$f(x) = x(x^2 - a^2)^{-1/2}$ $\int f(x)\,dx = (x^2 - a^2)^{1/2} + c$

$f(x) = x(a^2 - x^2)^{-1/2}$ $\int f(x)\,dx = -(a^2 - x^2)^{1/2} + c$

$f(x) = x(x^2 + a^2)^{-1}$ $\int f(x)\,dx = (1/2) \ln (x^2 + a^2) + c$

$f(x) = x(x^2 - a^2)^{-1}$ $\int f(x)\,dx = (1/2) \ln (x^2 - a^2) + c$

$f(x) = x(a^2 - x^2)^{-1}$ where $|a| > |x|$ $\int f(x)\,dx = -(1/2) \ln (a^2 - x^2) + c$

$f(x) = x(x^2 + a^2)^{-2}$ $\int f(x)\,dx = (-2x^2 - 2a^2)^{-1} + c$

$f(x) = x(x^2 - a^2)^{-2}$ $\int f(x)\,dx = -(1/2)(x^2 - a^2)^{-1} + c$

$f(x) = x(a^2 - x^2)^{-2}$ where $|a| > |x|$ $\int f(x)\,dx = (1/2)(a^2 - x^2)^{-1} + c$

$f(x) = x^2(x^2 + a^2)^{1/2}$ $\int f(x)\,dx = (x/4)(x^2 + a^2)^{3/2}$
$- (1/8)a^2 x(x^2 + a^2)^{1/2}$
$- (1/8)a^4 \ln [x + (x^2 + a^2)^{1/2}] + c$

$f(x) = x^2(x^2 - a^2)^{1/2}$ $\int f(x)\,dx = (x/4)(x^2 - a^2)^{3/2} + (1/8)a^2 x(x^2 - a^2)^{1/2}$
$- (1/8)a^4 \ln [x + (x^2 - a^2)^{1/2}] + c$

$f(x) = x^2(a^2 - x^2)^{1/2}$ $\int f(x)\,dx = -(x/4)(a^2 - x^2)^{3/2}$
$+ (1/8)a^2 x(a^2 - x^2)^{1/2}$
$+ (1/8)a^4 \sin^{-1} (a^{-1}x) + c$

$f(x) = x^2(x^2 + a^2)^{-1/2}$ $\int f(x)\,dx = (x/2)(x^2 + a^2)^{1/2}$
$- (1/2)a^2 \ln [x + (x^2 + a^2)^{1/2}] + c$

$f(x) = x^2(x^2 - a^2)^{-1/2}$ $\int f(x)\,dx = (x/2)(x^2 - a^2)^{1/2}$
$+ (1/2)a^2 \ln [x + (x^2 - a^2)^{1/2}] + c$

$f(x) = x^2(a^2 - x^2)^{-1/2}$ $\int f(x)\,dx = -(x/2)(a^2 - x^2)^{1/2}$
$+ (1/2)a^2 \sin^{-1} (a^{-1}x) + c$

$f(x) = x^2(x^2 + a^2)^{-1}$ $\int f(x)\,dx = x - a \tan^{-1} (a^{-1}x) + c$

$f(x) = x^2(x^2 - a^2)^{-1}$ $\int f(x)\,dx = x + (a/2) \ln [(x + a)^{-1}(x - a)] + c$

$f(x) = x^2(a^2 - x^2)^{-1}$ where $|a| > |x|$ $\int f(x)\,dx = -x + (a/2) \ln [(a - x)^{-1}(a + x)] + c$

$f(x) = x^2(x^2 + a^2)^{-2}$ $\int f(x)\,dx = -x(2x^2 + 2a^2)^{-1}$
$+ (2a)^{-1} \tan^{-1} (a^{-1}x) + c$

$f(x) = x^2(x^2 - a^2)^{-2}$ $\qquad\qquad \int f(x)\, dx = -x(2x^2 - 2a^2)^{-1}$
$\qquad\qquad\qquad\qquad\qquad\qquad\qquad + (1/4)a^{-1} \ln\left[(x + a)^{-1}(x - a)\right] + c$

$f(x) = x^2(a^2 - x^2)^{-2}$ where $|a| > |x|$ $\quad \int f(x)\, dx = x(2a^2 - 2x^2)^{-1}$
$\qquad\qquad\qquad\qquad\qquad\qquad\qquad - (1/4)a^{-1} \ln\left[(a - x)^{-1}(a + x)\right] + c$

$f(x) = ax^n$ $\qquad\qquad\qquad\qquad\quad \int f(x)\, dx = ax^{n+1}(n + 1)^{-1} + c$
$\qquad\qquad\qquad\qquad\qquad\qquad\quad$ provided that $n \neq -1$

$f(x) = ag(x)$ $\qquad\qquad\qquad\qquad \int f(x)\, dx = a \int g(x)\, dx + c$

$f(x) = g(x) + h(x)$ $\qquad\qquad\quad \int f(x)\, dx = \int g(x)\, dx + \int h(x)\, dx + c$

$f(x) = h(x)g'(x)$ $\qquad\qquad\qquad \int f(x)\, dx = g(x)\, h(x) - \int g'(x)(x) + c$

$f(x) = e^x$ $\qquad\qquad\qquad\qquad\quad \int f(x)\, dx = e^x + c$

$f(x) = ae^{bx}$ $\qquad\qquad\qquad\qquad \int f(x)\, dx = ae^{bx}/b + c$

$f(x) = x^{-1}e^{bx}$ $\qquad\qquad\qquad \int f(x)\, dx = \ln x + c + bx + (2! \cdot 2)^{-1}b^2x^2$
$\qquad\qquad\qquad\qquad\qquad\qquad\quad + (3! \cdot 3)^{-1}b^3x^3 + (4! \cdot 4)^{-1}b^4x^4 + \cdots$

$f(x) = xe^{bx}$ $\qquad\qquad\qquad\qquad \int f(x)\, dx = b^{-1}xe^{bx} - b^{-2}e^{bx} + c$

$f(x) = x^2e^{bx}$ $\qquad\qquad\qquad\quad \int f(x)\, dx = b^{-1}x^2e^{bx} - 2b^{-2}xe^{bx} + 2b^{-3}e^{bx} + c$

$f(x) = \ln^{-1} x$ $\qquad\qquad\qquad\quad \int f(x)\, dx = \ln(\ln x) + \ln x + c$
$\qquad\qquad\qquad\qquad\qquad\qquad\quad + (2! \cdot 2)^{-1} \ln^2 x + (3! \cdot 3)^{-1} \ln^3 x + \cdots$

$f(x) = x^{-2} \ln x$ $\qquad\qquad\qquad \int f(x)\, dx = -x^{-1} \ln x - x^{-1} + c$

$f(x) = x^{-1} \ln x$ $\qquad\qquad\qquad \int f(x)\, dx = (1/2) \ln^2 x + c$

$f(x) = \ln x$ $\qquad\qquad\qquad\qquad \int f(x)\, dx = x \ln x - x + c$

$f(x) = x \ln x$ $\qquad\qquad\qquad\quad \int f(x)\, dx = (x^2/2) \ln x - (1/4)x^2 + c$

$f(x) = x^2 \ln x$ $\qquad\qquad\qquad\quad \int f(x)\, dx = (x^3/3) \ln x - (1/9)x^3 + c$

$f(x) = \ln^2 x$ $\qquad\qquad\qquad\qquad \int f(x)\, dx = x \ln^2 x - 2x \ln x + 2x + c$

$f(x) = \sin x$ $\qquad\qquad\qquad\qquad \int f(x)\, dx = -\cos x + c$

$f(x) = \cos x$ $\qquad\qquad\qquad\qquad \int f(x)\, dx = \sin x + c$

$f(x) = \tan x$ $\qquad\qquad\qquad\qquad \int f(x)\, dx = \ln |\sec x| + c$

$f(x) = \csc x$ $\qquad\qquad\qquad\qquad \int f(x)\, dx = \ln |\tan (x/2)| + c$

$f(x) = \sec x$ $\qquad\qquad\qquad\qquad \int f(x)\, dx = \ln |\sec x + \tan x| + c$

$f(x) = \cot x$ $\qquad\qquad\qquad\qquad \int f(x)\, dx = \ln |\sin x| + c$

$f(x) = \sin ax$ $\qquad$ $\int f(x)\,dx = -a^{-1}\cos ax + c$

$f(x) = \cos ax$ $\qquad$ $\int f(x)\,dx = a^{-1}\sin ax + c$

$f(x) = \tan ax$ $\qquad$ $\int f(x)\,dx = a^{-1}\ln(\sec ax) + c$

$f(x) = \csc ax$ $\qquad$ $\int f(x)\,dx = a^{-1}\ln[\tan(ax/2)] + c$

$f(x) = \sec ax$ $\qquad$ $\int f(x)\,dx = a^{-1}\ln[\tan(\pi/4 + ax/2)] + c$

$f(x) = \cot ax$ $\qquad$ $\int f(x)\,dx = a^{-1}\ln(\sin ax) + c$

$f(x) = \sin^2 x$ $\qquad$ $\int f(x)\,dx = (1/2)\{x - [(1/2)\sin(2x)]\} + c$

$f(x) = \cos^2 x$ $\qquad$ $\int f(x)\,dx = (1/2)\{x + [(1/2)\sin(2x)]\} + c$

$f(x) = \tan^2 x$ $\qquad$ $\int f(x)\,dx = \tan x - x + c$

$f(x) = \csc^2 x$ $\qquad$ $\int f(x)\,dx = -\cot x + c$

$f(x) = \sec^2 x$ $\qquad$ $\int f(x)\,dx = \tan x + c$

$f(x) = \cot^2 x$ $\qquad$ $\int f(x)\,dx = -\cot x - x + c$

$f(x) = \sin^2 ax$ $\qquad$ $\int f(x)\,dx = (x/2) - (1/4)a^{-1}(\sin 2ax) + c$

$f(x) = \cos^2 ax$ $\qquad$ $\int f(x)\,dx = (x/2) + (1/4)a^{-1}(\sin 2ax) + c$

$f(x) = \tan^2 ax$ $\qquad$ $\int f(x)\,dx = a^{-1}\tan ax - x + c$

$f(x) = \csc^2 ax$ $\qquad$ $\int f(x)\,dx = -a^{-1}\cot ax + c$

$f(x) = \sec^2 ax$ $\qquad$ $\int f(x)\,dx = a^{-1}\tan ax + c$

$f(x) = \cot^2 ax$ $\qquad$ $\int f(x)\,dx = -a^{-1}\cot ax - x + c$

$f(x) = x\sin ax$ $\qquad$ $\int f(x)\,dx = a^{-2}\sin ax - a^{-1}x\cos ax + c$

$f(x) = x\cos ax$ $\qquad$ $\int f(x)\,dx = a^{-2}\cos ax + a^{-1}x\sin ax + c$

$f(x) = x^2\sin ax$ $\qquad$ $\int f(x)\,dx = 2a^{-2}x\sin ax$
$\qquad\qquad\qquad\qquad + (2a^{-3} - a^{-1}x^2)\cos ax + c$

$f(x) = x^2\cos ax$ $\qquad$ $\int f(x)\,dx = 2a^{-2}x\cos ax$
$\qquad\qquad\qquad\qquad + (a^{-1}x^2 - 2a^{-3})\sin ax + c$

$f(x) = (\sin x\cos x)^{-2}$ $\qquad$ $\int f(x)\,dx = 2\cot 2x + c$

$f(x) = (\sin x\cos x)^{-1}$ $\qquad$ $\int f(x)\,dx = \ln(\tan x) + c$

$f(x) = \sin x\cos x$ $\qquad$ $\int f(x)\,dx = (1/2)\sin^2 x + c$

$f(x) = \sin^2 x\cos^2 x$ $\qquad$ $\int f(x)\,dx = (x/8) - (1/32)\sin 4x + c$

$f(x) = (\sin ax \cos ax)^{-2}$ $\qquad$ $\int f(x)\,dx = 2a^{-1}\cot 2ax + c$

$f(x) = (\sin ax \cos ax)^{-1}$ $\qquad$ $\int f(x)\,dx = a^{-1}\ln(\tan ax) + c$

$f(x) = \sin ax \cos ax$ $\qquad$ $\int f(x)\,dx = (1/2)a^{-1}\sin^2 ax + c$

$f(x) = \sin^2 ax \cos^2 ax$ $\qquad$ $\int f(x)\,dx = (x/8) - (1/32)(a^{-1})\sin 4ax + c$

$f(x) = \sec x \tan x$ $\qquad$ $\int f(x)\,dx = \sec x + c$

$f(x) = \arcsin x = \sin^{-1} x$ $\qquad$ $\int f(x)\,dx = x\sin^{-1} x + (1 - x^2)^{1/2} + c$

$f(x) = \arccos x = \cos^{-1} x$ $\qquad$ $\int f(x)\,dx = x\cos^{-1} x - (1 - x^2)^{1/2} + c$

$f(x) = \arctan x = \tan^{-1} x$ $\qquad$ $\int f(x)\,dx = x\tan^{-1} x - (1/2)\ln(1 + x^2) + c$

$f(x) = \operatorname{arccsc} x = \csc^{-1} x$ $\qquad$ $\int f(x)\,dx = x\csc^{-1} x - \ln[x + (x^2 - 1)^{1/2}] + c$
when $-\pi/2 < \csc^{-1} x < 0$
$\int f(x)\,dx = x\csc^{-1} x + \ln[x + (x^2 - 1)^{1/2}] + c$
when $0 < \csc^{-1} x < \pi/2$

$f(x) = \operatorname{arcsec} x = \sec^{-1} x$ $\qquad$ $\int f(x)\,dx = x\sec^{-1} x - \ln[x + (x^2 - 1)^{1/2}] + c$
when $0 < \sec^{-1} x < \pi/2$
$\int f(x)\,dx = x\sec^{-1} x + \ln[x + (x^2 - 1)^{1/2}] + c$
when $\pi/2 < \sec^{-1} x < \pi$

$f(x) = \operatorname{arccot} x = \cot^{-1} x$ $\qquad$ $\int f(x)\,dx = x\cot^{-1} x + (1/2)\ln(1 + x^2) + c$

$f(x) = \sinh x$ $\qquad$ $\int f(x)\,dx = \cosh x + c$

$f(x) = \cosh x$ $\qquad$ $\int f(x)\,dx = \sinh x + c$

$f(x) = \tanh x$ $\qquad$ $\int f(x)\,dx = \ln|\cosh x| + c$

$f(x) = \operatorname{csch} x$ $\qquad$ $\int f(x)\,dx = \ln|\tanh(x/2)| + c$

$f(x) = \operatorname{sech} x$ $\qquad$ $\int f(x)\,dx = 2\tan^{-1} e^x + c$

$f(x) = \coth x$ $\qquad$ $\int f(x)\,dx = \ln|\sinh x| + c$

$f(x) = \sinh ax$ $\qquad$ $\int f(x)\,dx = a^{-1}\cosh ax + c$

$f(x) = \cosh ax$ $\qquad$ $\int f(x)\,dx = a^{-1}\sinh ax + c$

$f(x) = \tanh ax$ $\qquad$ $\int f(x)\,dx = a^{-1}\ln|\cosh ax| + c$

$f(x) = \operatorname{csch} ax$ $\qquad$ $\int f(x)\,dx = a^{-1}\ln|\tanh(ax/2)| + c$

$f(x) = \operatorname{sech} ax$ $\qquad$ $\int f(x)\,dx = 2a^{-1}\tan^{-1} e^{ax} + c$

$f(x) = \coth ax$ $\qquad$ $\int f(x)\,dx = a^{-1}\ln|\sinh ax| + c$

$f(x) = \sinh^2 x$ $\int f(x)\,dx = (1/2)\sinh x \cosh x - x/2 + c$

$f(x) = \cosh^2 x$ $\int f(x)\,dx = (1/2)\sinh x \cosh x + x/2 + c$

$f(x) = \tanh^2 x$ $\int f(x)\,dx = x - \tanh x + c$

$f(x) = \operatorname{csch}^2 x$ $\int f(x)\,dx = -\coth x + c$

$f(x) = \operatorname{sech}^2 x$ $\int f(x)\,dx = \tanh x + c$

$f(x) = \coth^2 x$ $\int f(x)\,dx = x - \coth x + c$

$f(x) = \sinh^2 ax$ $\int f(x)\,dx = (1/2)a^{-1}\sinh ax \cosh ax - x/2 + c$

$f(x) = \cosh^2 ax$ $\int f(x)\,dx = (1/2)a^{-1}\sinh ax \cosh ax + x/2 + c$

$f(x) = \tanh^2 ax$ $\int f(x)\,dx = x - a^{-1}\tanh ax + c$

$f(x) = \operatorname{csch}^2 ax$ $\int f(x)\,dx = -a^{-1}\coth ax + c$

$f(x) = \operatorname{sech}^2 ax$ $\int f(x)\,dx = a^{-1}\tanh ax + c$

$f(x) = \coth^2 ax$ $\int f(x)\,dx = x - a^{-1}\coth ax + c$

$f(x) = (\sinh x)^{-1}$ $\int f(x)\,dx = \ln |\tanh (x/2)| + c$

$f(x) = (\cosh x)^{-1}$ $\int f(x)\,dx = 2\tan^{-1} e^x + c$

$f(x) = (\sinh ax)^{-1}$ $\int f(x)\,dx = a^{-1}\ln |\tanh (ax/2)| + c$

$f(x) = (\cosh ax)^{-1}$ $\int f(x)\,dx = 2a^{-1}\tan^{-1} e^{ax} + c$

$f(x) = (\sinh x)^{-2}$ $\int f(x)\,dx = \coth x + c$

$f(x) = (\cosh x)^{-2}$ $\int f(x)\,dx = \tanh x + c$

$f(x) = (\sinh ax)^{-2}$ $\int f(x)\,dx = a^{-1}\coth ax + c$

$f(x) = (\cosh ax)^{-2}$ $\int f(x)\,dx = a^{-1}\tanh ax + c$

$f(x) = (\sinh x \cosh x)^{-2}$ $\int f(x)\,dx = -2\coth 2x + c$

$f(x) = (\sinh x \cosh x)^{-1}$ $\int f(x)\,dx = \ln |\tanh x| + c$

$f(x) = \sinh x \cosh x$ $\int f(x)\,dx = (1/2)\sinh^2 x + c$

$f(x) = \sinh^2 x \cosh^2 x$ $\int f(x)\,dx = (1/32)\sinh 4x - (1/8)x + c$

$f(x) = (\sinh ax \cosh ax)^{-2}$ $\int f(x)\,dx = -2a^{-1}\coth 2ax + c$

$f(x) = (\sinh ax \cosh ax)^{-1}$ $\int f(x)\,dx = a^{-1}\ln |\tanh ax| + c$

$f(x) = \sinh ax \cosh ax$ $\int f(x)\,dx = (1/2)a^{-1}\sinh^2 ax + c$

$f(x) = \sinh^2 ax \cosh^2 ax$ $\int f(x)\, dx = (1/32)a^{-1} \sinh 4ax - (1/8)x + c$

$f(x) = \text{arcsinh } x = \sinh^{-1} x$ $\int f(x)\, dx = x \sinh^{-1} x - (x^2 + 1)^{1/2} + c$

$f(x) = \text{arccosh } x = \cosh^{-1} x$ $\int f(x)\, dx = x \cosh^{-1} x + (x^2 - 1)^{1/2} + c$
when $\cosh^{-1} x < 0$
$\int f(x)\, dx = x \cosh^{-1} x - (x^2 - 1)^{1/2} + c$
when $\cosh^{-1} x > 0$

$f(x) = \text{arctanh } x = \tanh^{-1} x$ $\int f(x)\, dx = x \tanh^{-1} x + (1/2) \ln (1 - x^2) + c$

$f(x) = \text{arccsch } x = \text{csch}^{-1} x$ $\int f(x)\, dx = x \, \text{csch}^{-1} x - \sinh^{-1} x + c$
when $x < 0$
$\int f(x)\, dx = x \, \text{csch}^{-1} x + \sinh^{-1} x + c$
when $x > 0$

$f(x) = \text{arcsech } x = \text{sech}^{-1} x$ $\int f(x)\, dx = x \, \text{sech}^{-1} x - \sin^{-1} x + c$
when $\text{sech}^{-1} x < 0$
$\int f(x)\, dx = x \, \text{sech}^{-1} x + \sin^{-1} x + c$
when $\text{sech}^{-1} x > 0$

$f(x) = \text{arccoth } x = \coth^{-1} x$ $\int f(x)\, dx = x \coth^{-1} x + (1/2) \ln (x^2 - 1) + c$

$f(x) = \text{arcsinh } ax = \sinh^{-1} ax$ $\int f(x)\, dx = x \sinh^{-1} ax - (x^2 + a^{-2})^{1/2} + c$

$f(x) = \text{arccosh } ax = \cosh^{-1} ax$ $\int f(x)\, dx = x \cosh^{-1} ax + (x^2 - a^{-2})^{1/2} + c$
when $\cosh^{-1} ax < 0$
$\int f(x)\, dx = x \cosh^{-1} ax - (x^2 - a^{-2})^{1/2} + c$
when $\cosh^{-1} ax > 0$

$f(x) = \text{arctanh } ax = \tanh^{-1} ax$ $\int f(x)\, dx = x \tanh^{-1} ax$
$+ (1/2)a^{-1} \ln (a^{-2} - x^2) + c$

$f(x) = \text{arccsch } ax = \text{csch}^{-1} ax$ $\int f(x)\, dx = x \, \text{csch}^{-1} ax - a^{-1} \sinh^{-1} ax + c$
when $x < 0$
$\int f(x)\, dx = x \, \text{csch}^{-1} ax + a^{-1} \sinh^{-1} ax + c$
when $x > 0$

$f(x) = \text{arcsech } ax = \text{sech}^{-1} ax$ $\int f(x)\, dx = x \, \text{sech}^{-1} ax - a^{-1} \sin^{-1} ax + c$
when $\text{sech}^{-1} ax < 0$
$\int f(x)\, dx = x \, \text{sech}^{-1} ax + a^{-1} \sin^{-1} ax + c$
when $\text{sech}^{-1} ax > 0$

$f(x) = \text{arccoth } ax = \coth^{-1} ax$ $\int f(x)\, dx = x \coth^{-1} ax$
$+ (1/2)a^{-1} \ln (x^2 - a^{-2}) + c$

Suggested Additional Reading

Bluman, A. *Math Word Problems Demystified*. New York: McGraw-Hill, 2005.

Bluman, A. *Pre-Algebra Demystified*. New York: McGraw-Hill, 2004.

Gibilisco, S. *Everyday Math Demystified*. New York: McGraw-Hill, 2004.

Gibilisco, S. *Geometry Demystified*. New York: McGraw-Hill, 2003.

Gibilisco, S. *Statistics Demystified*. New York: McGraw-Hill, 2004.

Gibilisco, S. *Technical Math Demystified*. New York: McGraw-Hill, 2006.

Gibilisco, S. *Trigonometry Demystified*. New York: McGraw-Hill, 2003.

Huettenmueller, R. *Algebra Demystified*. New York: McGraw-Hill, 2003.

Huettenmueller, R. *College Algebra Demystified*. New York: McGraw-Hill, 2004.

Huettenmueller, R. *Pre-Calculus Demystified*. New York: McGraw-Hill, 2005.

Krantz, S. *Calculus Demystified*. New York: McGraw-Hill, 2003.

Krantz, S. *Differential Equations Demystified*. New York: McGraw-Hill, 2004.

Olive, J. *Maths: A Student's Survival Guide,* 2nd ed. Cambridge, England: Cambridge University Press, 2003.

Prindle, A. *Math the Easy Way,* 3rd ed. Hauppauge, New York: Barron's Educational Series, 1996.

Shankar, R. *Basic Training in Mathematics: A Fitness Program for Science Students*. New York: Plenum Publishing Corporation, 1995.

Index